Foundations of Physics

ROBERT BRUCE LINDSAY
*Hazard Professor of Physics Emeritus
Brown University*

AND

HENRY MARGENAU
*Eugene Higgins Professor of Physics and Natural Philosophy Emeritus
Yale University*

OX BOW PRESS
Woodbridge, Connecticut 06525
1981

1981 Reprint published by
Ox Bow Press
P.O. Box 4045
Woodbridge, Connecticut 06525

Copyright©, 1936 by Robert Bruce Lindsay and Henry Margenau.
Copyright©, 1957, by Robert Bruce Lindsay and Henry Margenau.

All rights reserved. No part of this book may be
reproduced in any form without written authorization
from the publisher.

ISBN 0-918024-18-8 (Hardcover)
ISBN 0-918024-17-X (Paperback)

Library of Congress Card Number 80-84973

Printed in the United States of America

TO
R. T. L.
AND
L. M.

Preface to 1981 Reprint

This book had its origin somewhat more than fifty years ago in discussions between the authors when they were colleagues in the Department of Physics of Yale University. Active writing on the book began after the senior author had left Yale to go to Brown University in 1930. When the manuscript was offered to John Wiley and Sons, Inc. early in 1934, the publisher raised the somewhat embarrassing question: How will this book fit into the usual college curriculum and specifically, for what courses in physics would it be suitable as a text? The answer was that it was not intended as a text for standard courses in physics as they then existed, but that it might well stimulate the development of courses stressing the philosophical aspects of physics and thus, in the opinion of the authors, serve to strengthen materially the students' grasp of the essential meaning of physical laws and theories. This feeling was strongly expressed in the original preface.

When in 1957 the book was reprinted by Dover Publications Inc., it was clear that the book had indeed served a useful purpose; by this time the philosophy of physics had become a subject of general interest and was actually being taught in college and university courses. The flattering offer of the Ox Bow Press to return the book to the market in 1981 bespeaks a further growth of that interest, not only among sophisticated theorists but among discriminating teachers as well.

R. Bruce Lindsay
Henry Margenau

1981

PREFACE

One of the striking scientific developments of the past decade or two has been an increasing interest in the philosophical aspects of physical science, an interest shared alike by physicists and philosophers and to a certain extent as well by the lay public. The stimulus to this has come largely from modern physics, particularly the quantum theory with its application to atomic structure and the theory of relativity, the concepts of which seem revolutionary to most of those whose physical thinking has been done in terms of nineteenth-century science. Physicists have become more than ever interested in the fundamental concepts on which their subject rests and are beginning to appreciate more than formerly the hypothetical nature of their theories; at the same time, philosophers have found new tasks in the re-interpretation of their views of the universe as a whole in terms of modern physical ideas.

One by-product of this interest has been the production of a number of books endeavoring to explain in simple language and with little or no technical rigor just what the physicist seeks to do. The widespread success of these popular and semi-popular works attests to the growth and strength of the liking for such discussion. Nevertheless, many have felt that the treatment adopted in books of this kind, though undoubtedly entertaining, is not sufficiently searching and, by the attempt to avoid analysis, often becomes misleading. The earnest reader, to gain a deeper understanding, naturally turns to the more elaborate treatises on theoretical physics. Here, however, he usually finds that the fundamental concepts are taken for granted and overlooked in the maze of the application of analysis to special problems—analysis which constantly makes larger demands on his mathematical powers. The result is apt to be bewilderment and a complaint that physics is becoming too technical and elaborate to be a worth-while field of endeavor.

The present volume has been written to satisfy the needs of those who keenly feel the gap between the popular discussions and the rigorous theoretical textbooks. The authors have tried to approach and discuss the fundamental problems of physics from a philosophical point of view; at the same time they have not hesitated to use mathematical analysis where it is essential for clear understanding, since analysis is the language of physics. Much of the book is thus superficially similar in appearance to a text on theoretical physics. It must be emphasized, however, that it has been no part of the authors' plan to write such a text, and the book as written must not be confused with one. In all cases the emphasis has been laid on the methods of physical description and the construction of physical theories rather than on special and elaborate applications. It follows that the present volume lacks completeness with respect to subject-matter: the particular subjects selected for study reflect the authors' opinion as to suitable material for exemplifying the basic structure of physical thought. Another divergent feature is the smaller amount of attention paid to strictly pedagogical requirements: the book is not intended for a single class of readers, and it is hardly to be expected that everyone will find all parts of equal ease or value. No scientific treatise is ever read with uniform interest by all its readers.

The attempt has been made, however, to keep the early chapters mathematically simpler than the later ones. For a full appreciation of the former, little more than a good general knowledge of elementary physics and the essentials of the calculus is required. The chapters on relativity and quantum mechanics naturally draw more heavily on the student's background, although even here the treatment has been kept as simple as is consistent with the degree of generality essential to the purpose of the book. Descriptive and reflective passages alternate freely with the analytical parts throughout. The authors would like to believe that, among other things, they have made clear, to put it crudely, how mathematics gets into physics and how, moreover, certain mathematical methods are peculiarly appropriate for particular physical theories. In this connection they hope that they have played a part, if only a small one, in encouraging modern philosophers to a more intensive study of the language of physics, so as to make the growing cooperation of physicists and philosophers more decisive for the progress of thought. Perhaps it is not out of place to say here that the authors have tried to be terse where terseness is an asset, viz., in the analytical parts. On the other hand, they have not scrupled to be discursive in the elucidation of the meaning of concepts.

In any critical review there always lurks a very real danger: the writers may either displease by adopting too dogmatic a tone in the presentation of their own viewpoint to the exclusion of possible alternatives, or they may go to the other extreme and incur the accusation of feebleness and indecision by presenting many diverse aspects without decisive analysis. The present authors have tried to steer a middle course: fairness to the common points of view, but unhesitating emphasis on what has seemed to them most reasonable.

The order of presentation of topics reflects the fundamental purpose of the authors. Thus an analysis of the meaning of a physical law and of the construction of physical theories in general, together with a discussion of the basic categories of space and time in terms of which all theories are expressed, is followed by a detailed treatment of specific theories beginning with the mechanical and continuing with the statistical. This sets the stage for a presentation of the more recent theories of relativity and quantum mechanics. The book closes with an examination of what is probably the most ardently discussed topic of contemporary physico-philosophical thought, viz., causality.

It is the authors' hope that the book will be of value and interest to an ever-widening circle of readers, including physicists, philosophers, mathematicians, and all others who feel an attraction to physical science in any of its phases. In its inception it grew out of courses in the foundations of physics which the authors have given for the past few years at their respective institutions. They trust that it will serve as a basis for similar courses elsewhere and possibly as a stimulus to the introduction of such courses. Certain chapters have been found by the writers to be suitable material for teaching advanced courses in theoretical physics. This should render the book valuable as collateral reading, even where it is not used as a regular text.

In a work like the present it is difficult to make specific acknowledgment of indebtedness to those whose thoughts have been drawn upon. Frequent footnote references throughout the book partially serve to discharge the authors' obligations in this respect. Deep gratitude is due those who have read the manuscript in advance of publication and who have contributed materially to accuracy and clarity of expression. Both authors wish particularly to express their sincere thanks to Professor Leigh Page of Yale University for his valuable advice and encouragement. The first-named author desires further to acknowledge his indebtedness to his wife, Rachel T. Lindsay, for considerable assistance with the manuscript and proof.

R. B. L.

December 2, 1935.

H. M.

CONTENTS

Chapter I
THE MEANING OF A PHYSICAL THEORY

		PAGE
1.1.	The Data of Physics	1
1.2.	Experiment and Measurement	4
1.3.	Symbolism	6
1.4.	Symbolism in Physics	10
1.5.	What is a Physical Law?	14
1.6.	What is a Physical Theory?	20
1.7.	The Mathematical Development of Physical Theories. The Method of Elementary Abstraction	29
1.8.	Further Illustrations of the Method of Elementary Abstraction. Wave Motion	36
1.9.	The Significance of Boundary Conditions in Physical Theories	48
1.10.	Integral Equations and Difference Equations in the Development of Physical Theories	55

Chapter II
SPACE AND TIME IN PHYSICS

2.1.	The Nature of the Problem	59
2.2.	Psychological Space or the Space of the Individual	59
2.3.	Public or Physical Space	61
2.4.	The Transition to Geometrical Space	62
2.5.	The Significance of Time in Physics	72

Chapter III
THE FOUNDATIONS OF MECHANICS

3.1.	Introduction	79
3.2.	Kinematical Concepts. Specification of Position	79
3.3.	The Concepts of Displacement, Velocity, and Acceleration	82
3.4.	Dynamical Concepts. The Newtonian Postulates for Mechanics	85
3.5.	The Concept of Mass	91
3.6.	The Concept of Force. Résumé of the Laws of Motion	94
3.7.	Interpretation of the Equation of Motion	98
3.8.	D'Alembert's Principle and the Principle of Virtual Displacements	102
3.9.	Gauss's Principle of Least Constraint	112

CONTENTS

		PAGE
3.10.	Hertz's Mechanics	118
3.11.	The Concept of Energy	120
3.12.	Hamilton's Principle and Least Action	128
3.13.	Generalized Coordinates and the Method of Lagrange	136
3.14.	The Canonical Equations of Hamilton	144
3.15.	The Transformation Theory of Mechanics	149

Chapter IV
PROBABILITY AND SOME OF ITS APPLICATIONS

4.1.	The Meaning of Probability	159
4.2.	Analytical Nature of Probability Aggregates	163
4.3.	Some Useful Concepts	167
4.4.	Bernoulli's Problem	169
4.5.	The Emission of α Rays as a Probability Problem	172
4.6.	Laplace's Formula	174
4.7.	Density Fluctuations	177
4.8.	Poisson's Formula	179
4.9.	Theory of Errors	181

Chapter V
THE STATISTICAL POINT OF VIEW

5.1.	Dynamical and Statistical Theories	188
5.2.	The Dynamical Theory of Electric Currents	201
5.3.	Application of Dynamical Theory to Thermodynamics	206
5.4.	The Fundamental Facts of Thermodynamics	212
5.5.	Statistical Mechanics. The Method of Gibbs	218
5.6.	Survey and Critique of Gibbs' Method	250
5.7.	The Method of Darwin and Fowler	252

Chapter VI
THE PHYSICS OF CONTINUA

6.1.	Concept of the Continuum in Physics	280
6.2.	Scalar and Vector Fields	287
6.3.	Deformable Media and Tensor Fields	292
6.4.	The Electromagnetic Field	302

Chapter VII
THE ELECTRON THEORY AND SPECIAL RELATIVITY

7.1.	Electron Theory of Electrodynamics	319
7.2.	Invariance of the Field Equations of Electron Theory	323
7.3.	The Special Theory of Relativity	330

Chapter VIII
THE GENERAL THEORY OF RELATIVITY

		PAGE
8.1.	The Fundamental Idea of Relativity	356
8.2.	Motion of a Particle in a Gravitational Field	364
8.3.	General Relativistic Mechanics	377

Chapter IX
QUANTUM MECHANICS

9.1.	Introduction	387
9.2.	Axiomatic Foundation of Quantum Mechanics	401
9.3.	Some Useful Theorems	413
9.4.	Remarks on Prop. 6. Uncertainty	420
9.5.	Hermitean Operators	422
9.6.	Special Forms of Schrödinger's Equation	426
9.7.	General Properties of Solutions and Eigenvalues of Schrödinger's Equation	440
9.8.	Matrices	448
9.9.	Formal Structure of Matrix Mechanics and Its Relation to the Operator Theory	453
9.10.	Perturbation Theory	460
9.11.	Quantum Dynamics	469
9.12.	Review and Orientation	477
9.13.	The Electron Spin	478
9.14.	The Exclusion Principle	488
9.15.	Atomic Structure	493
9.16.	Quantum Statistics	495
9.17.	Dirac's Theory of the Electron	501

Chapter X
THE PROBLEM OF CAUSALITY 515

SELECTED READINGS 529

INDEX 534

NOTE CONCERNING THE NUMBERING OF EQUATIONS

In each section, equations are numbered consecutively, bearing no reference to the section number, e.g., (9) or (10), etc. Any such equation number mentioned in the text of a particular section refers to the corresponding equation in that section. When reference is made in a particular section to equations occurring in other sections, the section number is placed before the equation number, e.g., (4.15–6) refers to eq. (6) in Sec. 4.15.

FOUNDATIONS OF PHYSICS

CHAPTER I

THE MEANING OF A PHYSICAL THEORY

1.1. The Data of Physics. Physics is concerned with a certain portion of human experience. From this experience the physicist constructs what he terms the physical world, a concept which arises from a peculiar combination of certain observed facts and the reasoning provoked by their perception. While most textbooks of physics set themselves the task of presenting to the student in the most adequate pedagogical manner the features of the physical world as they have been historically developed, it is the principal purpose of the present volume to conduct an inquiry into the logical constituents of these features, analyzing them into such elements as observed facts, definitions, postulates, and deductions.

What, then, is experience, and how much of it concerns the physicist? A detailed answer to the first part of this question would lead us into the depths of philosophy, to which this book is not primarily devoted. Nevertheless, a few remarks on this point are necessary to avoid possible misconceptions. We shall at once assume the possibility of experience and knowledge as the metaphysical basis upon which any science fundamentally rests. Moreover, we shall accept the genuineness of the sense-perceptions of normal people and abstain from quarreling about the meaning of normality in this connection. Naturally, the physicist like all scientists must be forever on his guard against abnormal perceptions, but he has, we assume, objective criteria for detecting them. We must also grant the possibility of the exchange of knowledge, that is, the understanding of another's sense-perceptions and reflections in terms of one's own.

A further assumption, however, must be made about experience before science as we know it becomes possible. Not only must there be agreement about it among normal people, but also it must exhibit a certain *uniformity*. That is, many sense-perceptions do not occur completely at random, but certain more or less well-defined groups of perceptions repeat themselves over and over again with only slight and often insignificant variations. The alternation of day and night with their attendant phenomena illustrates this point, and the reader will be able to supply other examples from all branches of science. There thus appears to be a kind of order in natural phenomena which is often expressed as the law of cause and effect: if a certain group of sense-perceptions appears a number of times associated with a second group, we expect that a repetition of the first group will *always* be associated with that of the second. This is not the place for a detailed discussion of the significance for physics of the assumption of uniformity in its various forms. We shall have occasion later (Chapter X) to investigate these matters more closely and are here introducing them merely as a means of initiating our discussion.

Now, clearly, physics is not concerned with the whole body of human experience. How, then, are the data of physics selected? Before we come to this question, however, let us make quite plain that the task of physics as of all science is found in the coherent *description* of experience. Physics has nothing to say about a possible real world lying behind experience. It is extremely important to realize that, although physics rests upon the assumptions of the possibility of knowledge and the uniformity of experience, the other philosophical problems connected with realism and idealism have no significance for it. One often hears the statement that the task of physics is to describe or explain the behavior of the material world. We cannot help feeling that this is meant to imply the existence of such a world, though what the adjective "material" here means is by no means clear. The physicist has been striving for years to attach a clear meaning to the term *matter*, and undoubtedly we have reason to believe that the concept means much more to us today than to the physicists of fifty years ago. However, to have a clear understanding of a physical concept like matter is not at all equivalent to the assumption that there is a "real" world behind our sense-perceptions which is responsible for the existence of matter. There is perhaps no harm in such an assumption—in fact, certain minds may find that it enables them more firmly to grasp and feel confidence in physical theories; yet it must be stressed that the assumption is no necessary part of physics, and that in a logical development of the subject the safest course is to omit it entirely. It is possible, indeed, to take the view that adopting such an assumption as part of the physicist's stock-in-trade involves a handicap. There is scarcely any use in believing in a real material world behind physical phenomena unless it is a permanent, unchanging affair toward the knowledge of which we progress with slow but certain steps. But our knowledge of it is based naturally on the prevailing physical theories. It seems as if the belief in such a world may tend to encourage too close adherence to reasonably successful physical theories with too small allowance for their necessary revision to meet the demands of new experience. Thus, the term "physical world" used in the introductory paragraph is not to be construed as being identical with real world.

We must now return to the problem proposed at the beginning of the second paragraph above. How are the data of physics selected? Experience is manifold, and physics is concerned with but certain aspects of it. If we take a thought journey back to the beginnings of science, we find the Greek philosophers and other early thinkers trying to understand many common natural phenomena such as the motions of the heavenly bodies, the motions of rivers, the tides, the rainbow, the action of fire, etc. It is hard to assess their interest in any other way than as curiosity about the world in which they found themselves and a desire to find some order in it.[1] Gradually there arose some method in their investigations. In particular it developed that certain phenomena lent themselves more readily to understanding because of their obviously more stable nature: those in which the behavior of living things entered only in a subjective way, i.e., as observation of the non-living world. On the other hand, the phenomena exhibited by living things themselves were early seen to lie in a different category marked by a lack of the definiteness and stability so characteristic of the phenomena of non-living nature. Hence we have the basis for the subdivision of the sciences into physical and biological. This distinction is here pressed as historical and not logical. We shall later have reason to see that there is probably but one method of science, and that is the method of physics. Modern progress in all branches amply confirms this view. However, for the moment let us consider the subject-matter of physics confined to inorganic nature.

The early scientists restricted their attention largely to things they could observe about them. The first physics was therefore largely observational in nature and did not involve the control which was later introduced through the medium of *experiment*. There is a considerable difference, for example, between the crude observation that a piece of metal increases its size on being heated and the experimental investigation of precisely how this expansion takes place, including the answer to the question: howmuch? Similar statements could be made about celestial phenomena like eclipses of the sun and

---

[1] To be sure, this curiosity often took a practical turn: the Chaldeans utilized their observations of the positions of the planets to make astrological predictions.

moon, optical phenomena like the rainbow, etc. The distinction we here have in mind is that between qualitative and quantitative aspects of experience. That almost all solid objects fall to the ground when released by the hand or other support is a statement of experience common to all normal persons and hence is a physical fact. The physicist, however, is not content with propositions of this type, though to be sure there was a time when the whole subject consisted of such. The physicist feels that these are too vague, that a deeper insight is to be gained, and that description can be pushed much further. It is the function of *experiment* and *measurement* to carry out this program, and we must turn to them for a further study of the data of physics.

1.2. Experiment and Measurement. What is an experiment? It is hardly sufficient to call it *controlled* sense-perception without some explanation. It will be found upon examination that this concept involves a great many notions and is altogether a rather complicated affair. In the first place, the performance of an experiment implies a focusing of attention on a certain relatively small portion of experience—an abstraction of a restricted region from the totality of physical phenomena. This means, in the second place, that the experimenter already has previous ideas concerning this group of sense-perceptions. These ideas result presumably from previous observations and reflections about his observations; by having these ideas the experimenter is stimulated to ask questions which he believes only experiment can answer. We are often told that an experimenter should go at his work with an open, unprejudiced mind. But this can never mean that his mind is to be bare of thoughts connected with the realm of sense-perceptions to be investigated by the experiment. In the third place, the notion of experiment implies a certain directed activity on the part of the experimenter. He has decided to go through certain *operations* in the hope that he may thus bring to pass experiences which he has not had in the passive state. Such operations in the first instance may be of purely manual character. Two stones of different weight may be dropped from the same height to ascertain whether they reach the ground at the same time. (Note, of course, how ideas of space and time enter fundamentally even into such a simple operation as this; we shall later discuss this fact fully.) Or a piece of glass may be formed in the shape of a triangular prism and sunlight after passing through a small hole be allowed to pass through the prism in order that the appearance of the illumination on a screen placed on the other side of the prism may be examined. The list of illustrations can be easily extended; the point we wish to emphasize is the *operational* character of the experiment   The same characteristic is to be found attached to the most sophisticated experiments of modern times in which complicated apparatus of glassware and metal is constructed and the operations thereafter are mostly a matter of closing and opening switches.

As distinguishing features of experiment we therefore have so far: (1) the abstraction from the totality of experience of a small domain for special investigation; (2) the presence in the mind of the experimenter of ideas relating to this domain leading him to frame certain questions the answers to which he seeks; (3) the operational character of the experiment involving well-defined activity on the part of the experimenter.

This procedure is common to science in general. In physics, however, another element enters in very decisive fashion. This is the quantitative aspect which finds its expression in *measurement*, perhaps the most important feature of physical science. Here a more highly sophisticated mode of thought makes its appearance, namely the *mathematical*, involving the idea of number.[1] The association of numbers with physical experiences may at first be on a rather simple level; e.g., one may inquire and ascertain the number of easily distinguishable colors in the solar spectrum, or the number of planets in the sky, etc. The elementary use of number in the mere counting of physical objects and qualities is of course far transcended in actual experiments. It brings out only one aspect, i.e., the so-called *cardinal* number. The concept of order and its representation by *ordinal* number is also of great importance, as in arranging a group of objects of equal size but different material in the order of their increasing weights (the primitive notion of density). However, even here we feel the crudeness of the association. To get further we must recognize the fact that scientific description is expressed in terms of *space* and *time*, and that physical measurement is ultimately concerned with the association of number with these concepts.

A rod with some equally spaced marks upon it (i.e., a scale) may be at first looked upon simply as a means of measuring length, though even this operation when analyzed closely appears very complicated, based as it is on certain gratuitous but ingrained ideas about space. However, every experiment in physics in which measurement plays a part reduces essentially to an operation involving the observation of the coincidence of a point or pointer with a mark on some scale, or the comparison of sets of such coincidences; and the symbolic expression

---

[1] See T. Dantzig, "Number, the Language of Science" (Macmillan Co., New York, 1930).

of the observation is made by means of numbers associated with the marks.

It is at once clear, then, that any discussion of physical measurement must involve a critique of the physical meaning of space, time, and number. Geometry enters at every turn in experiment; the measurement of time is fundamental in all experiments having to do with *motion*; and the number symbolism is the language by which we represent our results.

Evidently, measurement is not a simple matter. Yet the physicist uses it as his method of experiment in the laboratory, presumably because he feels that through it he obtains a much more detailed, precise, and therefore more satisfying description of the phenomena observed. For one man it may be enough to note the fall of a free body toward the earth; to the physicist it is much more pleasing to establish a numerical correspondence between the various positions assumed by the falling object on a vertical scale and the coexistent positions of the hands on a suitable clock.

The description of measurement is made in terms of symbolism which we must now examine.

1.3. Symbolism. In its widest sense, a symbol is any perceivable object representative of a definite conceptual or emotional situation. The use of symbolism is widespread indeed; it may take the form of music or of religious rites, and language itself is but a symbolic expression of human thoughts and feelings. The symbolism employed in physics or any other science, however, is of a more restricted kind.

Here a symbol may be defined as a mark of characteristic shape which is taken to represent a certain idea or group of ideas. Its simplest and most important use is encountered in connection with the notion of a real number, which, as we have seen before, plays an essential rôle in the procedure called measurement. Number in itself is a concept, the origin of which is still to some extent a matter of controversy among mathematicians and psychologists alike. As a concept, it may enter into combination with other concepts and appear in propositions according to rules of thought which are but vaguely defined. To remedy this elementary vagueness the scientist employs a remarkable procedure: he associates symbols with the initially conceived numbers, and attempts to state the rules of thought in terms of permissible *operations* upon these number symbols.

From this procedure he gains a threefold advantage. First he has transformed an abstract situation into a group of perceptual facts and is thereby enabled to enjoy the visual clarity which distinguishes this class from logical abstractions. The gain is largely psychological but nevertheless of enormous value, as is evident upon reflection on the curious fact that the drawing of figures, for instance, materially lessens the burden of solving difficult problems. Second, the rules regarding symbols can be fixed with a degree of uniqueness of which mere rules of thought are incapable. It is mainly on this account that logic itself makes use of a symbolism, which avoids the ambiguities inherent in non-symbolic or the less sophisticated symbolic type of linguistic reasoning. Last, not least, is the economy derived from the use of symbols, a point which is so apparent as to require no comment.

Symbolism in logic. An example which suitably illustrates the scientific use of symbolism may be found in logic. We are acquainted with the celebrated syllogism that runs:

All men are mortal
Socrates is a man
Therefore Socrates is mortal.

Let us now introduce some symbols. We let anyone in the class of men be denoted by A and anyone in the class of mortals be denoted by B. Let Socrates be X. The syllogism above then becomes

All A are included in B
X belongs to A
Therefore X belongs to B.

This formulation, indeed, can be made even more concise by representing symbolically the properties of A, X, etc., which above are expressed in words. Thus, let a certain class of objects or elements with respect to a certain property or group of properties (e.g., mortal beings) be denoted by B. Then the class of elements which are common to both classes at once may be denoted by AB (i.e., the juxtaposition or "product" of the two symbols). We may then represent the first statement of the syllogism by the relation

$$A = AB.$$

This is, of course, not to be interpreted as a mathematical relation in the simple algebraic sense. It is a shorthand way of expressing that the whole of the class A is the common part of the classes A and B. The reasoning involved in the syllogism may now be represented symbolically as follows:

$$A = AB$$
$$X = XA$$
$$\therefore \ X = XAB = XB,$$

whence we draw the conclusion that X is the common part of the class X (in our simple example, a single individual) and the class B; i.e., X belongs wholly to the class B. We have obtained this result, to be sure, only by virtue of certain assumptions about our symbolism, viz., certain rules of the game. To obtain the third line from the first two we assume that we may substitute for A in the second line that to which A is said to be equal in the first. With this rule the result follows at once. It is evident that this analysis is capable of very general application since the symbols may represent any classes satisfying the assumptions. The deduction represented is thus no longer a special but a general one, and we are led to attach to it greater significance. Though our illustration has been chosen from logic, the tendency to attach greater meaning to general than special propositions and deductions is just as weighty in science, and the ease with which such propositions are expressed in formal symbols adds greatly to the value of symbolism.[1]

Symbolic representation of relations and operations. We have not yet exhausted the sum of its merits. Very general abstract properties of relation and operation may be assigned to certain symbols, and from these properties a wealth of new relations among the symbols may be deduced. This is the method of mathematics. By the suitable interpretation of the symbols it becomes the method of theoretical physics. For the sake of definiteness, let us illustrate.

Among the properties of relation which may be assigned to symbols are those of *equality* and *order*. We say that the symbol A is equal to the symbol B (written in the form $A = B$) when we mean that A and B are so far alike or equivalent that wherever A occurs in any symbolic proposition B may replace it without in any way altering the validity of the proposition. If A stands for "hydrogen" and B for "lightest element," the equality means that in any proposition containing the word hydrogen we may just as well replace that word with the term "lightest element." The second important relation, namely, that of order, implies the possibility of arranging a set of symbols in a row so that one can be said to come before or after another. Thus if A and B are two symbols in such a set, if they are not equal in the sense mentioned above, we must have $A > B$ or or $A < B$; i.e., either A comes after B in the set or A comes before B.

---

[1] Reference may be made here to W. S. Jevons, "The Principles of Science" (Macmillan, London, 2d ed., 1900). An even more elaborate discussion of symbolic logic in which symbols are used more generally to represent whole propositions and not merely component parts is to be found in D. Hilbert and W. Ackermann, "Grundzüge der theoretischen Logik" (Springer, Berlin, 1928).

As an illustration, consider the symbols representing the points on a straight line. In this case one may indeed add to the basic notion of order that of *magnitude*, so that one symbol may be said to be *greater* or *less* than another, and this is in fact the usual meaning of the signs $>$ and $<$ when the symbols A and B are *numbers*.

In addition to the relations there exists the notion of *operating* on symbols by *combining* them in various ways. Thus, taking again two symbols of a set, A and B, we can imagine a combination of these two, written symbolically $A \circ B$, which is also a symbol of some set. As illustrations of the general operation $\circ$ we may cite addition, subtraction, multiplication, and division ($+$, $-$, $\times$, and $\div$), which of course apply when the symbols are numbers. The definition of the operation may or may not leave open such questions as the significance of the order of the symbols. Thus we may have $A \circ B = B \circ A$, in which case the symbols are said to *commute* under the given operation; or $A \circ B \neq B \circ A$, in which the symbols are non-commutative under the operation. If the symbols are integers the commutative rule holds for the operations of addition and multiplication but fails for subtraction and division.

Classes and groups. The use of symbols is of great value in representing the elements or objects belonging to a *class*. By a class we mean a collection of things (crudely speaking, objects or ideas) of such a nature that a rule exists by which we can tell of any given thing whether or not it belongs to the class. We may speak of the books in a library as forming a class, for example. Likewise the rotational movements which transform a given crystal into itself form a class. It is important to note that not only objects but also operations or events may be members of a class. Thus again we may speak of total eclipses of the sun as forming a class. Even abstract ideas may be the elements of a class, as when we refer to the class of mechanical theories of physical phenomena. Talking about the elements of a class in terms of symbols, i.e., allowing certain symbols to stand for these elements, immensely simplifies and generalizes thought about such a class. This is exemplified, for instance, in the important notion of *group*.

A set of symbols representing the elements of a class is said to form a group with respect to an operation $\circ$ when the following assumptions are satisfied:

1. If A and B belong to the set of symbols, $A \circ B$ belongs to the set.
2. The set contains a symbol I such that $A \circ I = I \circ A = A$, where A means any symbol of the set. I is called the idem or unit symbol.
3. Corresponding to any symbol A of the set, there belongs to the

set another symbol A' such that $A \circ A' = I$. A' is called the symbol inverse to A.

4. The *associative* law holds for operations on the symbols. That is, if A, B, C are three symbols in the set, $A \circ (B \circ C) = (A \circ B) \circ C$. This means that if first $B \circ C$ be formed and A be combined with it from the left, the resulting symbol of the set is equal to that formed by first forming $A \circ B$ and operating with it on C from the left.

The group concept is of great importance throughout mathematics and physics. For example, it can be shown that the set of movements of a rigid body in space forms a group. Group theory has recently assumed considerable significance in the new quantum mechanics.

We have now seen how, by the use of symbols, such an elaborate structure as a group has been developed. It is of greatest interest to observe the manner in which the rules of group theory may be employed to clarify the meaning of the apparently so simple concept of a number system, or rather its symbolic representative.

A set of symbols is called a number system if the symbols may be operated on by the two undefined operations $+$ and $\times$ and the following assumptions are made:

1. The set is a group with respect to the operation $+$. The idem symbol here is called 0 (zero).

2. The set is a group with respect to the operation $\times$ except that there is no inverse symbol to 0. The idem symbol for $\times$ is called 1 (unity).

3. The distributive law holds; viz., if A, B, C are any three symbols of the set there results

$$A \times (B + C) = A \times B + A \times C$$
$$(B + C) \times A = B \times A + C \times A.$$

Since the size of the set is not mentioned, it is clear that one may may visualize various number systems, some with a finite number of symbols and others with an infinite number. The reader who is interested in the foundations of mathematics will find a good discussion in the book of T. Dantzig above referred to and also in J. W. Young, "Fundamental Concepts of Algebra and Geometry" (Macmillan Co., New York, 1917).

1.4. Symbolism in Physics. We now want to pay some attention to the symbols used in physics. We have already noted that in performing an experiment the physicist goes through certain operations in the laboratory. In describing these operations and the resulting observations it is possible, of course, to use ordinary language, describ-

1.4 SYMBOLISM IN PHYSICS

ing in as minute detail as words will allow all that happens. Much early physical investigation was actually recorded in this way, particularly in connection with qualitative properties of light, heat, sound, electricity, and magnetism. For example, in the elementary physics laboratory we still ask the student to take a hard-rubber rod, rub it on fur, and examine what happens when the rubber rod is brought close to a suspended pith ball and then finally touched to it. All this may be described in ordinary language which will be satisfactory enough for certain purposes. However, the physicist is not wholly content with such a description as an ultimate part of physics. In the first place, we have seen that he insists on making measurements by assigning numbers to the results of his operations, i.e., his pointer-readings. Each experiment would then eventuate in various sets of number symbols, a comparison of which would constitute our knowledge of the particular physical phenomena being investigated. This already marks a considerable use of symbolism; but more than this is really implied, for symbols are also more intimately attached to the operations.

Symbolic description of a physical experiment. Consider for a moment the experiment on the properties of gases which is called verifying Boyle's law. The reader is familiar with the usual laboratory procedure, which involves such precautions as allowing the apparatus to come to thermal equilibrium, changing the mercury levels in the glass tubes very slowly, etc. The whole process is called measuring the volume and corresponding pressure of the gas in the closed tube. A symbol V is introduced for volume and a symbol P for pressure. The numbers corresponding to the observed coincidences are appropriately attached to the symbols, and the final result of the experiment is expressed in the symbolic relation $PV = $ constant for constant temperature. But this is too cursory a sketch of what is indeed a very highly involved process, even though the operations themselves may appear simple enough. Let us look rather more deeply if possible.

There is a geometrical concept of volume based on the space concept (which we have agreed to discuss in more detail later). The whole of solid geometry is a symbolic study of the space relations of ideal bodies built up from certain simple assumptions about undefined terms. We might suppose therefore that the physicist knows what volume is. Yet a question remains: how can we be sure that the physicist's method of assigning numbers to the symbol V is such as to make it mean the same thing as it does to the mathematical geometer? The only answer is that we cannot be sure. We must assume this and trust that our results will not discredit the assumption. Putting the matter baldly, the physicist from his previous geometrical knowledge is led to expect that, if he goes through certain operations with his tube and mercury column and obtains certain coincidences, he may justly call this the measurement of the volume of the enclosed gas and denominate the numbers associated with the coincidences the values of this volume under differing conditions. He does this with the expectation that any symbolic use he may make of the volume V so determined will not be inconsistent with the theorems of geometry.

Because of our more or less intuitive geometrical sense the measurement of volume does not seem to involve so much that is not clear. But what does the physicist do when he measures pressure? Actually, in the experiment we are describing, he notes the scale readings of the tops of the mercury columns in the closed and open tubes respectively, subtracts the corresponding numbers, and calls the result the difference between the pressure of the air inclosed in the tube and the pressure of the atmosphere. This introduces a new quantity, namely, the pressure of the atmosphere. However, if you ask the physicist how he gets this, he will reply that a single tube, about one meter in length and closed at one end, is filled completely with mercury, and inverted carefully, allowing no air to enter, into a dish of mercury. If the tube so treated is fixed vertically, and a scale (of the same character as that used in the previous experiment) is attached, the difference between the numbers associated with the scale readings for the level of the mercury in the dish and for the top of the column in the tube is what he means by atmospheric pressure, and this number is the one which when used in connection with the previous experiment gives the actual pressure of the air in the closed tube or the quantity represented by the symbol P.

The meaning of a physical symbol. It thus appears that the symbol here is but a shorthand expression for the results of a given operation leading to the assignment of a number value to the symbol. Instead of describing in words the entire series of acts involved in the setting of the tubes and the reading of the scale, the whole matter is summed up in the one phrase: measurement of P. Is this then all that there is to the meaning of symbolism? If it were necessary to associate a symbol with the results of every single physical operation, the description of these operations might indeed be simplified, but it would not constitute what we now consider theoretical physics. The real power of symbolism in physics first becomes clear when we envisage the possibility of letting a symbol stand for a concept which is, so to speak, the synthesis of the results of a whole set of operations which may

1.4 SYMBOLISM IN PHYSICS

appear to be superficially dissimilar, but are assumed by the physicist to have a common element. We feel that there is *something* common to all operations of the set, and this something is the concept to which we give symbolic representation. For example, in the illustration we have been·using, the concept of pressure and its symbolic representation P must not be too immediately associated with the specific set of operations mentioned. The student will at once think of the various methods of measuring pressure besides that involving the mercury column, viz., the aneroid barometer, the Bourdon pressure gauge, the ionization gauge, etc. Now these are superficially different, yet we do not hesitate to use such operations as a means of assigning numbers to the symbol P. Why is this? It must be because we feel that there is so close a connection between these operations that a single concept represented by a single symbol is competent to describe their essential nature. Or we might put it this way: the behavior of fluids can best be described in terms of a certain concept known as *pressure*, represented by the symbol P in such a way that in every operation involving fluids the physicist is in no doubt what number to assign to the symbol.

It will already be clear to the thoughtful reader that it is the aim of the physicist to work with the smallest possible number of concepts and their representative symbols, and to try to describe the most diversified phenomena in terms of these symbols. In this program he is very materially aided by the fact that the symbols lend themselves readily to mathematical manipulation. Thus, beginning with a few fundamental symbols and introducing certain assumed relations among them, viz., mathematical equations, it is possible by purely analytical methods to deduce a great many new relations. If the original relations can be identified with the result of some known experiment, the same should be true of the deduced relations, which can then be tested in the laboratory. In this way the symbolic method leads to the further development of the whole subject. In using it in the way indicated the physicist has not only a shorthand description of actually observed phenomena but also a powerful tool for the *prediction* of new and hitherto unobserved phenomena.

However, we are here venturing on matters which really transcend the limits of the present section, whose purpose is merely to sketch the importance of symbolism without giving a detailed discussion of the precise ways in which it is used in physics. The preceding paragraph gets us into deep water, for it contains in condensed form the essential content of the method of theoretical physics. Before going further with our treatment of symbolism, therefore, we must necessarily face two important questions. What is the nature of symbolic relations like PV = constant, the so-called physical *laws*? What, moreover, is the nature of the *assumed* relations among symbols which have been mentioned as the starting point of physical prediction? These questions take us into the very heart of the problem of physical theory.

1.5. What is a Physical Law? The word *law* is a good illustration of the difficulty associated with the use of everyday language in the statement of scientific propositions. It has too many meanings, and this fact has led to great confusion in much of the popular writing about science. The dictionary gives roughly ten principal meanings of the word, and some of these are further subdivided. For our purposes, perhaps, we need only call attention to the fact that the term law may signify a rule or way of action imposed on the members of society by society itself either as tradition or as enactment by legislation. In this sense the term has the implication of imposition and compulsion. On the other hand, we sometimes speak of divine law, using the term in its theological sense and implying the notion of necessity. A different use again appears in the phrase moral law. It is essential to emphasize most strongly that these aspects of the word have nothing to do with physical law. What, then, do we mean by a law in physics?

Symbolic representation of the results of laboratory operations. Let us go back to the consideration of experiments or operations performed in the laboratory. Let us imagine that a group of such operations leads to a certain result, essentially, as we have seen, a set of pointer-readings. For example, in the experiment on the properties of gases discussed above, the result will be a set of numerical values attached to the volume symbol and an associated set of values for the pressure symbol, with a one-one correspondence between the two sets. Now this experiment may be repeated—at least, we make the more or less unconscious assumption that it is possible for us to go through *essentially* the *same* set of operations, using perhaps a different piece of glass tubing, a different amount of mercury, and a different volume of air at atmospheric pressure, and perhaps a different temperature. The result is again two associated sets of numbers for P and V, respectively. Now it is unlikely that precisely the same values of P and V will occur in both tables, and a superficial examination might not lead to anything more striking than the apparent fact that, in both cases, as the numerical values of P increase, those of V decrease.

To deal with the situation more accurately the physicist may avail himself of the graphical representation of mathematics. That is, he will establish rectangular axes in a plane, represent values of V along

1.5 WHAT IS A PHYSICAL LAW?

one axis and values of P along the other axis, and plot each associated pair of numerical values of P and V as a point in the plane (Fig. 1.1). Thus the first experiment might be represented by the array of points marked I, and the second by the points II. Already the appearance of the points in the two cases reveals a considerable similarity, and this is heightened by the next step which consists in drawing the smoothest possible curve connecting the set of points in I and performing the same operation for II.

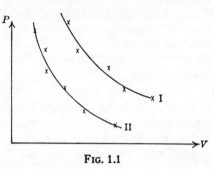

FIG. 1.1

At this stage in the process, an important question arises: just *why* should a curve be drawn anyway, and just *how* should it be drawn if there is any justification for it? Obviously an infinite number of possible curves might pass through all the points, some of which would look very queer indeed. But why draw a curve at all? The actual experiment yields us only the discrete set of points. If physics is a description of experience, what right have we to replace the actual discrete set of experimental points by a continuum? The answer, of course, is to be found in the meaning which the physicist attaches to the word *description*. Baldly speaking, he replaces the discrete set by the continuum because he believes that the resultant description will be simpler and more useful. This opens up the larger question as to what constitutes simplicity and utility in science; and we must ultimately face these questions. For the moment, however, let us postpone this embarrassment, and say that the physicist simply makes the assumption that a discrete set of symbolic data may be profitably replaced by a continuous one.

Now this may be translated into other and more familiar terms by employing the mathematical notion of *function*. From the variation of P and V in the two cases (and we are naturally supposing that many similar experiments have been performed with apparently similar results) the physicist draws the conclusion that P and V are *functionally* related; that is, to each value of P is definitely associated one or more (in this case only one for a given experiment keeping outside conditions the same) values of V. As a matter of fact, the related set of values of P and V represented by the points already constitute a function, but it is a highly complicated type of function, and not at all easy to handle by mathematical analysis. If the physicist wishes to make use of the large body of more simple mathematics at his disposal, he is practically forced to assume that the functional relation of P and V is a *continuous* one. This assumption proves to be extremely significant, for it has guided practically the whole development of physical theory for many years. Incidentally it is a good illustration of the rôle of *hypothesis* in physics, a matter to be discussed presently.

Description and prediction. The hypothesis that P is a continuous function of V leads at once to the search for a simple functional relationship which will represent analytically the continuous curve drawn among the experimental points. In the present case, of course, this results in the well-known relation

$$P = K/V, \qquad (1)$$

where K is a constant. In geometrical terms, the curve is most closely simulated by a rectangular hyperbola. A precisely similar procedure in the experiment labeled II yields a similar result, i.e., eq. (1), with, however, a different numerical value of the constant. We have then a symbolic description of two experiments of like nature though with different initial conditions. Both are summed up in a single type of formula. We must be careful to observe, however, that this description is so far strictly limited. It is only when we have performed a large number of similar experiments in widely separated places, at widely separated intervals, with widely varying initial conditions (e.g., varying initial volumes of gas at atmospheric pressure and different constant temperatures), and using gases of different chemical composition, that our description begins to be felt as having a greater range of application and validity. If in every such case the symbolic description is most closely that represented by (1) we say that we have established a *physical law*. The statement of the law now represents our confidence in a certain routine of experience. We feel assured that a repetition of a certain experiment will yield a quite definitely predictable result. We have no hesitation in predicting this result, if furnished with the initial conditions.

But let us suppose that someone presents us with a gas which has never been experimented upon previously in this way. We are, of course, tempted to predict what will happen on the basis of our "law." The experiment is performed, and the prediction, unfortunately, comes out wrong. What shall we do? Now this is a circumstance which frequently arises. For example, if someone were to ask us to experiment on carbon dioxide at room temperature, we should find the law represented symbolically by (1) a very poor description indeed.

1.5 WHAT IS A PHYSICAL LAW?

What then becomes of the law? Temporarily we might be justly worried as to its value, and should probably wish to extend the scope of our experiments considerably. But ultimately we should find that, by repeating the experiment on CO_2 at higher temperature, the law of Boyle is once again verified.

The conclusion is that we must recognize strict limitations in the applicability of a physical law. It is only a symbolic description of a limited range of phenomena, and we must always be wary about extrapolation. Because a particular law furnishes an accurate shorthand description of a given group of operations we must not immediately assume that similar operations under different initial conditions will be equally well subsumed by the same law. This emphasizes again the extreme importance in every experimental operation of noting carefully all apparently subsidiary conditions, such as temperature, state of atmosphere, proximity of neighboring bodies, time of day or night, etc. Naturally it is necessary to grant the practical impossibility of reproducing again and again the same conditions in the performance of a given experiment. The best we can hope for is close approximation, and we may well expect that the law chosen as the best description of a routine of experience will have its form determined to a certain extent by the degree of approximation of the routine. The more exactly the experiment is done and the more carefully controlled the various factors that enter into it, the more precise will be the resulting law.

Thus for many purposes the law describing the behavior of a simple pendulum, viz., $T = 2\pi\sqrt{l/g}$, connecting the period T of the motion with the length l of the string and g, the acceleration of gravity, will be an adequate description of the phenomenon in question. However, for more precise purposes this is not a good description. A much more accurate version is $T = 2\pi\sqrt{l/g}(1 + \ell_0^2/16)$, where ℓ_0 is the small initial angle the string makes with the vertical. It is seen at once that if this angle is very small the new law reduces to the old "for all practical purposes." Other illustrations of the necessity for the more exact formulation of simple laws will be found in the equation of state of gases, in the temperature dependence of resistance of conductors, in the anharmonic oscillator, etc.

The above discussion will serve to discredit partly the feeling of the early physicists that there is an underlying simplicity about nature and that the more carefully and thoroughly one performs experiments the more simple in form will be the laws which describe them. A difficulty about this is inherent in the word "simplicity." Just what does it mean? One criterion of simplicity might be the number of necessary symbols in the law. Another, the type of mathematical analysis used—e.g., the nature of the functions involved. On either or both of these tests the primitive feeling mentioned is quite certainly false as far as our experience goes. The more thoroughly a given experimental operation is carried out the more difficult it is in general to fit the results with a law of simple form. We appear to have traveled considerably beyond the view at one time held that the very simplicity of a physical law was a guarantee of its truth.

The criterion of simplicity. It will perhaps be wise at this point to introduce a few further remarks on the question of simplicity. Most of us will undoubtedly maintain that we know when a concept or a law is simple—that we have an "intuitive feeling" for such a property. Yet we should probably be forced to admit on closer inspection that this feeling is based largely on our individual background of familiarity with the subject under consideration. The differential calculus can hardly be called simple by one who has not mastered the rudiments of algebra, though it may be second nature to the mind thoroughly versed in advanced analysis. It seems that the criterion of simplicity must remain forever arbitrary, depending, as it does, largely on familiarity. Physical theories that seem simple enough to us in our day would undoubtedly have appeared far from simple to physicists of an earlier time. However, for our future discussion it will probably be wise for us to adopt some kind of criterion, no matter how arbitrary. We shall agree to seek simplicity in a minimum number of distinct concepts in terms of which physical laws describing particular phenomena may be expressed. This criterion will be of some further use in our discussion of the development of physical theories.

So far then we have been led to look upon a physical law as a symbolic description (in the "simplest" form) of an observed routine in a limited field of phenomena. It is well to emphasize again its descriptive nature. It never purports to give a reason for any of the phenomena described, in the metaphysical sense. That is, it does not constitute what is popularly called an *explanation*. Newton's law of gravitation is not an explanation of gravitation in the sense that it explains *why* particles attract each other. It merely seeks to give an exact description of the observed attraction. The physical law attempts to answer the question "how" rather than the question "why."

Cause and effect. This brings us to another interesting question connected with the nature of a physical law. Does a physical law always state a relation of cause and effect? The proper definition

1.5 WHAT IS A PHYSICAL LAW? 19

of the term *cause* has been a controversial subject among philosophers for a long time, and it is not intended to enter this controversy here. We must recognize, however, that there has been a tendency in the past to separate a physical happening into two parts, of which the first is said to be the cause of the second and the second the effect of the first. Thus, in the extension of a wire by a weight or other force (Hooke's law), the force has popularly been called the *cause* of the extension, and the extension then figures as the effect of the force, the idea being that with the force in operation the extension takes place, and that without it the extension would not take place. Or to go back to our much-used illustration of Boyle's law, the increase in pressure on the gas is popularly looked upon as the cause of the decrease in volume, the decrease in volume being the effect of the increase in pressure. Now there may be a certain advantage to this popular version; certainly, it is an instinctive stand taken in everyday life. But on careful examination it must be confessed that the view has little value for physics. Thus, in the example of Hooke's law, as far as the law itself is concerned in its symbolic form, it is just as sensible to call the extension the cause and the force the effect as vice versa. Similarly with the pressure and volume in Boyle's law. All that the laws state is a *relation* among symbols which represent well-defined operations in the laboratory, and no notion of precedence or antecedence, or dynamic enforcement, is involved in them. But in the cause-effect idea another suggestion is implied, namely, that of necessity or certainty. We say that the given cause *always* results in the same effect. This was mentioned in Sec. 1.1 as a possible way of stating the general assumption of the uniformity of nature. The *assumption* is clearly a valuable one—without it we could do little business in science; but it is quite another thing to state dogmatically that the idea of *necessity* or *certainty* is always implied in every physical law. Our presentation of the meaning of a physical law has clearly shown the limitations to which such a law is subject. To be sure, it predicts; but its prediction is always surrounded by a nebula of uncertainty. All we can say is that there is a certain degree of *probability* that its prediction will be verified. This at once introduces the important concept of probability. We shall hear much more of it later in connection with theories in physics and with errors in measurement (Chapter IV).

What has been said above about cause and effect in their relation to laws should by no means be interpreted as denying the importance of *a principle of causality* in physics. This principle has a very definite significance independent of the popular ideas about cause and effect.

It is in itself a *hypothesis* concerning the behavior of physical systems. Reference should be made to Chapter X for an extended discussion.

Are qualitative propositions physical laws? Before closing this brief examination of the nature of physical law, we might note one further point. Shall we classify in the list of physical laws such purely qualitative observations as the propositions: copper conducts electricity, the melting point of ice is lowered by pressure, a grating forms a spectrum, etc., or such numerical statements as that there are three states of matter, there are two types of waves possible in an elastic medium—compressional and distortional, etc.? It is conceivable that a difference of opinion may be honestly entertained here. Some authors [1] take the view that these really are examples of physical laws. No one doubts that they express physical facts, i.e., routines of experience. Nevertheless, they do appear in a different category from the mathematical relations among symbols to which numbers may be attached by means of operations performed in the laboratory. These relations are the ones which we shall consider as forming the class of real physical laws in the sense in which the term is used in the present text. It is rather significant, indeed, that there *are* physical laws (in our sense) *connected* somehow with all such qualitative propositions of the kind illustrated above. These laws serve to carry the description of our knowledge far beyond the frontier set by such propositions themselves. Herein, of course, lies a large part of their value.

1.6. What is a Physical Theory? In the discussion of physical laws included in the previous section the impression might momentarily have arisen that as a brief symbolic description of a limited range of phenomena each law might well stand as complete in itself and independent of all other laws. That such is not the case, however, becomes clear when we recall what has been said about the nature of the symbolism in which the laws are expressed; the very idea of symbolism implies, as we have seen, that a symbol must represent a concept transcending the particular operation which it is used to represent. If the symbol P in Boyle's law represented *only* the one particular operation of taking the difference between the levels of two mercury columns, it would have very little meaning for the many other operations which may be carried out on fluids. Because P somehow represents the synthesis of a large number of operations, we may expect that physical laws involving P have some connection with one another. The possibility indeed lies open that, among

[1] For example, Norman R. Campbell in "Physics—The Elements." (Cambridge Univ. Press, 1920.)

certain symbols representing primitive concepts, a few mathematical relations may exist from which can be deduced, by purely mathematical operations, a whole group of physical laws involving the concept of pressure. Now when such a state of affairs exists we say that we have a *physical theory*. Let us analyze more closely what this means.

Mechanics as an illustration of physical theory. It will be desirable in this connection to refer to some simple example. The reader will recall the many "formulas" relating to motion of various kinds—uniform, uniformly accelerated, simple harmonic, motion in a central field, etc. Each of these formulas is a real physical law in the sense of our previous discussion. Each is expressed in terms of symbols to which numbers may be attached by means of laboratory operations. These are symbols with which we are familiar under the names: displacement, time, velocity, acceleration, mass, force, etc. Now all the relations among these symbols can be derived from a limited number of fundamental assumptions. We begin with certain primitive concepts closely joined in our minds with laboratory operations. The first of these is distance or interval, a space concept. The second is the concept of time. Both of them demand extensive investigation (see Chapter II). However, for the present take them for granted. From these it is possible to construct or define the concepts of velocity and acceleration by admitting suitable mathematical operations performed on representative symbols. Now from the various experiments performed with moving bodies we are led to make two simple hypotheses about the accelerations of pairs of particles which influence each other. (For the details consult Chapter III.) These assumptions lead at once to the definition of a new concept, that of *mass*, with a symbol to represent it. Finally, we make the last assumption that for any system of particles there exists a certain set of functions of the position coordinates and velocity components of the particles, such that every possible motion of any of the particles, say the ith one, is described by a differential equation of the form $F_i = m_i \dfrac{d^2 r_i}{dt^2}$.

From the integration of this equation or its component Cartesian equations (a purely mathematical process), a whole set of symbolic relations among position, velocity, and time for the ith particle will be obtained. These will be actual physical laws describing the possible motions of the particle, and will be found to be verified by experiments in which the position, velocity, and time symbols are associated with definite laboratory operations whereby numbers may be attached to the symbols.

The building of a theory. This then is the essence of a theory of mechanics. Naturally, many special points need to be looked after in a detailed study. But the method is clear. We start with intuitive space-time concepts. We then construct new symbols which are, by the very method of construction, defined in terms of the primitive ones, so that we have no doubt as to the proper laboratory operation to carry out in order to assign a number to each one. This is what is meant, for example, when we say that we understand how to measure acceleration and mass. We are then ready for the next step—the choice of hypotheses or assumed fundamental relations among the symbols by logical deduction from which all the special results of the theory, viz., the *laws*, are to be obtained.

The origin of these hypotheses requires a word of comment. Of course it *would* be perfectly possible to assume any set of self-consistent relations among the symbols which have been introduced. This is precisely the method of the mathematician in looking for new results; but it will hardly do for physics, where the deductions from the postulates must predict actual laboratory results. Hence in framing postulates for a physical theory we must consult our previous experience of the phenomena which the theory is to describe. In mechanics, for example, there are at our disposal a large number of observations on the motions of particles; these are indeed the *laws* of moving bodies. They supply the hints toward the proper construction of the hypotheses, the principal one of which is that, through the placing of the product of mass and acceleration of a particle equal to an appropriate function of its position and velocity, and those of neighboring particles, we may obtain a complete description of the possible motions of the particle.

Relation of law and theory. Several points remain to be cleared up. One may ask: what is the relation of the fundamental assumption of mechanics to Newton's laws of motion? The answer is that Newton's laws, however phrased, constitute the hypotheses at the basis of mechanics. In other words, they are not physical laws in the sense in which we have used this term. They are not laws like Boyle's law or the law which connects the range of a projectile with its initial velocity. They are really the basis of the theory of mechanics (perhaps one should better say "a" theory of mechanics, as we shall see presently). Of course, it is possible to phrase them in a variety of ways, as will be made clear in Chapter III; but fundamentally they remain hypotheses, suggested indeed by experiment but never verified directly by experiment. Their value lies in the fact that together with appropriate assumptions about forces they predict all the phenomena connected with the motions of gross bodies that have ever been observed. They therefore provide a comprehensive description of a large field of experience—a great deal in a small package.

Incidentally, it will be found that many so-called "laws" in physics are really hypotheses of theories and not laws in the sense of the present text. For example, the symbolic expression known commonly as Newton's law of gravitation, stating that any two particles attract each other with a force varying directly as their masses and inversely as the square of their distance apart, is the fundamental hypothesis of the Newtonian theory of gravitation, from which, together with the fundamental hypotheses of mechanics, many observed motions may be adequately described. It is probable that this looseness of nomenclature is hardly to be avoided. We have noted, indeed, in the preceding section that a distinct hypothetical element enters into the statement of every physical law. Moreover, usually the very statement of a physical law implies the existence of a theory in which that law is included in a deductive sense. For example, Boyle's law implies really the existence of a theory of fluid behavior in which one of the chief symbolic concepts is pressure, and the laws of motion of the heavenly bodies imply the existence of a theory of motion in which one of the chief symbols is mass, etc. It may be desirable from the logical standpoint, however, to distinguish as sharply as possible between a law and a theory. Perhaps the leading hypothesis of a theory might properly be called a principle. Thus one might better refer to Newton's "laws" as the principles of mechanics rather than laws of motion, leaving that term to express the actual laws immediately descriptive of the motions of bodies. Similarly, the leading hypothesis of the theory of relativity might from this point of view be referred to as the principle of relativity. Other illustrations will occur to the reader. Of course matters of nomenclature are bound to be arbitrary.

Hypothesis framing. Certain tendencies in modern physics provoke an important question about the construction of a physical theory. We have just emphasized that in making the fundamental definitions and postulates of a theory we are always guided by background and previous experience. This has certainly been true in the development of most physical theories. For example, the theory of mechanics has been so remarkably successful that there has been a tendency in the description of other phenomena to frame hypotheses based directly on mechanical pictures, i.e., implying the validity of the principles of mechanics. One thinks immediately of the kinetic theory of gases, the dynamical theory of acoustics, the dynamical theory of electric currents, etc. (Some of these mechanical theories of natural phenomena are discussed in Chapter V.) These are important and well-established theories and appear to justify our faith in the essentially mechanical concepts which are so closely related to our intuitive notions of space and time. Nevertheless, the attempt to use these concepts in framing hypotheses for theories of atomic structure has by no means been so successful. So the necessity has arisen for seeking postulates outside the fold of mechanics. What have we here to guide us? We still have a considerable background to call upon, but the results of modern research show that free mathematical construction has a great place after all, and that postulates which appear to have little connection with anything anybody has ever experienced in space and time may prove remarkably apposite in the formulation of theories.

Of course, atomic physics is still in a state of flux; but the present indication is that the postulates of its theories will continue to grow more abstract, having less immediate connection with or plausibility derived from experiment, at least experiment with macroscopic phenomena. Inevitably there arises the feeling that postulates will become more arbitrary and that physical theories will eschew easy visualization and the picturesqueness characteristic of models based on large-scale experience. Does this mean that physics is really drawing away from experience or, to use the word that induced shivers of horror in the physicists of the past two centuries, is becoming metaphysical? It is not likely that we need be worried over this question. Physical theories will always be based on hypotheses, but the deductions from these hypotheses must always have a readily identifiable relation with experience. If they do not meet this test, then, of course, they must be discarded. To ask that the hypotheses at the basis of a physical theory should *always* be expressed in terms of the most primitive concepts (e.g., those of space and time) may actually impose a handicap on the handling of the situation now being encountered in recent atomic theories. No theory of atomic structure, for example, in which the fundamental assumptions are based on primitive space-time concepts has as yet proved successful. Hence the temptation has been strong to build an entirely new and necessarily more abstract point of view, the deductions from which appear to be in good accord with experiment, and which is able to predict successfully the results of new experiments. This is the quantum mechanics of Heisenberg, cast into very finished and elegant form by Dirac, Weyl, v. Neumann, and others.

We must be careful not to give the impression that no atomic theory will ever be developed based on primitive concepts, i.e., where the fundamental postulates are definitely suggested by our experience of the behavior of matter in the large. We mean simply that the attempt to carry out such a program has been accompanied hitherto by such great difficulty as to seem for the present unwise. There are fashions in the development of physics as in any other field of knowledge, and at present it cannot be denied that the trend is away from models. As a matter of fact, we shall later see that although the new quantum mechanics is abstract, the postulates are not, of course, so completely arbitrary as they look at first sight. To sum up our present discussion: we have emphasized the importance of experimental and theoretical background in the framing of new hypotheses, but at the same time have stressed the danger involved in allowing prejudices produced by a too close adherence to this background to govern completely the choice of principles.

In this brief general analysis of the meaning of a physical theory we have laid so much stress on its hypothetical character that it will be only fair to speak a little further of the types of hypothesis of use to the physicist. As we have used the term in this section we have so far had chiefly in mind an actual assumed analytical relation among symbols. The function of hypothesis is much wider that this, however, as has already been suggested in previous sections, for in the actual business of developing theories we constantly make hypotheses without realizing that they are being made. Such, for example, are what Poincaré has called "unconscious" or "natural" hypotheses—a type which one hardly ever challenges, for it seems too unlikely that we could make progress without them. Nevertheless it should be the endeavor of the physicist always to drag them out into the light of day, so that it may be perfectly clear what we are actually doing. We have already noted the tendency of physicists to use *continuous* functions in the statement of physical laws and in the mathematical manipulations incidental to the transition from theories to laws. This is an unconscious hypothesis. It is hardly necessary to add that, although it is an extremely valuable one and does seem natural, it has an element of danger in it, since in many branches of physics functions with one or more points of singularity have proved necessary, and in quantum mechanics physicists have introduced improper functions (e.g., Dirac's δ function; cf. Chapter IX). Hence we must not allow the unconscious leaning toward continuous functions to render us unwilling to employ other types, even though this entails more involved mathematical technique.

Other illustrations of unconscious hypotheses which have been definitely recognized are: the use of a particular kind of geometry, e.g., Euclidean; the use of the idea of *symmetry*, e.g., the method of Archimedes in establishing the principle of the lever; the feeling that all physical effects should vanish at sufficiently great distance of separation of particles, i.e., at mathematical infinity; the argument from *analogy*, e.g., the feeling that the same method which works for the description of macroscopic effects should work in the microscopic domain. The reader should be readily able to supply other examples.

Poincaré has also stressed the essentially *indifferent* nature of certain hypotheses. This is a particularly important point in the building of physical theories. Thus it may happen that two or more superficially different hypotheses lead on deduction to the same physical laws on appropriate identification of the symbols. We may say that it is a matter of indifference which form of hypothesis we use since the ultimate result is the same. Of course, it may happen that the mathematical technique involved in the one type is somewhat more direct for certain kinds of problems than that involved in the other. This is the situation, for example, in mechanics, where there are many different formulations of the fundamental principles, all leading, however, in any particular problem to the same final result. Thus we have, besides the Newtonian formulation, the principle of D'Alembert, Gauss's principle of least constraint, Hamilton's principle, etc. Some of these prove much more direct than the others for the handling of certain problems. Yet they are essentially indifferent hypotheses in the sense of Poincaré. Other illustrations are afforded by electricity and heat. As far as the laws of electrostatics are concerned it makes little difference whether we base our theory on the assumption of a single electrical fluid, two fluids, or the existence of discrete smallest charges, viz., electrons and protons. Similarly, as far as the laws of calorimetry are concerned we may indifferently begin with the assumption of heat as a material substance, or the mechanical point of view which looks upon heat as a mode of motion. Again, as far as many phenomena are concerned, the statistical theory of gases leads to the same results as thermodynamics.

It must be confessed that the existence of indifferent hypotheses is not considered an unmitigated good by physicists, and the endeavor is constantly being made to reduce their number by rendering hypotheses more and more precise and making the experimental demands on theories more exacting. In this way it is hoped to be able to make a decision for one mode of the framing of hypotheses as against another.

1.6 WHAT IS A PHYSICAL THEORY?

The discovery by Foucault in 1850 that the velocity of light is greater in air than in water definitely decided against the corpuscular theory of light in the form suggested by Newton in favor of the wave theory, which indeed had previously been more successful than the former in describing the phenomena of interference and diffraction. This seemed to remove the "indifference" and to enthrone the wave theory as the best description of the phenomena of optics. However, it is well known that a new corpuscular theory has come back in the guise of the photon theory of the past three decades. We shall not enter here upon any discussion of the apparently anomalous situation produced by this "reincarnation," but shall merely point out that hypotheses constantly ebb and flow and change their form so that it is very difficult to be sure that any one type has definitely been discarded for good and all.

Need for precise definition of concepts. One of the great difficulties associated with the framing of hypotheses for any theory lies in the definition of the symbolic concepts which are to be used in the hypotheses. It goes without saying that the more precisely this definition is carried out the more definite will be the results of the theory, and the more valuable it will be as a description of phenomena. It would hardly be necessary to stress this matter at all were it not for the fact that in many physics texts—popular and professional—definitions are cited which are quite meaningless or positively misleading. Among the most glaring illustrations is the carelessness in the definitions of mechanical concepts such as mass and force. One still occasionally sees the wholly unnecessary absurdity: mass is the amount of matter in a body. Perhaps its uselessness, i.e., the fact that it is merely stated for the sake of form, is the reason why it is not so baneful in its effect on students as one might actually expect. Of course it is not at all necessary, for there is a perfectly satisfactory definition of mass in mechanics (cf. Chapter III). Much the same confusion exists in connection with force, though here it is true that the physicists are perhaps less at fault than the engineers, many of whom persist in using a purely anthropomorphic concept of force which has no significance for physical theory, and of course becomes wholly meaningless in any discussion of atomic forces, etc. Force as used by the physicist has one perfectly definite meaning, and once the concept is settled the name is ever after preëmpted. In no other way can logical thinking in physics be possible. The precision with which physical concepts are formulated is obviously one of the criteria of the success of the theory based on these concepts.

Criteria for the success of a theory. This brings us to the last point in our general discussion of physical theory. We must face the question: when is a theory to be judged *successful*? This is not a simple subject, and we shall hardly be able to do it justice in our short treatment. At first, however, one would be tempted to ask: is it not after all rather obvious? If the conclusions reached by mathematical analysis from the postulates of the theory are found to be in agreement with experiment, is not the theory successful? Unfortunately a simple consideration shows that this view is illusory. We have already seen that there are usually different ways of formulating a theory to describe the same range of physical experience, and it is small esthetic comfort or enlightenment to know that if theory A is successful, theory B is also successful. Still worse, it is easy to convince ourselves that, once given a set of physical laws which seem best to represent a certain range of experience, we can always construct a physical theory, and indeed a great many theories, to subsume these laws. This is true no matter how complicated the phenomena may be. That a physical theory is able "merely" to generalize known experience and boil it down into a few postulates of rather arbitrary character is then hardly a test of the theory's success. But, one may observe, of all the theories descriptive of a given range of phenomena, surely one is *simpler* than all the other easily conceivable ones! Is not this the really successful one, is not *simplicity* the genuine criterion for success? We have already made the acquaintance of the idea of simplicity in the previous section, and we there came to the conclusion that it is after all a rather arbitrary yardstick by which to judge anything. Unless one sets up a very arbitrary standard, our preconceived tastes and prejudices are likely to determine that which we shall call simple in a physical theory. And these prejudices themselves are largely fixed by the *familiarity* we have acquired of the concepts in question. To a person familiar by long thought with the theory of quantum mechanics it seems as elegantly simple as the theory of mechanics, whereas the person approaching it for the first time is repelled by what he terms its abstractness and wide deviation from the "common-sense" ideas of classical physics. Further thought will probably convince the reader that the criterion of simplicity is a doubtful one.

Where then shall we look for a more satisfactory criterion for the success of a theory? We have so far, indeed, overlooked one point, and that is the power of a physical theory to *predict* physical observations previously unknown. This is a remarkable faculty. Suppose an experimentalist is about to perform an experiment. He describes to the theoretical physicist exactly what operations he intends to make and asks what he may expect to find. We may well agree that if the

1.7 MATHEMATICAL DEVELOPMENT OF PHYSICAL THEORIES 29

theoretician is able as a deduction from his theory to tell the experimentalist exactly what he should find and then the experimentalist proceeds to find it, there must be something in the theory. It has met a most searching test, and the more of such it is able to meet, the more successful we shall be inclined to consider it, even though it is based on concepts which are difficult to understand in terms of those of other well-established physical theories. It is easy to sympathize with modern theoretical physicists [1] who are inclined to emphasize the power of prediction as the really crucial test of a theory. This does not indeed relieve a theory of the necessity of subsuming within itself all the already *known* facts in the range it is intended to describe, but it can serve to pick out of a number of theories which describe with apparently equal facility a certain range, the one which is most worth following up, the best gamble, as it were, for future investigation. Nor must we forget the influence that this has on the course of experimental investigation. A theory already successful in meeting the test of prediction usually suggests many new lines of research for the discovery of new relations, new laws, new physical knowledge. This is a tremendous stimulus to development, and it is in this way that the rapid advance in physics in certain periods has taken place. The tremendous progress in quantum theory during the last two decades is a very good example of this fact.

In this section we have been discussing certain general aspects of the meaning of physical theory. In our further study it will be desirable to become more concrete. The general outline of the subject should be clear, but there still remains the problem of the *special methods* employed in the building of theories, especially the type of mathematical analysis used. An examination of this will be our task in the next sections.

1.7. The Mathematical Development of Physical Theories. The Method of Elementary Abstraction. Our general discussion of the previous section needs to be amplified from the standpoint of concreteness. The reader recalls that the great principles of physics are symbolically expressed in terms of *differential equations*. As a matter of fact, the differential notation was historically invented for the purpose of physical description. The question arises: why has this particular form of mathematics been adopted?

Perhaps we can best introduce the subject by quoting from a celebrated parable of the philosopher Schopenhauer.[2] " Two China-

---
[1] See, for example, P. A. M. Dirac, " Principles of Quantum Mechanics " (Oxford, 1930).
[2] "Studies in Pessimism."

men traveling in Europe went to the theater for the first time. One of them did nothing but study the machinery, and he succeeded in finding out how it was worked. The other tried to get at the meaning of the piece in spite of his ignorance of the language. Here you have the Astronomer and the Philosopher." Now this may be somewhat rough on the philosopher, but everyone at once recognizes the profound distinction in method that is implied. A certain group of sense-impressions were experienced by the two observers, who sought to describe the experience by two totally different ways: the one tried to appreciate the experience as a *whole*; the other, foregoing this, picked out of the whole one small part which he thought he might understand and successfully describe. It is this process of abstraction from the totality of physical phenomena which has undoubtedly been a leading feature in the success of the physical theorist. It is precisely this method of abstraction pushed to its logical conclusion that leads to the use of the differential calculus in physics. We shall call it the method of elementary abstraction.

Elementary abstraction in fluid motion. To consider a simple illustration, let us imagine that a physicist wishes to build a theory of fluid motion. Does he merely go down to the river and watch the moving water in its almost infinitely complex course past muddy bends and over rocks and stones? In other words, does he confine his attention to the large-scale phenomena of fluids and then proceed to build hypotheses designed to account for this behavior of fluids considered in the large? Of course, he does not. To be sure, he might come back with valuable information, with hints, with suggestions. But when he starts to theorize on the motion of fluids he really proceeds quite differently. If in his imagination he visualizes a large quantity of fluid he abstracts from the whole a very small volume element for his special consideration, and he considers its behavior over very short intervals of time. He makes the assumption that the elements and intervals may be symbolically represented by mathematical infinitesimals. The symbolic expression of his further hypotheses concerning the behavior and properties of the fluid element will then contain differentials and derivatives, and hence be a differential equation. The order and degree of the equation will depend on assumptions as to the variation of the fluid properties over the extent of the element. It will be wise before proceeding further to illustrate these points concretely by studying one differential equation in the theory of fluids, namely, that commonly designated as the *equation of continuity.*

Let us discuss the behavior of the fluid in terms of the rectangular

1.7 MATHEMATICAL DEVELOPMENT OF PHYSICAL THEORIES

coordinates x, y, z. That is to say, the position of every point in the fluid with respect to a given frame of reference is indicated by its perpendicular distances respectively from three mutually perpendicular planes, forming a frame of reference. When we further agree to allow x, y, z to take on *all* possible values in any given interval, i.e., all values which can be put into one-one correspondence with the points on a line (a continuum in the mathematical sense), we imply the important physical hypothesis that the fluid is continuous in the sense that throughout the whole region under investigation we never encounter a volume element no matter how small in which there is no fluid. We make the final fundamental assumption that at no place in the whole volume is there any creation or destruction of fluid. It now remains to give symbolic form to these hypotheses.

We shall take as our element the rectangular parallelepiped indicated in Fig. 1.2 with sides, dx, dy, dz, respectively. The

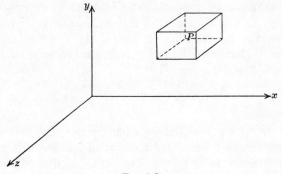

Fig. 1.2

point P with coordinates x, y, z is its midpoint. Now imagine that the box remains stationary with respect to the reference system while the fluid flows through it. The instantaneous *velocity* of the fluid at time t at the point P is assumed to be the vector **q**. This quantity itself is a fruit of the method of elementary abstraction, as we shall later have occasion to note in Chapter III. Here we shall content ourselves with the reminder that through every point in the fluid it is possible to pass a curve the tangent to which at that point gives the *direction* of flow at the point; the instantaneous rate of flow (a purely mechanical quantity defined similarly to the rate of progression of a material particle) at this point is called the magnitude of the instantaneous velocity at the point. The vector **q** has components along the three axes which we shall denote by u, v, w respectively. These are called the component velocities of flow of the fluid. What are the component velocities at the *faces* of the volume element dx, dy, dz? Since in general we must suppose that u, v, w vary with x, y, z, i.e., are *functions* of these variables, it is clear that the velocity components at the faces will not be the same as at the midpoint. Let us concentrate our attention first on the component velocity in the x direction. Its value at the face parallel to the yz plane and nearer to it will be given in terms of that at P by means of the well-known Taylor's theorem, viz.,

$$u(x - dx/2) = u(x) - \frac{\partial u}{\partial x}\frac{dx}{2} + \frac{\partial^2 u}{\partial x^2}\frac{dx^2}{4\cdot 2!} + \ldots, \qquad (1)$$

where the derivatives are *partial* derivatives, since u is a function of x, y, z and t in general. Now the question at once arises: how far do we wish to push our use of mathematical analysis? Do we wish to use differentials of order higher than the first? If so, it is clear that our resulting relations will rapidly become complicated. We therefore make the assumption that a satisfactory description will be attained by the use of first order differentials only. This leaves us only the first two terms on the right of eq. (1). In similar fashion the x component velocity at the face parallel to the yz plane and *farther* from it than P will be given by

$$u(x + dx/2) = u(x) + \frac{\partial u}{\partial x}\frac{dx}{2}) \ldots, \qquad (2)$$

neglecting as before the higher order terms. The *volume* of fluid flowing with velocity u in unit time normally across the area S is uS if u is the average velocity over the area. And similarly the *mass* flow per unit time, or mass discharge rate, will be $\rho u S$, if ρ is the density or mass per unit volume of the fluid. Both ρ and u may vary from place to place in the fluid and also at any given place may vary with the time.

With this and the preceding results in mind we may write for the rate of mass flow per unit time *into* the one face parallel to the yz plane and *out* of the other, the two expressions, respectively,

$$\left(\rho u - \frac{\partial(\rho u)}{\partial x}\frac{dx}{2}\right)dy\, dz$$

and

$$\left(\rho u + \frac{\partial(\rho u)}{\partial x}\frac{dx}{2}\right)dy\, dz.$$

We notice that ρu is differentiated as a whole. This is because both ρ and u are in general functions of x. The excess of mass flow in the

1.7 MATHEMATICAL DEVELOPMENT OF PHYSICAL THEORIES 33

x direction out of the element over the flow into the element is the difference between the second expression and the first, viz., $\frac{\partial(\rho u)}{\partial x} dx\, dy\, dz$. It is an easy matter to derive the corresponding expressions for the excess of efflux over influx in the y and z directions, obtaining the expressions $\frac{\partial(\rho v)}{\partial y} dx\, dy\, dz$ and $\frac{\partial(\rho w)}{\partial z} dx\, dy\, dz$, respectively. Hence the *total* excess of efflux over influx in mass of fluid per unit time is the sum

$$\left(\frac{\partial(\rho u)}{\partial x} + \frac{\partial(\rho v)}{\partial y} + \frac{\partial(\rho w)}{\partial z}\right) dx\, dy\, dz. \tag{3}$$

Before proceeding further we must discuss briefly a point which may have occurred to the reader. The velocity $u - \frac{\partial u}{\partial x}\frac{dx}{2}$ is certainly not exactly the velocity of the fluid over the *whole* nearer face parallel to the yz plane, nor is $u + \frac{\partial u}{\partial x}\frac{dx}{2}$ the exact velocity over the *whole* farther parallel face, for these quantities in general will vary with y and z. It might then seem that our result above is vitiated. Closer inspection, however, shows that if we take account of the variation of $u - \frac{\partial u}{\partial x}\frac{dx}{2}$, etc., with y and z we ultimately get the above with additional second order terms. But since we have already agreed to neglect all beyond the first our result is shown to be sound.

The equation of continuity. The assumption has been made that the fluid is continuous and moreover is indestructible. Hence whatever fluid flows into the element must flow out again, or if there is an excess of efflux over influx, it must be compensated by a decrease in the mass of fluid inside the element. The time rate of change of mass in the box is, however, $\frac{\partial \rho}{\partial t} dx\, dy\, dz$, and the statement we have just made is represented symbolically by the equation (obtained by dividing out the common factor dx, dy, dz)

$$\frac{\partial(\rho u)}{\partial x} + \frac{\partial(\rho v)}{\partial y} + \frac{\partial(\rho w)}{\partial z} = -\frac{\partial \rho}{\partial t}. \tag{4}$$

This is the famous *equation of continuity*, fundamental for the physical theory of continuous media. It is a *partial differential equation of the first order* and as such is a symbolic expression of the continuity and indestructibility of the medium with which the quantities ρ, u, v, w are associated. It is a partial differential equation since the fundamental quantities u, v, w, ρ are functions of both space and time variables. This will in general be the case in the use of the method of elementary abstraction in physics. As stated in general form in (4), this particular equation then applies to any medium in which ρ and u, v, w may vary both in time and space. Indeed, it is not restricted to a *fluid* medium, but is also descriptive of the behavior of a deformable, continuous, and indestructible solid medium, as a study of elasticity shows. It is thus of very general significance. We shall later note an important application to statistical mechanics (Sec. 5.5).

To appreciate its meaning more fully, let us consider the special case of an *incompressible* medium. By this is meant one in which the density remains constant in time at every point, though it may still vary from point to point, i.e., the medium may be *non-homogeneous*. In this case eq. (4) becomes

$$\frac{\partial(\rho u)}{\partial x} + \frac{\partial(\rho v)}{\partial y} + \frac{\partial(\rho w)}{\partial z} = 0; \qquad (5)$$

that is to say, the total excess of efflux over influx in any volume element vanishes, the rate of influx being precisely equal to the rate of efflux. In the further event of a homogeneous medium (i.e., ρ the same everywhere) the equation becomes even simpler, viz.,

$$\frac{\partial u}{\partial x} + \frac{\partial v}{\partial y} + \frac{\partial w}{\partial z} = 0. \qquad (6)$$

The equation of continuity may thus be looked upon as a symbolic statement of a fundamental hypothesis concerning continuous indestructible media. It is an integral part of the *theory* of such media. Although from it alone it is not possible by direct mathematical deduction to obtain symbolic description of *all* the observed properties of such media, this description may result from the combination of the equation of continuity with certain other differential equations which express the motion of a small element of the medium (the so-called equations of motion). This is the task of hydrodynamics and elastokinetics. The important point for us to recognize here is that the theory that professes to describe the behavior of continuous media is based on certain definable mechanical concepts (like time, space, velocity, and density) and certain hypotheses whose ultimate symbolic expression takes the form of partial differential equations like (4), precisely be-

1.7 MATHEMATICAL DEVELOPMENT OF PHYSICAL THEORIES

cause of the use of the method of elementary abstraction in the development and statement of the theory.

Physical meaning of integration. Now that we have our differential equation, what shall we do with it? From it we can derive new equations which are in the nature of physical laws describing possible laboratory operations. Just how is this transition carried out? The mathematical process of passing from the differential equation to the physical law is known as *integration*. From the physical point of view the word almost literally conveys the meaning of the method, for just as the differential equation is a symbolic description based on the method of elementary abstraction, the resulting so-called solution is an algebraic equation describing the large-scale operations in the laboratory. Hence in a very true sense the passage from the differential equation to its solution involves a symbolic integration of small-scale phenomena into large-scale phenomena. Let us try to illustrate this with a simple case, namely, that of an incompressible, homogeneous fluid, for which the equation of continuity takes the form (6). What does this really tell us about the possible large-scale motion of the fluid? It certainly assures us that not *all* conceivable values of u, v, w are actually possible. For example, if we conceive of a fluid flowing so that there is no component motion in the y or z directions, i.e., $v = w = 0$, the equation tells us that

$$\frac{\partial u}{\partial x} = 0,$$

which on integration gives immediately

$$u = f(y, z) + C,$$

where C is a constant and $f(y, z)$ an *arbitrary* function of y and z. That is, u must be a function of y and z only and not of x. It must be constant as far as x is concerned—a considerable restriction. As a matter of fact, the same result will follow if v is a constant or a function of x and z alone and w a constant or a function of x and y alone; neither need be zero, as inspection shows. A more specialized case would be that in which $v = c_1 y + f_1(y, z)$, $w = c_2 z + f_2(x, y)$, where f_1 and f_2 are arbitrary functions. Then eq. (6) becomes

$$\frac{\partial u}{\partial x} = -(c_1 + c_2),$$

or on integration

$$u = -(c_1 + c_2)x + f_3(y, z).$$

Here, then, u also appears as a linear function of x. Numerous other special cases will occur to the reader, and he will already recognize an important point: the solutions of a partial differential equation like (6) are always very *general*, involving in every case, as we have seen, *arbitrary* functions as well as arbitrary constants. One would then be inclined to ask: does not this seriously restrict the usefulness of the method which leads to such equations? The answer to this question involves a consideration of the important initial or boundary conditions which serve to limit the degree of arbitrariness of the solution. We must also heed the fact already mentioned that many valuable physical laws result from the combination of two or more such equations as that of continuity. Before looking into these, however, let us first note one more point concerning eq. (6). Suppose that u, v, w are so related that there exists a function of x, y, z, say ϕ, such that

$$u = \frac{\partial \phi}{\partial x}, \quad v = \frac{\partial \phi}{\partial y}, \quad w = \frac{\partial \phi}{\partial z}. \tag{7}$$

The function ϕ, if it exists, is called the *velocity potential*. Substitution into (6) yields the equation

$$\frac{\partial^2 \phi}{\partial x^2} + \frac{\partial^2 \phi}{\partial y^2} + \frac{\partial^2 \phi}{\partial z^2} = 0. \tag{8}$$

We started with a first order partial differential equation in the three dependent variables u, v, w. We are led by our assumption (7) to a *second* order partial differential equation with a single dependent variable ϕ. Eq. (8) is called *Laplace's equation* and is one of the most important in all theoretical physics, playing a significant rôle in the dynamics of particles, hydrodynamics, elasticity, electricity and magnetism, and optics. We shall not embark here on a discussion of its solutions. We merely remark that they are called *harmonics*, are very numerous, and involve arbitrary functions which have to be further specified to meet the needs of particular problems. These questions, however, are perhaps better illustrated for our purposes by the *wave equation*, which we proceed to discuss in the next section.

1.8. Further Illustrations of the Method of Elementary Abstraction. Wave Motion. A brief study of *wave motion* will enable us to give particular attention to the nature of the solutions of the partial differential equations encountered in the building of physical theories.

Equations of motion of a fluid. We are going to discuss more fully the motion of a continuous medium, more specifically, a fluid. What more can we say about this motion than is already expressed in the

1.8 METHOD OF ELEMENTARY ABSTRACTION

equation of continuity? The assumption can be made that, if we concentrate on a small volume element such as the parallelepiped in Fig. 1.2, we can describe the motion of the fluid inside by means of the principles of mechanics. In a later chapter we shall give a critique of the foundations of this subject. But for the present it will be sufficient to say that the use of mechanics implies merely this: we shall construct equations stating that the mass of the fluid element multiplied by the component acceleration in the x, y, or z direction is equal to the force acting on the element in that particular direction. That is, we shall merely use the celebrated equation $\mathbf{F} = m\mathbf{a}$. It is necessary, of course, to exercise care in constructing the expression for the acceleration. Going back to Fig. 1.2, assume again that at the point $P(x, y, z)$ the velocity components at the time t are u, v, w, respectively, while at the nearby point $Q(x + dx, y + dy, z + dz)$ the corresponding quantities at the later time $t + dt$ are $u + du$, $v + dv$, $w + dw$. Now du, dv, dw will clearly depend on x, y, z and t. Hence we may write

$$\left. \begin{aligned} du &= \frac{\partial u}{\partial t} dt + \frac{\partial u}{\partial x} dx + \frac{\partial u}{\partial y} dy + \frac{\partial u}{\partial z} dz \\ dv &= \frac{\partial v}{\partial t} dt + \frac{\partial v}{\partial x} dx + \frac{\partial v}{\partial y} dy + \frac{\partial v}{\partial z} dz \\ dw &= \frac{\partial w}{\partial t} dt + \frac{\partial w}{\partial x} dx + \frac{\partial w}{\partial y} dy + \frac{\partial w}{\partial z} dz \end{aligned} \right\} \quad (1)$$

Suppose that the particle of fluid (a portion of fluid not larger than the volume element above considered) which was at P at time t is at Q at the time $t + dt$. We must then have

$$dx = u\,dt, \quad dy = v\,dt, \quad dz = w\,dt,$$

and du, dv, dw will be the increments in the velocity components of the particle in time dt. Therefore the components of the acceleration of this particle are:

$$\left. \begin{aligned} \frac{du}{dt} &= \frac{\partial u}{\partial t} + u \frac{\partial u}{\partial x} + v \frac{\partial u}{\partial y} + w \frac{\partial u}{\partial z} \\ \frac{dv}{dt} &= \frac{\partial v}{\partial t} + u \frac{\partial v}{\partial x} + v \frac{\partial v}{\partial y} + w \frac{\partial v}{\partial z} \\ \frac{dw}{dt} &= \frac{\partial w}{\partial t} + u \frac{\partial w}{\partial x} + v \frac{\partial w}{\partial y} + w \frac{\partial w}{\partial z} \end{aligned} \right\} \quad (2)$$

Careful distinction must be made between du/dt and $\partial u/\partial t$, etc. The latter refers to the rate of change of u with time at a *particular* place; the former gives the actual rate of change of u for a particle moving from place to place.

Concentrate attention once more on the element of volume in Fig. 1.2. Let the pressure at P at any instant be p. We may now use precisely the method employed in the previous section to show that the components of force on the fluid in the box due to the pressure of the surrounding medium are $-\dfrac{\partial p}{\partial x}\,dx\,dy\,dz$, $-\dfrac{\partial p}{\partial y}\,dx\,dy\,dz$, $-\dfrac{\partial p}{\partial z}\,dx\,dy\,dz$ in the x, y, z directions, respectively. This follows from the definition of pressure as force per unit area. The other assumptions involved are precisely those which were discussed *in extenso* in Sec. 1.7. If no external forces in addition to that due to pressure act on the fluid we may get the equations of motion in the usual fashion by setting the mass times acceleration (in the three component directions) equal to the corresponding force components (due to pressure). Denoting the density by ρ, as before, and dividing both sides of the resulting equations by $dx\,dy\,dz$, we have finally for the equations of motion of a fluid subject to no external forces

$$\left.\begin{aligned}\frac{\partial u}{\partial t} + u\frac{\partial u}{\partial x} + v\frac{\partial u}{\partial y} + w\frac{\partial u}{\partial z} &= -\frac{1}{\rho}\frac{\partial p}{\partial x} \\ \frac{\partial v}{\partial t} + u\frac{\partial v}{\partial x} + v\frac{\partial v}{\partial y} + w\frac{\partial v}{\partial z} &= -\frac{1}{\rho}\frac{\partial p}{\partial y} \\ \frac{\partial w}{\partial t} + u\frac{\partial w}{\partial x} + v\frac{\partial w}{\partial y} + w\frac{\partial w}{\partial z} &= -\frac{1}{\rho}\frac{\partial p}{\partial z}.\end{aligned}\right\} \quad (3)$$

It will again be noted that we have to do with partial differential equations of the first order, as in the equation of continuity. Their integration under various special conditions yields many important physical laws of fluid motion, such as Bernoulli's celebrated theorem, for example. At present, however, we are more interested in another situation.

The wave equation. Take the special case of a fluid confined in a right circular cylindrical tube whose axis lies along the x axis of coordinates. We shall suppose that the only possible motion of the fluid is that along the x axis. This is, of course, a highly idealized situation. Nevertheless it will be worth our while to envisage the consequences. Assume further that initially the fluid in the tube is at rest. Suppose then a disturbance is created at some point

1.8 METHOD OF ELEMENTARY ABSTRACTION

in the tube involving the displacement of the particles of fluid in the immediate neighborhood of this point from their equilibrium positions. We make the additional assumption that these displacements, the corresponding velocities, and rates of change of velocity are small enough so that their powers higher than the first and their products may be neglected. What becomes of the equations of motion (3)? Only the first, of course, is applicable, since we are confining all motion to the x direction. Since $v = w = 0$, this becomes

$$\frac{\partial u}{\partial t} + u\frac{\partial u}{\partial x} = -\frac{1}{\rho}\frac{\partial p}{\partial x}. \tag{4}$$

But our order of magnitude hypothesis involves the result that $u\dfrac{\partial u}{\partial x} << \dfrac{\partial u}{\partial t}$, since both u and $\partial u/\partial x$ are small quantities. Hence the equation has the approximate form

$$\rho\frac{\partial u}{\partial t} = -\frac{\partial p}{\partial x}. \tag{5}$$

Recall the equation of continuity (1.7-4), which in the present case reduces to

$$\frac{\partial(\rho u)}{\partial x} = -\frac{\partial \rho}{\partial t}. \tag{6}$$

We shall suppose that the fluid before being disturbed has everywhere a constant density ρ_0 and a constant pressure p_0. Owing to the disturbance the pressure is altered by an amount p_e (the "excess" pressure, so called, which is a function of time and position in the tube), and the density by an amount ρ_e, the "excess" density. Thus we may write in general

$$\rho = \rho_0 + \rho_e \tag{7}$$

$$p = p_0 + p_e \tag{8}$$

It must be noted that our postulate of small displacements and velocities is to be supplemented by the assumption that $p_e << p_0$ and $\rho_e << \rho_0$.

Denote the displacement of fluid along the tube by ξ. Then $u = \partial\xi/\partial t$ by definition. Substituting for ρ the value given by (7) in eq. (6), we have

$$\frac{\partial}{\partial x}\left[(\rho_0 + \rho_e)\frac{\partial \xi}{\partial t}\right] = -\frac{\partial \rho_e}{\partial t}.$$

From the relative magnitudes of ρ_0 and ρ_e, this may be written to a good approximation

$$\rho_0 \frac{\partial}{\partial x}\left(\frac{\partial \xi}{\partial t}\right) = -\frac{\partial \rho_e}{\partial t}. \tag{9}$$

We integrate both sides with respect to t, noting that we may interchange the order of differentiation of ξ, and get

$$\rho_0 \frac{\partial \xi}{\partial x} = -\rho_e + C(x), \tag{10}$$

where $C(x)$ is a function of x only and independent of the time. By proper choice of the origin (which is arbitrary), it can be made zero. Now in every fluid there is a relation between the excess pressure and excess density. Assume that this is in our case a linear relationship,

$$p_e = c^2 \rho_e, \tag{11}$$

where c^2 is the proportionality constant. Substituting into (10), with $C(x) = 0$, gives us

$$\rho_0 \frac{\partial \xi}{\partial x} = -p_e/c^2. \tag{12}$$

Differentiating both sides with respect to x further yields

$$\frac{\partial p_e}{\partial x} = -c^2 \rho_0 \frac{\partial^2 \xi}{\partial x^2}. \tag{13}$$

But, from the equation of motion (5), we have

$$\frac{\partial p_e}{\partial x} = -\rho_0 \frac{\partial^2 \xi}{\partial t^2}, \tag{14}$$

where we neglect the small term $\rho_e \frac{\partial^2 \xi}{\partial t^2}$ as compared with the much larger term $\rho_0 \frac{\partial^2 \xi}{\partial t^2}$. The combination of (13) and (14) finally gives

$$\frac{\partial^2 \xi}{\partial t^2} = c^2 \frac{\partial^2 \xi}{\partial x^2}. \tag{15}$$

This is a second order partial differential equation the integration of which is to tell us how ξ must depend on x and t, that is, how the displacement of the fluid from equilibrium depends on time and position along the tube. It is very instructive to note once more how we got this equation. We obtained it from the combination of the equation of continuity (reducing in this special case to the form (6)), which expresses the effective continuity and indestructibility of the medium, and the equation of motion (5), which describes how limited portions of the fluid move in accordance with the principles of mechanics, provided it is assumed that the motion in this case is that involved in small displacements from equilibrium, as recognized in the approximation from which (5) follows from (4). The final differential equation (15), thus implies the two physical ideas of *continuity* and *mechanical description*. It is worth while to note that (15), besides being a second order equation, is also *linear* in that neither the dependent variable nor its derivatives enter in any form higher than the first power. The reason for this, of course, is closely connected with the approximations involved in neglecting terms like $u \dfrac{\partial u}{\partial x}$ in eq. (4), etc. We shall later see that the linearity of the equation has great physical significance.

Eq. (15) is a special form of what is usually called the " wave equation "—another of the predominantly important equations of theoretical physics. It will be worth our while to examine its solution to understand the significance of the name attached to it. The discussion of the last section has already forced upon us a realization of the extremely general and arbitrary nature of the solution of a partial differential equation. Indeed, it is probable that the reader has often wondered how the physicist could ever hope to get anything definitely physical out of such an equation. It will be our task in the rest of this section to answer this query, at least from the standpoint of the equation of wave motion.

Meaning of a wave. This is not the place to give a complete mathematical treatment of an equation like (15). We note merely that its general solution may be expressed in the form

$$\xi = f_1(x - ct) + f_2(x + ct), \qquad (16)$$

where f_1 is *any arbitrary* function of the argument $x - ct$ and f_2 is an arbitrary function of the argument $x + ct$. The reader may verify by direct substitution that this expression does satisfy the equation. Let us examine the physical meaning of the first part of the sum, viz., $f_1(x - ct)$. Choosing an arbitrary function, we plot $\xi = f_1(x - ct_0)$, where t_0 is a particular instant of time. The result is depicted graphically in Fig. 1.3 by the curve marked I. Pick out the value of

ξ for the point x_0; thus $\xi_0 = f_1(x_0 - ct_0)$. Now plot the curve for $\xi = f_1(x - ct_1)$, where t_1 is a later instant of time than t_0. The curve is represented in the figure by II. (Note that c is taken as a positive constant.) It is now clear that the ordinate or value of ξ of the second curve at $x = x_1$, viz., $f_1(x_1 - ct_1)$, will be equal to the value of ξ for the first curve at the point $x = x_0$, provided

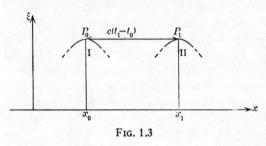

FIG. 1.3

$$x_1 - ct_1 = x_0 - ct_0,$$
or
$$x_1 - x_0 = c(t_1 - t_0). \qquad (17)$$

This means that the value of the displacement ξ corresponding to a particular place and instant of time x_0, t_0, respectively, is reproduced at the place and time x_1, t_1, respectively, where the interval $x_1 - x_0$ is precisely that which would be traveled in the time interval $t_1 - t_0$ by a particle moving with constant velocity c.[1] Applying the same reasoning to every point on the curve I, we see that $f_1(x - ct)$ is a function moving along the positive x direction with constant velocity c. The disturbance represented by $f_1(x - ct)$ is propagated along the x axis with velocity c. Such a propagated disturbance in a medium is called a *wave*, and the type of motion represented by a function of the

---

[1] Consulting eq. (11) we note that the velocity $c = \sqrt{p_e/\rho_e}$, where p_e is the excess pressure and ρ_e the excess density in the fluid. The expression may be put in terms of the actual absolute pressure p and density ρ. If the fluid is a liquid of volume V, a small change in volume ΔV is associated with a change in pressure Δp given by the well-known relation

$$\Delta p = -\frac{\Delta V}{V} E,$$

where E is the modulus of volume elasticity (bulk modulus). But from the definition of density $\Delta V/V = -\Delta\rho/\rho$. Hence $\Delta p/\Delta\rho = E/\rho$. But $\Delta p = p_e$ and $\Delta \rho = \rho_e$ in our above definitions. Therefore for a liquid the velocity of the elastic wave being considered is $c = \sqrt{E/\rho}$. It may be proved that, if the fluid is a gas, $c = \sqrt{p/\rho}$ or $\sqrt{\gamma p/\rho}$ (where γ is the ratio of the specific heat of the gas at constant pressure to that at constant volume) according as Boyle's law or the adiabatic law is chosen for the relation between p and ρ. Experiment indicates that the adiabatic law is the correct choice.

1.8 METHOD OF ELEMENTARY ABSTRACTION

form $f_1(x - ct)$ is *wave motion*. It may be shown in similar fashion that $f_2(x + ct)$ represents a wave moving in the *negative* x direction with velocity c. We must emphasize that the form or shape of the wave is purely arbitrary—all that the function implies is that a shape of arbitrary magnitude moves through the medium, preserving its shape as it moves. The complete solution (16) represents waves moving in the positive and negative x directions, respectively.

Introduction of boundary conditions. This is what happens when we create a small disturbance at any point of the fluid in the tube. From one point of view the result is not very startling and hardly illuminating, for it is not definite and precise enough for the purpose of the physicist. It is necessary to see how the observed properties of waves may be deduced from the mathematical description just obtained. We are again faced with the problem of deducing physical laws which may be tested in the laboratory. For this purpose it is clear that the mathematical description must become more concrete. Now it is important to observe that we can do this only by the introduction of *boundary conditions*, which in the case at hand serve to specify the values of the displacement in the fluid at certain chosen places at certain designated times. To understand this, let us suppose that the fluid is contained in a finite tube of length l with ends so closed that no motion is possible at either end at any time. This at once imposes on the general solution (16) the conditions

$$\left. \begin{array}{l} f_1(-ct) + f_2(ct) = 0 \\ f_1(l - ct) + f_2(l + ct) = 0, \end{array} \right\} \qquad (18)$$

where for simplicity the value of x for one end of the tube is taken to be zero and that for the other end l. Now what do these relations tell us about the nature of the functions f_1 and f_2? From the first we have

$$f_2(ct) = -f_1(-ct),$$

and therefore using the second

$$f_2(ct + l) = -f_1(l - ct) = f_2(ct - l).$$

Consequently if we represent $l + ct$ by z we have

$$f_2(z) = f_2(z - 2l), \qquad (19)$$

which means that, considered as a function of z, $f_2(z)$ repeats its value when z changes by $2l$. That is, $f_2(z)$ is a *periodic* function of z with period equal to $2l$. We may, however, write (19) in the form

$$f_2(ct + l) = f_2(c(t - 2l/c) + l). \qquad (20)$$

This states that f_2 is a *periodic* function of the *time* with period equal to $2l/c$. What we have shown for f_2 can also be proved for f_1. We see then that the mere imposition of *boundary conditions* has led to the limiting of the arbitrariness of the general solution (16) by making it necessary for the functions entering into the solution to be periodic functions of time with period equal to $2l/c$; this is the time required by the wave to travel from one end of the tube to the other and back. This extremely important result shows clearly the great power of boundary conditions in introducing concreteness and definiteness into the solution of a partial differential equation. Moreover, from the purely physical point of view the result is significant. It seems most natural to interpret it thus: waves pass along the tube in both directions and are *reflected* at the ends, so that a particular wave after passing a given point P will after two reflections at the ends again pass P in the same direction with the same magnitude of displacement as before.

Harmonic waves. This periodicity introduced by the boundary conditions at once suggests that periodic functions will be of great utility in the study of wave motion.[1] The simplest type of periodic function is found in the circular functions, i.e., the sine and cosine. Waves of this form are called *harmonic*. It will later be indicated that it is possible to express *any* periodic wave in terms of harmonics.[2]

Let us then confine our attention to harmonic waves. The displacement ξ at any point in the fluid at any time may be denoted by

$$\left. \begin{array}{l} \xi = A \sin k(x - ct) + B \sin k(x + ct) \\ + C \cos k(x - ct) + D \cos k(x + ct) \end{array} \right\} \quad (21)$$

where A, B, C, and D are arbitrary constants. It can be shown by differential equation theory that (21) is the most general simple harmonic solution of eq. (15). It may be readily proved that the constant k, inserted as a factor of the arguments $(x - ct)$ and $(x + ct)$, is equal to $2\pi/\lambda$, where λ is the wave length of the harmonic wave. To consult Fig. 1.4, which represents the shape of a harmonic wave at a given instant of time, i.e., a snapshot picture, the distance between two successive crests (or any two successive points in the same phase) is called the wave length and is designated by λ. Suppose that the single wave $A \sin k(x - ct)$ is considered at the time t_0. If the dis-

---

[1] The reader must take care to note that there is *no* idea of periodicity in the wave concept itself. Waves may be either periodic or non-periodic. As a matter of fact, however, it is the periodic type which is of greatest use in physical theory.

[2] The great theorem of Fourier. See below.

placement has its maximum at $x = x_0$ (i.e., the displacement there equals the wave *amplitude*, A), we have

$$\sin k(x_0 - ct_0) = 1 = \sin [k(x_0 - ct_0) + 2\pi n],$$

where n is any integer. But at $x_0 + \lambda$ we must also have by definition

$$\sin k(x_0 + \lambda - ct_0) = 1.$$

Therefore
$$k(x_0 + \lambda - ct_0) = k(x_0 - ct_0) + 2\pi,$$

since $n = 1$ from the definition of λ; it follows that

$$k = 2\pi/\lambda, \qquad (22)$$

as was stated. As the wave progresses in the positive x direction, the

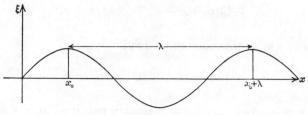

FIG. 1.4

time it takes for the crest to move through one wave length is called the *period*, usually denoted by T. This is related to the wave length by the relation

$$\lambda = cT, \qquad (23)$$

so that we may also express k as $2\pi/cT$. Finally the reciprocal of the period, or the number of waves which pass any point per unit of time, is the *frequency*, usually denoted by $\nu = 1/T$. In terms of frequency, k appears as $2\pi\nu/c$. Often we write $\omega = 2\pi\nu$ and then $k = \omega/c$. Matters of notation are of course merely questions of convenience.

What physical information do we expect to obtain from our solution (21)? We wish to be able ultimately to predict the displacement ξ at any place in the tube at any time, for if we are able to do this we may truly say that we know the state of motion of the fluid, and all other physical laws concerning its motion will also follow automatically. In order to carry out this program it is necessary to evaluate the four arbitrary constants A, B, C, and D. As before, we confine our attention to a finite tube of length l. To take a special case, suppose that both ends of the tube are rigidly closed so that $\xi = 0$ at $x = 0$ and $x = l$ for all t. Writing the condition for $x = 0$ we have for all t

$$(B - A) \sin kct + (C + D) \cos kct = 0. \qquad (24)$$

Now if this is to be true for all t

$$\left.\begin{array}{l} B = A \\ C = -D, \end{array}\right\} \qquad (25)$$

which immediately reduces the number of independent arbitrary constants to *two*. Our solution (21) takes the form

$$\xi = A[\sin k(x - ct) + \sin k(x + ct)]$$
$$+ C[\cos k(x - ct) - \cos k(x + ct)],$$

which after simple trigonometric reduction becomes

$$\xi = (2A \cos kct + 2C \sin kct) \sin kx. \qquad (26)$$

We are now ready to apply the boundary condition at $x = l$. From (26) it follows that

$$\sin kl = 0, \qquad (27)$$

whence

$$kl = n\pi, \qquad (28)$$

where n is any integer or zero. This interesting relation states that waves of not every frequency are allowed in the tube, but only those waves whose frequencies can satisfy this condition. Using the value of k in terms of the frequency and velocity, viz., the relation (28) becomes

$$\nu = nc/2l. \qquad (29)$$

Thus the allowed frequencies are integral multiples of the quantity $c/2l$, called the fundamental frequency. The permitted frequencies are the so-called "harmonics" of the fundamental. The striking nature of this result will be appreciated when it is noted that the boundary conditions have introduced *discreteness* into the problem. This function of boundary conditions will be further emphasized in the next section. We have shown that for each integral value of n

1.8 METHOD OF ELEMENTARY ABSTRACTION

there is a definite mode of vibratory motion of the fluid in the closed tube,[1] the symbolic expression for which is obtained from eq. (26) by substituting for k the value $n\pi/l$ according to (28). We thus get a whole set of possible solutions. Now it will be recalled that the original differential equation (15) is a *linear* equation. It is an important mathematical property of such an equation that, if one has any number of individual solutions, their sum is also a solution. This may be readily proved by inspection for eq. (15). To get a complete harmonic solution, therefore, it is necessary to add the individual solutions corresponding to the various values of n. The result may be represented by the following summation,

$$\xi = \sum_{n=1}^{\infty} \left(A_n \cos \frac{n\pi c}{l} t + C_n \sin \frac{n\pi c}{l} t \right) \sin \frac{n\pi x}{l}, \qquad (30)$$

where we have replaced $2A$ and $2C$ by A_n and C_n, respectively, to indicate that the constants for each particular solution need not be the same. The fact that the solutions representing particular modes of oscillation can be added together in this way to get the complete solution is known as the *principle of superposition*. Daniel Bernoulli first enunciated it in 1755 as the principle of the " coexistence of small motions."

It will hardly do to leave the result in this form, however. We still have on our hands the double set of arbitrary constants A_n, C_n. It is clear that their evaluation involves the introduction of further boundary conditions, though these should presumably now be conditions in *time* rather than *space*. Let us imagine that the displacement of the fluid at time $t = 0$ is denoted by ξ. This is, of course, a function of x and gives what may be called the initial state of the fluid. Thus we have

$$\xi_0 = \sum_{n=1}^{\infty} A_n \sin n\pi x/l. \qquad (31)$$

---

[1] The reader may show that if both ends of the tube are open, as might be the case if the fluid is a gas, the allowed frequencies or modes of oscillation are also given by (29). On the other hand, if one end is open and the other closed, $\nu = \dfrac{2n+1}{4} \cdot c/l$, where n is again integral. For the boundary condition at the open end, it is customary to assume that the excess pressure there is zero. This leads to $\partial \xi / \partial x = 0$ there, from eq. (12). As a matter of fact, experiment indicates that this is only a first approximation.

Suppose that the *velocity* of displacement in the fluid is initially given by $\left(\frac{\partial \xi}{\partial t}\right)_0$. Thus

$$\left(\frac{\partial \xi}{\partial t}\right)_0 = \sum_{n=1}^{\infty} C_n \cdot \frac{n\pi c}{l} \cdot \sin n\pi x/l. \tag{32}$$

We now have the problem of getting A_n and C_n out of limbo. Multiplying both sides of (31) by $\sin n'\pi x/l$, where n' is a given integer, and integrating from 0 to l, it is easily verified that

$$\int_0^l \xi_0 \sin n'\pi x/l \cdot dx = A_n \cdot l/2,$$

the reason being that all terms save one in the integral on the right vanish since they are of the form

$$\int_0^l \sin n_1\pi x/l \cdot \sin n_2\pi x/l \cdot dx,$$

with $n_1 \neq n_2$. We have a similar situation with respect to (32) and hence may finally write

$$\left.\begin{array}{l} A_n = 2/l \cdot \int_0^l \xi_0 \sin n\pi x/l \cdot dx \\[1em] C_n = 2/n\pi c \cdot \int_0^l \left(\frac{\partial \xi}{\partial t}\right)_0 \sin n\pi x/l \cdot dx. \end{array}\right\} \tag{33}$$

The immediately preceding analysis means that, if we know the initial displacement and displacement velocity for every point in the fluid, we can compute the displacement and velocity at any point for all subsequent times (or alternatively for all past times, for that matter). This has an interesting correlative mathematical result. It is conceivable that we can make the initial state of the fluid in the tube anything we like compatible with the imposed condition. That is, specifically, ξ_0 can be any continuous function of x defined in the interval from zero to l. What we have shown (without mathematical rigor, to be sure) is the possibility of expanding ξ_0 in the interval $(0, l)$ in terms of a series of circular trigonometric functions, the coefficients in the expansion being the constants A_n. This is termed a *Fourier series*; such series are of immense importance in theoretical physics.

1.9. The Significance of Boundary Conditions in Physical Theories. We have had occasion to note in the previous section the

1.9 BOUNDARY CONDITIONS IN PHYSICAL THEORIES 49

important rôle played by boundary conditions in the deduction of physical laws from physical theories whose fundamental hypotheses are expressed in terms of differential equations and in particular partial differential equations. The solutions of such equations involve both arbitrary functions and arbitrary constants and hence are at first of little concrete physical value. They obtain concreteness, as we have seen, only through the imposition of special conditions specifying the type of function of greatest utility, and evaluating the arbitrary constants. It is thus evident that the boundary condition concept is one of the utmost importance in theoretical physics, and is worthy of more extended study.

Specific boundary conditions. In this section we shall try to unify our ideas concerning such conditions and see what light they throw over the development of the whole subject. We begin by dividing boundary conditions into two classes, denoted as the *specific* and *general*, respectively. A *specific* boundary condition is simply a postulated event in space and time expressed by the statement that a symbol, representing a certain physical quantity, shall have a definite value or set of values throughout a specified region of space within a specified interval of time. From the mathematical standpoint there seems to be nothing very striking about this type of boundary condition. When we specify such a set of events in space and time we get a set of equations containing the arbitrary constants; from these equations the constants can be evaluated. Thus in the discussion of fluid motion in the preceding section we were able to evaluate certain of the arbitrary constants by assuming a knowledge of the initial displacement and velocity of every part of the fluid. This information led to a set of equations involving the constants.

Another illustration may help further to fix the idea. Using Newton's law of gravitation and the fundamental principles of mechanics we are able to compute the position and velocity of a particle moving in a gravitational field (such as a planet moving about the sun, for example) at all times past or future as soon as we know its position and velocity at a single instant. This information is just sufficient to evaluate for us the two arbitrary constants that occur in the second order ordinary differential equation expressing the application of the principles of mechanics to the problem. But we may note that boundary conditions like these have a physical significance as well. The necessity for such a condition tells us that, in order to predict the future behavior of any physical system (not necessarily restricted to a *dynamical* system), not only must we know the differential equation governing its changes, but also we must have definite knowledge of its state at a certain specified time or at least some property of the system expressible by a set of physical events. This may seem trivial to those accustomed to the purely mathematical point of view, but it appears to impose a definite restriction on the possibility of describing physical phenomena. It is perhaps unnecessary to add that these specific boundary conditions are customarily in physical texts referred to as *initial* conditions, since they usually specify events happening at some time chosen as the initial time. This fact, however, should not prevent us from recognizing them as limiting or restrictive and hence as boundary conditions in the sense in which we desire to use this term. In other words, in the present section we are not using the term *boundary* in its purely spatial significance. For us a boundary condition is any condition which limits or restricts the generality of a relation among physical symbols. The use of the term in this wider sense enables us to discuss together a number of very interesting questions which might otherwise seem disconnected.

Thus there are many physical problems peculiarly noteworthy because of our ignorance of the appropriate boundary conditions. One of the best illustrations is to be found in the kinetic theory of gases, which assumes that a gas consists of a large number of small particles called molecules moving in straight lines in every direction with varying velocities and colliding frequently with one another and the walls of the confining vessel. It is further postulated, at least in the simplest presentations of the theory, that each molecule is a rigid sphere and that its motions are to be described by the ordinary laws of collisions in classical mechanics (viz., conservation of momentum, etc.). From the strictly mechanical point of view, then, the problem of the description of the large-scale properties of a gas would reduce to the calculation of the position and velocity of every molecule at every instant of time. What does this imply? It implies the knowledge of boundary conditions, viz., the initial position and velocity of every molecule. This, however, is a hopeless task, for the number of entities is too great. What do we do in such a quandary? We invariably fall back on *probability* considerations. We cease to fix attention on the individual members of the aggregate and concentrate on the average, employing the principle of large numbers, that is, the method of statistics. In another part of this text this method is discussed in great detail (Chapters IV and V). All we wish to emphasize here is its connection with the ignorance of initial or boundary conditions where systems with many degrees of freedom (i.e., independent coordinates) are involved. This ignorance renders the strict use of the dynamical method impracticable for such problems and is

1.9 BOUNDARY CONDITIONS IN PHYSICAL THEORIES 51

thus closely tied up with the transition from a dynamical to a statistical theory.

Measurement and indeterminacy. It is also clear that specific boundary conditions are closely associated with the possibility of measurement. Fixing an event for physical purposes implies experiment and measurement. The calculation of the orbit of a planet involves, as we have seen, certain observations of the planet's position at different instants of time, equivalent to a knowledge of its position and velocity at some definite instant. The corresponding problem for a particle like an *electron* has recently been receiving considerable attention, the result of which has been the enunciation by W. Heisenberg of a principle of *uncertainty* or *indeterminacy* (the latter name appearing to be the more firmly established). This states that one cannot hope to carry out to indefinite precision the measurements fundamental to the fixation of the boundary conditions for particles in motion, no matter what the particles are, nor how small their number.

Putting the matter more concretely, suppose we wish to measure the position of an electron with great exactness. The principle asserts that we must then sacrifice precision in measuring its velocity, that is, we must remain content merely with the probable value of the velocity, or with the probability that the electron will have a velocity lying in some assigned interval. More precisely still, if we denote by x and v the position coordinate and velocity of the particle, respectively, by Δx the limit to which x can be measured (i.e., the *possible* error in fixing x) and by Δv the corresponding limit for velocity, the principle of indeterminacy may be expressed symbolically in the form

$$\Delta x \cdot \Delta v \geq h/m,$$

where h is Planck's constant of action with magnitude 6.55×10^{-27} erg sec and m is the mass of the particle. The product $\Delta x \cdot \Delta v$ is at least as great as h/m, though it is to be observed that its actual magnitude is not stated, the implication being, however, that the order of magnitude of the product is that of h/m. It is then clear that, for particles of ordinary mass, the uncertainty in measurement predicted by the principle is negligible. However, this is by no means true for a particle like an electron, whose mass in grams is numerically comparable with h in erg seconds. Thus if Δv is very small, Δx may be relatively enormous, indicating complete inability to localize the electron if the velocity measurement is to be precise, and vice versa. We shall deal more adequately with the principle in Chapter IX and attempt to assess its place in the modern quantum physics. What we wish to do here, however, is to stress its importance for the boundary condition problem. It puts our ignorance of the necessary boundary conditions in the atomic domain on a quite different basis from the probability considerations just discussed in connection with the kinetic theory of gases. There the disregard of the boundary conditions and the reversion to the study of average behavior were dictated by the desire for economy and convenience. In *principle*, the boundary conditions could be written down; in actual fact, this would prove so difficult that the exact dynamical method is dispensed with. However, the principle of indeterminacy is much more thoroughgoing, for it converts a question of convenience into a matter of *necessity*. It expresses the conviction ultimately that we shall *never* in *any* case be able to carry out even in principle the measurements necessary for the exact determination of boundary conditions in any physical problem. It is evident that such a drastic restriction demands more careful study.

General boundary conditions. We encounter a type of boundary condition different from the specific type so far discussed when we come to physical problems which involve the flow of energy across the interface of two media. It is necessary to postulate at such an interface a certain continuity for the functions representing the disturbance which is being propagated. It is not meant that we specify the exact values at the boundary; what we have to do is to set up certain relations among the physical quantities descriptive of the disturbance in the two media at the boundary. We may thus look upon this type of condition as more general. A simple illustration is provided by the transmission of a sound wave through a tube (see the previous section) in which there is an abrupt change in cross-section. We require that at the place where the change occurs there shall be *continuity* in the pressure and volume displacement of the air disturbance. This assumption when applied to the transmission equations leads to a value for the fraction of incident sound energy transmitted across the junction which is in agreement with that experimentally observed. Situations not essentially different are encountered in the behavior of light and all electromagnetic radiation at an interface. The imposition of the continuity conditions serves to fix the distribution of energy in the reflected and transmitted beams. Sometimes even more general considerations are involved. The familiar law of refraction (i.e., Snell's law) for an obliquely incident beam proves on the electromagnetic theory of light to be independent of the *precise* form of the boundary conditions: all that is demanded is that the quantities to be compared on the two sides of the boundary have the

1.9 BOUNDARY CONDITIONS IN PHYSICAL THEORIES

same functional form in the time and the coordinates of points on the bounding surface.

In the above illustrations, then, we see the significance of general boundary conditions: they impose fundamental restrictions on the type of activity possible for the system considered. The very presence of a spatial boundary in a wave field profoundly modifies the whole wave pattern. We have, indeed, seen this very clearly in the discussion of the motion of fluid in a tube in the preceding section. The condition of no motion at the closed ends is precisely a continuity condition at a boundary. It is of peculiar importance since, as we have noted, it serves to introduce discreteness into the problem by limiting the possible frequencies of periodic vibrations to a certain definite series of values, the so-called " harmonic " or " characteristic values." The general problem of which this is a special case has long had a particular charm for mathematicians. We have already emphasized this in connection with the classical physics of the motion of a continuous medium. In a later chapter we shall give further attention to this matter. However, it is worth pointing out here that the importance of boundary conditions has assumed a new aspect with the recent advent of the wave mechanics theory of atomic structure. In this theory the possible states in which an atom can exist are represented by the solutions of a certain wave equation (in form not unlike that of a continuous vibrating medium like a fluid or a string, but differing from it in the physical interpretation of the quantities involved—cf. Chapter IX) satisfying certain boundary conditions. As a result of these conditions it develops that the principal parameter of this equation, representing the energy of the atom (corresponding to the frequency in the analogous problem of the vibrating medium) is in general restricted to a limited set of values. In the cases so far studied these values are the correct energy values indicated by experiment for the possible stationary states of the atom. It is difficult to resist the feeling that the discreteness introduced by the above boundary condition is intimately connected with the discreteness which we ought to expect to be physically characteristic of an atomic mechanism and which is indeed the most striking experimental fact about all atomic phenomena.

The reader who is familiar with the history of recent atomic theory will recall that the classical Bohr theory of atomic structure employs a boundary condition which achieves essentially the same end as the condition of quantum mechanics with, of course, distinctive differences of detail which appear to favor the latter. Speaking specifically, the Bohr theory postulates that in an atom (constructed according to the

Rutherford nuclear model) the electrons move in accordance with dynamical laws, but from the whole continuum of possible mechanical orbits it selects as actually existent only those for which certain path integrals of the component radial and angular momenta are equal to integral multiples of h, the fundamental constant of Planck (already mentioned above in connection with the indeterminacy principle). Once more the function of the boundary conditions (here called *quantum* conditions) is to introduce discreteness into the problem. The energies of the allowed or "stationary" states of the atom are the energies of the orbits which are permitted by the quantum conditions; these are thus just as much boundary conditions as those involved in quantum mechanics.

The present subject takes on added interest from a study of the theory of relativity, for through it we may see our way to an even more general view of the nature of boundary conditions. Concomitant with the development of relativity has arisen the conviction that physical laws in general must be of such a character that they retain their form intact when the coordinate system in which the various quantities are expressed is transformed to another system in accordance with the now celebrated Lorentz-Einstein transformation equations, the latter being only the formal mathematical way of expressing the impossibility of detecting absolute motion by any means and the fact that the velocity of light has the same value for any two coordinate systems moving with respect to each other at constant velocity. In a certain sense we may interpret the invariance with respect to the Lorentz-Einstein transformation to be just as truly a boundary condition as is the quantum condition. It has, to be sure, a more general scope than the quantum condition, for it serves to select the type of physical law which is best adapted to describe physical phenomena successfully. Of still greater generality because of its larger realm of applicability is the fundamental postulate of general relativity that all physical laws must be expressed in the form of tensor equations. Here again the restriction is on the type of general law and not merely on the particular kind of behavior resulting as a special case of a given law. In this sense boundary conditions like the tensor requirement are more general than the differential equations in which most of the laws of physics are embodied. In truth, it would seem as if physicists were coming to realize more and more the importance of ascertaining the general boundary conditions which whole groups of physical laws must satisfy. This puts us in a position where the possible type of law is of greater significance than the specific laws themselves. From the esthetic point of view this

1.10 INTEGRAL EQUATIONS AND DIFFERENCE EQUATIONS

marks a tremendous gain for science as a whole. Nor is it without utility, as is illustrated by Einstein's use of it in the discovery of the general law of gravitation.

The reader should be able to suggest many other illustrations; but, to render the matter as clear as possible, it is well to point out that in our present view many so-called physical laws are really boundary conditions in disguise. Thus the law of the conservation of mechanical energy appearing as the first integral of the equations of motion of a conservative dynamical system is a boundary condition in the sense that it is the mathematical expression of the fact that the system under consideration is conservative; it thereby delimits or fixes a boundary for the class of systems considered and separates them from all others. The importance of this step becomes clear when we find that many of the specific laws obeying this general condition are found to be satisfied approximately in the world of physical phenomena. And the notion of the separation of dynamical systems into conservative and non-conservative systems has been of prime importance in the advancement of theoretical mechanics.

In summing up the whole matter we may say that closely associated with every physical law are to be found important boundary conditions, of both a general and a specific nature, forming a very significant part of the physical meaning of the law as well as its practical application. The specific conditions show us how to use the law to predict physical events, and ignorance of them forces us back on probability considerations; the general conditions fix the possible types of laws or the possible kinds of functions which enter into them. The mind can conceive countless forms of differential laws. The general boundary conditions serve to pick out the useful ones. Our search should forever turn in the direction of these conditions, the discovery and clear statement of which represent much of the progress of mathematical physics.

1.10. Integral Equations and Difference Equations in the Development of Physical Theories. Before leaving our general survey of the meaning of a physical theory we ought to consider one or two more special mathematical methods of physical interest. The first and more important of these is that involved in the *integral* equation, encountered in problems in which the idea of *heredity* enters. For example, consider a fiber which is twisted from its normal equilibrium configuration by the application of torque. When released it displays the familiar phenomena of elastic fatigue and hysteresis. This means that a knowledge of the state of twist and angular velocity of the fiber at any instant is not sufficient for a prediction of its state and motion at any subsequent time. Rather we need for this purpose the whole history of the fiber since first it began to move at all; that is, we must know its *heredity*. The name hereditary mechanics has been given to the field of problems into which there enter what are essentially boundary conditions extending over continuous intervals of space and time and demanding integrals for their representation.

To see this, let us examine the problem of the fiber more closely. The relation between the strain (i.e., the angular twist) and the torsional stress can now no longer be written in terms of the simple law of Hooke. Rather we must have for the strain

$$\delta = kX + \theta, \tag{1}$$

where X is the stress, k an elastic constant, and θ a quantity which depends on all the values which X has assumed from the time of application of the initial stress to the time t being considered. Now a quantity which depends on a whole range of values of another quantity is known as a *functional*. The area under a curve and lying between two ordinates provides a simple illustration. If we denote the equation of the curve by $y = f(x)$, the area under this and included by $x = a$ and $x = b$ clearly depends on the whole set of values of $f(x)$ between a and b. Coming back to (1) we shall assume that unit torsional stress applied to the fiber during a time interval of $d\tau$ from τ to $\tau + d\tau$ not only produces a certain twist in the fiber at this instant but also contributes a twist at the future time t of amount $\theta(t, \tau)d\tau$. The resultant angle of twist δ at the time t will then be given by $kX(t)$ plus the sum of all the residual terms $\theta(t, \tau)X(\tau)d\tau$. This sum of course must be an integral over all the time elapsing from the initial application of stress at t_0 to t. We can then write for the twist the generalized Hooke's law

$$\delta(t) = kX(t) + \int_{t_0}^{t} \theta(t, \tau)X(\tau)d\tau. \tag{2}$$

A knowledge of the previous torsional stress at any instant coupled with a knowledge of the function $\theta(t, \tau)$ would then allow us to compute the resultant twist at time t. Usually in (2) we are more interested in evaluating $X(t)$ which occurs under the integral sign, when we know $\delta(t)$ and $\theta(t, \tau)$, the latter of which is called the *coefficient of heredity*. Eq. (1) is then called an *integral equation*, and its solution demands special mathematical technique into which we shall not go.[1]

---

[1] The interested reader should consult, for example, Volterra, "Theory of Functionals" (Blackie, London, 1930).

1.10 INTEGRAL EQUATIONS AND DIFFERENCE EQUATIONS

We have merely desired to emphasize the existence of problems in which the boundary conditions are of such a nature as to call for this more powerful technique.

Difference equations. Our emphasis in the preceding sections on the principle of elementary abstraction and the resulting expression of physical principles and laws in terms of differential equations should not blind us to the fact that there are in physics very fundamental equations which are not in differential form and which do not appear to be readily deducible from differential equations. Such a one, for example, is the celebrated equation of quantum mechanics

$$h\nu = E_1 - E_2, \qquad (3)$$

which gives the frequency of the radiation emitted or absorbed by an atom in the transition from a stationary state of energy E_1 to one of energy E_2. This is strictly a *difference* equation, for although the energy difference $E_1 - E_2$ is usually very small compared with macroscopic energies, it cannot be made as small as we please.

It is very possible that finite difference equations will play an important rôle in the development of atomic physics. Indeed, it has been suggested [1] that even in classical physics we ought to use the calculus of finite differences rather than the differential method, since no real physical intervals ever become strictly infinitesimal. A suggestion of this kind implies that physical functions are defined only for *discrete* values of the independent variables, so that only finite differences of variables and functions have meaning. Thus in the simple example of the motion of a free particle along the x axis the differential equation of motion is

$$\frac{d^2x}{dt^2} = 0, \qquad (4)$$

leading to the integral

$$x = c_1 t + c_2. \qquad (5)$$

The corresponding difference equation is

$$\frac{\Delta^2 x_i}{\Delta t^2} = 0, \qquad (6)$$

where $i = 1, 2, \ldots n, \ldots$, and $x_1, x_2 \ldots x_n, \ldots$ are the values of x at the respective instants $t_1, t_2 \ldots t_n, \ldots$ with $t_k - t_{k-1} = \Delta t$ for all k. The solution of (6) is in the form

$$x = At + B, \qquad (7)$$

[1] See, for example, A. E. Ruark, *Phys. Rev.*, **37**, 315, 1931.

where A and B are functions of t arbitrary in all respects save that they are periodic with period equal to Δt. The distinction between (5) and (7) is clearly brought out by the following graph (Fig. 1.5), in which the straight line represents (5) and the dotted curve represents (7). The two cut each other at the points $P_0, P_1, \ldots, P_n \ldots$, which correspond to the only places where the position of the particle is definitely known. The solution (5) tells us too much. Eq. (7) in its very arbitrariness is more in accord with the *actual* physical situation. However, as soon as one attempts to generalize the method to the point where more complex cases can be handled, serious mathematical difficulties arise. It is likely that these will stand in the way of any considerable development of the point of view here noted.

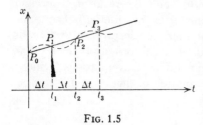

Fig. 1.5

We now bring our brief introductory review of the meaning of physical theory to a close.[1] After the foregoing general considerations it is well to particularize on certain special matters in more thorough fashion. We need to examine more closely the fundamental concepts of *space* and *time* in which all physical theories are expressed. We wish further to make a careful study of the theory of mechanics as the most highly developed example of a physical theory. We need also to pay attention to the important part which the concepts of *probability* and *statistics* have played and will continue to play in the development of physics. These will form the material of the next few chapters, which will then lead on directly to the fundamental questions connected with the growth of what may be called contemporary physics.

---

[1] No attempt has been made to provide a complete bibliography of the extensive material bearing on this subject. However, in addition to the older works of H. Poincaré, "Foundations of Science" (Science Press, N. Y., 1913), and N. R. Campbell, "Physics—The Elements" (Cambridge University Press, 1920), the reader is referred to the more recent books by V. F. Lenzen, "Nature of Physical Theory" (John Wiley and Sons, N. Y., 1931), and H. Jeffreys, "Scientific Inference" (Cambridge University Press, 1931). Consult also the articles by H. Reichenbach and H. Thirring in "Allgemeine Grundlagen der Physik," vol. IV of the "Handbuch der Physik" (Springer, Berlin, 1929).

CHAPTER II

SPACE AND TIME IN PHYSICS

2.1. The Nature of the Problem. The whole of classical physics is based on the concepts of space and time, and it is essential that before proceeding further we conduct a critique of them as they are used by the physicist. It is only thus that we can really appreciate the ultimate basis on which physical description has rested in the past and can understand the meaning of the statement which is now often made that space-time description is inadequate for the needs of modern physical theory, e.g., for a consistent theory of atomic structure and radiation.

On asking the question, what is space? we are at once confronted with what has been one of the major problems of philosophy through the ages. It is impossible for us to consider, within the limits of this book, all that philosophers have had to say about this matter. What should our attitude be toward the problem? The first thing that must be observed, apparently, is that strictly there is no such thing as *space*, but only *spaces*. By this is meant that connected with the notion of space are many quite different ideas. From the psychological standpoint, space is only a mode of sense-perception, a way in which we perceive things. In this sense, each individual possesses his own space which we may call *private* or *individual*. Its properties we shall discuss in a moment. But for the sake of human intercourse we agree to forego the emphasis of each on his own private space and create a mental abstraction which we may call *public* or *physical* space. This is an extrapolation or abstraction of the space of experience. Finally, the mathematician creates purely conceptual spaces possessing properties unconditioned by physical observations. These are the various spaces of geometry. It shall be our first task to distinguish between these different kinds of space and so justify the point of view which recognizes the distinction.

We shall see in a moment (Sec. 2.5) that much the same type of situation exists with respect to time.

2.2. Psychological Space or the Space of the Individual. We agreed in the beginning of Chapter I to look upon human experience as made up of sense-perceptions and the reasoning associated with these perceptions. For purposes of convenience in description we say that our perceptions are due to objects localized in space. Our sense-perceptions are not a confused jumble but display a certain order—it is this order which we denominate by the term space. Following the philosopher Leibniz one may define space as the "order of coexistent sense-impressions." The word space therefore appears as a symbolic representation of our ability to perceive things apart. Now it is clear that, as the sense-perceptions of the individual are of various kinds, to each type there corresponds a special space. Thus there are visual, tactile, motor, olfactory, auditory, and gustatory spaces corresponding to the different senses. Of these it seems clear that the first three are more important than the last three. A person who cannot and never has been able to see or feel or move will surely have a very limited concept of space. It is, to be sure, probably rather artificial to separate the various spaces, and some might well hold that space to each individual is the synthesis of the ordering of experience in its broadest sense. However, this is not so important from the standpoint of our present discussion. What is more to the point is the question: what are the properties of psychological space?

To answer, each individual needs only to consult his own experience. His space is *finite*, for his range of sense-perceptions is finite. This is the only meaning which we can give to the word finite here. His space is non-homogeneous; that is, it does not possess the same properties in every part. It has frequently been remarked that a cavity in a tooth feels very much larger to the tongue than to the finger, and the pricks of a pair of compasses placed close together are easily noted as two by the finger but appear as one prick on the back. We describe this physiologically and psychologically by saying that the nerve centers in tongue, finger, and back are spaced differently. This, however, already implies a more abstract concept of space. We are here dealing with raw experience. Coming to visual space, a house at a distance may appear smaller than a pencil close at hand, railroad tracks appear to converge in the distance, etc.

Private space is also non-isotropic; that is, it does not possess the same properties in every direction. The sense of sight, for example, is not equally good in every direction—we say that the nerve endings at the edge of the retina are not so sensitive as those near the yellow spot; the sense of hearing is most acute when the source of sound is on the line joining the two ears. Private space is discontinuous, as the existence of the "blind" spot in the eye indicates for the visual sense. The dimensionality of private space is highly uncertain. Visual space is strictly two dimensional in so far as the action of lens and retina alone are concerned. Yet the muscular power of accommodation gives depth and distance and thus effectively adds a third dimension. It is difficult to think of dimensionality at all in connection with audition, whereas motor space may appear to have an indefinite number of dimensions. The reader must recall that all this discussion is based on pure sense-perception. The beginning of conceptualization at once introduces a new situation wherein the space of the individual is abstracted into what may be called " public " space.

2.3. Public or Physical Space. Public or physical space is an abstraction by the mind of the aggregate of the various modes of sense-perception. One may think of it also as a kind of idealization of the space of the individual in which for purposes of social convenience many properties lacking in that space are supplied. Thus physical space is homogeneous, continuous, isotropic, three-dimensional, and Euclidean (in its geometry—of which see more further on). It is this space which the physicist uses to describe his laboratory operations and experiments, and it is this space whose properties play such a dominant rôle in the development of his theories. The evolution of the concept of physical space is an interesting topic in psychology, but lies somewhat outside our present field.

From the standpoint of physics, however, some factors must be stressed. When a physicist does an experiment, if it is of that quantitative type which we have learned to think of as the most significant for the development of physics, he invariably ends with sets of pointer-readings. The very possibility of such pointer-readings implies the existence of scales, i.e., rods with marks on them bearing some orderly relation to one another. The construction and use of a scale are therefore matters of the utmost importance to experimental physics and hence to the whole development of the subject. In the construction of a scale there is already involved an operational notion of a *space interval*. It is based, of course, on operations carried out with actual physical objects. Associated with each pair of marks (ultimately points) on a *rigid*, i.e., experimentally not easily deformable, physical object, there is a certain space interval, to which numerical magnitude may be given as soon as agreement is reached as to when two such intervals are equal. The convention is made that, if it is possible to *superpose* the two pairs of points so that when one of each pair coincide the others also coincide, the associated intervals are *equal*.

It will be noted that this and other conventions employed in measurement may be set up and discussed without reference to abstract space; it is only necessary to talk about actual operations with carefully made apparatus. The properties thus specified are, strictly speaking, properties of the apparatus and the operations employed. Some of those properties or conventions are indeed very far reaching. For example, it is assumed that the interval associated with any two points on a rigid rod (used as a scale) is independent of any motion of the rod; that is, no matter how the rod is moved about as a measuring instrument, the interval remains the same. This is a natural assumption; it is difficult to see how measurement could be carried out at all simply without it. Yet we must recognize its postulational nature. As soon as we have transferred it from its status as a postulate about operations with physical objects to a postulate about "space," it becomes equivalent to the assumption of isotropy and homogeneity (or free mobility, as we shall later call it) which are then considered characteristics of physical space.

How then may we sum up the matter? In order to carry out experimental measurements and describe them with the greatest convenience the physical scientist decides upon certain conventions with respect to his measuring apparatus and the operations performed with it. These are, strictly speaking, conventions with regard to physical objects and physical operations. However, for practical purposes it is convenient to assume for them a generality beyond any particular set of objects or operations. They then become, as we say, properties of space. This is what is meant by physical space, which we may define, in brief, as the abstract construct possessing those properties of rigid bodies that are independent of their material content. Physical space is that on which almost the whole of physics is based, and it is of course the space of everyday affairs.

2.4. The Transition to Geometrical Space. Why is it necessary to go further than physical space? Why should we not remain content with this abstraction from actual physical experiments? The answer to these questions is to be found in the following considerations. By means of experiments made with physical objects certain results of general validity are obtained. It is found, for example, that if rods are joined in various ways to make what may be called figures, the figures have properties which appear to be independent of the particular nature of the rods. Certain rods are called "straight," such a rod being one for which the interval associated with its extremities has the smallest numerical measure in terms of unit intervals of arbitrary size —at least, this is one possible definition. When one end of one straight rod is joined to one end of another straight rod the difference in the directions of the two rods gives rise to the concept of angle, and angular measure is invented. When three straight rods are joined to form a closed figure it is called a triangle. Measurement then dis-

2.4 THE TRANSITION TO GEOMETRICAL SPACE

closes a number of properties of such a figure, e.g., the sum of its three angles is equal to two right angles.

It must be emphasized that we are here considering these results as obtained directly by experiment on physical objects. They are therefore subject to the same uncertainties that surround all physical experiment. Nevertheless it is found possible to build a theory in which the concepts are points, lines, planes, etc., which are abstractions from rods and sheets, etc., and by the assumption of certain postulates which appear operationally reasonable, to deduce all the results of measurement on rigid rods and rigid bodies. This theory is geometry, and the particular type which best appears to fit the experiments on actual bodies is the geometry of Euclid. We may take this geometry as defining a *space* which is an idealization of the physical space of rods and other measuring apparatus. This is what we mean by *geometrical space*. It is, of course, a construction of the mind and bears the same relation to physical space that the world of physical theories (i.e., what we have termed " the physical world ") bears to the external world of phenomena. Like physical theories in general, geometry employs symbols and consists ultimately of those relations among these symbols which can be derived from the initial postulates.

Euclidean geometry. Because of its highly complete nature and the accuracy with which it has met all experimental tests, there was in the past a tendency to regard Euclidean geometry as forced on the mind in *a priori* fashion, as conditioning " real " space, that in which all things actually happen. Thus if somewhat more than 100 years ago anyone had raised the question as to the uniqueness of the transition from physical space to geometrical space, there would have been no doubt of the answer: all experiment, all experience in actual space, point directly to *one* geometrical space, namely, that of Euclid. So strong was the feeling for this uniqueness of geometrical space that philosophers like Kant took the stand that the axioms and postulates of Euclid are *a priori* synthetic judgments, a necessary condition for knowledge, something without which man cannot hope to describe experience. It is an interesting fact that a conclusive answer to this point of view was provided by mathematicians who were seeking to settle another question. For many centuries one of the outstanding problems of mathematics had been the attempt to reduce the number of Euclid's postulates. It will be recalled that Euclid began his treatment of geometry with *definitions*, *axioms*, and *postulates*. For convenience we shall list some of these here.

    A. Definitions, e.g.:
        1. A *point* is that which has no parts.

2. A *line* is length without breadth.
3. A *straight line* is one which lies evenly between two of its points.

B. Axioms, five in number, as follows:
   1. Things equal to the same thing are equal to each other.
   2. If equals be added to equals the results are equal.
   3. If equals be subtracted from equals, the remainders are equal.
   4. The whole is greater than any one of its parts.
   5. Things that coincide are equal.

C. Postulates, also five in number, as follows:
   1. It is possible to draw a straight line joining any two points.
   2. A terminated straight line may be extended without limit in either direction.
   3. It is possible to draw a circle with given center and through a given point.
   4. All right angles are equal.
   5. If two straight lines in a plane meet another straight line in the plane so that the sum of the interior angles on the same side of the latter straight line is less than two right angles, the two straight lines will meet on that side of the latter straight line. (This is often called the *parallel postulate*, since Euclid found it necessary to assume it in order to prove that through a point not on a given straight line there is one and only one line parallel to the given line.)

It is interesting to note the arrangement into axioms and postulates, the axioms presumably referring to self-evident rules of thought, and the postulates to necessary geometrical assumptions. It was a highly natural procedure to endeavor to show that some of these postulates are merely deductions from the definitions and the axioms. Particular attention was paid to the fifth postulate, and many so-called " proofs " were concocted. For a good review of the history of this matter the student is referred to J. W. Young, " Fundamental Concepts of Algebra and Geometry " (Macmillan Co., New York, 1917), pp. 26 ff. We need only point out that all attempts were failures. In particular, the *reductio ad absurdum* method of proof which consisted in assuming that the postulate was not true and then endeavoring to show that in the derived theorems a contradiction ensued, merely indicated that the denial of the postulate led to no inconsistencies.

Non-Euclidean geometries. Finally, in 1832 the Hungarian *Bolyai* and in 1835 the Russian *Lobatchevsky* definitely announced the discovery of a non-Euclidean geometry which is just as complete and consistent as that of Euclid. They denied the parallel postulate (while retaining all the other postulates) by assuming in effect that it is possible to pass through a point not on a given straight line an infinite number of lines parallel to the given line. The theorems of Lobatchevsky geometry naturally sound a bit strange. Thus the sum of the angles of a triangle is *less* than two right angles, and the

2.4 THE TRANSITION TO GEOMETRICAL SPACE

defect (i.e., difference from two right angles) varies directly as the area of the triangle. Another important result of this geometry is that it is impossible to construct two similar figures of different dimensions, i.e., it is impossible to magnify or diminish a figure without altering its shape. We shall consider shortly the physical importance of this. Though the theorems are different from those of Euclid, no one has ever detected any trace of inconsistency in them. It is now quite certain that no one ever will.

In 1854 the mathematician Riemann suggested another type of non-Euclidean geometry based on an extension of the first postulate of Euclid and a denial of the fifth. He assumed that through a point not on a given straight line *no* parallels can be drawn to the given line, and that there exist infinitely many pairs of points through each pair of which an infinite number of straight lines can be drawn. Here again many interesting theorems may be derived without ever running into inconsistency. Among these, for example, is the result that the sum of the angles of a triangle is *greater* than two right angles, with the excess proportional to the area of the triangle.

Riemann's geometry is of particular interest because it was through it that mathematicians were led to the conviction that a non-Euclidean geometry can be just as consistent as that of Euclid. It develops that, for the case of two dimensions, Riemann's geometry reduces to that of the sphere. It will be recalled that the analogue of the straight line on a sphere, i.e., the shortest distance between two points, is the arc of a great circle. Since every great circle intersects every other one we see that it is impossible to draw parallel " straight lines " on the sphere. Moreover, in spherical triangles the sum of the angles is always greater than $180°$, etc. For every theorem in Euclidean plane geometry there is a corresponding theorem in spherical geometry which can be translated into the former by the appropriate definition of the terms employed. If there were an inconsistency in the spherical geometry there would have to be a corresponding inconsistency in plane Euclidean geometry. If, therefore, we believe in the consistency of the Euclidean geometry we are compelled to believe in that of the spherical geometry. But for every theorem in the Euclidean geometry of the sphere there is a corresponding theorem in the three-dimensional analogue, which is precisely Riemann's geometry. The mere increase in the number of dimensions cannot entail loss in logical consistency. In this way, then, we convince ourselves of the consistency of the Riemann geometry.

That a similar state of affairs exists for the geometry of Lobatchevsky and Bolyai was clearly shown by the Italian mathematician

Beltrami. He proved that this type of non-Euclidean geometry is the three-dimensional analogue of the Euclidean geometry of a surface formed by rotating a *tractrix*, or curve with the parametric equations $x = \cos \phi$, $y = \log \tan (\pi/4 + \phi/2) - \sin \phi$, about the y axis. The shape of the tractrix is indicated in the accompanying figure (Fig. 2.1). It is tangent to the x axis and approaches the y axis asymptotically. This surface of revolution has been called a "pseudo-sphere," and the Lobatchevsky type is therefore sometimes referred to as "pseudo-spherical" geometry.

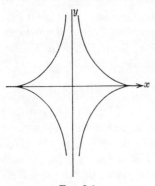

Fig. 2.1

Space curvature. Undoubtedly the clearest view with respect to the physical meaning of the various kinds of geometries or geometrical spaces is supplied by Riemann's generalization of Gauss's important mathematical theory of curved surfaces. It will be recalled that if we express the equation of a surface (in three-dimensional space) in the form

$$z = f(x, y), \tag{1}$$

the so-called *total curvature* of the surface at any point is given by the expression

$$C = \frac{1}{R_1 R_2} = \frac{rt - s^2}{(1 + p^2 + q^2)^2}, \tag{2}$$

where the symbols have the following significance:

$$p = \frac{\partial z}{\partial x}, \quad q = \frac{\partial z}{\partial y}, \quad r = \frac{\partial^2 z}{\partial x^2}, \quad t = \frac{\partial^2 z}{\partial y^2}, \quad s = \frac{\partial^2 z}{\partial x \partial y}.$$

R_1 and R_2 are the so-called *principal* (i.e., maximum and minimum) radii of curvature at the point in question. It can be shown that C has a definite value at every point of the surface independently of the way in which the surface is represented. This value will, of course, in general vary from point to point. Without going into detail, let us merely illustrate by means of the sphere, whose equation in the form (1) is

$$z = \sqrt{a^2 - x^2 - y^2}, \tag{3}$$

where a = radius. If we form the various derivatives indicated above and substitute into (2) we find $R_1 = R_2 = a$ and

$$C = 1/a^2. \tag{4}$$

This result might of course have been obtained much more simply from the fact that for a sphere the principal radii of curvature R_1 and R_2 at any point (which are here the radii of curvature of the circles formed by cutting the sphere by perpendicular diametral planes) are each equal to a, the radius of the sphere. The important fact, in any case, is that the total curvature of the sphere is constant and greater than zero. A plane may be treated as a sphere of infinite radius, and hence for it also the total curvature is constant and indeed equal to zero. On the other hand, there are surfaces for which the total curvature is negative, i.e., R_1 of the opposite sign from R_2. Such is true, for example, at certain points on a torus, though it is clear that the curvature is not constant for a torus. However, the pseudo-sphere of Beltrami is a surface of constant negative curvature.[1]

Now what is the physical significance of constant curvature? Consider any surface, and imagine that a perfectly flexible but inextensible piece of cloth with a figure on it is placed on the surface so that the contact between cloth and surface is perfect, i.e., there are no wrinkles. Suppose now that we wish to slide the cloth over the surface without altering the dimensions of the figure. In general this cannot be done. For most surfaces the cloth would have to leave the surface to satisfy the condition or alternatively would have to be extensible to stay in perfect contact with the surface. However, for a surface of *constant* curvature the indicated operation *can* be carried out. Thus a two-dimensional being living on such a surface could move about without changing his dimensions or those of his various pieces of apparatus.

If now we pass from two to three dimensions, as Riemann did, the problem becomes more complicated. But Riemann showed that in a space of n dimensions there exist $n(n-1)/2$ expressions having definite value at each point of the space. These characterize the nature of the space about every point just as the total curvature characterizes the nature of a surface in the neighborhood of a given point. There are certain spaces for which these expressions are identical, reducing to a single expression which in turn remains constant for all points. In this case the space is referred to as space of " constant curvature,"

[1] See, for example, G. Scheffers, "Anwendung der Differential und Integral Rechnung auf Geometrie" (Veit & Co., Leipzig, 1910), Vol. II, pp. 139 ff.

and the characteristic constant is called the absolute "space curvature."

This is, indeed, a rather unfortunate nomenclature, since in the generalization the physical idea of curvature is lost. Perhaps it would be better to refer to this quantity as the *space-constant*, as suggested by Hobson.[1] However, in any case, Riemann showed that, if this space-constant or "space curvature" is zero, the geometry of the corresponding three-dimensional space is Euclidean. If it is less than zero the geometry is Lobatchevskian or "pseudo-spherical," whereas if it is greater than zero, the geometry is that of Riemann or "spherical" (these terms here, of course, all refer to three dimensions). The expression for the "space curvature" or space-constant is too complicated for us to go into. However, it may be remarked that it is a measure of the deviation between the sum of the angles of a triangle and two right angles. Thus if Δ represents this deviation for a triangle of area A, we have Δ/A = space-constant. The more highly "curved" the space, the greater is Δ for a given triangle.

It is customary to refer to three-dimensional space with Euclidean geometry as *flat* or *homaloidal* space. The other two varieties are non-homaloidal or "curved." These three types of space, homaloidal, spherical, and pseudo-spherical, are the only three which satisfy the condition of *free mobility*, that is, the possibility of moving physical objects from place to place without altering their dimensions. This is, of course, the most fundamental criterion for physical measurement, for all such measurement implies the superposition of scales and hence of physical objects. Hence if objects were to change their dimensions on displacement, measurement would become meaningless. If, therefore, physical experimentation involving measurement is to have any meaning, it would seem that we are restricted to geometrical space of constant curvature. Even this allows considerable leeway. However, it can be shown that, in spherical and pseudo-spherical space, magnification or diminution of a geometrical figure alters its shape; that is, it is impossible to have in "curved" space *similar* figures of different dimensions. These can exist only in flat space.

Poincaré's non-Euclidean world. We have seen in the preceding paragraphs that Euclidean geometry is not the only logically consistent one. Indeed, there exist three geometrical spaces of constant "space curvature." The question may now be raised: granted the truth of what has been said, what is its significance for physics? Even if the "curved" spaces are perfectly logical, does this alter the fact that,

[1] E. W. Hobson, "The Domain of Natural Science" (The University Press, Cambridge, England, 1923).

2.4   THE TRANSITION TO GEOMETRICAL SPACE

after all, flat space is the one best fitted to describe all physical phenomena? It is just at this point, however, that doubt arises. This doubt can be made most clear by an illustration due to Poincaré. He asks us to imagine a universe in the form of a great sphere (as we look upon it from the outside). The temperature is taken to be a maximum at the center and decreases toward the periphery where it attains the value absolute zero. More specifically, the radius of the sphere being R and the absolute temperature being T at distance r from the center, we assume with Poincaré that $T = K(R^2 - r^2)$, where K is a constant proportionality factor. At the center, then, $T = KR^2$, its maximum value, while $T = 0$ for $r = R$, i.e., at the boundary. Poincaré further assumes that a decrease in temperature is accompanied by a contraction in size, so that any object moving outward from the center toward the periphery shrinks and becomes indefinitely small as the periphery is approached. The assumption is also made that these alterations in size take place instantaneously, that is, any object at a given place takes on at once the dimensions corresponding to the temperature at that place. We must, of course, grant the possibility of motion for physical objects from point to point in this hypothetical world.

What, then, are some of the space properties that we, looking in from the outside, should associate with this universe? We, of course, should consider it a finite world. An inhabitant, however, would naturally look upon it as infinite, if he had any conception of infinity, for since in moving toward the boundary he contracts in size, his steps or whatever rhythmic movements he makes in motion get smaller and smaller and he would have to take an infinite number of them to reach the boundary. He would doubtless conclude that there is no boundary. Would he be apt to get a clue to his difficulty by detecting any change in his size? No, because all his measuring rods would change with him, and relative dimensions would be preserved.

What kind of geometry would the inhabitants of Poincaré's world invent? Would it be Euclidean? To answer this question consider what the inhabitant would call a straight line. If by a straight line he meant a *geodesic*, that is, the shortest path or that involving the smallest number of steps, then it may be shown that a straight line is the arc of a circle which intersects the sphere orthogonally. Indeed it can be proved that through any two points inside the sphere one and only one " straight line " can be drawn.[1] Of course, this is only one physical aspect of a straight line. A critic might object that,

---

[1] See, for example, Young, "Fundamental Concepts of Algebra and Geometry," p. 18.

though the shortest path is a circle, to the hypothetical inhabitant it would not look "straight," that is, light would not necessarily travel along such a path. To obviate this difficulty let us assume that the sphere is filled with some gas whose index of refraction varies from point to point so that light travels along curved paths and indeed along the geodesics above described. In this case the inhabitant would look upon his geodesics precisely as we look upon our straight lines.

Further study shows that in this universe the parallel postulate will not be true. Rather, through a given point, not on a given orthogonal circle, it is possible to draw infinitely many orthogonal circles which do not intersect the given circle inside the sphere. All these lines would presumably be considered parallels by the inhabitants of Poincaré's hypothetical world. These parallels, to be sure, lie in an angular region about the point which becomes smaller as the distance to the center of the sphere becomes smaller. Very close to the center the inhabitant might not be able by his measuring instruments to separate out the various possible parallels and recognize them as distinct.

Now if our own earth were actually situated in such a universe as has been described, and if it were very near the center, we should probably not be able to detect our presence in it by any measurement, since the orthogonal circles very near the center are so close to what we call straight lines that no measurements of ours could hope to show the difference. More important still, any endeavor to detect the various distinct lines through a given point parallel to a given line would fail since the lines are too close together for separation. The same situation would exist for the various other geometric properties. In other words, we may be living in a non-Euclidean world (i.e., non-homaloidal space) without being able to detect the fact. This must then be taken into consideration in any attempted decision on the question as to whether flat space is after all the one best adapted to describe all our experiences. It is perfectly conceivable that a physical phenomenon may be discovered which can be adequately described only in terms of a non-homaloidal space such as has just been discussed. In other words, we have no right to feel certain that homaloidal space will always remain the one most useful for our purposes. As a matter of fact, we shall see a little further on (Chapter VIII) that the theory of general relativity actually implies that the space of the physical universe is non-homaloidal, and various types of "curved" spaces have played an important rôle in the recent developments of the theory.

It would seem that we must conclude that experiment can afford

2.4 THE TRANSITION TO GEOMETRICAL SPACE

no valid criterion for a choice of geometry. Experiments are performed with actual physical objects, and we have just seen that the transition from the experimental results to geometrical space is by no means unique. Choice of geometry is therefore dictated by reasons of convenience; for most purposes Euclidean geometry works well. When it fails to satisfy, we do what we ordinarily do with other physical theories in a similar predicament: frame a new geometry. It is well to emphasize again that we are taking toward spatial ideas the same postulational attitude we have all along taken toward physical theories in general. We have nothing to do with any possible "real" space which lies behind phenomena, as it were, or in which phenomena appear to be embedded, and towards a knowledge of which we may be supposed to be slowly but surely progressing.

Other geometries. The question may be asked: are there other geometries besides the three which correspond to space of constant "curvature"? As one might expect, the answer to this question is in the affirmative. There are indeed an infinite number of geometries. Restrictions may be placed on the number only by the introduction of subsidiary conditions. Thus Sophus Lie[1] has proved that if we assume: (1) that space has a finite number of dimensions n, (2) that free mobility is possible, (3) that a finite number, p, of conditions are necessary to determine the position of a figure in this space, then the total number of geometries compatible with these requirements is finite. These geometries, of course, contain theorems which sound very queer to one accustomed to Euclidean geometry. For example, Poincaré has investigated a geometry in which restrictions are placed on the possible rotation of a straight line in a plane about a point on the line. Thus to consult Fig. 2.2, in Euclidean geometry one usually assumes implicitly (or unconsciously) not only that the line AC may be rotated about A so as finally to coincide with AB, but also that AC may be further rotated so as to coincide with AB produced beyond A. Poincaré has showed that admission of the first assumption and denial of the second leads to a geometry quite as consistent as Euclidean. One of the theorems of this geometry is that a straight line may be perpendicular to itself. Of course, the notion of perpendicularity in this geometry will be different from that in Euclidean geometry.

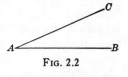

FIG. 2.2

Attention may also be called to the non-Archimedean geometry of Hilbert, so called because it is based on a denial of the assumption of

---

[1] See Poincaré, "Foundations of Science" (Science Press, New York, 1913), p. 63.

Archimedes which is equivalent to the axiom of Euclid that two magnitudes are said to have a ratio if they are such that a multiple of either may exceed the other. More specifically, from a geometric point of view, if we have any segment of a straight line, no matter how small, it is always possible to find an integer n such that n times the given segment is larger than any previously assigned segment. It is clear that this assumption is fundamental for measurement as we understand it in physics. If now all the axioms and postulates of Euclid save this alone are assumed we get an entirely new geometry. Those interested in these considerations should refer to the book by J. W. Young already mentioned.

Further discussion of the significance of space in physics will be postponed to Chapters VII and VIII.

2.5. The Significance of Time in Physics. It is certainly one of the striking features of human thinking that we have not one mode of grouping sense-impressions, but two. In addition to space, which we have interpreted in its most general sense as the mode of perceiving things apart, there is time, which is our way of referring to the perception of events one after the other, i.e., succession. It is, indeed, perfectly possible to imagine a universe in which space alone has significance, but it would be one in which nothing ever happens, or rather in which there occurs a single event, namely, the observation on the part of some observer of a static perception pattern. As soon as we admit the possibility of more than one event in our universe we imply the possibility of *order* for these events. This is what we mean by the term *time*. At least this is the first meaning.

Private time. There is, however, an additional mode of perception denoted by the word "duration." As human beings we not only introduce the idea of order into the description of events; we also attempt to assign a metric to this order. We are not satisfied with noting that event B follows event A and event C follows event B. We feel the urge to assign a numerical estimate to the time " interval " from A to B as compared with that from B to C. We are doubtless stimulated to this judgment by the observation of certain periodicities in vital phenomena, such as recurring hunger and fatigue, as well as similar rhythms in external nature, such as the reappearance of day after night, the rise and fall of the tides, etc. The faculty of estimating duration is probably universal, but at the same time it is subject to wide fluctuations in judgment between different individuals and between judgments of the same individual at different times. This is recognized in the familiar remark that time passes slowly in childhood and more rapidly with increasing age. Time appears to fly when we are deeply engrossed in mental or physical activity, whereas it drags when we have nothing to do.[1] We can sum up by saying that each individual appears to have his own *private* time which is to a certain extent analogous to his private space; i.e., it is conditioned by characteristic peculiarities in his mental and physical constitution.

Public time. For the sake of forwarding mutual activities mankind has found it necessary to establish arbitrarily a public time based on the recurrence of some simple phenomenon evident to all so-called normal persons and agreed upon as exhibiting a type of regularity in consonance with which they are willing to adjust their affairs. Such is the passage of the sun or the stars across the meridian. The invention of mechanisms (viz., clocks) with periods closely approximating such phenomena naturally follows, and the measurement of public time has been a subject of increasing precision through the ages.

As with space, however, human thought has not remained content with either the intuitive feeling for duration or its empirical measure by a conventional mechanism: it has striven for a more abstract concept to serve as a foundation for logical thinking about physical events—a *time* which would be independent of individual consciousness or the operation of any external mechanism. This conceptualization goes back, indeed, to the time of the Greek philosophers, who puzzled themselves greatly over the question of the divisibility of time intervals. In this connection the paradoxes of Zeno come to mind with their assumption that an infinity of instants necessarily corresponds to an infinite time interval, leading to results very disturbing to theorists, which have been made clear only by the relatively recent mathematical studies of the infinite and infinitesimal.

Physical time. The first clear statement of the nature of the concept of time as used in the development of physical theories is undoubtedly that of Newton in his " Principia." Distinguishing between absolute time (the abstract concept for use in physics) and common time (which is equivalent to the " public " time discussed above), Newton adopted this assumption: " Absolute, true and mathematical time or duration flows evenly and equably from its own nature and independent of anything external; relative, apparent and common time is some measure of duration by means of motion (as by the motion of a clock) which is commonly used instead of true time." Everyone can appreciate Newton's feeling for the desirability of setting up a definition independent of the common notion of time. Nevertheless, the most cursory inspection shows that the definition is useless since

[1] Reference may here be made to William James, "Psychology, Briefer Course" (Holt, New York, 1892), chapter entitled "The Sense of Time."

it defines time in terms of the thing itself, without which the words "evenly" and "equably" have no meaning.

The modern interpretation of Newton's conceptual time would probably run somewhat as follows. Abstract time, as it appears in the equations of physics, is merely a parameter which serves as a useful independent variable and whose range of variation is the real number continuum. Thus every time interval is put into a one-one correspondence with the interval between two real numbers. In this sense the time parameter possesses all the continuity and other properties of the real number continuum, and its analytical usefulness in physical theories may be said to rest on this fact. This interpretation does not of course explain why physics places so much weight on the time parameter. Actually when we have defined our concepts (e.g., the mechanical concepts of velocity, acceleration, etc.) in terms of this kind of symbolic time, have set up our physical principles, have derived the physical laws, and by the solution of the equations suitable to the description of the behavior of a given system have expressed the state of the system as a function of t, we have still left on our hands the problem of associating the result with experimental observation. The t in our equation then becomes time as measured on a clock, viz., public time. Now a clock is only a special type of physical system, so that after all what we are doing resolves itself into a comparison of one physical system with another. Experience has shown it to be particularly convenient to conduct this comparison with respect to some *standard* physical system rather than to compare *any* two physical systems directly. Thus it appears simpler to relate the behavior of a falling particle and a vibrating spring both to the same system, viz., a clock, than to express the behavior of the two former systems in terms of each other. Nevertheless, transformations of t are perfectly feasible if desired.[1] From the point of view taken here, the use of the time parameter in physics may be looked upon as a matter of convenience and nothing more. All measurements reduce in the last analysis to *space* measurements; consequently it should be possible to describe all physical phenomena in terms of symbols directly representing the results of such operations only. The selection of certain standard systems for comparison suggests, however, the use of a certain parameter referring specifically to such a system. This parameter we call the time, and its use renders our equations in general mathematically much simpler than they would be without it. It may be that some day we shall decide to forego its use and not assign such

[1] See, for example, Poincaré, "Science and Hypothesis," German edition (Teubner, Leipzig, 1914), p. 296.

2.5 THE SIGNIFICANCE OF TIME IN PHYSICS 75

outstanding importance to any physical system like a clock. This might conceivably help in the removal of certain embarrassing difficulties which are associated with the use of the time parameter. However, it would also most certainly introduce new complication into physical laws.

Simultaneity. One of the most serious problems connected with time in physics is that of simultaneity—particularly for events separated by spatial distance. The time for events taking place at a certain point in a reference system may be measured directly by a clock fixed at that point, and there arises no question concerning such events which happen at the same time. However, what meaning can be assigned to the question: if it is 10 o'clock here, what time is it *now* on the moon? When we look through a telescope at a distant star—so distant that light takes many years to reach our eyes—what meaning is there in the statement that the light which enters our eyes at a given instant left the star at a time which was simultaneous with some past event on earth, perhaps before our birth? The statement may, indeed, be dismissed as meaningless; nevertheless the question assumes considerable importance as soon as one is faced with the problem of assigning t to two physical systems moving with respect to each other in some reference system. This is not an academic problem, as we shall see in the discussion of electrodynamics in Chapter VII.

A natural scheme for securing simultaneity would be to set clocks on different systems by means of signals. If these signals were transmitted with infinite velocity, no difficulty would arise. The finite velocity of light, however, removes the two systems from contact, so to speak, during the time of passage of the signal. Hence any method of synchronizing the one clock with the other must be based on an arbitrary convention. In the chapter on relativity (Sec. 7.3) we shall discuss the one proposed by Einstein. We shall also postpone to that chapter the whole question concerning the assignment of t to moving systems and the importance of space-time.

Directionality of time. A fundamental distinction may be pointed out, namely, that the private time of the individual and the public time of measurement are unidirectional, whereas the abstract time of physical theory, strictly speaking, is not. The time of our consciousness certainly seems closely bound to memory which in turn concerns itself with exclusively past events and not with the future, save in persons who claim the power of prevision. The strong feeling also that experience having once been lived cannot be relived attaches a uniqueness to every event in human consciousness which is most simply interpretable in terms of one-way time. On the other hand, with the symbol t, as it occurs in physical equations, the matter stands quite otherwise. As we shall see in our discussion of mechanics, the equations are just as valid for negative values of t as they are for positive values. If the equations predict future events, they predict past ones as well. Time in mechanics is then completely reversible or two-way time.

However, there is a good deal of physics which, as we shall see in Chapter V, is not easily describable in mechanical terms. In thermodynamics, which is most readily handled by statistical reasoning, we meet the famous second law with its associated concept of entropy. This looks upon the whole course of physical experience as being accompanied by thermodynamical degeneration involving the *increase* of entropy (in *no* case its decrease, although in special cases it remains constant).

Eddington [1] has seized upon this property of entropy as fixing the direction of time—"time's arrow," as he so picturesquely puts it. Others, notably G. N. Lewis,[2] have argued that since the increase in entropy is only *probable* and never *certain* one should not associate unidirectional time with it. Lewis holds that non-directional or two-way conceptual time is satisfactory in the description of all physical phenomena. If one could consider the whole universe as a single physical system, its state at any instant could *never* recur if the law of the increase in entropy is absolute. If this law is interpreted statistically, however, there is always a probability, small though finite, that *any* given state will recur if one waits long enough. This, however, breaks down the unidirectionality argument as far as thermodynamics is concerned. The question can hardly be considered as settled yet. We can, at any rate, safely say that, in the actual use of the time concept in physics, the physicist will consult his convenience with respect to reversibility or irreversibility. The situation here is precisely that encountered in connection with other concepts. If the demands for clarity and simplicity cannot be met by the more intuitive notions, the physicist has no hesitation in modifying these notions and subliming them into more abstract concepts.

Time: continuous or discrete? The last point which we wish to examine in our brief survey of the conceptual time of physics is the question of *continuity*. Although psychological time is certainly discrete, in classical physics, as we have noted above, the time parameter is assumed to vary continuously over the range of all real numbers. This justifies the use of the time differential element dt in the applica-

---

[1] "The Nature of the Physical World" (Macmillan, London and New York, 1929).
[2] G. N. Lewis, "The Symmetry of Time in Physics," *Science*, 71, 569, 1930.

2.5 THE SIGNIFICANCE OF TIME IN PHYSICS

tion of the principle of elementary abstraction, and in so far as differential equations constitute a satisfactory way of expressing the fundamental principles of physical theories, the continuity of time would seem to be a convenient postulate. However, the recent stress on the element of *discreteness* involved in the quantum theory and its application to atomic structure has focused attention on the possible value of associating discreteness with conceptual time or space or both. The quantum concept of stationary state can be reconciled with a continuous time parameter only through the medium of quantum mechanics (see Chapter IX). If one tries to interpret this concept in terms of the usual space-time picture of classical physics, one is faced with the uncomfortable inconsistency that, although the behavior of a system in a stationary state can be described in terms of classical mechanics with its continuous t, the transitions from one state to another correspond to nothing in classical mechanics. This breakdown of the classical method has been stressed by Bohr ever since his founding of the quantum theory of atomic structure in 1913. As we have said, only the introduction of quantum mechanics with its new concept of *state* seems to be able to solve the difficulty, and this only by departing considerably from our primitive notions. The quantum theory lays great stress on the concept of frequency, related to the radiated quantum energy by the celebrated relation $E = h\nu$, where h is the Planck constant of action. It has been suggested [1] that there exists a maximum energy quantum corresponding to the largest possible radiation frequency which can exist. This would in turn correspond to a minimum period of time $T_{min.} = 1/\nu_{max.}$, which has been taken as a time-atom and called by some the *chronon*. The highest frequency radiation so far observed is that of cosmic rays, viz., approximately $1/4.5 \times 10^{24}$ sec^{-1}, which yields, for $T_{min.}$, 4.5×10^{-24} sec. Now the theory of relativity (Chapter VII) suggests that the energy associated with any mass m is $E = mc^2$, where c is the velocity of light (3×10^{10} cm/sec). If one assumes that all this energy can be changed into radiation energy the frequency will clearly be mc^2/h. The substitution of the mass of a *proton* or *neutron* for m yields a value very close to $\nu_{max.}$ above. If the proton is the most massive atomic particle which can be, so to speak, annihilated with the change of its mass into energy in accordance with the Einstein relation $E = mc^2$, some meaning might be attached to the chronon as just defined. Since it is conceivable, however, that still more massive particles (i.e., atomic nuclei) may be subject to the same process, one is hardly justified at the present time in taking the chronon hypothesis too seriously.

[1] See, for example, G. I. Pokrowski, *Nature*, 127, 667, 1931.

It is of interest to observe, in any case, that if there exists a time quantum, all spectra must be line spectra and "continuous" spectra are only those in which the lines are too close together to be resolved by ordinary optical instruments. The future alone can decide the fate of the assumption of discrete conceptual time. It may be in order, however, to remark briefly the profound alteration the introduction of discrete time would necessitate in all physical theories. Continuous conceptual time provides, so to speak, a continuous background against which to describe both discrete and continuous phenomena. If time itself is assumed discrete, this background is lost and the whole question of the use of time in physical description must be examined anew.

CHAPTER III

THE FOUNDATIONS OF MECHANICS

3.1. Introduction. Mechanics is the science of motion. Since motion of some kind is at the basis of the conceptual description of practically all physical phenomena, the fundamental importance of mechanics hardly needs to be emphasized. We shall here be concerned with a study of the methods by which it achieves its aim. These methods involve the construction of theories as indicated in Chapter I for physical theories in general. This remark would scarcely be necessary were it not for the fact that, owing to their slow historical development and the common use to which they are every day put in applied science, the principles of mechanics seem to possess an obviousness not shared by other physical theories. This mistaken view needs correction by a careful examination of the postulates on which mechanics rests and the various points of view from which the principles may be developed. It has probably been the experience of most teachers of physics to note the astonishment of the elementary student over the fact that the principles of mechanics may be stated in a great variety of ways. This surprise is usually dissipated when it is made clear that mechanics is a *description* of motion, and differing methods of description may well serve different purposes, emphasizing different aspects of the phenomena. This is particularly well illustrated in the application of mechanical concepts to the understanding of acoustical, electrical, optical, and, more recently, atomic phenomena, as we shall have occasion to note later.

We wish now to make a critical analysis of the logical content of mechanics as perhaps the most important illustration of the construction of a physical theory. From time to time we shall advert to the historical development.

3.2. Kinematical Concepts. Specification of Position. At the outset we are confronted with the question as to the meaning of motion. It is hardly desirable to enter into too philosophical a discussion of this matter, which is one of great difficulty. We can certainly say this much: in the world of our experience we are often confronted by *changes* in our sensations. Certain of these changes we are able to compensate by appropriate muscular reactions—turning the eyes, etc.—and we have grown accustomed to the association of the word motion to such compensable changes. This in brief may be looked upon as the physiological-psychological meaning of motion. However, it does not answer the question: what moves? The usual definition that it is *matter* which moves is not very helpful since we can hardly say that we know exactly what matter is. Strictly speaking, it might well be possible to describe motion wholly in terms of the compensating changes in the observer. But this may be criticized on the grounds of anthropomorphism and still more on those of inconvenience. We shall therefore at once introduce the assumption that all motion may be symbolically described in terms of the motion of *particles*. A *particle* is an entity assumed to possess several properties the most fundamental of which are as follows:

1. A particle has *position* (we shall define this more carefully below) but no *extension*; in this respect it is analogous to a geometrical point.

2. A particle has *inertia*, the measure of which is *mass*. This likewise will be described more fully later.

3. A particle will have certain relations to other particles—gravitation, etc.

It must be remarked with respect to the above that the particle is a *symbolic* conception like all conceptions in physics. It is, of course, suggested by our experience but represents an idealization of that experience. The statement of the above properties is at this stage necessarily vague, implying a more careful consideration of each point. Nevertheless it will be evident that the properties mentioned are the necessary and sufficient ones for the classical mechanical picture.

We therefore begin our study of mechanics with the discussion of the motion of a particle. It is a very interesting fact that the very definition of a particle necessarily implies to a certain extent the principles of mechanics themselves. It would be difficult for a physicist to introduce a definition of a particle which was entirely independent of and had no connection with the principles fundamental to a description of motion.

Frames of reference. The statement that a particle has *position* and the corollary that it may change its position imply that we shall develop mechanics in terms of certain notions of *space* and *time*. These notions and their varieties have been discussed in Chapter II. Hence all we need to state here is that every particle is conceived to be placed in a three-dimensional Euclidean space. Its position in this space will be given with respect to a system or frame of reference. We therefore imply that the meaning of position amounts to the

3.2 KINEMATICAL CONCEPTS. SPECIFICATION OF POSITION

specification of a *method* of reaching the particle. Hence position is purely relative and depends on the reference system chosen, which may vary depending on the purpose required. For many problems the earth furnishes a suitable frame. The most general reference system of use in mechanics is that formed by the average positions of the " fixed " stars. This has been called the *primary inertial* system and plays an important part in the theory of relativity.

The position of a particle with reference to a certain frame of reference may be symbolized in an infinite variety of ways. Certain of these are of sufficiently general use to receive special mention in this discussion of foundations. The position of a point P may be specified by the perpendicular distances from P to three mutually perpendicular planes intersecting in a point O. (See Fig. 3.1.) The three perpendicular distances are respectively symbolized as the rectangular coordinates x, y, z of the point. We have already used such coordinates (e.g., Sec. 1.7). Obviously there are possible variations in which the axes are oblique instead of mutually perpendicular.

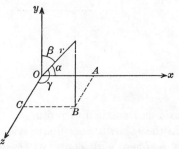

FIG. 3.1

The position of P may also be specified by the straight-line distance r from P to O and the direction of OP with respect to the coordinate axes as measured by the three angles α, β, γ which OP makes with the x, y, and z axes respectively. The four quantities r, α, β, γ, which may be called *polar coordinates*, are not all independent, since we have the identical relation

$$\cos^2 \alpha + \cos^2 \beta + \cos^2 \gamma = 1. \tag{1}$$

This follows at once from

$$x = r \cos \alpha, \quad y = r \cos \beta, \quad z = r \cos \gamma. \tag{2}$$

The line OP from O to P is a physical quantity which has both magnitude and direction. It is a *vector* symbolized as **r**, the magnitude of which is r in appropriate units. The vector **r** completely determines the position of P with reference to the given reference system, and as such it is called a position vector. The x, y, and z coordinates of P are called the *components* of the vector **r** along the three coordinate axes, respectively. Vector representation is of great importance in mechanics and throughout physics. The reason for this will be made clearer in our discussion of relativity where it will be shown that every vector has the important property of invariance with respect to a *rotation* of the coordinate axes.

There are many other types of coordinate specification which are of value for special problems. Two of these will be mentioned for the sake of reference. In *cylindrical* coordinates (Fig. 3.2a) the position of P is given on the surface of a right circular cylinder with the z

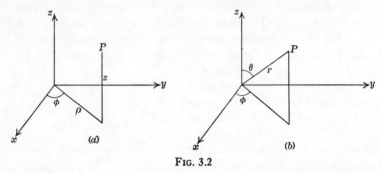

FIG. 3.2

axis as its axis and radius ρ. The coordinates are then ρ, ϕ, and z. In the *spherical* coordinate system (Fig. 3.2 b) P is given on the surface of a sphere with center at O and radius r. The longitude is given by ϕ, and the co-latitude by θ. The relation between these systems and rectangulars is given by the following formulas

CYLINDRICALS   SPHERICALS

$x = \rho \cos \phi$ $x = r \sin \theta \cos \phi$

$y = \rho \sin \phi$ $y = r \sin \theta \sin \phi$

$z = z$ $z = r \cos \theta$

In more complicated problems it may be found necessary to use parabolic or elliptic coordinates. These, however, will not be discussed here.

Specification of position by distance implies, of course, the introduction of a *unit* of length. We need here only call attention to the purely arbitrary nature of this unit, which in the so-called metric system is called the *centimeter* (strictly the *meter*, which is 100 centimeters) and in the English system is the *foot* (strictly the *yard*, which is three times as great).

3.3. The Concepts of Displacement, Velocity, and Acceleration. When a particle changes its position in a given reference frame it is said to undergo a *displacement*. Thus, to consult Fig. 3.3, suppose that the particle is at one instant (say time t) at P_1, in which its posi-

3.3   DISPLACEMENT, VELOCITY, AND ACCELERATION

tion is given by the vector $\mathbf{r}_1$, and at the later instant $t + \Delta t$ is at P_2, represented by the vector $\mathbf{r}_2$. We shall denote the displacement by the vector $\mathbf{P}_1\mathbf{P}_2$. This implies that the particle has moved from P_1 to P_2 in a straight line, which may of course not be true at all. It is conceivable that in moving from P_1 to P_2 the particle may actually wander over a very large region. This, however, would render the concept of displacement meaningless. We therefore make the important assumption, viz., that it is possible to describe the motion of a particle during a sufficiently short time interval by the displacement $\mathbf{P}_1\mathbf{P}_2$. This will be referred to as an *elementary* displacement. All actual displacements during finite time intervals will be assumed to be vector sums of elementary displacements. This significant assumption implies that the path of the particle during displacement shows no discontinuities. It is very similar to the hypothesis made in connection with the discussion of the construction of

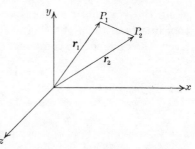

FIG. 3.3

physical laws as descriptions of physical operations (Chapter I). Experimental observation always yields a discrete set of points. We introduce the hypothesis of representation by continuous functions by drawing a smooth curve through the points. Similarly all we really know about a particle may be said to be a discrete set of points representing its successive positions in a given reference frame. Our assumption is that the particle has traversed a *continuous* path which is to be represented as the limit of the vector sum of elementary displacements.

Vector representation of displacement and velocity. The elementary vector displacement $\mathbf{P}_1\mathbf{P}_2$ may be most conveniently represented as the *difference* between the two position vectors $\mathbf{r}_2$ and $\mathbf{r}_1$. We may write the vector equation

$$\mathbf{r}_2 - \mathbf{r}_1 = \mathbf{P}_1\mathbf{P}_2 = \Delta\mathbf{r}, \tag{1}$$

where $\Delta\mathbf{r}$ denotes an increment in $\mathbf{r}$. The ratio $\Delta\mathbf{r}/\Delta t$ is then defined as the *average velocity* of the particle during the displacement. The next assumption is that this ratio has a unique limit as Δt approaches zero, implying that $\mathbf{r}$ is a continuous function of t, a variable assuming values which can be put into one-to-one correspondence with the set of all real numbers (cf. Sec. 2.5). This limit, $d\mathbf{r}/dt$, is called the *instantaneous* velocity of the particle at the instant in question. We write

$$\mathbf{v} = \lim_{\Delta t \doteq 0} \frac{\Delta \mathbf{r}}{\Delta t} = \frac{d\mathbf{r}}{dt} = \dot{\mathbf{r}}, \tag{2}$$

if we use Newton's dot notation to indicate differentiation with respect to the time. Velocity is, of course, a vector. For many purposes it is convenient to represent $\mathbf{r}$ as the sum of the three mutually perpendicular vectors $\mathbf{i}x + \mathbf{j}y + \mathbf{k}z$ where $\mathbf{i}, \mathbf{j}, \mathbf{k}$ are vectors of unit magnitude along the x, y, and z axes respectively and x, y, z are the rectangular coordinates of the point P of which $\mathbf{r}$ is the position vector or alternatively the rectangular components of $\mathbf{r}$. Thus we have

$$\mathbf{r} = \mathbf{i}x + \mathbf{j}y + \mathbf{k}z, \tag{3}$$

and (2) becomes

$$\mathbf{v} = \mathbf{i}\dot{x} + \mathbf{j}\dot{y} + \mathbf{k}\dot{z}, \tag{4}$$

$\dot{x} = v_x,\ \dot{y} = v_y,\ \dot{z} = v_z$ appearing as the rectangular components of the velocity $\mathbf{v}$. All this implies that, during the motion, the axes remain fixed. We shall assume this during most of the subsequent discussion although moving axes have proved very convenient in the description of certain motions.

Acceleration. The case in which $\mathbf{v}$ is a vector constant in time is an unusual one in mechanics. It is therefore important to consider the change of velocity with the time, introducing the concept of *acceleration*. If the velocity of a particle is $\mathbf{v}_1$ at a given instant and $\mathbf{v}_2$ at time Δt later, we shall define $(\mathbf{v}_2 - \mathbf{v}_1)/\Delta t = \Delta \mathbf{v}/\Delta t$ as the *average acceleration* of the particle during the interval Δt. It is also a vector, and its formation implies assumptions precisely similar in character to those made above in connection with $\Delta \mathbf{r}/\Delta t$. The instantaneous acceleration is defined as

$$\mathbf{a} = \lim_{\Delta t \doteq 0} \frac{\Delta \mathbf{v}}{\Delta t} = \frac{d\mathbf{v}}{dt} = \frac{d^2\mathbf{r}}{dt^2} = \ddot{\mathbf{r}}. \tag{5}$$

Here again it is often convenient to set

$$\mathbf{a} = \mathbf{i}\ddot{x} + \mathbf{j}\ddot{y} + \mathbf{k}\ddot{z}, \tag{6}$$

where $\ddot{x}, \ddot{y}, \ddot{z}$ appear as the rectangular components of the instantaneous acceleration of the particle. Whenever the words velocity and acceleration are used hereafter, unless otherwise specified, they shall be taken to mean the instantaneous values.

3.4 DYNAMICAL CONCEPTS. THE NEWTONIAN POSTULATES

From the logical point of view there is no reason why we should not go further and introduce the time rate of change of acceleration $\dot{a} = \dddot{r}$ and the higher derivatives. But our aim is to build a foundation for mechanics with the fewest possible fundamental concepts. Strictly speaking, it is conceivable that we might have stopped at displacement and velocity. The actually observed motions of bodies, however, do not appear to be easily described by these two concepts alone. Hence we introduce the further concept of acceleration. Will this suffice? The only way to answer this question is to make some postulates relating these concepts and then see with what success the theory so constituted describes observed motions.

As a matter of fact, the attempt to carry out such a program suggests the desirability of introducing certain additional concepts of a nature apparently rather different from the kinematical concepts discussed in this section. These are *mass* and *force*.

3.4. Dynamical Concepts. The Newtonian Postulates for Mechanics. The basis for mechanics presented in most elementary textbooks of physics is that expressed in the form of Newton's so-called "laws" of motion with accompanying "definitions." We shall therefore begin by stating and analyzing these famous pronouncements. This procedure may, indeed, be criticized from the logical standpoint: one might prefer to consider at once a logical foundation for mechanics independent of the historical background. This method, of course, may be found in several treatises, notably that of Hertz.[1] In the present instance, however, there is probably more to be gained at the moment by an examination of what will doubtless remain for a long time, in spite of defects, the classical formulation. This will in addition serve admirably to introduce us to the real problem involved in the fundamental concepts of the subject.

Newton's definitions and laws. Newton in the "Principia Mathematica" began his enunciation of the principles of mechanics with several definitions. The more important of these follow (in somewhat abbreviated form).

1. The quantity of matter in a body will be measured by its density and volume conjointly. This quantity of matter will be known by the name of *body* or *mass*; it will also be known by the *weight* of the body in question. That mass is proportional to weight has been found by very careful experiments on pendulums.

2. Quantity of motion is measured by the velocity and the quantity of matter (mass) conjointly.

3. The innate force of matter is a power of resisting by which every body, as much as in it lies, endeavors to persevere in its present state, whether it be of rest or of moving uniformly forward in a straight line.

---
[1] H. Hertz, "Principles of Mechanics" (Leipzig, 1894). See Sec. 3.11.

4. An impressed force is an action exerted upon a body in order to change its state either of rest or of moving uniformly forward in a straight line.

There are four other definitions, concerning centripetal force, which we need not consider at this place.

The laws of motion placed after the definitions were then stated in the following form:

I. Every body continues in its state of rest or of uniform motion in a straight line, except in so far as it is compelled by forces to change that state.

II. Change of motion is proportional to the force and takes place in the direction of the straight line in which the force acts.

III. To every action there is always an equal and contrary reaction; or, the mutual actions of any two bodies are always equal and oppositely directed along the same straight line.

Let us try to understand what the laws mean, using Newton's definitions and our experimental knowledge as a guide. Our critique will be conducted with special reference to the laws themselves rather than to the definitions.

The first law contains several words descriptive of concepts which must be made clear before it can be understood. These are " body," " state of rest," " uniform motion," " straight line," " forces." By the word " body " Newton without doubt meant a particle or an aggregate of particles which do not change their mutual distances (i.e., a rigid body). These ideas involve no difficulties beyond those already discussed in Sec. 3.2. From a critical standpoint it would doubtless be better to state all fundamental rules in terms of the single material particle, and leave properties of an aggregate of particles a matter of further deduction. For the sake of clarity, therefore, we may for the time being interpret " body " to mean " particle " in the sense of the earlier section. The " straight line " is a geometrical concept, and the various questions involved in it have already been taken up in Chapter II. " State of rest " and " uniform motion " are purely kinematical expressions and are to be interpreted as in Secs. 3.2 and 3.3. This leaves only the word " force " or " forces." It is just here that great difficulties begin to be encountered. Clearly the first law depends for its meaning on an understanding of the concept of force. Suppose that we were to approach the law with no notion whatever of this concept. Just what would we make of it? We might say this: There is a kind of motion which is to be distinguished for its simplicity, the implication being that this kind is a preferred one and would be universal were it not for the intervention of some outside influence which prevents it from being always attained. But from this point of view, what is left of

3.4 DYNAMICAL CONCEPTS. THE NEWTONIAN POSTULATES

the law save the statement (made notorious by Eddington[1]) that "every particle continues in its state of rest or uniform motion in a straight line, except in so far as it doesn't"? Granted our ignorance of force, the law becomes a pious statement calling attention to a possible state of motion which may sometimes be realized in nature and sometimes not.

However, this is not the only interpretation of the law. Newton certainly meant to convey something very definite. What he wished to express was doubtless the feeling that the idea of force is intuitively known and does not need further explanation. This at first strikes one as a reasonable point of view. When the word force is mentioned, nearly everyone associates *some* meaning to it. An unabridged dictionary yields some ten common meanings of the term ranging all the way from "strength" or "vigor" to "moral power" and a "body of men trained for action in any way, as a *police force*." It is also true, however, that most people associate with the word a kind of technical significance, as when it is said that to move an object one must exert a force on it. This is distinctly implied in Newton's third definition. In this sense the word conveys the notion of muscular push or pull; most people seem to have no trouble in extending the notion to inanimate objects: one speaks of the force with which the locomotive pulls on the train or the force with which a house pushes against its foundation. With this sort of notion in mind (and this is probably that which Newton had) one can get a qualitative sense for the first law. One visualizes a particle which somehow wants to keep going straight, but a push or a pull interferes with its intentions. All this is very much tinged with anthropomorphism and, inasmuch as science tries to get as far away from anthropomorphic ideas as possible, is of doubtful value for mechanics. Indeed, its futility is made very clear as soon as curved-line motions are encountered without the obvious agency of any push or pull, as in the very simple act of throwing a ball into the air. One may talk about the force of gravitation, but naturally one hesitates to think of the earth as acting humanly with muscles, etc. At any rate, this is the feeling of the modern scientist. Naturally there is no sense in saying that anthropomorphism is wrong. It was good enough for our ancestors. Perhaps it ought to be good enough for us. Certainly we can never get away from it in at least a diluted form. Science is made by the observer and the observer is human, no matter how impersonal he tries to make his observations. However, modern science is definitely

[1] A. S. Eddington, "Nature of the Physical World" (Macmillan, New York, 1930), p. 124.

committed to removing the human element as far as possible. The limitations inherent in this attempt are being well brought out in the recent discussions of quantum mechanics concerning the principle of indeterminacy of Heisenberg (see Chapter IX) or the more thoroughgoing principle of complementarity of Bohr. In this connection it may also be well to add that there is a vast difference between having some kind of understanding of a physical concept and giving a precise definition of it by which (if it is a quantity) it may be measured.

There is, however, certainly one further attitude one can take toward the first law and escape the embarrassments just outlined. This is the now rather common view that the law is really a *definition of force*. In other words it does not make a statement about what happens to a particle when a force (supposed known and understood) acts on it; rather it purports to say that a force is anything which acts to change the state of a particle from rest or uniform motion in a straight line. This would mean that, whenever we note a particle which is not in this particular ideal state, we are to say that it is acted on by a force. There can be no logical criticism of this attitude, which is essentially that of Newton's fourth definition. Its utility, however, is open to doubt, for it conveys no notion how to measure force and thus has little quantitative significance. This is presumably to be looked for in the second law, to which the first is then to be considered a kind of qualitative introduction. It would indeed not be wholly correct to say that the first law has *no* quantitative significance, for it does describe the meaning of *zero* force. No force acts in the ideal state indicated. This might lead to certain apparent perplexities. For example, a book resting on a table might then be taken as an illustration of zero force. This contradicts the common-sense feeling that the book actually exerts a force on the table. Of course we clear up the difficulty at once by deciding that the state of rest or equilibrium involves the action of more than one force; we meet here the notion of force pairs and resultants.

Another drawback to the utility of the law as a definition of zero force comes from the experimental fact that uniform motion in a straight line is the exception and not the rule. Indeed we may go so far as to say that it is never encountered in large-scale motions save approximately (as in the case of rain drops, parachutes, etc.). This is possibly the reason why the Greek philosophers paid so little attention to it, preferring instead to look upon uniform motion in a circle as the "perfect" motion. They would have been astonished to see so much emphasis laid on a kind of motion that is practically never observed! Many elementary textbook writers content themselves

3.4 DYNAMICAL CONCEPTS. THE NEWTONIAN POSTULATES 89

with observing that when a hockey puck slides on ice, the smoother the ice the farther the puck travels with a given blow before coming to rest. They then ask us to imagine that the ice becomes in the limit perfectly smooth—an ideal surface which has no effect on the puck. The assertion is then made that the puck would continue indefinitely in a straight line with constant velocity. As a suggestive illustration one can hardly criticize this, though it is well to point out that the surface must be idealized beyond the limit suggested, i.e., it must be made infinite in extent and, more important still, must be flat, i.e., cannot be on the surface of the earth. A perfectly smooth sheet of ice level with the earth will not answer, for in this case the path will not be a straight line at all. It would not even be a great circle, owing to the effects of gravity coupled with the rotation of the earth. In other words, the illustration which sounds at first not bad proves very unfortunate on closer inspection. Probably much the same thing would be true of any attempted large-scale phenomenal illustration of the first law of Newton. It is questionable procedure to try to clear up the meaning of a fundamental physical law by giving an illustration which breaks down badly on the slightest questioning.

To continue our critique, let us examine the meaning of the *second* law of motion. It will be noted that, like the first, this introduces both the kinematical concept of motion and the dynamical idea of force. To understand the statement we must recall Newton's second definition in which he uses the term " motion " in a technical sense, namely that now commonly referred to as *momentum*, or mass times velocity. " Change in motion " may be translated into " change of momentum." In modern texts this law usually appears in the form of the vector equation of motion

$$\mathbf{F} = \frac{d(m\mathbf{v})}{dt}, \tag{1}$$

where $\mathbf{F}$ is the force (a vector) and m is the mass. For most purposes outside of the field of relativity this is equivalent to $\mathbf{F} = m\mathbf{a}$. Now in interpreting the law or its equivalent equation we are faced with a new problem, namely, the meaning of *mass*. Of course, if we deal directly with motion as *momentum*, the issue is dodged, but we have on our hands the equally difficult task of defining momentum. On the whole it seems simpler to deal with mass. It is probable that Newton considered this notion, like that of force, as more or less intuitively known. Nevertheless, he introduced the name in his first definition, stating essentially that the mass of a body is the product of its volume and its density. To us who are accustomed to define density as mass per unit volume, this definition is of course circular. It throws the burden on the definition of density independent of mass. Newton evidently looked upon density as the more fundamental thing, though it is not clear just how he proposed to measure it, that is to compare the density of one material with that of another, unless it be by weight, which has the disadvantage of varying from point to point on the earth's surface.

Successive ages have seen the difficulty inherent in the concept of mass but have been inclined to avoid it until comparatively recently, i.e., until approximately the time of the appearance of Ernst Mach's "Mechanics." In most nineteenth-century textbooks of mechanics and general physics one finds with very little significant variation the statement that "mass is the amount of matter in a body" or "mass is somehow the body itself or the matter of which it is composed." It would doubtless be considered supererogatory at this date to dignify such statements with attention were it not for the fact that one still occasionally encounters them in texts in current use. One wonders just what meaning is attached to the word "definition" by those who would seriously use such a form of words. If, following Mill, we consider that a definition is a proposition declaratory of the meaning of a word, it is hard to see how statements like the above ever could have been considered in the category of definitions, for what meaning do they convey or wherein do they enlighten? Matter is still a mystery which is being only gradually unraveled, and the words "amount of matter" convey nothing unless an actual method for ascertaining "amount" is specified. Even if such statements were intended to convey merely an impression such as, for example, is conveyed by the definition of force as a push or pull, it must be confessed that they fail utterly, for the concept of mass as actually used in dynamics involves the idea of inertia and no such notion is even implicit in the above statements.

One sometimes meets the definition of mass as inertia; when this is coupled with a discussion of a concrete experiment it has the merit of at least providing a qualitative notion. But it has already been emphasized that no definition of a physical quantity has any value unless it describes a process by which the quantity may be measured.

Turning back to eq. (1) or the equivalent $\mathbf{F} = m\mathbf{a}$, one may see the possibility of defining mass as represented by the symbol m in terms of the force and acceleration. This, however, implies that force has already been independently defined, which is by no means the case. Let us, however, examine for a moment the possibility of defining force in a satisfactory manner independently of Newton's second law.

We have already touched briefly upon this question earlier in this section. Almost inevitably our first thought is of rods and strings with weights, pushes, and pulls. We wish, however, to keep as clear as may be from anthropomorphism. What is the actual situation? We have a number of particles that are observed to be moving with various accelerations. We wish to describe their motions by saying that certain forces act on the particles. How shall we tell what force acts on a given particle? Here an idea suggests itself. Let us apply a push or a pull to the particle in question until its acceleration is reduced to zero. The push or pull which accomplishes this will then be equal in magnitude and opposite in direction to the original force acting on the particle. This, however, merely shifts the problem to that of measuring a push or pull. Such is, of course, always applied through a material medium, and any attempt to measure it will have to be made in terms of the observed properties of such media. Since media differ widely, this is by no means a satisfactory state of affairs. Surely we want a definition which is ultimately independent of the properties of special media! We might think to get around this difficulty by the use of gravity. That is, we might balance every push or pull against the weight of some object and so express ultimately all forces in terms of weights. But this again is unsatisfactory, for weight differs from point to point on the surface of the earth in no very simple way. It looks as if the attempt to define force independently of the second law is fraught with too many and too great difficulties to be worth while.[1]

Enough has probably been said to emphasize the fact that the common interpretation of Newton's second law is by no means logically sound and that it leaves many questions open. It is incumbent on us to provide some connected theoretical basis with which to replace the Newtonian laws and which will lead to the results which Newton and his successors have sought to draw from these laws.

3.5. The Concept of Mass. There are without doubt many perfectly logical definitions of mass, depending on the point of view which is chosen. At the same time it should be equally emphasized that all involve numerous assumptions which make the definitions highly idealized. This is a procedure not to be deprecated. We wish our analysis to be searching enough to reveal as many critical points as possible. In this sense we are not concerned with merely "practi-

---

[1] See H. Poincaré, *op. cit.*, p. 104, for a rather complete discussion of this point, to which we are here considerably indebted. In particular, note his critique of the attempt of Andrade and his "thread school" to define force in terms of the deformation of material bodies.

cal" considerations like the fact that for most purposes the mass of an object can be determined by means of a beam balance. This amounts essentially to a definition of mass in terms of weight, and since the experimental procedure involved is susceptible of extremely great precision one might suppose that this definition is perfectly satisfactory: one would choose a certain piece of material as a standard of mass and then compare all others with it by balancing them against known fractions or multiples of the standard. From the logical point of view, however, this is not a satisfactory definition. It fails to emphasize the leading qualitative idea involved in mass, namely, *inertia*; more important still, it has the drawback of making mass effectively a function of the properties of the earth. The definition under discussion depends on the postulate that mass is directly proportional to weight. But weight, of course, involves the idea of force. It therefore amounts to defining mass in terms of force. This puts us back in the position of having to give a definition of force independently of mass and the second law of Newton. The dilemma is obvious.

Definition of mass. Let us therefore proceed as carefully as possible. We shall deal entirely with particles. Suppose we have two of them, A and B, which are very far away from all others in the universe; they may be considered effectively isolated. In order to build a physical theory concerning these two particles which has any chance of agreeing with the facts it is necessary to assume that the particles influence each other mutually. We shall assume that this mutual action takes the form of accelerated motion. The acceleration of A with respect to the primary inertial system described in Sec. 3.2 will be denoted by $\mathbf{a}_{AB}$ and that of B by $\mathbf{a}_{BA}$. We shall now make the fundamental assumption suggested by experiment that the ratio

$$\frac{\mathbf{a}_{AB}}{-\mathbf{a}_{BA}}$$

is a positive scalar constant which is independent of the relative position of the particles (within the limits of the assumption just made concerning their isolation), of their relative velocity (in so far as it is small compared with the velocity of light—in the contrary case our definition must give way to one satisfying the relativity criterion), and of the time and place. As a concrete illustration consider two small solid spheres attached to each other by a rubber band or spring.

We shall state the fundamental assumption as follows:

$$\frac{\mathbf{a}_{AB}}{-\mathbf{a}_{BA}} = M_{BA}. \tag{1}$$

In seeking to give an interpretation to the constant M_{BA} we take another particle C and let it interact with A and B respectively in turn. We then have relations analogous to (1), viz.:

$$\frac{a_{AC}}{-a_{CA}} = M_{CA}, \quad \frac{a_{BC}}{-a_{CB}} = M_{CB}. \tag{2}$$

It is perfectly conceivable that the constants M_{BA}, M_{CA}, M_{CB} may be independent. However, experiment indicates that this would be a bad assumption. Actually we shall postulate on the basis of experience that

$$M_{CB} = M_{CA}/M_{BA} \tag{3}$$

for *any* three particles. On substitution of (3) into the second relation in (2) there results

$$M_{BA}a_{BC} = -M_{CA}a_{CB}. \tag{4}$$

The relation (4) is at once subject to a very interesting interpretation, for it shows that by choosing A as a *standard* particle we are able to associate with every other particle (such as $B, C \ldots$) a constant whose value does not depend on the particle with which the one in question interacts. This constant when multiplied by the acceleration of the given particle due to its interaction with another gives a quantity which is, so to speak, anti-symmetrical with respect to the interaction. We agree to give to the constant or some function of it a name. We shall call M_{BA} the *mass* of B relative to A. If it is understood that A is to be taken as a standard reference particle, we then have a whole set of values of $M_{BA}, M_{CA}, M_{DA} \ldots$ for the masses of $B, C, D \ldots$. These may more conveniently be written $m_B, m_C, m_D \ldots$, where we have omitted the subscript A, since that is now understood.

It should be emphasized that the above definition of mass is arbitrary in the further sense that any function of the constant M_{BA} might equally well be chosen for this title. For example, one might prefer to call $1/M_{BA}$ the mass of B. One could not criticize this step logically, though it must be confessed that it would be playing fast and loose with words, since under it the body which opposed greater resistance to push and pull (in the anthropomorphic sense) would then have the smaller mass, thus destroying the qualitative significance of mass as a measure of inertia.

The definition of mass given in this section is essentially that of Mach, although the truly postulational nature of the definition is more clearly brought out than in Mach's treatment. The question of the choice of a unit of mass is of course of practical importance but hardly concerns us here in a logical discussion.

The justification for the measurement of mass in terms of weight can be better appreciated when we have discussed the definition of force in the next section.

3.6. The Concept of Force. Résumé of the Laws of Motion.

Let us return to a consideration of the equation (3.5–4). We have already noted that it expresses the fact—a deduction from the assumptions (3.5–2) and (3.5–3)—that the product of mass and acceleration is a quantity antisymmetric with respect to the interaction of the two particles B and C. We shall now make the hypothesis that the value of this quantity in any given case depends on the relative positions of the particles and sometimes on their relative velocities as well as the time. We express this functional dependence by introducing a vector function $\mathbf{F}_{BC}(\mathbf{r}, \dot{\mathbf{r}}, t)$, where $\mathbf{r}$ is the position vector of B with respect to C and $\dot{\mathbf{r}}$ is the relative velocity. We then write

$$m_B \mathbf{a}_{BC} = \mathbf{F}_{BC}, \tag{1}$$

and define the function $\mathbf{F}_{BC}$ as the *force* acting on the particle B due to the particle C. For the latter we have similarly

$$m_C \mathbf{a}_{CB} = \mathbf{F}_{CB}. \tag{2}$$

The relation (3.5–5) becomes

$$\mathbf{F}_{BC} = -\mathbf{F}_{CB}, \tag{3}$$

which may be stated in words: the force on B due to C is equal in magnitude but opposite in direction to the force on C due to B. It will be noted at once that this is the mathematical formulation of the third law of Newton, which thus appears as a *deduction* from our previous assumptions and the definitions of mass and force. Of course it is already implicit in the relation (3.5–4). We shall discuss in greater detail the meaning of the third law a little later. For the moment, however, it is worth while to stress the significance of the definition of force presented here. It will be noted that no merely anthropomorphic notion of push or pull is involved. The definition satisfies the accepted criterion in that a method is provided for the attachment of a number to the magnitude of the symbol $\mathbf{F}$ as soon as a unit has been decided on. It has the additional interest that it introduces an actual equation which can be used in the study of all mechanical problems. For any particle of mass m, we can write (1) in the form of the vector differential equation of the second order

$$m\ddot{\mathbf{r}} = \mathbf{F}, \tag{4}$$

where **F** may now be looked upon as the *resultant* force acting on the particle, i.e., the vector sum of all the individual forces. This last step is, to be sure, an additional assumption to be justified by its success in practice. It expresses our feeling that each force acts independently of the others. This need not be true, but works satisfactorily in many cases encountered in mechanics. It is an illustration of a general principle called the principle of superposition. Eq. (4) is then called the "equation of motion" of the particle, stating that the product of mass and acceleration, usually known as the *kinetic reaction*, is equal to the *force*. The substitution of a particular functional form of **F** in (4) yields a differential equation or rather a set of three scalar differential equations in terms of the three components of **r** along the rectangular axes. If these can be solved, the position and velocity of the particle in the given reference system at any time can be found (subject, of course, to the existence of certain initial or boundary conditions—see Sec. 1.9). Eq. (4) is thus equivalent to the generally accepted mathematical formulation of Newton's second law, eq. (1), in which, however, all terms are now explicitly defined.

The relation between *mass* and *weight* follows very simply from (4) or its equivalent. At any one place on the surface of the earth the acceleration of free fall in vacuo is approximately the same for all bodies (as both Galileo and Newton demonstrated experimentally). Hence for two particles of masses m_1 and m_2, respectively, we have

$$m_1/m_2 = W_1/W_2, \qquad (5)$$

where W_1 and W_2 are the magnitudes of the respective weights. This justifies the use of the balance for the measurement of mass. An interesting variation of this method for the demonstration of the equality of the masses of two particles is provided by Atwood's machine.

What is the relation between the concept of force developed in this section and that used in statics? In the latter field we are accustomed in practice to deal with weights, pushes, and pulls. It is well to point out, however, that there is no fundamental distinction between static and dynamic force. In statics one deals merely with balanced forces: a particle which is acted on by *two* forces equal in magnitude but opposite in direction will have zero acceleration and will be said to be in equilibrium. A book resting on a table is acted on by the force of gravity and would have the usual downward acceleration g. Balancing the weight there is, however, the upward force of the table on the book. We must therefore think of the book as being acted on by two forces whose vector sum is zero. It will be noted that we have here used the principle of the mutual independence of effects, i.e., superposition.

Our definition of force makes no use of the common idea of causation. In the view exemplified by Newton's definitions and laws, force is looked upon as something which somehow causes change in the state of motion of a body, so that without the action of the force the change in state would not take place. We shall not at this point examine in detail the notion of causation or causality (see Chapter X), but it will suffice if we emphasize that the use of the word force in this sense has no particular value in the solution of mechanical problems. Here again one encounters the distinction between a precise definition on which actual theoretical developments can be based and the vague forms of words which convey at best only qualitative ideas. One gets, to be sure, some picture of force by looking at it in the traditional causative sense and some people may feel that it means something to them, but it is only necessary to try to put the notion to definite use in a concrete example in order to find it illusory. Doubtless no harm is done in saying that the force of gravitation is that influence which causes the planets to pursue their orbits about the sun, but this is of little value in an exact description of the actual motion. The word force is a good illustration of a much-overworked expression. This, however, is no reason why it should not have a precise meaning in physics. We recall in this connection Clerk Maxwell's humorous commentary on Herbert Spencer's appearance before Section A of the British Association at the memorable meeting in Belfast in 1874. Among other references Maxwell says:

"Mr. Spencer in the course of his remarks regretted that so many members of the Section were in the habit of employing the word Force in a sense too limited and definite to be of any use in a complete theory of evolution. He had himself always been careful to preserve that largeness of meaning which was too often lost sight of in elementary works. This was best done by using the word sometimes in one sense and sometimes in another and in this way he trusted that he had made the word occupy a sufficiently large field of thought."

The introduction of this quotation is not intended as a diatribe against the use of the word force in any other sense than the purely physical one. This would be ridiculous. The important point is simply that, whenever the term is employed in philosophical discussions based on physical ideas and assumptions, the meaning should be restricted to the precise physical definition. Failure to do this is attended by utter confusion, and all attempted analogy becomes meaningless. One is tempted to believe that much quasi-physical speculation nowadays indulged in comes to grief primarily for this reason.

Third law of motion. We have already adverted to the significance of our definition of force for the third law of Newton. Its formulation is considered by Mach to be Newton's most important contribution to the principles of mechanics. On the view taken here, it is not a postulate but a theorem following from the previous definitions and principles. This fact, however, robs it of none of its importance. It states that accelerations never occur singly but in pairs. There is no meaning to be attached to the acceleration of a single particle in a reference system; whenever one body is accelerated there must be one or more others accelerated also. This is its kinematical meaning. In terms of force we may say that, if a force acts on a given body, the body itself exerts an equal and oppositely directed force on some other body. The simplest illustrations are those drawn from push and pull forces: the book pushes down on a table with a certain force, and the law then states that the table pushes up on the book with precisely the same force. Newton called the two aspects of the force the *action* and the *reaction*, whence the usual statement of the law.

It is a somewhat curious though perhaps understandable fact that, although elementary students find no difficulty in grasping the significance of this static type of illustration, most of them almost instinctively disbelieve in its application to dynamical cases. They commonly ask: " If action and reaction are always equal and opposite, how is it that bodies ever move? " If the wagon pulls back on the horse with precisely the same force with which the horse pulls forward on the wagon, how can the wagon move? Is it not necessary for the latter force to be just slightly greater than the former, at any rate momentarily, etc.? The confusion is easily cleared up by the reminder that the acceleration of any body, such as a wagon, depends on the forces (to use the conventional manner of speaking) which act on it and not on the forces with which it acts on other bodies. That is, in eq. (4) the force F on the right-hand side is the resultant force acting *on* the particle of mass m and is not concerned with the forces with which the particle may act on other particles.

Further interesting illustrations of the third law will be found in Mach's " Mechanics." It is, however, perhaps not out of place here to emphasize one important application of the law. This is the deduction of the law of the conservation of momentum. We shall deal here with a very simple aspect, namely, the case of two colliding spheres. Let us suppose that the two spheres have masses m_1 and m_2 and initial velocities u_1 and u_2, respectively, in the same straight line. After impact the velocites will be supposed to be v_1 and v_2, respectively. Let the time of impact be τ, and let the force with which the mass m_2 acts on the mass m_1 at any time t during this period be F (variable, in general). The impulse of the force during this interval is $\int_0^\tau F dt$, and it may be very easily shown from eq. (4) that the impulse is equal to the change in momentum of the body on which the force acts. We have

$$\int_0^\tau F dt = m_1(v_1 - u_1)$$

Now from the third law it follows that the force with which m_1 acts on m_2 during the impact is $-F$. We therefore have

$$\int_0^\tau F dt = m_2(u_2 - v_2).$$

Comparison of these two equations gives at once

$$m_1 u_1 + m_2 u_2 = m_1 v_1 + m_2 v_2,$$

which states that the total momentum of the system composed of the two spheres remains constant during the collision. This is of course a special case. There is no particular difficulty in making a similar application to a whole aggregate of particles subject only to their mutual actions.

It might appear possible to use the conservation of momentum to provide a definition of mass and hence of force. However, closer inspection shows that the same general postulates are necessary as have been used in the definitions established in this and the previous section, and hence little appears to be gained by the change.

3.7. Interpretation of the Equation of Motion. With the precise definitions of mass and force established, every mechanical problem treated on the basis of the Newtonian principles reduces to the solution of a vector equation of the type of (3.6–4). Though it is not our purpose to enter into detailed discussion of special problems, it is worth while to consider a little more closely the meaning of this so-called " equation of motion " and in particular the methods by which one is enabled to write it. If for the moment we confine our attention to a single particle, the question that immediately arises is this: how do we know in any particular case what to put for F on the right-hand

3.7 INTERPRETATION OF THE EQUATION OF MOTION

side of the equation? The situation appears to reduce to this: experience or experiment makes us familiar with a certain type of motion, e.g., planetary motion. This knowledge is at first purely kinematical, i.e., we note the positions and velocities of the bodies in question in their dependence on the time. Our task in giving a mechanical description of the observed phenomena consists in deducing symbolically from the equation or equations of motion the kinematical observations and in predicting new ones. Ultimately what we do is to *guess* an **F**, which, when inserted into the equation, leads on solution to the observations and in connection with appropriate boundary conditions makes predictions of future observations which are later found to be correct. It is indeed true that some of the apparent arbitrariness of this procedure is mitigated by the fact that certain types of forces appear to be more reasonable than others.

Common types of forces. For example, in discussing the two-body problem (the simplest case in planetary motion), it seems most natural to introduce the assumption of a central force, i.e., one always directed along the line joining the particles and a function only of their distance apart. This is essentially what Newton did, taking the functional relation to be that of the inverse square of the distance. Again in the case of simple harmonic motion a purely mathematical acquaintance with the equation $\frac{d^2x}{dt^2} = -kx$ (k positive) suggests the choice of a linear function of distance for the force to describe this type of motion. In the case of a spring, indeed, we may feel that we have a guide in Hooke's law, but this is restricted to a small range of displacement values and can manifestly be looked upon only as a suggestion. Naturally in the growth of mechanics certain types of forces become well known, and our familiarity with them undoubtedly tends to infuse in us a certain confidence in their metaphysical significance. There grows a tendency to think of certain forces as especially favored by nature and hence as always to be invoked in the attack on new problems. We tend to forget that it is after all the physicist who invents the function which is to be inserted in the equation of motion. It is of course conceivable that this standardization of force assists one to a better grasp of the meaning of mechanics; it connects closely with our physiological feeling of push and pull and muscular sensations and our urge to endow nature with similar feelings. However, it has its dangers also. If for example in the case of the motion of a particle attached to the end of a spring we follow Hooke's law blindly, we do indeed discover simple harmonic motion, but we are annoyed to find that the particle does not really perform this type of motion after all, for instead of continuing its vibrations indefinitely, it more or less rapidly comes to rest. It is clear that to describe the actual motion we must modify the equation. Something different from $-kx$ must appear on the right-hand side. What this something is must again be found by the process of judicious guessing. Ultimately we observe that the addition of $R\dot{x}$ leads to a solution representing a damped oscillation and hence is a better description of the observed motion. If, however, the motion in question is that of an eardrum, for example, our equation will still be found to be insufficient, and it is only when a term of the form cx^2 is added to the force side that the solution agrees with the observed facts. We say that the eardrum is an *asymmetrical* oscillator. These remarks are not intended to suggest that in handling the equation of motion we should throw away all our previously acquired knowledge of mechanics. It only means that one must be careful not to be led astray by preconceptions but must be willing to experiment with the right-hand side of the equation of motion without attaching too sacred a significance to **F** from a metaphysical point of view.

Of all the types of forces considered in mechanics, central forces have appeared to be the most useful, and this has undoubtedly been an important stimulus to the feeling that mechanics can be developed entirely in terms of such forces. This notion received perhaps its greatest emphasis in the work of Hertz which has already been mentioned and which will be further considered in Sec. 4.11. It has an important bearing on modern physics since it leans strongly on the assumption of an atomic constitution for all bodies, i.e., that all bodies are in the last analysis composed of ultimate particles the forces between which control the physical properties of substances. Of course the modern view of the electrical nature of the particles has altered somewhat the older classical mechanical view, since electrodynamic forces are no longer always central. Nevertheless, central forces still play an overwhelmingly significant rôle in quantum mechanics.

The importance of boundary conditions in the solution of physical problems in general has been emphasized previously. It is sufficient to remind the reader that since the equation of motion is of the second order the solution always involves two arbitrary constants which must be specified in order to provide the final solution in any given problem. Knowledge of the position and velocity for every particle in the system at a given instant is sufficient to fix these constants, and these are the usual boundary conditions (" specific " type in the sense of Sec. 1.9) involved in particle dynamics. Continuous

3.7 INTERPRETATION OF THE EQUATION OF MOTION

media, of course, demand special attention and will be taken up in Chapter VI.

Invariance. In the use of the equation of motion a question of considerable importance arises. Does the equation retain its form invariant with respect to changes in the coordinate system? This is not the case. The reader will readily convince himself that there is nothing particularly sacred about rectangular coordinates, and hence in the procedure used in setting up the equation the acceleration components can be expressed in terms of spherical or cylindrical coordinates, etc., by the use of transformation equations such as those given in Sec. 3.2. The component equations of motion are then not as symmetrical as the rectangular form, viz.

$$m\ddot{x} = F_x, \quad m\ddot{y} = F_y, \quad m\ddot{z} = F_z, \tag{1}$$

but they are often much more useful. More important than this type of change in coordinates is that to a system *moving* with respect to the primary inertial system in which the equation of motion was set up. Here it is comparatively easy to show that there *is* a type of moving system in which eqs. (1) retain their form. This is one moving with constant velocity with respect to the primary inertial system. Thus if the two systems are moving with velocity v along their common x axis, the transformation equations become

$$\left. \begin{array}{l} x' = x - vt - x_0 \\ y' = y - y_0 \\ z' = z - z_0 \end{array} \right\} \tag{2}$$

where x_0, y_0, z_0 are the initial coordinates of the origin of the primed system in the unprimed system. Since v is constant, we have

$$\ddot{x}' = \ddot{x}, \quad \ddot{y}' = \ddot{y}, \quad \ddot{z}' = \ddot{z},$$

and the form of the equations is unaltered. This neglects the transformation of the F's; they would not in general have the same form in x', y', z' as in x, y, z. This difficulty disappears when we recall that the coordinates appearing in F_x, F_y, F_z are not *absolute* position coordinates but always differences in the coordinates of two particles denoted, let us say, by subscripts 1 and 2 (cf. Newton's third law of motion). Since we always have $x_1 - x_2 = x_1' - x_2'$, etc., the complete invariance is proved.

If v were not constant this result would not follow and the equations in the primed system would be of quite different form. The system moving with constant velocity with respect to the primary inertial system is called a secondary inertial system, and what we have shown is that the Newtonian equations of motion preserve their form in all inertial systems. This is sometimes referred to as the Newtonian principle of relativity. It is worth mentioning here, however, that inertial systems are not so common in mechanics as one might hope. For example, a system of axes rotating with the earth is not an inertial system, and hence the investigation of the motion of particles with respect to the earth using equations of the form (1) is logically wrong. As a matter of fact, owing to the relatively small angular velocity of the earth the error committed is in practice very small.

Summary. We may summarize the discussion of the present section by saying that the solution of every problem in particle dynamics by the Newtonian method consists in the solution of sets of equations of the form (1) (one set for each particle) in which F_x, F_y, F_z are appropriately chosen functions of coordinates and velocities (and perhaps the time) subject to certain constraints symbolized by relations among the coordinates of the particles, and the solutions rid of arbitrary constants by the choice of suitable initial conditions. The system of reference must, moreover, be an inertial system; otherwise special investigation is necessary.

We are now ready to consider another question: is the Newtonian method the only basis for mechanics? We shall see that the answer to this is in the negative.

3.8. D'Alembert's Principle and the Principle of Virtual Displacements. Although the Newtonian principles have long been the most useful simple approach to mechanics, many other formulations have arisen in the attempt to simplify the solution of problems of various types. It is important that we should devote some attention to these alternative formulations, not so much to follow them into the detailed study of their applications as to note them as illustrations of various methods of constructing physical theories and of the value of new points of view in the attack on old problems. Mechanics has the advantage of being the oldest branch of knowledge to which the method of physical science has been applied. It is only natural that many attempts have been made to introduce simplification by the introduction of broader concepts making necessary fewer postulates in the development of the theory.

It is important to recognize at the outset that these alternative formulations differ considerably in their methods of making use of the concepts of mass and force. Some of them are hardly more than restatements of the content of Newton's laws with a shift of emphasis; others introduce new concepts and state new principles in terms of them. In the first class belong the principle of D'Alembert which seeks effectively to reduce dynamics to statics but which employs the concepts of mass and force essentially as they are contemplated in Newton's laws, and the Gaussian principle of least constraint which, though it avoids the use of the force concept, does employ that of mass. In the second class belong the various principles employing the concept of energy, which so far we have not mentioned. The attempt of Hertz to formulate an entirely new basis for mechanics renouncing the use of force, though closely related to the Gaussian principle, was such an ambitious undertaking that it should be considered in a class by itself. This we shall do in Sec. 3.11. No one of these reformulations is wholly independent of the Newtonian basis. Since they must all lead to the same observed results they must all be somehow connected. The incautious observer might then be tempted to remark: if all mechanical problems can be solved by Newtonian mechanics, is it really economical to introduce a flock of differently stated principles which after all can accomplish no more? To this we make a threefold rejoinder. In the first place, the ease of solving a given problem generally depends on the way in which it is stated, and a method which solves it when it is stated one way may be vastly simpler than that which handles it when the statement is made in another form. In the second place, we can fairly say that every restatement of the fundamental principles deepens our appreciation of and feeling for the whole subject: two methods of solving the same problem mean more in our understanding than the solution of two problems by the same method. More important in many respects than these answers is, however, the third: it is by no means sure that the Newtonian principles are actually competent to describe *all* phenomena in which motion occurs. Indeed, it seems certain that they break down in attacking the problem of atomic structure. It seems plausible that alternative points of view may themselves suggest fundamental modifications in mechanical principles which will lead to successful attacks on the newer problems. As a matter of fact, precisely this has happened, first in connection with the development of the Bohr theory and then more recently in quantum mechanics.

D'Alembert's principle. Since there appears to this day to be some confusion over the meaning of this principle, it will be desirable to examine the statement as given by its celebrated author in his " Traité de dynamique " first published in 1743 and reprinted with revisions and additions in 1758. It is here entitled, " A general principle for finding the motions of several bodies which react on each other in any fashion." D'Alembert begins by observing that bodies act on each other in three different ways, viz.: (1) by collisions; (2) by means of some body placed between them to which they are attached (i.e., strings, springs, or rods, etc.); and (3) by mutual attraction, as the earth and sun. We should naturally enlarge the last class to include such things as magnetic and electrostatic attractions and repulsions. D'Alembert was not particularly interested in the third class, however, considering probably that gravitational attraction had already at his time received sufficient attention (though the classic work of Laplace had not yet been done). His attention was concentrated on what he thought were the more or less neglected problems of the first two types, although he intended his principle to be of perfectly general application.

The statement of the problem which the principle is to solve is given in the following form. Let us imagine a system of bodies (equivalent to particles in our notation) which interact with each other in any fashion whatever. The attempt is made to impress on each particle a certain motion (by the word " motion " here D'Alembert explains that he means strictly " velocity " in the vector sense). However, because of the influence of the other particles, the one in question is unable to perform the impressed motion. The problem then is to determine the actual motions of all the particles of the system.

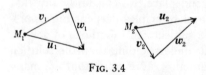

FIG. 3.4

Denote the particles of the system by $M_1, M_2, M_3, \ldots$, and let the *impressed* motions be $u_1, u_2, u_3, \ldots$ (cf. Fig. 3.4). The *actual* motions are to be represented by $v_1, v_2, v_3 \ldots$. Now let us think of the motion u_1 as compounded of the actual motion v_1 and another motion which we shall call w_1. That is,

$$u_1 = v_1 + w_1.$$

Similarly for the other particles,

$$u_2 = v_2 + w_2$$

$$u_3 = v_3 + w_3$$

$$\vdots \quad \vdots \quad \vdots \quad \vdots \quad \vdots$$

It follows that the motions of the particles would be the same if instead of impressing $u_1, u_2, u_3, \ldots$, we were to impress the equivalent combinations $v_1 + w_1, v_2 + w_2 \ldots$. But the motions actually per-

3.8 D'ALEMBERT'S PRINCIPLE

performed are $v_1, v_2, v_3 \ldots$ Hence the motions $w_1, w_2, w_3 \ldots$ must be of such a character that they alter nothing in the motions $v_1, v_2, v_3 \ldots$ Otherwise expressed, if the particles had been impressed originally only with the motions $w_1, w_2, w_3, \ldots$ the latter would have mutually compensated each other and the particles if originally at rest would have remained at rest. The solution of every problem involving the motion of interacting particles is then to be handled by this program: the impressed motions $u_1, u_2, u_3 \ldots$ are to be resolved into pairs, viz.: $v_1 + w_1$, etc., of such a nature that if the motions $v_1, v_2, v_3 \ldots$ were the impressed ones the particles would perform them without influencing each other, whereas if the motions $w_1, w_2, w_3 \ldots$ had been impressed the system would remain at rest. The motions $v_1, v_2, v_3 \ldots$ are those which solve the problem.

Illustration. It is interesting to observe that in the enunciation of the principle the word "force" nowhere appears, nor, for that matter, does the word "mass." However, a moment's thought will convince us that the concepts are really after all implicit in the principle, for any attempt to apply the principle to the solution of problems involves their use. Thus the notion of force is involved in the operation of "impressing" a velocity on a particle, and it is impossible to express the condition of equilibrium necessary for the application of the principle without employing mass explicitly. This is most readily brought out by examining one of the problems which D'Alembert proposed and solved. Fig. 3.5 represents a rigid rod AB to which there are attached mass particles $C, D,$ and B of masses $m_1, m_2,$ and m_3, respectively. The rod is supposed to be fastened to a rigid support at A and to be able to move about an axis normal to the plane of the diagram. It is assumed that if the particles were not attached to the rod (supposed for convenience to be of negligible mass) they would suffer in some infinitesimal time interval dt the infinitesimal displacements $CC', DD',$ and BB', respectively, all perpendicular to the rod. The problem is to determine the motion of the rod. Suppose that in the time dt the rod moves from AB to AB'' with C moving to C'' and D moving to D''. For simplicity we denote the impressed displacements as $d\mathbf{r}_1, d\mathbf{r}_2,$ and $d\mathbf{r}_3$, respectively, so that the impressed velocities are $d\mathbf{r}_1/dt = u_1, d\mathbf{r}_2/dt = u_2, d\mathbf{r}_3/dt = u_3$, respectively. Similarly the *actual* velocities are $d\mathbf{r}_1'/dt = v_1, d\mathbf{r}_2'/dt = v_2, d\mathbf{r}_3'/dt = v_3$. The principle says that the particles and hence the rod to which they are attached would remain at rest if the velocities $u_1 - v_1, u_2 - v_2, u_3 - v_3$ were to be given to the masses

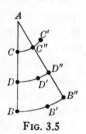

FIG. 3.5

m_1, m_2, m_3, respectively. This equilibrium condition is expressed by D'Alembert essentially in the following form

$$m_1(u_1 - v_1)\overline{AC} + m_2(u_2 - v_2)\overline{AD} + m_3(u_3 - v_3)\overline{AB} = 0, \quad (1)$$

where we have simplified by omitting the vector signs, since $\mathbf{u}_1$ and $\mathbf{v}_1$ are approximately collinear vectors, etc. Now we have the kinematic relations

$$v_2/v_1 = \overline{AD}/\overline{AC}; \ v_3/v_1 = \overline{AB}/\overline{AC}.$$

Therefore solving for v_1 yields

$$v_1 = \frac{(m_1 u_1 \overline{AC} + m_2 u_2 \overline{AD} + m_3 u_3 \overline{AB})\overline{AC}}{m_1 \overline{AC}^2 + m_2 \overline{AD}^2 + m_3 \overline{AB}^2}, \quad (2)$$

which provides the solution for the problem, since if we know v_1 we can readily calculate v_2 and v_3 from the kinematic relations. We are interested, however, in the interpretation of the equilibrium equation (1). It is apparently clear to D'Alembert from his previous consideration of equilibrium, for he gives no explicit justification of it. On its understanding, however, rests our whole grasp of the method involved in his principle. Fortunately it is not difficult to interpret it in terms of familiar concepts. The difference $\mathbf{u}_1 - \mathbf{v}_1$ is the negative of the *change* in velocity brought about by the constraint, i.e., by the fact that the particle is not free to move but is attached to the rod. This change in velocity divided by the time interval dt may therefore appear as the *acceleration* due to the constraint and when multiplied by the mass becomes the *force* on the particle due to the constraint. Under the totality of these forces the system of particles must be in equilibrium. Eq. (1) expresses the condition that the total torque due to these forces about the point A is zero, and is the familiar equilibrium condition for rotation.

There is indeed another possible interpretation based on *the principle of virtual displacements* or *virtual work*. Let $\delta\theta$ represent the angle through which the rod rotates when B moves to B'', etc. If we multiply (1) by $\delta\theta$ we obtain

$$\left. \begin{array}{c} m_1(u_1 - v_1)\overline{AC}\cdot\delta\theta + m_2(u_2 - v_2)\overline{AD}\cdot\delta\theta \\ + m_3(u_3 - v_3)\overline{AB}\cdot\delta\theta = 0. \end{array} \right\} \quad (3)$$

But $\overline{AC}\cdot\delta\theta = \overline{CC''}$, $\overline{AD}\cdot\delta\theta = \overline{DD''}$, $\overline{AB}\cdot\delta\theta = \overline{BB''}$. Hence the condition (3) may be interpreted in the following way: the sum of the products obtained by multiplying the constraint force on each particle

3.8   D'ALEMBERT'S PRINCIPLE

by an elementary displacement of this particle compatible with the constraints is equal to zero. The more modern interpretation would of course be that the work done by the constraint forces during any set of elementary displacements compatible with the constraints is zero.[1] This is the principle of virtual displacements, first enunciated by John Bernoulli in 1717. From it the well-known vector conditions of equilibrium follow, and it can indeed be deduced from these. Since it plays an important rôle in the application of D'Alembert's principle, it will be worth while to stop and consider it briefly here, using more familiar notation.

Virtual displacements. The principle of virtual displacements may be stated in the general form: if there are n particles $1, 2, 3, \ldots, n$ acted on by forces $\mathbf{F}_1, \mathbf{F}_2, \ldots, \mathbf{F}_n$, respectively, and if these are given arbitrary (virtual) displacements [2] $d\mathbf{r}_1, d\mathbf{r}_2, \ldots, d\mathbf{r}_n$ respectively compatible with the constraints, the condition for equilibrium under the action of the forces is

$$\mathbf{F}_1 \cdot d\mathbf{r}_1 + \mathbf{F}_2 \cdot d\mathbf{r}_2 + \ldots + \mathbf{F}_n \cdot d\mathbf{r}_n = 0. \qquad (4)$$

The products in each case are *scalar* or *dot* products, that is, $\mathbf{F}_j \cdot d\mathbf{r}_j = F_j dr_j \cos \theta_j$, where θ_j is the angle between the vectors $\mathbf{F}_j$ and $d\mathbf{r}_j$. This principle may indeed be deduced from the more common vector equilibrium rule, but we shall not concern ourselves here with that.[3] Rather we wish to give an illustration of its application which will assist us in our understanding of D'Alembert's principle.

Let us consider the case of two particles P_1 and P_2 of masses m_1 and m_2, respectively, tied to the ends of an inextensible string which is wrapped around a smooth peg at the vertex of two smooth inclined planes of angles θ_1 and θ_2, respectively (Fig. 3.6). The impressed forces can be considered to be

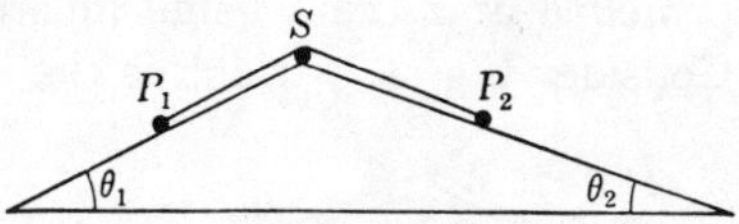

FIG. 3.6

$$\mathbf{F}_1 = m_1 \mathbf{g} + \mathbf{R}_1 + \mathbf{T} \quad \text{and} \quad \mathbf{F}_2 = m_2 \mathbf{g} + \mathbf{R}_2 + \mathbf{T},$$

where $\mathbf{T}$ is the tension in the string, $\mathbf{R}_1$ is the reaction of the plane against P_1, and $\mathbf{R}_2$ is that against P_2. Allow the system virtual dis-

---

[1] Note that in our illustration the constraint force and the corresponding displacement are in the *same* direction.

[2] $\mathbf{r}_1, \mathbf{r}_2, \ldots, \mathbf{r}_n$ are the position vectors of the particles respectively. See Secs. 3.2 and 3.3.

[3] See Mach, "Mechanics," for a discussion of the principle.

placements $d\mathbf{r}_1$ and $d\mathbf{r}_2$, which for the sake of convenience we shall take *along* the two planes respectively. The principle says that for *equilibrium*

$$(m_1\mathbf{g} + \mathbf{R}_1 + \mathbf{T}) \cdot d\mathbf{r}_1 + (m_2\mathbf{g} + \mathbf{R}_2 + \mathbf{T}) \cdot d\mathbf{r}_2 = 0, \qquad (5)$$

with the condition that in order to be compatible with the constant length of the string

$$dr_1 = -dr_2, \qquad (6)$$

where positive displacement always means *down* the plane and negative *up* the plane. Recalling that $\mathbf{R}_1 \cdot d\mathbf{r}_1 = \mathbf{R}_2 \cdot d\mathbf{r}_2 = 0$ since the vectors are normal to each other, we have from (5)

$$m_1 g \sin \theta_1 = m_2 g \sin \theta_2. \qquad (7)$$

It is interesting to note that neither the tension nor the reaction plays an effective rôle in the problem. Of course (7) is equivalent to the result of applying the usual static equilibrium condition. Nevertheless our example is instructive in pointing out the value of the principle in telling us how to proceed in general without having to pay too much attention to the details of each special problem.

The problem just discussed was a case of equilibrium. Let us now consider the corresponding dynamical problem, i.e., suppose that the system may be accelerated. We shall use D'Alembert's principle and the scheme illustrated in D'Alembert's first problem. Consider Fig. 3.7, which is the same as Fig. 3.6 save that the vector velocities $\mathbf{u}_1, \mathbf{u}_2$ and $\mathbf{v}_1, \mathbf{v}_2$ are indicated. The former are the velocities which the two particles would have respectively if they were allowed to move freely, i.e., they are the velocities of *free fall*. The latter are the actual velocities, directed along the plane. The principle states that the vectors $m_1(\mathbf{u}_1 - \mathbf{v}_1)$ and $m_2(\mathbf{u}_2 - \mathbf{v}_2)$, which are respectively proportional to the constraint forces on m_1 and m_2, form a system in equilibrium. Applying the principle of virtual displacements we have

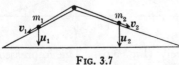

FIG. 3.7

$$m_1(\mathbf{u}_1 - \mathbf{v}_1) \cdot d\mathbf{r}_1 + m_2(\mathbf{u}_2 - \mathbf{v}_2) \cdot d\mathbf{r}_2 = 0. \qquad (8)$$

For convenience we choose $d\mathbf{r}_1$ and $d\mathbf{r}_2$ *along* the planes, whence the constraint condition that the string is inextensible becomes as before

3.8　D'ALEMBERT'S PRINCIPLE

$dr_1 = -dr_2$. The same condition gives us $v_1 = -v_2$. Eq. (8) then yields

$$v_1 = \frac{m_1 u_1 \sin \theta_1 - m_2 u_2 \sin \theta_2}{m_1 + m_2}. \tag{9}$$

If the motion takes place in time t we have

$$u_2 = u_1 = gt, \quad v_1 = at, \tag{10}$$

where g is the numerical value of the acceleration of gravity and a the acceleration of m_1 (the same numerically as that of m_2) along the plane. Then (9) becomes

$$a = \frac{(m_1 \sin \theta_1 - m_2 \sin \theta_2)g}{m_1 + m_2}, \tag{11}$$

the usual result.

Mach's formulation of D'Alembert's principle. The modes of stating D'Alembert's principle have been many and various. Above we have tried to remain as close as possible to the original formulation. That used by Mach is worth considering, since he adheres definitely to the concept of force in his enunciation. Thus Mach speaks of a system of n particles being acted on by certain *impressed* forces $\mathbf{F}_1{}^i, \mathbf{F}_2{}^i, \ldots, \mathbf{F}_n{}^i$, which, if the particles were perfectly free to move, would give them accelerations $\mathbf{a}_1 = \mathbf{F}_1{}^i/m_1, \ldots, \mathbf{a}_n = \mathbf{F}_n{}^i/m_n$. Owing to the constraints, however, the actual accelerations are such as would be produced in free particles by the forces $\mathbf{F}_1{}^e, \mathbf{F}_2{}^e, \ldots, \mathbf{F}_n{}^e$, which are called the *effective* forces. One can then form the vector differences

$$\mathbf{F}_1{}^i - \mathbf{F}_1{}^e = \mathbf{V}_1, \ldots, \mathbf{F}_n{}^i - \mathbf{F}_n{}^e = \mathbf{V}_n. \tag{12}$$

The $\mathbf{V}$'s may be looked upon as the parts of the impressed forces which are, so to apeak, not effective. Hence they must form a system of forces under the influence of which the system of particles must remain in equilibrium. This is the content of D'Alembert's principle, according to Mach. The $\mathbf{V}$'s are sometimes referred to as the "lost" forces. This change of notation introduces no essential difference into the principle. For one can look upon the $\mathbf{F}_1{}^i$ as proportional to the mass times the impressed velocity gained in a given infinitesimal time interval, while the $\mathbf{F}_1{}^e$ is proportional to the mass times the actual velocity gained in the same time interval. Let us solve the previous problem from this viewpoint. The impressed forces are $\mathbf{F}_1{}^i = m_1\mathbf{g}$ and $\mathbf{F}_2{}^i = m_2\mathbf{g}$. The effective forces are $\mathbf{F}_1{}^e = m_1\mathbf{a}_1$ and

$F_2^e = m_2 a_2$, where a_1 and a_2 are the actual accelerations along the planes respectively. The principle then says that

$$m_1(\mathbf{g} - \mathbf{a}_1)\cdot d\mathbf{r}_1 + m_2(\mathbf{g} - \mathbf{a}_2)\cdot d\mathbf{r}_2 = 0. \qquad (13)$$

From the constraints, $dr_1 = -dr_2$ and $a_1 = -a_2$, so that we finally have

$$a_1 = a = \frac{(m_1 \sin \theta_1 - m_2 \sin \theta_2)g}{m_1 + m_2}, \qquad (14)$$

a result identical with (11). It is of interest to note that it is unnecessary to include the tension in the string and the reactions of the surfaces since the latter always disappear from the final result in the process of applying the principle of virtual displacements. This is a great help in a more complicated problem.

A word of caution is advisable. When one reads (as the statement is sometimes made) that the $\mathbf{V}$'s form a system in equilibrium one must not interpret this as meaning that the system of vectors $\mathbf{V}_j$ has a zero resultant. This is not the meaning. Rather $\Sigma \mathbf{V}_j \cdot d\mathbf{r}_j = 0$, where the $d\mathbf{r}_j$ satisfy the constraints. Actually we may, if we wish, look upon the matter thus: each $\mathbf{V}_j$ is balanced by a constraint force $\mathbf{R}_j$ which we may think of as acting on the jth particle, so that $\mathbf{V}_j + \mathbf{R}_j = 0$. Then $\Sigma(\mathbf{V}_j + \mathbf{R}_j)\cdot d\mathbf{r}_j = 0$ identically. Therefore $\Sigma \mathbf{V}_j \cdot d\mathbf{r}_j = -\Sigma \mathbf{R}_j \cdot d\mathbf{r}_j$. But by virtue of their nature the constraint forces by themselves can do no work on the system during displacements which are compatible with them. Hence $\Sigma \mathbf{R}_j \cdot d\mathbf{r}_j = 0$, and therefore $\Sigma \mathbf{V}_j \cdot d\mathbf{r}_j = 0$, which is the content of D'Alembert's principle in this interpretation.

Another illustration of D'Alembert's principle. A somewhat more elaborate illustration of D'Alembert's principle is provided by the motion of two particles of masses m_1 and m_2, respectively, attached to a flexible string which is rigidly fixed at the point A (see Fig. 3.8). The distance from A to m_1 is l_1, and the length of string connecting m_1 with m_2 is l_2. The string with attached masses, which may be termed a double pendulum, is supposed at first to be hanging vertically in equilibrium. It is then drawn aside through a small angle and let go. Let the displacements of m_1 and m_2, respectively, from the vertical along circular arcs be denoted by x_1 and x_2. The effective forces on m_1 and m_2, respectively, are $m_1 a_1$ and $m_2 a_2$. The impressed forces are, respectively, $m_1 \mathbf{g} + \mathbf{T}$ and $m_2 \mathbf{g}$, where $\mathbf{T}$ is the force on m_1 due to the weight of m_2. The principle says that the

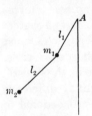

FIG. 3.8

system is in equilibrium under the action of the forces $m_1\mathbf{a}_1 - m_1\mathbf{g} - \mathbf{T}$ and $m_2\mathbf{a}_2 - m_2\mathbf{g}$. Hence if we give the two particles virtual displacements $d\mathbf{r}_1$ and $d\mathbf{r}_2$, respectively (compatible with the constraints, of course), we must have

$$(m_1\mathbf{a}_1 - m_1\mathbf{g} - \mathbf{T})\cdot d\mathbf{r}_1 + (m_2\mathbf{a}_2 - m_2\mathbf{g})\cdot d\mathbf{r}_2 = 0. \qquad (3)$$

It is simplest to take $d\mathbf{r}_1$ and $d\mathbf{r}_2$ along the respective arcs, with $|d\mathbf{r}_1| = dx_1$ and $|d\mathbf{r}_2| = dx_2$. Then

$$m_1\mathbf{a}_1 \cdot d\mathbf{r}_1 = m_1\ddot{x}_1 dx_1; \quad m_2\mathbf{a}_2 \cdot d\mathbf{r}_2 = m_2\ddot{x}_2 dx_2,$$

while

$$m_1\mathbf{g}\cdot d\mathbf{r}_1 = -\frac{m_1 g x_1}{l_1} dx_1,$$

and

$$m_2\mathbf{g}\cdot d\mathbf{r}_2 = -\frac{m_2 g(x_2 - x_1)}{l_2} dx_2.$$

Finally

$$\mathbf{T}\cdot d\mathbf{r}_1 = m_2 g\left(\frac{x_2 - x_1}{l_2} - \frac{x_1}{l_1}\right) dx_1.$$

Therefore eq. (3), reduced to scalar form, becomes

$$\left\{m_1\ddot{x}_1 + \frac{m_1 g x_1}{l_1} - m_2 g\left(\frac{x_2 - x_1}{l_2} - \frac{x_1}{l_1}\right)\right\} dx_1$$
$$+ \left\{m_2\ddot{x}_2 + \frac{m_2 g(x_2 - x_1)}{l_2}\right\} dx_2 = 0. \qquad (4)$$

Now within the limit of small displacements, dx_1 and dx_2 are *arbitrary* and to satisfy (4) we must therefore have

$$m_1\ddot{x}_1 = -\frac{(m_1 + m_2)g x_1}{l_1} + m_2 g\frac{(x_2 - x_1)}{l_2} \qquad (5)$$

and

$$m_2\ddot{x}_2 = -\frac{m_2 g(x_2 - x_1)}{l_2}. \qquad (6)$$

These are the equations of motion of the system (neglecting the very small vertical motions). This is as far as D'Alembert's principle carries us; the rest of the problem involves the mathematical integration of the equations to find x_1 and x_2 as functions of t and four arbitrary constants. (For the complete solution consult, for example, Lamb, "Dynamical Theory of Sound," second edition, 1925, p. 39.)

Summary. We have now examined enough examples of the use of the principle to make its meaning clear. It must be evident that every one of the illustrative problems can be readily solved by the direct application of Newton's laws. Indeed in most cases the solution in that manner follows more directly. It is true that at the time when the principle was enunciated such problems involving connections and constraints in general had hardly been attacked at all and there was every reason for the introduction of the unified procedure of D'Alembert. Yet it must be admitted that the subsequent high development of the Newtonian method has rendered the principle less necessary. Why then have we paid so much attention to it? Because it focuses our vision on another aspect of mechanics and by that much cannot fail to deepen our understanding of the subject. The idea contained in it that all problems in dynamics may be treated from the standpoint of equilibrium is certainly of importance in unifying mechanical theory. Moreover, it is not amiss to seek in mechanics an esthetic element lending it grace and elegance of form, and at the same time incorporating within it economy of thought. Judged from this standpoint D'Alembert's principle is worthy of close attention. The success of mechanics depends not only on the utility of the method for the solution of concrete problems but also on the conciseness and simplicity of the statement of the general principles and the ease with which the laws of mechanics flow from them. The principle of D'Alembert provides a single fundamental formula from which may be derived all other mechanical laws. It was undoubtedly this feature of universality which impressed Lagrange when in his celebrated treatise " Mécanique Analytique " (1788) he used the principle as the starting point for the development of his powerful general methods. In a later section we shall have occasion to note this in detail.[1]

3.9. Gauss's Principle of Least Constraint. In the year 1829, Gauss, who had been pondering on the principles of mechanics, was led to enunciate a principle which he hoped would reduce the subject to a single generalization. It was intended to be particularly adapted to the study of systems subject to specific constraints.

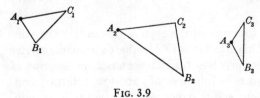

Fig. 3.9

Let us again imagine that we have a system of n particles of masses $m_1, m_2, \ldots, m_n$. Three of them are indicated in Fig. 3.9 as occupying at the same instant the positions A_1, A_2, A_3. If the particles were perfectly free to

---

[1] For further interesting detailed applications of D'Alembert's principle see P. Appell, "Traité de mécanique rationnelle," second edition (Paris, 1904), p. 290.

3.9   GAUSS'S PRINCIPLE OF LEAST CONSTRAINT

move under the action of certain impressed forces, we shall suppose that in the infinitesimal time Δt they would undergo the elementary displacements $\overline{A_1B_1}, \overline{A_2B_2}, \overline{A_3B_3}, \ldots, \overline{A_nB_n}$. Because of the constraints the actual displacements in this time interval are $\overline{A_1C_1}, \overline{A_2C_2}, \overline{A_3C_3}, \ldots, \overline{A_nC_n}$, the effect of the constraints being to produce a change in the motion represented by the vector differences $\overline{B_1C_1}, \overline{B_2C_2}, \ldots, \overline{B_nC_n}$. Gauss's principle states that the actual displacements of the particles in the time Δt are determined by the condition that the sum

$$m_1 \overline{B_1C_1}^2 + m_2 \overline{B_2C_2}^2 + \ldots + m_n \overline{B_nC_n}^2 \qquad (1)$$

shall be a minimum, that is, shall be *less* for the actual motion of the system than for any other motion compatible with the constraints. Gauss termed the sum indicated in (1) the total *constraint*, and hence the principle has come to be known as that of *least constraint*.

The formulation reminds one at once of D'Alembert's principle, with its emphasis on free motions and actual motions, and indeed it will shortly be shown that it can be deduced from the latter principle. However, Gauss no doubt intended it to serve as an independent postulate.   Hence to get a clearer notion of its meaning it will be worth while to apply it to a special problem, viz., the case already discussed earlier by the method of D'Alembert, in which two particles of masses m_1 and m_2 are on two smooth inclined planes (see Fig. 3.6). In time Δt, if the particles were perfectly free, instead of being tied and constrained to move on the planes, they would each descend vertically downward through a displacement

$$\Delta \mathbf{r} = \tfrac{1}{2}\mathbf{g}(\Delta t)^2.$$

Actually, however, the displacements of the particles are, respectively,

$$\Delta \mathbf{s}_1 = \tfrac{1}{2}\mathbf{a}_1(\Delta t)^2,$$

$$\Delta \mathbf{s}_2 = \tfrac{1}{2}\mathbf{a}_2(\Delta t)^2.$$

The application of the principle now leads to the minimizing of the expression

$$\left. \begin{array}{l} \dfrac{m_1}{4}\left|\mathbf{g} - \mathbf{a}_1\right|^2(\Delta t)^4 + \dfrac{m_2}{4}\left|\mathbf{g} - \mathbf{a}_2\right|^2(\Delta t)^4 \\[6pt] = \dfrac{m_1}{4}(g^2 + a_1^2 - 2ga_1\sin\theta_1)(\Delta t)^4 + \dfrac{m_2}{4}(g^2 + a_2^2 - 2ga_2\sin\theta_2)(\Delta t)^4, \end{array} \right\} \qquad (2)$$

subject to the auxiliary condition $a_1 = -a_2$. Utilizing this, differentiating the above expression with respect to a_1, and setting the result equal to zero, we obtain the usual answer

$$a_1 = (m_1 \sin \theta_1 - m_2 \sin \theta_2)g/(m_1 + m_2) = -a_2. \tag{3}$$

It is of interest to observe that in getting this result we have not used the factor $(\Delta t)^2$ in any essential way. Actually we might simply have recalled that

$$\Delta s_1 = k a_1, \quad \Delta s_2 = k a_2, \quad \Delta r = k g; \tag{4}$$

i.e., the displacements in the same time interval are in the same ratio as the corresponding accelerations. We may therefore treat the total constraint in the Gaussian sense as being simply proportional to

$$m_1 |\mathbf{g} - \mathbf{a}_1|^2 + m_2 |\mathbf{g} - \mathbf{a}_2|^2. \tag{5}$$

Applying the condition for a minimum at once without further reduction we have

$$2m_1 |\mathbf{g} - \mathbf{a}_1| d |\mathbf{g} - \mathbf{a}_1| + 2m_2 |\mathbf{g} - \mathbf{a}_2| d |\mathbf{g} - \mathbf{a}_2| = 0 \tag{6}$$

subject to the condition $a_1 + a_2 = 0$. It is seen that

$$d |\mathbf{g} - \mathbf{a}_1| = -|d\mathbf{a}_1| \cos \{(\mathbf{g} - \mathbf{a}_1), d\mathbf{a}_1\}, \tag{7}$$

where the argument of the cosine is the angle between the vectors $d\mathbf{a}_1$ and $\mathbf{g} - \mathbf{a}_1$. Hence from the definition of the dot product

$$|\mathbf{g} - \mathbf{a}_1| d |\mathbf{g} - \mathbf{a}_1| = (\mathbf{g} - \mathbf{a}_1) \cdot d(\mathbf{g} - \mathbf{a}_1). \tag{8}$$

The minimum condition then becomes effectively

$$m_1(\mathbf{g} - \mathbf{a}_1) \cdot d\mathbf{a}_1 + m_2(\mathbf{g} - \mathbf{a}_2) \cdot d\mathbf{a}_2 = 0, \tag{9}$$

and since $d\mathbf{a}_1$ and $d\mathbf{a}_2$ are proportional to the *virtual* displacements $d\mathbf{r}_1$ and $d\mathbf{r}_2$ of the two particles, respectively, we can write the above in the form

$$m_1(\mathbf{g} - \mathbf{a}_1) \cdot d\mathbf{r}_1 + m_2(\mathbf{g} - \mathbf{a}_2) \cdot d\mathbf{r}_2 = 0, \tag{10}$$

which is precisely (with the condition $d\mathbf{r}_1 + d\mathbf{r}_2 = 0$) the equation to which D'Alembert's principle led us in the solution of this problem.

Deduction of Gauss's principle. The close connection between the two principles can be established in general without difficulty by a

3.9 GAUSS'S PRINCIPLE OF LEAST CONSTRAINT

method exemplified by Mach. Consulting Fig. 3.10 we note that $\overline{A_jB_j}$ is the displacement which the jth particle would undergo in time Δt if free to do so under the action of the *impressed* force $\mathbf{F}_j{}^i$; the actual displacement is $\overline{A_jC_j}$, which is that which would be produced in a free particle by the *effective* force $\mathbf{F}_j{}^e$. The displacement $\overline{B_jC_j}$ corresponds to the *constraint* force $\mathbf{V}_j = \mathbf{F}_j{}^i - \mathbf{F}_j{}^e$. The measure of the constraint on the jth particle is $m_j \overline{B_jC_j}^2$. If the particle had gone to D_j instead of C_j under

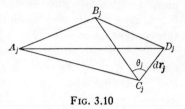

Fig. 3.10

the same constraints, the measure of the constraint would have been $m_j \overline{B_jD_j}^2$. We can show that for every motion compatible with the constraints

$$\Sigma m_j \overline{B_jC_j}^2 < \Sigma m_j \overline{B_jD_j}^2. \tag{11}$$

$\overline{B_jD_j}$ can be obtained from $\overline{B_jC_j}$ by adding the vector $d\mathbf{r}_j$ which makes the angle θ_j with $\overline{B_jC_j}$. In fact, we have

$$\Sigma m_j \overline{B_jD_j}^2 = \Sigma m_j \overline{B_jC_j}^2 + \Sigma m_j dr_j{}^2 - 2\Sigma m_j \overline{B_jC_j}\, dr_j \cos \theta_j. \tag{12}$$

But $m_j \overline{B_jC_j}$ is proportional to $\mathbf{V}_j$, the constraint force, and therefore $m_j \overline{B_jC_j}\, dr_j \cos \theta_j$ is proportional to $\mathbf{V}_j \cdot d\mathbf{r}_j$, where $d\mathbf{r}_j$ now figures as a possible virtual displacement of the system. The principle of D'Alembert, in the form in which we have used it, says that $\Sigma \mathbf{V}_j \cdot d\mathbf{r}_j = 0$. Therefore $\Sigma m_j \overline{B_jC_j}\, dr_j \cos \theta_j = 0$ for alternative motions compatible with the constraints, and (12) becomes

$$\Sigma m_j \overline{B_jD_j}^2 = \Sigma m_j \overline{B_jC_j}^2 + \Sigma m_j\, dr_j{}^2. \tag{13}$$

But $\Sigma m_j dr_j{}^2$ is always positive. Therefore we have proved (11) and have shown that $\Sigma m_j \overline{B_jC_j}^2$ is a minimum for the actual motion.

The minimal concept. The demonstrated connection between the principles of D'Alembert and Gauss tends to rob the latter of some of its attraction. But we must remember that all general methods of solving mechanical problems must be interrelated so that any one may be derived from any other, unless the latter possesses a lesser degree of generality. The relation between the two principles we have been discussing is indeed a mutual one, and there is no problem which can be solved by the one which cannot be solved by the other. Yet the principle of least constraint does introduce one feature which merits somewhat closer attention. It is an important illustration of the use of the *minimal* idea in mechanics, the assignment of meaning to the minimizing of some expression or function. The fact that we shall meet this again in the principles of Hamilton and least action does not diminish its significance here, for the use of the minimal concept by Gauss does not involve the introduction of the *energy* concept: from this conceptual point of view it is probably the simplest of the minimal principles in mechanics. It represents the more or less unconscious feeling that of all the possible motions of a mechanical system that is the actual one which approaches as closely as possible free motion. To be sure, this notion has an anthropomorphic basis and in this respect resembles the principle of least action for which Maupertuis had so cleverly argued many years before Gauss. In the present case the anthropomorphism in no way diminishes the physical significance of the principle since the latter is a perfectly definite result of well-established mechanical principles. Moreover, it is allowable to believe that the idea of least constraint does provide a deeper insight into mechanics than would be possible without it. We cannot help having a lively interest in the "greatest or least" in every department of knowledge. Heinrich Hertz generalized the principle and embodied it in his scheme of mechanics as we shall note presently. Although they are so closely connected, the two principles of D'Alembert and Gauss stress quite different physical aspects. D'Alembert emphasizes the *equilibrium* notion inherent in every mechanical problem, whereas Gauss stresses the extent to which a mechanical system endeavors to reduce the effect of the constraints to a minimum and move as a free system under the action of the impressed forces. The method of D'Alembert in the form usually employed makes free use of the concept of force, while that of Gauss in all strictness makes the use of force wholly unnecessary. It was perhaps this feature of it which appealed to Hertz, who had a dislike for the metaphysical causational view of force prevailing in his day.

Gauss's principle and least squares. In connection with the statement of his principle, Gauss made the interesting observation that there is an analogy between least constraint in mechanics and least squares in the adjustment of observations. It will be shown in Sec. 4·9 that if one has a series of observations of a certain physical quantity the most probable value for the quantity (on the basis of the Gauss error law) is that which makes the sum of the squares of the deviations of the observed values from this value a minimum. The arithmetical mean satisfies this criterion. Thus in a certain sense we may look upon the circumstances attending a physical measurement as imposing a constraint on our ability to assign a value to the physical quantity in question. We like to choose the value which will make this constraint as measured by the sum of the squares of the deviations between measured values and actual value a minimum, just as the mechanical system on its part likes

3.9 GAUSS'S PRINCIPLE OF LEAST CONSTRAINT

to move in such a way as to make the effect of the constraint on it a minimum. Yet we should perhaps refrain from pushing the analogy, interesting though it may be, too far, for while the mechanical principle is universal in the sense that all mechanical laws can be derived from it, the principle of least squares has been sometimes looked upon as of questionable utility in the theory of measurements.

System subject to constraint. It may be of some interest to use the Gaussian principle in the deduction of the equations of motion of a dynamical system subject to constraint. Imagine such a system composed of n masses $m_1, m_2 \ldots m_n$. Let them be acted on by impressed forces $\mathbf{F}_1{}^i, \mathbf{F}_2{}^i \ldots \mathbf{F}_n{}^i$, and let the connections be such as to be represented by one equation of the form

$$\phi(x_1, y_1, z_1, \ldots, x_n, y_n, z_n) = 0 \qquad (14)$$

where x_s, y_s, z_s are the rectangular coordinates of the sth particle. For example, the particles may all be attached to the same closed string. The principle of least constraint states that the sum

$$\sum_s m_s \,|\, \mathbf{a}_s - \mathbf{F}_s{}^i/m_s \,|^{\,2} \qquad (15)$$

is a minimum subject to the condition (14).[1] Here $\mathbf{a}_s$ is the actual acceleration of the sth particle. This may be expressed in Cartesian form as follows:

$$\sum_s m_s \{(a_{sx} - F_{sx}{}^i/m_s)^2 + (a_{sy} - F_{sy}{}^i/m_s)^2 + (a_{sz} - F_{sz}{}^i/m_s)^2\}, \qquad (16)$$

where the x, y, z subscripts serve to represent the respective rectangular components of the vector in question. In making (16) a minimum we employ Lagrange's method of undetermined multipliers.[2] We must have

$$\sum_s m_s \{(a_{sx} - F_{sx}{}^i/m_s)\delta a_{sx} + (a_{sy} - F_{sy}{}^i/m_s)\delta a_{sy} + (a_{sz} - F_{sz}{}^i/m_s)\delta a_{sz}\} = 0 \qquad (17)$$

still subject to the condition (14), which yields on the variation of δx_s, etc.,

$$\sum_s \left(\frac{\partial \phi}{\partial x_s}\delta x_s + \frac{\partial \phi}{\partial y_s}\delta y_s + \frac{\partial \phi}{\partial z_s}\delta z_s\right) = 0. \qquad (18)$$

Now since the variations all correspond to the same time interval, we can set

$$\delta a_{sx} = k\,\delta x_s, \quad \delta a_{sy} = k\,\delta y_s, \quad \delta a_{sz} = k\,\delta z_s,$$

where k is a constant factor. Hence (17) can be written

$$\sum_s \{(m_s\ddot{x}_s - F_{sx}{}^i)\delta x_s + (m_s\ddot{y}_s - F_{sy}{}^i)\delta y_s + (m_s\ddot{z}_s - F_{sz}{}^i)\delta z_s\} = 0 \qquad (19)$$

subject to the condition (18). Multiplying the latter through by the at first unde-

---

[1] In writing $\mathbf{F}_s{}^i/m_s$ for the "free" acceleration of the sth particle, we are, of course, assuming that Newton's second law applies to the "free" motion.

[2] See, for example, L. Page, "Introduction to Theoretical Physics" (D. Van Nostrand Co., New York, 1928), pp. 18 and 284.

termined multiplier λ and adding to (19), we get on equating to zero the coefficients of the arbitrary variations, the equations

$$m_s\ddot{x}_s = F_{sx}{}^i + \lambda \frac{\partial \phi}{\partial x_s}, \quad m_s\ddot{y}_s = F_{sy}{}^i + \lambda \frac{\partial \phi}{\partial y_s}, \quad m_s\ddot{z}_s = F_{sz}{}^i + \lambda \frac{\partial \phi}{\partial z_s}. \quad (20)$$

These constitute a set of $3n$ equations (obtained by letting s run from 1 to n) and together with (14) give us $3n + 1$ equations for the determination (on integration) of the $3n + 1$ quantities $\lambda, x_1, y_1, z_1, \ldots, x_n, y_n, z_n$. This completes the physical set up of the problem. The generalization to the case where there are k (any integer $< n$) conditions of type (14) is not difficult.

3.10. Hertz's Mechanics. In 1894 Heinrich Hertz published "Principles of Mechanics," which is of considerable significance for our survey of the foundations in spite of the fact that it has actually obtained rather slight recognition in the common discussions on the subject. The reason is doubtless to be found in the fact that the greatest modern emphasis in mechanics is placed on its utility in application rather than its fundamental meaning for physics; Hertz's work has indeed little of a utilitarian nature to recommend it. Nevertheless it was an earnest and thoughtful attempt to rid the science of much of the metaphysical uncertainties and vagueness with which mechanics was still befogged during the late nineteenth century. To be sure, Mach and others had already brought to bear searching criticism on the usual Newtonian presentation, but Hertz doubtless felt that this criticism was not sufficiently thoroughgoing and that a complete re-establishment of the principles was desirable for the sake of logical consistency.

D'Alembert sought to unify the theory of mechanics by reducing dynamics to statics; Hertz endeavored to achieve a similar result by reducing dynamics to kinematics, thus avoiding the use of concepts like mass, force, and energy which he regarded as undesirable because of the contemporary looseness of their interpretation. He proposed to do this by the invocation of concealed motions of particles which introduce the apparently obvious complexity into what would otherwise be the simple motions of observed bodies. The idea is a tempting one. A mechanical system (i.e., an aggregate of particles with certain connections) is observed to move in some more or less complicated fashion; what is more natural than to suppose that if this system were assumed to be a part of a larger system the whole could be described as a "free" system moving in a more simple fashion? However, to understand Hertz's ideas more clearly, it is essential to look into his fundamental concepts more closely.

Hertz's dynamical system. We shall not try to present in full the mathematical details of Hertz's formulation. Fortunately this is not

3.10 HERTZ'S MECHANICS

necessary for an understanding of his theory. His conception of a dynamical system is like that presented earlier in this chapter, i.e., an aggregate of material particles in Euclidean space of three dimensions. The *position* of such a system is the aggregate of the positions of its component particles. The *path* of a system is the totality of positions (each considered simultaneously) occupied by the system in time. For two successive positions infinitely near to each other, the path is called a path element. The magnitude of the path of a system is defined to be the quadratic mass-weighted mean of the corresponding displacements of the individual particles. Specifically, if the coordinates are the $3n$ quantities $x_1, x_2, \ldots x_j, \ldots x_{3n}$ and the masses are denoted by m_j (where it is to be understood that not all the m_j are unequal, since there are three coordinates for each particle) the path between the positions $x_1', x_2' \ldots x_{3n}'$ and $x_1, x_2, \ldots x_{3n}$ has the magnitude s' where

$$ms'^2 = \sum_{j=1}^{3n} m_j(x_j' - x_j)^2. \qquad (1)$$

Here m is the total mass of the system. Hertz considers every material particle as in turn made up of what we may call infinitesimal mass corpuscles which are all alike, invariable, and indestructible. The mass of a given particle is then defined as the ratio of the number of corpuscles in the particle to the number in some standard particle. He later assumes this number to be given practically by the use of the balance. If two paths run from $x_1 \ldots x_{3n}$ to $x_1' \ldots x_{3n}'$ and $x_1 \ldots x_{3n}$ to $x_1'' \ldots x_{3n}''$, respectively, the angle θ between the two path directions is defined by

$$ms's'' \cos \theta = \sum_{j=1}^{3n} m_j(x_j'' - x_j)(x_j' - x_j). \qquad (2)$$

This defines effectively the path direction of a system. The definitions (1) and (2) are arbitrary but sufficiently suggestive.

The path of a system is said to be *straight* if it has the same direction in all its positions; the path is called *curved* if the direction changes with position. Path curvature is defined as the limit of the ratio of the difference in direction at the two ends of a path element to the magnitude of the element. One path element of a system is called straighter than another if it has a smaller curvature. That possible path element which is straighter than all other possible ones having the same position and direction is called the *straightest* path element for this position and direction. A *straightest* path is then one all of whose elements are straightest elements.

Principle of the straightest path. Among the particles of a system connections may exist, analytically specified by certain relations among their coordinates. If these relations are independent of the time and depend only on the mutual positions of the particles the system is said to be *free*. The fundamental principle of Hertzian mechanics is: *Every free system remains either in a state of rest or in uniform motion along a straightest path.* The aim of mechanics is to investigate all the consequences of this principle. It might be objected that it is too narrow since most systems encountered in practice are non-free. Hertz meets such an objection by assuming that every non-free system is part of a free system. Every motion of a free system or its non-free parts which obeys the fundamental principle is called a *natural* motion, and Hertzian mechanics is concerned with natural motions only. The treatment of every system as either a free system or part of a free system clearly brings to the fore the necessity for the admission of concealed motions. Thus, even in the simple case of the system consisting of a single particle in a gravitational field, the principle certainly does not hold, for if the particle were reduced to rest and then set free it would not stay at rest. In order to make the principle work it is necessary to invoke the existence of other particles in the universe, which may not be immediately discernible. Later the originally " concealed " motions may be associated with actually observed motions, as was indeed the case in the solar system. We shall have more to say about the invocation of concealed masses and motions in the latter part of this chapter, particularly in the section on energy. From the illustration just given it is clear that the notion of force does not play any necessary part in Hertzian mechanics. If it is introduced it appears in the mutual action of the non-free parts of a free system; that is, one part of a free system can be considered as exerting a force on the other part. This is a matter of convenience.

The close relation between Hertzian mechanics and the Gaussian principle of least constraint is obvious. It must be admitted that Hertz has created a logically beautiful structure, even if it lies outside the realm of utilitarian mechanics.

3.11. The Concept of Energy. It may be thought strange that we have brought our discussion of the foundations of mechanics so far with hardly a mention of *energy*, which now holds such an important place among the concepts of physics. We have indeed had no desire to slight the significance of this concept, but have wished to emphasize that the logical development of mechanics is quite possible without it.

3.11 THE CONCEPT OF ENERGY

All the problems of classical mechanics can be solved without reference to it. The question at once arises: why then should it have been introduced at all? This is what we wish to discuss. We must first remark, however, that the *idea* of energy is historically much older than the name. Without doubt it goes back at least to Galileo on his observation with respect to machines that "what is gained in power is lost in speed," referring to the fact that the force required to lift a weight (by means of a pulley system) multiplied by the distance through which the force has to be applied remains constant, though either factor in itself may vary. The concept of work is involved here. Its importance was, however, overshadowed by Galileo's epoch-making discoveries of the laws of motion, and it was not until the time of Huyghens that it again became prominent in the concept of "vis viva" or *living force*, i.e., a quantity varying as the mass multiplied by the square of the velocity. The attribution of the term *energy* to the concept of "vis viva" did not come until the nineteenth century.

Space integral of force and kinetic energy. Leaving the historical and returning to the logical aspect of the question, we get the simplest illustration of the fundamental meaning of energy in mechanics by considering the *space integral* of the force acting on a particle. If $\mathbf{F}$ is the force and $d\mathbf{r}$ an elementary displacement of the particle, we inquire what meaning may be attached to $\int_{P_0}^{P_1} \mathbf{F} \cdot d\mathbf{r}$, where the integration extends from the initial position P_0 of the particle to its final position P_1. Substituting for $\mathbf{F}$ the kinetic reaction $m\ddot{\mathbf{r}}$ from the equation of motion,

$$m \int_{P_0}^{P_1} \ddot{\mathbf{r}} \cdot d\mathbf{r} = \frac{m}{2} \int_{P_0}^{P_1} \frac{d}{dt}(\dot{\mathbf{r}} \cdot \dot{\mathbf{r}}) dt \tag{1}$$

from simple vector properties.[1] Now $\dot{\mathbf{r}} \cdot \dot{\mathbf{r}} = v^2$, where v is the *magnitude* of the velocity of the particle. The integral then becomes

$$\frac{m}{2} \int_{v_0}^{v_1} d(v^2) = mv_1^2/2 - mv_0^2/2, \tag{2}$$

and we have the result that the space integral of the force during a given displacement is equal to the difference between the value of the quantity $\frac{1}{2}mv^2$ at the end of the displacement and its value at the beginning. The integral $\int_{P_0}^{P_1} \mathbf{F} \cdot d\mathbf{r}$ is called the *work* done by the

---

[1] Or cf. Lindsay, "Physical Mechanics," pp. 84, 85.

force on the particle during the displacement, while the quantity $\frac{1}{2}mv^2$ is one-half that which used to be known as "vis viva." We now call it the *kinetic energy*. Though we have made the above deduction for a single particle, it may be extended to a system of particles, viz.,

$$\sum_{i=1}^{n} \int \mathbf{F}_i \cdot d\mathbf{r}_i = \frac{1}{2}\sum m(v^2 - v_0^2),$$

where the sum is extended over all n particles of the system. What is the physical content of this result? Why are we so interested in it that we give names to the quantities which appear? It would seem that the answer must be similar to that which was given to our earlier question concerning the place of such ideas as those of D'Alembert and Gauss; viz., we introduce a concept, name it, and work with it because we feel in the first place that it is useful, i.e., it deepens our understanding of the whole subject and enables us to solve certain problems more directly than other methods would permit; and, in the second place, that it adds a further esthetic quality to the development of the theory, a note of completeness which helps to satisfy our demand for elegance. In the present case it is not difficult to trace both these influences. Once there has been developed the notion of *force* as something associated with the accelerated motion of a particle, we cannot help feeling interested in the *integrated* effect of the force on a particle. This integration may be, of course, either *temporal* or *spatial*. It at once develops that the time integral of the force is the change in momentum of the particle, a concept of great value in the principle of conservation of momentum, as has already been seen. The space integral is the change in another quantity somehow characteristic of the motion of the particle, and this is what has been named the kinetic energy. Both integrals provide interesting aspects of force. It is not surprising that during the late seventeenth and early eighteenth century there should have been wordy argument as to the proper measure of force, Descartes and his followers insisting on the time integral as the only true measure, and Leibniz combating this view vigorously in favor of the space integral. This famous Cartesian-Leibnizian controversy was finally settled when D'Alembert pointed out that both integrals are perfectly legitimate, though different aspects of the effect of force, and that the problem is one involving nomenclature only.

The usefulness of the new concept in solving problems arises chiefly from the fact that in many cases we are interested mainly in the velocities of the particles in question. It will suffice to consider a simple example, and we may as well choose that which we have already studied by means of the principles of D'Alembert and Gauss (see Fig. 3.6). In the motion of P_1 down its plane a distance x and the corresponding motion of P_2 up its plane the same distance, the space-integrated force is

$$m_1 g x \sin \theta_1 - m_2 g x \sin \theta_2.$$

Placing this equal to the total kinetic energy gained, we have the equation

$$(m_1 g \sin \theta_1 - m_2 g \sin \theta_2) x = \tfrac{1}{2}(m_1 + m_2)v^2 - \tfrac{1}{2}(m_1 + m_2)v_0^2, \quad (3)$$

which enables us to compute the velocity gained by the system in any displacement. Mathematically, of course, the equation merely amounts to a first integration (a spatial one) of the two equations of motion, with subsequent addition. Actually there is a certain economy in being able to write at once the result in so far as the right-hand side of the equation is concerned. Other simple examples will occur to the reader, e.g., a sphere rolling down a plane, where the total kinetic energy gained is made up jointly of kinetic energy of translation of the center of mass and kinetic energy of rotation about the center of mass.

Potential energy. If the concept of energy had gone no further it is doubtful whether it would have proved of great significance for mechanics, to say nothing of physics as a whole. The following consideration indicates how the concept became enormously enlarged in meaning and fruitfulness. Contemplating once more the integral $\int_{P_0}^{P} \mathbf{F} \cdot d\mathbf{r}$, let us write the vectors in terms of their Cartesian components, viz. (see Sec. 3.3),

$$\mathbf{F} = \mathbf{i}F_x + \mathbf{j}F_y + \mathbf{k}F_z; \quad d\mathbf{r} = \mathbf{i}dx + \mathbf{j}dy + \mathbf{k}dz.$$

Our result above becomes

$$\int_{P_0}^{P_1} (F_x dx + F_y dy + F_z dz) = \tfrac{1}{2} m(v_1^2 - v_0^2). \quad (4)$$

The force components F_x, F_y, F_z are functions of x, y, z and the expression $F_x dx + F_y dy + F_z dz$, a linear differential form. Suppose it happens to be a perfect differential, i.e., suppose there exists a function $V(x, y, z)$ such that

$$F_x dx + F_y dy + F_z dz = -dV, \qquad (5)$$

with

$$F_x = -\frac{\partial V}{\partial x}, \quad F_y = -\frac{\partial V}{\partial y}, \quad F_z = -\frac{\partial V}{\partial z}. \qquad (6)$$

Then (4) becomes, on carrying out the integration,

$$V_0 - V_1 = \tfrac{1}{2}mv_1^2 - \tfrac{1}{2}mv_0^2$$

or

$$\tfrac{1}{2}mv_1^2 + V_1 = \tfrac{1}{2}mv_0^2 + V_0, \qquad (7)$$

where V_0 is the value of the function $V(x, y, z)$ at the point $P_0(x, y, z)$ and V_1 that at the point $P_1(x, y, z)$. Now the result (7) is remarkable in the fact that for any displacement the quantity $\tfrac{1}{2}mv^2 + V$ keeps the same value as the particle moves from one point to the other. In other words, we have discovered something which remains *constant* during the motion. It is not surprising that we should like to attach considerable significance to such a quantity when it is realized that all the previous concepts connected with motion involve quantities which may and usually do vary with the time. An invariant of the motion is sure to appeal to our esthetic sense. The dimensions of this quantity, and thus of V alone, are those of work or kinetic energy. It is therefore natural to look upon V as a kind of energy whose value depends on the position of the particle only. We call it the *potential* energy, and the sum the *total mechanical energy* of the particle. It is important to note that these definitions depend for their meaning on the mathematical existence of the function V. If $F_x dx + F_y dy + F_z dz$ is not a perfect differential, the whole procedure ceases to have the meaning we have associated with it. Nevertheless, experience tells us that there are many F functions which describe observed motions reasonably well (at least for short time intervals) for which the function V exists. This suffices to fix firmly the definition of mechanical energy as a valuable mechanical concept, and we immediately begin to give it significance even for cases for which it never could have been logically defined. Thus we begin to divide mechanical systems into two classes: one the so-called *conservative* systems in which V exists and the total energy is constant or "conserved," and the other the *non-conservative* systems in which the energy does not remain constant during the motion. This is in strict logic an exceedingly curious procedure, for we have just seen that the very definition of mechanical energy rests upon the discovery of something which remains constant during motion and now we wish to apply the concept to the case where this constancy does not exist! Why are we not more impressed by the illogical nature of this situation? It is probably because we feel that the idea of energy is too valuable to be dispensed with; so we save it in difficult situations by introducing the possibility of its changing with the time and endow *every* mechanical system with energy whether constant or not. A very simple historic example will help to make clear this extrapolation.

Conservative and non-conservative systems. It was Galileo who visualized the experiment of allowing a ball to roll down one inclined plane and up another. If gravitational force is one for which a function V exists, it follows that the kinetic energy gained at the bottom is equal to the potential energy lost in falling through the height of the plane and that the ball should travel up the other plane to a vertical height equal to that from which it originally fell. Now when the experiment is performed, this result never actually happens: the ball does not quite reach the same height even when the bottom angle is smoothed over. What is the conclusion? Shall we say that there is no meaning to the concept of energy for gravitational force? At first sight this appears to be a troublesome problem. The approximation which considers the force of gravitation to be constant in magnitude and vertical in direction near the surface of the earth (an approximation which we know from other experiments to be a very good one) insures that the expression $\mathbf{F} \cdot d\mathbf{r}$ is a perfect differential and hence to this approximation the gravitational potential exists; there must be a meaning to the total energy of a particle moving in such a gravitational field of force. Rather than look upon this disagreement between theory and experiment as fatal to the theory we seize upon the fact of the motion along the planes and through the air as providing a reason for the discrepancy, and conclude that the ball would have a constant total energy if the air and the planes were not there. The air and the planes are then assumed to rob the ball of some of its energy; or as we may say, the energy is partly *dissipated*. In this simple example we have already decided upon a far-reaching extrapolation of the original notion of mechanical energy. We begin to look upon it as a definite property of a mechanical system rather than merely another and somewhat more abstract way of characterizing the state of such a system than by its position and velocity, a property of which the system can have more or less and which is therefore *extensive* rather than *intensive*. Thence arises the somewhat unfortunate tendency to endow the concept with a "substantial" foundation.

Dissipative system. From the standpoint of mechanics the motion of so-called *dissipative* systems can be readily handled in general by the introduction of a function of the velocity on the right-hand side of the equation of motion (see Sec. 3.7). In this way it becomes a simple matter to compute the rate of dissipation of energy by the system. Let us take an illustration which has a great deal of significance in many branches of physics, viz., the linear harmonic oscillator. The equation of motion without dissipation is

$$m\ddot{x} = -kx, \qquad (8)$$

with m the mass and k the stiffness of the system. The energy equation is

$$\tfrac{1}{2}m\dot{x}^2 + \tfrac{1}{2}kx^2 = E, \qquad (9)$$

where E appears as the (here) constant total energy. In order to make the analysis agree with observation we rewrite eq. (8) in the form

$$m\ddot{x} = -kx - R\dot{x}, \qquad (10)$$

where R is the so-called *damping* or *dissipation coefficient*. This equation can at once be written in the form

$$\frac{d}{dt}(\tfrac{1}{2}m\dot{x}^2 + \tfrac{1}{2}kx^2) = -R\dot{x}^2,$$

and if we still agree to call the quantity inside the parenthesis the instantaneous *energy* of the system, we may write

$$\dot{E} = -R\dot{x}^2, \qquad (11)$$

which says that the rate of change (in this case, loss) of energy of the system is equal in magnitude to $R\dot{x}^2$, a quantity one-half of which has been called by Lord Rayleigh the *dissipation function*. A great variety of interesting special results can be obtained with this analysis as the starting point. Reference to texts on mechanics, acoustics, and electricity may be made for such developments. The important point for our present purpose is to note how the concept of energy has here been extended and a precise meaning given to its variation with time.

A natural corollary of the substantialization of energy is the question: if the energy of such a dissipative system as we have just been considering decreases with the time, what becomes of it? Once having conceived energy in the form of substance the mind dislikes to contemplate its annihilation and immediately seizes upon the possibility of transformation into another type with conservation of the whole. There must, however, be some phenomenon with which we can plausibly associate the transformation. This is found in the heat changes that are always observed in connection with dissipation in mechanical systems. The development of the concept of heat as a form of energy undoubtedly has served to increase enormously the importance of the mechanical concept and to extend its utility into every branch of physics. The principle of the conservation of energy became the first law of thermodynamics. In the mechanical theory of heat (cf. Chapter V) the idea of energy transformation is kept on a strictly mechanical basis in the sense that all energy is thought of as kinetic and potential energy of the particles which on the atomic hypothesis are considered to constitute gross matter. The unifying value of this generalization is evident, even if it is at times a source of confusion to persons who have the vague and naïve impression of energy as something "which makes everything go," "the source of all activity," etc. This is much to be deplored; energy is primarily a mechanical concept, and care should be taken to keep this in mind when referring to it. The question arises: what can one say about such a thing as electrical energy, for example? In a later section (Sec. 5.2) it is shown that the energy concept has meaning for electricity only to the extent to which electrical theory can be developed along mechanical lines.

When we come to the meaning of energy in atomic theory the situation is considerably more complicated since classical mechanics breaks down in the description of atomic structure. Nevertheless a formal use of the concept in connection with the quantum hypothesis has led to great success in interpreting atomic problems (cf. Chapter IX).

Use of the energy equation. We now revert to mechanical energy in the strict sense and consider how useful the energy principle may actually prove to be in the solution of mechanical problems. It might be thought that the application of the energy equation, i.e., eq. (7), would be a much more economical method of solving a problem than the use of the Newtonian equation of motion. As a differential equation it is of the first order and might appear to offer a simpler basis for mechanics. Of course, in using it, one encounters the same sort of trouble we have discussed in connection with the equation of motion, namely, the choice of the potential energy function V. The remarks made in Sec. 3.7. concerning force will apply here with equal weight to V. However, this is a trivial matter compared with the real difficulty with the energy equation, viz., the fact that it is but a single equation and hence is by itself alone incompetent to describe a mechanical system of more than *one* degree of freedom, i.e., one in which more than a single coordinate is necessary for a complete description of the motion. It gives, one may say, merely *one* integral of the equation of motion.

The question then arises: is there any way of stating the general principles of mechanics using the concept of energy alone and making no *a priori* use of the equation of motion? The answer is that we find such a formulation in *Hamilton's principle* and its various interpretations.

3.12. Hamilton's Principle and Least Action. We first consider this principle, enunciated by Hamilton in 1824, as a postulate. Associated with every mechanical system there is, as we have seen, a certain kinetic energy T and, if it is conservative, a potential energy V; the former is a function of the velocities and positional coordinates of the particles of the system, and the latter we shall first assume to be a function of their positional coordinates only. We define

$$L = T - V. \qquad (1)$$

The content of Hamilton's principle may then be stated in words as follows: the motion of the system between any two instants t_0 and t_1 takes place in such a way that the integral

$$\int_{t_0}^{t_1} L\, dt \qquad (2)$$

has a value which is either *less* than its value for any other possible motion of the system in the interval from t_0 to t_1 or *greater* than its value for any other such possible motion, it being understood that the initial and final states are the same respectively for all such motions. In other words, the integral has a *stationary* value for the actual motion as compared with all other possible motions between the same states. This is often referred to as a *minimum* principle, but although this turns out to be correct for the majority of cases encountered in practice, it is not so in general. In our use of the principle we need employ only the stationary property. Let us try to make the statement clearer by the use of a pictorial representation. In Fig. 3.11 we represent the state of the system at the instant t_0 by A and that at time t_1 by B. Let the path actually followed by

3.12 HAMILTON'S PRINCIPLE AND LEAST ACTION

the system be ACB, while ADB and AEB represent two "varied" paths—possible paths for the system and performed in the same time. It must be remarked that our representation is abstract. In certain simple cases of the plane motion of a single particle, ACB, ADB, and AEB may represent possible geometrical orbits from point A to point B in the plane, with the understanding that to be comparable they must of course all be traversed in the *same* time, viz., in the interval $t_1 - t_0$. However, the representation is in general purely schematic, i.e., A merely *represents* for convenience the state of a system of several particles and is a point in what may be called *configuration* space to distinguish it from ordinary three-dimensional physical space. Hence the paths we are speaking of are not in general paths in physical space. In any case the meaning of the principle is that the time integral $\int_{t_0}^{t_1} L dt$ along the actual path traversed, ACB, is either *greater* than that along *any other* paths such as ADB and ACB, or it is *less* than that along any such paths. The quantity L is termed the *Lagrangian* function or *kinetic potential*.

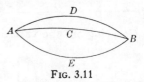

Fig. 3.11

Simple harmonic motion. The use of the principle in the setting up of mechanical problems can perhaps best be appreciated by considering first its application to a very simple case, namely, that of a single particle constrained to move along a straight line in such a way that there exists a potential function

$$V = \tfrac{1}{2}kx^2,$$

where x is the particle's distance from some chosen origin and k is a positive constant. The kinetic energy being denoted by $T = \tfrac{1}{2}m\dot{x}^2$, with m appearing as the mass of the particle, the principle now involves finding the condition that

$$\int_{t_0}^{t_1} (\tfrac{1}{2}m\dot{x}^2 - \tfrac{1}{2}kx^2)dt$$

shall have a stationary value. This means that if we vary the actual motion very slightly the corresponding variation in the integral will be zero. In the present case, of course, the geometrical shape of the physical orbit may not change, since the particle must always move on the straight line. Hence any variation in the motion implies only an altered dependence of x on t. In the usual notation of the calculus of variations, the stationary condition will be written

$$\delta \int_{t_0}^{t_1} (\tfrac{1}{2}m\dot{x}^2 - \tfrac{1}{2}kx^2)dt = 0. \tag{3}$$

Carrying out the variation yields [1]

$$\int_{t_0}^{t_1} \left[m\dot{x} \frac{d}{dt}(\delta x) - kx\delta x \right] dt = 0, \tag{4}$$

where we have used the fact (true in the present instance) that the operations of d and δ are interchangeable (i.e., $d/dt\,(\delta x) = \delta \dot{x}$). Integration of $\int_{t_0}^{t_1} m\dot{x}\frac{d}{dt}(\delta x)dt$ by parts yields $-\int_{t_0}^{t_1} m\ddot{x}\delta x dt$, and hence the principle leads to

$$\int_{t_0}^{t_1} (m\ddot{x} + kx)\delta x dt = 0. \tag{5}$$

Since within the limits imposed by the fact that the variation must be small and zero at $t = t_0$ and $t = t_1$, δx is arbitrary, the only way for the integral to be zero for *all* δx is to have the integrand vanish. Thus we are led to

$$m\ddot{x} + kx = 0, \tag{6}$$

which we recognize as the equation of motion for the linear harmonic oscillator with mass m and stiffness coefficient k.

The important feature of this illustration whose concrete result is so trivial rests simply in the fact that we have been led to an equation of motion, making no use of the Newtonian laws. Treating $T = \tfrac{1}{2}m\dot{x}^2$ as a concept defined independently of the Newtonian basis we have inquired what kind of linear motion corresponds to a positive potential energy varying quadratically with the distance from a given point. Hamilton's principle gives us the answer that it is simple harmonic motion. Note that in framing our question we used the energy concept throughout; there was no question about force. Our illustration suggests that we have here, therefore, an entirely new way in which to approach mechanical problems. The formal proof of this comes in showing that the principle is a straightforward deduction from the principle of D'Alembert.

[1] See, e.g., Lindsay, "Physical Mechanics," p. 304.

3.12 HAMILTON'S PRINCIPLE AND LEAST ACTION

Deduction of Hamilton's principle. We imagine a system with n particles of masses $m_1, m_2, \ldots, m_n$ and position vectors $\mathbf{r}_1, \mathbf{r}_2, \ldots, \mathbf{r}_n$. On the system there are impressed forces $\mathbf{F}_1^i, \ldots, \mathbf{F}_n^i$. The effective forces are $m_1\ddot{\mathbf{r}}_1, m_2\ddot{\mathbf{r}}_2, \ldots, m_n\ddot{\mathbf{r}}_n$. The motion of the system according to D'Alembert and the principle of virtual displacements (Sec. 3.8) is given by

$$\sum_{j=1}^{n} (m_j\ddot{\mathbf{r}}_j - \mathbf{F}_j^i) \cdot \delta\mathbf{r}_j = 0, \tag{7}$$

where $\delta\mathbf{r}_j$ is the virtual displacement of the jth particle. Now from the properties of δ,

$$\ddot{\mathbf{r}}_j \cdot \delta\mathbf{r}_j = \frac{d}{dt}(\dot{\mathbf{r}}_j \cdot \delta\mathbf{r}_j) - \tfrac{1}{2}\delta(v_j^2), \tag{8}$$

where we have set $\dot{\mathbf{r}}_j \cdot \dot{\mathbf{r}}_j = v_j^2$ (Sec. 3.11). Therefore

$$\frac{d}{dt}\sum m_j\dot{\mathbf{r}}_j \cdot \delta\mathbf{r}_j = \delta\sum \tfrac{1}{2}m_j v_j^2 + \sum m_j\ddot{\mathbf{r}}_j \cdot \delta\mathbf{r}_j. \tag{9}$$

The summation in all cases is taken over j running from 1 to n. Integrate both sides of this equation from $t = t_0$ to $t = t_1$. We get (with $T = \Sigma\tfrac{1}{2}m_j v_j^2$)

$$\sum m_j\dot{\mathbf{r}}_j \cdot \delta\mathbf{r}_j \Big]_{t_0}^{t_1} = \int_{t_0}^{t_1} (\delta T + \Sigma m_j\ddot{\mathbf{r}}_j \cdot \delta\mathbf{r}_j)dt. \tag{10}$$

Now let us assume that at times t_0 and t_1 the $\delta\mathbf{r}_j$ are all zero, i.e., we consider only those motions of the system for which the initial and final positions are the same. In this case the left-hand side of (10) vanishes and we have

$$\int_{t_0}^{t_1} (\delta T + \Sigma m_j\ddot{\mathbf{r}}_j \cdot \delta\mathbf{r}_j)dt = 0. \tag{11}$$

This is in itself an interesting and rather general result. If, however, we suppose in addition that there exists a function V of the rectangular coordinates of the n particles such that

$$\Sigma \mathbf{F}_j^i \cdot \delta\mathbf{r}_j = \Sigma m_j\ddot{\mathbf{r}}_j \cdot \delta\mathbf{r}_j = -\delta V, \tag{12}$$

eq. (11) takes the form

$$\delta\int_{t_0}^{t_1} (T - V)dt = 0, \tag{13}$$

i.e., the mathematical statement of Hamilton's principle. It applies, as we know, to a *conservative* system. But it is worthy of note that (11) presents an analogue to the principle which holds for both conservative and non-conservative systems. Hence the principle is really more general than our first enunciation of it implied. As one might suspect, however, its utility has been most marked in conservative systems.

Least action. At this point we ought to consider a principle closely related to that of Hamilton and even more celebrated historically, namely, that of *least action*. We shall discuss it for a conservative system of n particles. The total momentum of the system is $\Sigma m_j \dot{\mathbf{r}}_j$. Let us form the space integral of this between any two points P_0 and P_1,

$$\int_{P_0}^{P_1} \Sigma m_j \dot{\mathbf{r}}_j \cdot d\mathbf{r}_j, \tag{14}$$

which we shall call the *action* of the system. The principle of least action states that the actual motion of the system from P_0 to P_1 takes place in such a way that the action is *stationary*, i.e., an extremal compared with all other possible motions between the points P_0 and P_1 that correspond to the *same* value of the total energy. As in Hamilton's principle the integral has a stationary value and need not be a minimum, although this is the more common case encountered in practice. An important distinction between the two principles should be noted, however. In Hamilton's principle all the varied paths are supposed to be traversed in the *same* time; the variations in the coordinates are perfectly arbitrary save at the end points. In the case of least action this arbitrariness is restricted by the assumption that the total energy shall remain constant and the same over every varied path. Thus from $T + V = E =$ constant it follows that

$$\delta T = - \delta V. \tag{15}$$

We can no longer assume that all the varied paths are traversed in the same time. Keeping this in mind we may still transform (14) to a time integral by noting that

$$\Sigma m_j \dot{\mathbf{r}}_j \cdot d\mathbf{r}_j = \Sigma m_j(\dot{x}_j dx_j + \dot{y}_j dy_j + \dot{z}_j dz_j) = 2T dt.$$

We thus get for the principle

$$\delta \int_{t_0}^{t_1} 2T dt = 0, \tag{16}$$

3.12 HAMILTON'S PRINCIPLE AND LEAST ACTION

subject to (15), so that whereas t_0 may be kept constant, t_1 must be varied. The principle may be verified by actually carrying out the the variation indicated in (16), the process being similar to that employed in deducing Hamilton's principle above. Hamilton's principle is evidently more general than that of least action since it does not necessitate the assumption (15). Moreover, it can be shown that the extension to non-conservative systems is much more readily accomplished in the former than the latter. In general, then, Hamilton's principle seems to be much the more suitable starting point for advanced mechanical investigations. From the historical standpoint, however, least action is very interesting as it provides many sidelights on the development of minimal principles in physics.

Maupertuis and least action. The principle of least action was due apparently in the first instance to Maupertuis, who expressed in 1747 the belief that the wisdom of God is best exemplified in having the motions of bodies take place with the "minimum quantity of action." In this way he felt that there could be secured for mechanics a true metaphysical basis. Undoubtedly his instinct was sure, but his actual mathematical formulation was very crude. Thus as a measure of the action of a body he took the product of mass, velocity, and distance. This is so vague as to be almost useless, since distance depends on time and, if distance in unit time is meant, the quantity really becomes proportional to the kinetic energy. At any rate the dimensions are the same as those of action as defined in our previous discussion—the logical definition due to Lagrange (1760), and of course for a particle moving with constant velocity the action for a given path reduces to the Maupertuis definition. An examination of the way in which Maupertuis used his idea in the solution of problems indicates clearly that he had no really definite process which was the same in every case, but rather tried to find in each a quantity which when subjected to the minimizing condition led to the known result. As a matter of historical interest we consider a couple of examples.

The first is the inelastic collision of two particles of masses m_1 and m_2 moving initially in the same straight line with velocities whose magnitudes are u_1 and u_2, respectively. After collision they move off together with the common velocity v. The change that has taken place in the universe by this occurrence may be described by saying that instead of the mass m_1 moving with velocity u_1 through distance u_1 in unit time, it moves with velocity v through distance v in unit time. Similarly m_2 moves through v in unit time with velocity v instead of through u_2 in unit time with velocity u_2 (Maupertuis assumes at the outset that $u_1 > v > u_2$). He then points out that the same result would occur if, while m_1 was moving ahead with velocity u_1, etc., it were at the same time to be carried *backward* on an ideal plane with velocity $u_1 - v$ through space $u_1 - v$ in unit time, and m_2 while moving ahead with velocity u_2 were to be carried *forward* with velocity $v - u_2$ through space $v - u_2$ in unit time. The total action involved in the process must be a minimum. The latter measured from Maupertuis' point of view as the product of mass, velocity, and distance (presumably in unit time—at least that is his interpretation in this problem) becomes

$$m_1(u_1 - v)^2 + m_2(v - u_2)^2. \tag{17}$$

The minimum condition is

$$-m_1(u_1 - v) + m_2(v - u_2) = 0$$

or

$$v = \frac{m_1 u_1 + m_2 u_2}{m_1 + m_2}, \tag{18}$$

the usual expression. Maupertuis shows further how elastic collisions can be treated in similar fashion. However, we are here more interested in noting that this solution of the problem really reduces to the previously discussed Gaussian principle of least constraint and has very little to do with the principle of least action as understood at the present time.

Propagation of light. Maupertuis' reasoning may be further observed by an examination of his method of dealing with the propagation of light. If light travels from point A in medium I in which the velocity is v_1 to point B in medium II (see Fig. 3.12), in which the velocity is v_2, across the boundary SS', we desire to find the path pursued, i.e., essentially the point C, since it is tacitly assumed that the path lies in a plane and is a straight line in each medium. According to Maupertuis the path will be such that the total action is a minimum, where by action here he means the quantity

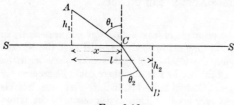

Fig. 3.12

$$v_1 \overline{AC} + v_2 \overline{CB}, \tag{19}$$

justified by the fact that here again a velocity is multiplied by a displacement. Since only one particle (light corpuscle?) is imagined, the mass factor is unnecessary. We have then to minimize the expression

$$v_1 \sqrt{x^2 + h_1^2} + v_2 \sqrt{(l - x)^2 + h_2^2}.$$

The usual rule yields

$$\frac{v_1 x}{\sqrt{x^2 + h_1^2}} - \frac{v_2(l - x)}{\sqrt{(l - x)^2 + h_2^2}} = 0,$$

or

$$\sin \theta_1 / \sin \theta_2 = v_2/v_1 = \text{constant}. \tag{20}$$

In so far as the ratio $\sin \theta_1/\sin \theta_2$ comes out to be a constant, this method gives the law of refraction of light in its empirical form. However, we see at once that, since the constant is equal to v_2/v_1, the result disagrees with experience, according to which the ratio is inverted. Eq. (20) could indeed agree with the observed results if the velocity of light in the optically denser medium (e.g., glass) were greater than that in the optically rarer medium (e.g., air). The contrary fact was not made plain by experiment until the middle of the nineteenth century, and hence Maupertuis may perhaps be pardoned for believing that he had given a rigorous derivation of the law of refraction of light. We now know that Fermat had the correct idea in his principle of least time.

Enough has been said to make clear the hazy nature of the ideas of Maupertuis. At the same time, credit is due him for his emphasis on the minimal or more general vari-

3.12 HAMILTON'S PRINCIPLE AND LEAST ACTION

ational concept in mechanics. This has long attracted the fancy of mathematicians and physicists. The idea is a very old one. It is well known that Hero (or Heron) of the Alexandrian school of Greek philosophers was able to deduce the law of the *reflection* of light from the assumption that in traveling from one point to another light always follows the *shortest* path. * The attempt to deduce the law of *refraction* (i.e., Snell's law) from the same principle naturally failed. It will be recalled that Descartes rediscovered the law and published it as his own. He tried to describe it in terms of the motion of corpuscles and was led (like Newton) to the necessity of assuming that the velocity increases with the density of the medium.

Fermat's principle. It is of great interest to note that Fermat considered this assumption wholly unlikely and in 1662 proceeded to give his famous derivation of the law on the basis of the postulate that the actual path taken by light in traveling from one point to another irrespective of the medium is such that the *time* is *less* than for any other path connecting the two points. Since we have already noted Maupertuis' treatment of this problem it will be of value to examine that of Fermat, recognizing its significance as one of the earliest successful uses of the minimal concept in physics. Referring again to Fig. 3.12 it is now assumed that the time taken for light to travel from A to B is a minimum. The expression

$$\frac{\overline{AC}}{v_1} + \frac{\overline{CB}}{v_2}$$

must now be minimized instead of $v_1 \overline{AC} + v_2 \overline{CB}$ in Maupertuis' method. Writing the above in the form

$$\sqrt{x^2 + h_1^2}/v_1 + \sqrt{(l-x)^2 + h_2^2}/v_2$$

the usual rule [1] yields

$$x/v_1\sqrt{x^2 + h_1^2} - (l-x)/v_2\sqrt{(l-x)^2 + h_2^2} = 0$$

or

$$\sin\theta_1/\sin\theta_2 = v_1/v_2, \tag{21}$$

the law of Snell in the form which agrees with observation. The principle of Fermat can be expressed in more general form as follows. If the element of path of the ray is denoted by ds, and v is the velocity (which may change continuously if the medium varies continuously), we have, using the notation of the first part of this section,

$$\delta \int \frac{ds}{v} = 0, \tag{22}$$

where the integral is taken between the two points between which the ray is desired. Clearly if the medium does not change and v remains constant, (22) reduces to the principle of the shortest path used by Hero. It is also clear that if the propagation of light is treated as being due to the motion of corpuscles moving with the velocity of light the corpuscles do not move in accordance with the principle of least action. This situation has been discussed by L. de Broglie in his wave mechanics.

Variational problems in physics attracted much attention during the late seventeenth and early eighteenth centuries. We merely mention such problems as that of the determination of the form assumed by a freely hanging chain or cord,

---

[1] Fermat did not, of course, use explicitly the calculus method. One of his great contributions to seventeenth-century mathematics was the development of geometrical methods for solving problems in maxima and minima.

which was solved by J. Bernoulli on the assumption that the center of mass of the cord must lie as *low as possible*; the general determination of the path of a material particle moving under *gravity* from one point to another in the *least time* (this being the so-called *brachistochrone*, investigated extensively by the Bernoullis, Leibniz, L'Hôpital, and Newton); and finally the general problems characteristic of what is now known as the *calculus of variations* investigated by Euler and Lagrange. By the beginning of the nineteenth century the use of the minimal or variational idea in physics had become firmly established and its elegance heartily appreciated.

Time averages. Before leaving the present discussion of Hamilton's principle we might note another interesting physical aspect. It will be recalled that $\int_{t_0}^{t_1} L dt$ is by definition $(t_1 - t_0)$ times the *time average* of the Lagrangian function over the interval in question. We may thus interpret the principle as stating that the time average of L for the actual motion has a stationary (usually a minimum) value. Alternatively since $\int_{t_0}^{t_1} L dt = \int_{t_0}^{t_1} T dt - \int_{t_0}^{t_1} V dt$, we may also state the principle in this form: the difference between the time averages of the kinetic energy and of the potential energy has a stationary value; i.e., for the actual motion the average kinetic energy either approaches the average potential energy more closely than for any other motion between the same states on the same time or diverges from it more than for any other such motion. Usually it is the former situation which is physically realized, and then we may say simply that the actual motion is that for which the average kinetic energy and average potential energy become as nearly equal as they can. The reader will readily recall the simple harmonic oscillator as a case where the averages actually are equal.

3.13. Generalized Coordinates and the Method of Lagrange. We have just seen how Hamilton's principle can provide an energetical basis for mechanics. The principle, indeed, is what may be called an *integral* principle. The question may quite logically arise: is there a method for the solution of mechanical problems in which the energy is basic but enters in a *differential* form? This question was probably first answered in the affirmative by Lagrange in his famous "Mécanique Analytique." A somewhat different answer was given later by Hamilton. We shall first consider the former. Lagrange proceeded from very general considerations, endeavoring to reduce mechanics to pure analysis and emancipate it from the connection with geometry which had been one of its outstanding characteristics as developed by Newton. With triumphant éclat Lagrange announced in the preface to his book that "there are no figures in this work,"

3.13 GENERALIZED COORDINATES—METHOD OF LAGRANGE

implying that all had been reduced to algebraic analysis (in the large sense).

To understand Lagrange's method it is essential to recall that so far our discussion of mechanics has been in terms of ordinary space coordinates; i.e., those specifying the positions of material particles. It may be, however, that this materially hampers us from generalizing the mechanical attack on general physical problems. There are many physical phenomena in which we do not directly observe the motions of material particles, e.g., many of the phenomena of optics, acoustics, and electricity. Yet the temptation is strong to attempt to describe these in mechanical terms, and this attempt has proved on the whole remarkably successful (cf. the early sections in Chapter V). This has been possible largely through the introduction of *generalized coordinates*. Let us suppose that we have a certain phenomenon which we wish to describe. We have made quantitative experiments concerning it. These have involved the tabulation of the values of certain parameters, for example, as functions of the time. Thus the value of the electric current in a wire is an illustration. We finally find it convenient to think of the phenomenon as associated with a certain *physical system* and consider the parameters as typifying the *state* of the system in the sense that when the values of the parameters are known at a given instant the state of the system is said to be known at that instant.

So far, nothing has been said about mechanics; we now, however, further suppose that the behavior of the system is really controlled by the motion of particles which remain intrinsically concealed from us. The values assumed by the descriptive parameters must then be connected with the position coordinates of the particles. This functional relation may be wholly unknown; nevertheless it provides the means for a mechanical description. The parameters are called *generalized coordinates*. If n of them are essential to fix the configuration of a system at a given instant we say that the system has n *degrees of freedom*. Call the n coordinates $q_1, q_2, \ldots, q_n$. The rectangular coordinates of the jth particle in the concealed aggregate may be written in terms of the q's as follows

$$x_j = x_j(q_1 \ldots q_n); \quad y_j = y_j(q_1 \ldots q_n); \quad z_j = z_j(q_1 \ldots q_n). \quad (1)$$

Similarly the velocity components of the jth particle will be

$$\dot{x}_j = \sum_{k=1}^{n} \frac{\partial x_j}{\partial q_k} \dot{q}_k; \quad \dot{y}_j = \sum_{k=1}^{n} \frac{\partial y_j}{\partial q_k} \dot{q}_k; \quad \dot{z}_j = \sum_{k=1}^{n} \frac{\partial z_j}{\partial q_k} \dot{q}_k \quad (2)$$

where the $\dot{q}_k$ are termed *generalized velocities*. Now the kinetic energy of the system will by assumption be the total kinetic energy of the particles, viz.,

$$T = \tfrac{1}{2}\Sigma m_j(\dot{x}_j{}^2 + \dot{y}_j{}^2 + \dot{z}_j{}^2), \qquad (3)$$

where the summation here is taken over the whole number of particles in the aggregate and m_j is the mass of the jth particle. Substitution from (2) and reduction lead to the following expression for T in terms of the generalized coordinates and velocities

$$T = \Sigma a_{ik}\dot{q}_i\dot{q}_k. \qquad (4)$$

Here the summation is a double one, i.e., both i and k vary from 1 to n, and the a_{ik} are functions of the q's. This result is important: the kinetic energy of the system is a *homogeneous quadratic* function of the generalized velocities.

Derivation of Lagrange's equations. It is next postulated that the system has potential energy, viz., that of the aggregate of concealed particles. If the latter depends, as will be first supposed, on the position coordinates alone, it will likewise be a function of the q's alone, from (1). We are now in a position to apply Hamilton's principle to the system. Let us therefore investigate

$$\delta \int_{t_0}^{t_1} \{T(q,\dot{q}) - V(q)\}dt = 0, \qquad (5)$$

where q represents for brevity $q_1 \ldots q_n$ and $\dot{q}$ represents $\dot{q}_1 \ldots \dot{q}_n$. We here carry out formally the variation much as in the simple example of the oscillator in the previous section, using $L = T - V$. Thus

$$\int_{t_0}^{t_1} \left\{ \sum_{j=1}^{n} \left(\frac{\partial L}{\partial q_j} \delta q_j + \frac{\partial L}{\partial \dot{q}_j} \frac{d}{dt}(\delta q_j) \right) \right\} dt = 0. \qquad (6)$$

The integration by parts of the second term in the integral yields

$$\int_{t_0}^{t_1} \sum_{j=1}^{n} \frac{\partial L}{\partial \dot{q}_j} \frac{d}{dt}(\delta q_j)dt = \sum_{j=1}^{n} \frac{\partial L}{\partial \dot{q}_j} \delta q_j \Big]_{t_0}^{t_1} - \int_{t_0}^{t_1} \sum_{j=1}^{n} \frac{d}{dt}\left(\frac{\partial L}{\partial \dot{q}_j}\right) \delta q_j dt. \qquad (7)$$

Since $\delta q_j = 0$ for all j at $t = t_0$ and $t = t_1$, the integrated part vanishes and there results simply from (6)

$$\int_{t_0}^{t_1} \sum_{j=1}^{n} \left\{ \frac{\partial L}{\partial q_j} - \frac{d}{dt}\left(\frac{\partial L}{\partial \dot{q}_j}\right) \right\} \delta q_j dt = 0. \qquad (8)$$

3.13 GENERALIZED COORDINATES—METHOD OF LAGRANGE

The arbitrary nature of the δq_j in the interval from t_0 to t_1 therefore leads at once to the n equations

$$\frac{d}{dt}\left(\frac{\partial L}{\partial \dot{q}_j}\right) - \frac{\partial L}{\partial q_j} = 0, \; j = 1, 2, \ldots n. \tag{9}$$

These are known as Lagrange's equations for the physical system. They are second order equations, and hence their solution involves $2n$ arbitrary constants. Their solution will express each q_j as a function of the time and the $2n$ constants, which must be fixed by the initial conditions. The state of the system at any instant will then be formally represented by the solution of (9). The utility of these equations in describing phenomena in which there are no obvious motions of particles will be made clear in a subsequent chapter (Chapter V).

Central field motion. We here illustrate the use of the equations in a clear-cut mechanical problem, namely, the motion of a single mass particle in a central field of force. Our purpose is to show how the general equations for such motion in *spherical coordinates* (defined in Sec. 3.2, Fig. 3.2b) can be written with great ease by the Lagrangian method.

The kinetic energy of a particle is

$$T = \tfrac{1}{2}m(\dot{x}^2 + \dot{y}^2 + \dot{z}^2) = \frac{m}{2}(\dot{r}^2 + r^2\dot{\theta}^2 + r^2\sin^2\theta\cdot\dot{\phi}^2), \tag{10}$$

and if the field is central the potential energy V is a function of r only, so that the Lagrangian function becomes

$$L = \frac{m}{2}(\dot{r}^2 + r^2\dot{\theta}^2 + r^2\sin^2\theta\cdot\dot{\phi}^2) - V(r), \tag{11}$$

the system being one with three degrees of freedom with $q_1 = r$, $q_2 = \theta$, and $q_3 = \phi$. The Lagrangian equations (9) are

$$\left.\begin{aligned}\frac{d}{dt}(m\dot{r}) + \frac{dV}{dr} - mr(\dot{\theta}^2 + \sin^2\theta\cdot\dot{\phi}^2) &= 0, \\ \frac{d}{dt}(mr^2\dot{\theta}) - mr^2\sin\theta\cos\theta\,\dot{\psi}^2 &= 0, \\ \frac{d}{dt}(mr^2\sin^2\theta\cdot\dot{\phi}) &= 0.\end{aligned}\right\} \tag{12}$$

From these the whole theory of central motion can be developed. It will be observed that the longitude ϕ does not appear explicitly in any of eqs. (12). Moreover, $\dot{\phi}$ is determined from the third equation in terms of r and θ, viz.,

$$\dot{\phi} = \frac{C}{mr^2 \sin^2 \theta}, \tag{13}$$

where C is a constant. This may be substituted into the first two equations, yielding

$$\left.\begin{array}{l} \dfrac{d}{dt}(m\dot{r}) + \dfrac{dV}{dr} - mr\left(\dot{\theta}^2 + \dfrac{C^2}{m^2 r^4 \sin^2 \theta}\right) = 0, \\[2mm] \dfrac{d}{dt}(mr^2\dot{\theta}) - \dfrac{C^2 \cos \theta}{mr^2 \sin^3 \theta} = 0. \end{array}\right\} \tag{14}$$

The problem is effectively reduced to one with two degrees of freedom. This means of course that central motion takes place in a *plane*. Since it is most simple to take this plane as the plane $\phi = 0$ (i.e., the xz plane), we have $\dot{\phi} = 0$ (i.e., $C = 0$, above), and consequently may write

$$\left.\begin{array}{l} \dfrac{d}{dt}(m\dot{r}) + \dfrac{dV}{dr} - mr\dot{\theta}^2 = 0, \\[2mm] \dfrac{d}{dt}(mr^2\dot{\theta}) = 0. \end{array}\right\} \tag{15}$$

The law of areas, viz.,

$$mr^2\dot{\theta} = \text{constant}, \tag{16}$$

follows at once from the second equation.

Non-holonomic systems. The above discussion of the Lagrangian method and equations is not completely general. When we passed from eq. (8) to eq. (9) we assumed that the δq_j are completely arbitrary in the time interval *between* t_0 and t_1, the only restriction on them being that they all vanish *at* t_0 and t_1. It may be that there are certain restrictions placed on the variations δq_j. Two cases should be considered. First, these restrictions may result from certain relations among the q_j, i.e., the generalized coordinates themselves. However, in this case, we can see at once that we are using too many coordinates, i.e., they are not all independent, for by the relations certain of them can be expressed in terms of the rest. We should then recognize that the number of degrees of freedom for our problem is really less than the number of coordinates being used and decrease that number accordingly and proceed with the Lagrangian method as before.

3.13 GENERALIZED COORDINATES—METHOD OF LAGRANGE

However, there is a second possibility. The restrictions in question may be expressible only in relations among the δq_j (not the q_j): the relations may be *non-integrable*, so that from them one cannot obtain corresponding relations among the q_j. When this is the case the system is said to be *non-holonomic* (the term *holonomic* being used to refer to the case where such non-integrable relations do not exist) and an altered method of attack must be employed. Suppose that the relations in question are $k(< n)$ in number and may be written in the form

$$\sum_{j=1}^{n} a_{1j}\delta q_j = \sum_{j=1}^{n} a_{2j}\delta q_j = \ldots = \sum_{j=1}^{n} a_{kj}\delta q_j = 0, \quad (17)$$

the a_{ij} being in general functions of the q_j. All we are now allowed to say is that the integrand in (8) must vanish, i.e.,

$$\sum_{j=1}^{n} \left\{ \frac{\partial L}{\partial q_j} - \frac{d}{dt}\left(\frac{\partial L}{\partial \dot{q}_j}\right) \right\} \delta q_j = 0, \quad (18)$$

subject to the k conditions (17). Following Lagrange we now multiply these relations by the parameters $\lambda_1, \lambda_2, \ldots, \lambda_k$, respectively, and add to (18). The result is

$$\sum_{j=1}^{n} \left\{ \frac{\partial L}{\partial q_j} - \frac{d}{dt}\left(\frac{\partial L}{\partial \dot{q}_j}\right) + \lambda_1 a_{1j} + \ldots + \lambda_k a_{kj} \right\} \delta q_j = 0. \quad (19)$$

The λ's are the so-called Lagrangian undetermined multipliers, which we have already used in Sec. 3.9. Write the k equations obtained by setting the coefficients of δq_j equal to zero for $j = 1 \ldots k$. Thus:

$$\left.\begin{aligned}\frac{\partial L}{\partial q_1} - \frac{d}{dt}\left(\frac{\partial L}{\partial \dot{q}_1}\right) + \lambda_1 a_{11} + \ldots + \lambda_k a_{k1} &= 0 \\ \cdots \cdots \cdots \cdots \cdots \cdots \cdots \cdots \cdots \cdots \cdots \\ \frac{\partial L}{\partial q_k} - \frac{d}{dt}\left(\frac{\partial L}{\partial \dot{q}_k}\right) + \lambda_1 a_{1k} + \ldots + \lambda_k a_{kk} &= 0.\end{aligned}\right\} \quad (20)$$

The remaining δq_j (in number $n - k$) are now perfectly arbitrary because there are only k equations restricting them. Hence we must have

$$\left.\begin{aligned}\frac{\partial L}{\partial q_{k+1}} - \frac{d}{dt}\left(\frac{\partial L}{\partial \dot{q}_{k+1}}\right) + \lambda_1 a_{1,k+1} + \ldots + \lambda_k a_{k,k+1} &= 0 \\ \cdots \cdots \cdots \cdots \cdots \cdots \cdots \cdots \cdots \cdots \cdots \\ \frac{\partial L}{\partial q_n} - \frac{d}{dt}\left(\frac{\partial L}{\partial \dot{q}_n}\right) + \lambda_1 a_{1n} + \ldots + \lambda_k a_{kn} &= 0.\end{aligned}\right\} \quad (21)$$

The k equations (17) and the n equations (20) and (21) are sufficient to calculate the k λ's and the n expressions $\frac{\partial L}{\partial q_j} - \frac{d}{dt}\left(\frac{\partial L}{\partial \dot{q}_j}\right)$. The Lagrangian equations for the non-holonomic system are therefore of the form

$$\frac{d}{dt}\left(\frac{\partial L}{\partial \dot{q}_j}\right) - \frac{\partial L}{\partial q_j} = \lambda_1 a_{1j} + \ldots + \lambda_k a_{kj}, \qquad (22)$$

where j goes from 1 to n.

A simple illustration of a non-holonomic system is a sphere rolling on a rough plane. One even more simple is a vertical wheel with a thin edge, rolling on a horizontal sheet of paper without slipping in any direction. The wheel always has one point in contact with the paper. If the coordinates of this point are x and y, and if the plane of the wheel makes the angle θ with the x axis, the motion is subject to the non-integrable relation

$$\delta y = \delta x \tan \theta$$

corresponding to the general relations (17).[1]

Non-conservative systems. The discussion of Lagrange's equations in the earlier part of this section lacks generality in another respect: the systems considered were *conservative*. This restriction is unnecessary. If we go back to Hamilton's principle in the general form (3.12–11), we easily obtain

$$\int_{t_0}^{t_1}\left\{\sum_{j=1}^{n}\left(\frac{\partial T}{\partial q_j}\delta q_j + \frac{\partial T}{\partial \dot{q}_j}\delta \dot{q}_j\right)\right\}dt = -\int_{t_0}^{t_1}\sum_{k=1}^{n}\mathbf{F}_k\cdot\delta\mathbf{r}_k dt. \qquad (23)$$

The right-hand side can be transformed into a function of the generalized coordinates by means of the transformation equations (1) and (2), etc. We write

$$\sum_{k=1}^{n}\mathbf{F}_k\cdot\delta\mathbf{r}_k = \sum_{j=1}^{n}Q_j\delta q_j, \qquad (24)$$

where the Q_j (functions of the q_j and $\dot{q}_j$ and possibly of t) appear in the guise of *generalized forces*, since $\sum_{j=1}^{n}Q_j\delta q_j$ represents the total work done on the system by the actual external forces $\mathbf{F}_k$ in the displacements

---

[1] For an elaborate treatment of non-holonomic systems, see Whittaker "Analytical Dynamics," Chapter VIII.

3.13 GENERALIZED COORDINATES—METHOD OF LAGRANGE

represented by δr_k. On substituting and carrying out the variations as usual there results the set of n equations

$$\frac{d}{dt}\left(\frac{\partial T}{\partial \dot{q}_i}\right) - \frac{\partial T}{\partial q_i} = Q_j. \quad (25)$$

These are Lagrange's equations of motion for the system and are perfectly general, valid for non-conservative as well as conservative systems as long as they are holonomic.

Eqs. (25) can be written in another form when, as is often the case, the force acting on a system is conveniently broken up into two parts—one intrinsically connected with the system (e.g., gravity, elastic restoring force of an oscillator, etc.) corresponding to potential energy V', and the other more definitely external (e.g., the driving force on an oscillator) which may or may not be conservative. If we call the latter Q'_j we can write in general

$$Q_j = Q'_j - \frac{\partial V'}{\partial q_i}. \quad (26)$$

V' being assumed to be a function of the q_i only, we take as our Lagrangian function $L = T - V'$, whence eqs. (25) become

$$\frac{d}{dt}\left(\frac{\partial L}{\partial \dot{q}_i}\right) - \frac{\partial L}{\partial q_i} = Q'_j. \quad (27)$$

If Q'_j is also conservative and equal to $-\frac{\partial V''}{\partial q_i}$, eqs. (27) reduce to the more familiar eqs. (9) with, however, $L = T - (V' + V'') = T - V$.

If the Lagrangian equations in the form (9) hold, it follows from the analysis in the first part of this section, eqs. (5), (6), (7), that

$$\delta \int_{t_0}^{t_1} L\, dt = 0, \quad (28)$$

i.e., Hamilton's principle in the usual form.

It is well to emphasize again, however, that this result is based on the assumption that the q_i do not vary at the ends of the dynamical path of the system. If this assumption is not made it is clear from eq. (7) that (28) must be replaced by

$$\delta \int_{t_0}^{t_1} L\, dt = \sum_{j=1}^{n} \frac{\partial L}{\partial \dot{q}_i} \delta q_i \Big]_{t_0}^{t_1} \quad (29)$$

This generalization of eq. (5) is sometimes called *Hamilton's principle of varying action*. It will be useful in the discussion in Chapter V (cf. Sec. 5.3).

3.14. The Canonical Equations of Hamilton. For most purposes the equations of Lagrange are a sufficiently powerful method of attack on mechanical problems. It is well to point out, however, that they are not the only differential equations of motion in which the energy enters. Consider once more a dynamical system of n degrees of freedom described by the generalized coordinates $q_1, \ldots q_n$. We shall now define the *momentum* p_i associated with or conjugate to q_i as follows:

$$p_i = \frac{\partial L}{\partial \dot{q}_i}, \qquad (1)$$

where L is the Lagrangian function. It is easy to see that in simple cases this abstract definition reduces to momentum as defined in elementary mechanics. Thus if $L = \frac{1}{2}m\dot{x}^2$ for a single free particle with $q = x$, we get $p = m\dot{x}$, or mass times velocity. However, we must be careful to note the generalized nature of the p_i. Just as the q's need not have the dimensions of length, so the conjugate p's may not have the dimensions of mass times velocity. They are "generalized" momenta conjugate to generalized coordinates. For some purposes it is extremely useful to write the dynamical equations in a form involving the q's and p's rather than the q's and $\dot{q}$'s. Consider the following function:

$$H = \sum_{j=1}^{n} p_j \dot{q}_j - L(q, \dot{q}). \qquad (2)$$

From the way it has been constructed it might appear to be a function of the q's, the $\dot{q}$'s, and the p's. As a matter of fact, it really is a function of the q's and p's alone. Let us form its total differential in the usual way, assuming that L does not depend on t explicitly, viz.,

$$dH = \sum_{j=1}^{n} p_j d\dot{q}_j + \sum_{j=1}^{n} \dot{q}_j dp_j - \sum_{j=1}^{n} \frac{\partial L}{\partial \dot{q}_j} d\dot{q}_j - \sum_{j=1}^{n} \frac{\partial L}{\partial q_j} dq_j. \qquad (3)$$

From the definition of p_i given in (1) it is plain that the first and the third terms on the right-hand side of (3) cancel each other, leaving only

$$dH = \sum_{j=1}^{n} \dot{q}_j dp_j - \sum_{j=1}^{n} \frac{\partial L}{\partial q_j} dq_j. \qquad (4)$$

3.14 THE CANONICAL EQUATIONS OF HAMILTON

From the mathematical definition of a total differential

$$dH = \sum_{j=1}^{n} \frac{\partial H}{\partial q_j} dq_j + \sum_{j=1}^{n} \frac{\partial H}{\partial p_j} dp_j. \qquad (5)$$

Consequently

$$\left.\begin{array}{l} \dfrac{\partial H}{\partial q_j} = -\dfrac{\partial L}{\partial q_j} = -\dot{p}_j \\[1em] \dfrac{\partial H}{\partial p_j} = \dot{q}_j. \end{array}\right\} \qquad (6)$$

The replacement of $\partial L/\partial q_j$ by $\dot{p}_j$ is of course a result of the Lagrangian equations (3.13-9) coupled with the definition of the conjugate momentum.

The Hamiltonian function. The $2n$ partial differential equations of the first order (6) are known as the *canonical equations* of motion. Since they were introduced by Hamilton they have also been termed the Hamiltonian equations of the dynamical system. The function $H(p, q)$ is called the *Hamiltonian* of the system. The contrast between the Hamiltonian equations and those of Lagrange is rather striking. In the first place the latter are n in number (one for each degree of freedom) and of the second order; the former are $2n$ in number but of the first order. In each case the total number of arbitrary constants of integration which must be fixed by the boundary conditions is $2n$. The Lagrangian equations when set up for a particular problem are usually integrated directly to find the state of the system; the Hamiltonian equations might be but actually are never in practice integrated directly; rather they serve as the starting point for a number of considerations usually grouped together as the "transformation" theory of mechanics (Sec. 3.15). The meaning of the canonical equations becomes clearer if we inquire further into the nature of the function H. From eq. (5) we see that the time rate of change of H is

$$\dot{H} = \sum_{j=1}^{n} \frac{\partial H}{\partial q_j} \dot{q}_j + \sum_{j=1}^{n} \frac{\partial H}{\partial p_j} \dot{p}_j. \qquad (7)$$

But the substitution for $\dot{q}_j$ and $\dot{p}_j$ from the Hamiltonian equations immediately gives

$$\dot{H} = 0 \quad \text{or} \quad H = \text{constant}. \qquad (8)$$

Thus H remains constant in time. We might at once suspect some connection with the energy. This suspicion is verified by reverting to the definition of H in eq. (2). If we suppose that the potential energy of the system does not depend explicitly on the $\dot{q}_j$, we may write effectively $p_j = \partial T/\partial \dot{q}_j$, and H becomes

$$H = \sum_{j=1}^{n} \frac{\partial T}{\partial \dot{q}_j} \dot{q}_j - L. \tag{9}$$

Now T (eq. (3.13-4)) is a homogeneous quadratic function of the generalized velocities. By an application of Euler's theorem on homogeneous functions it develops that

$$\sum_{j=1}^{n} \frac{\partial T}{\partial \dot{q}_j} \dot{q}_j = 2T, \tag{10}$$

whence we have at once

$$H = T + V. \tag{11}$$

The Hamiltonian function then has a numerical value equal to that of the total energy of the system. We must again emphasize, however, that in functional form it is expressed in terms of the p's and q's and not in terms of q's and $\dot{q}$'s. A few simple illustrations will suffice to make its meaning clearer. For a free particle of mass m, $H = p^2/2m$, and the first equation (6) is simply

$$\dot{p} = 0 \quad \text{or} \quad p = \text{constant}$$

or the ordinary Newtonian equation of motion, while the second is the relation between p and $\dot{q}$:

$$\dot{q} = p/m.$$

For the simple harmonic oscillator (one degree of freedom) $H = p^2/2m + kq^2/2$, where k is the stiffness of the system. The canonical equations become

$$\dot{p} = -kq,$$
$$\dot{q} = p/m.$$

Here again the first equation appears in the guise of the ordinary Newtonian equation of motion while the second is the relation between p and $\dot{q}$.

In the discussion above we have assumed that L and hence H are not explicit functions of the time. This is an unnecessary restriction. If L depends on the time explicitly, i.e., if the parameters entering into the expressions for T and V vary with the time (e.g., a pendulum in in which the length of the string changes with the time), eq. (3) merely needs to be supplemented on the right with the expression $\partial L/\partial t \cdot dt$ and similarly for eqs. (4, 5). But it is clear that the Hamiltonian equations (7) still hold. Only we now no longer have, of course, $\dot{H} = 0$. Rather

$$\dot{H} = -\frac{\partial L}{\partial t} \tag{12}$$

and its value will depend on the functional dependence of the parameters on the time. However, the procedure leading to eq. (11) still remains valid and the Hamiltonian still appears as the instantaneous (though varying) energy of the system. This case is often of value in statistical mechanics (Chapter V).

Derivation of the canonical equations. It is of interest to note that it is possible to obtain the canonical equations directly from Hamilton's principle. Thus writing $L = \sum_{j=1}^{n} p_i \dot{q}_i - H(p_i, q_i)$ we have the condition

$$\delta \int_{t_0}^{t_1} \left[\sum_{j=1}^{n} p_i \dot{q}_i - H \right] dt = 0, \tag{13}$$

which becomes, on taking the variation,

$$\int_{t_0}^{t_1} \left[\sum_{j=1}^{n} \delta p_i \dot{q}_i + \sum_{j=1}^{n} p_i \delta \dot{q}_i - \sum_{j=1}^{n} \frac{\partial H}{\partial q_i} \delta q_i - \sum_{j=1}^{n} \frac{\partial H}{\partial p_i} \delta p_i \right] dt = 0. \tag{14}$$

By partial integration we have

$$\int_{t_0}^{t_1} \sum_{j=1}^{n} p_i \delta \dot{q}_i dt = -\int_{t_0}^{t_1} \sum_{j=1}^{n} \dot{p}_i \delta q_i dt.$$

Therefore on substitution we get

$$\int_{t_0}^{t_1} \left[\sum_{j=1}^{n} \dot{q}_i \delta p_i - \sum_{j=1}^{n} \dot{p}_i \delta q_i - \sum_{j=1}^{n} \frac{\partial H}{\partial q_i} \delta q_i - \sum_{j=1}^{n} \frac{\partial H}{\partial p_i} \delta p_i \right] dt = 0. \tag{15}$$

This result must be true no matter what the variations are, as long as they vanish at the termini of the dynamical path. Hence we must have identically

$$\sum_{j=1}^{n} \frac{\partial H}{\partial q_i} \delta q_i + \sum_{j=1}^{n} \frac{\partial H}{\partial p_i} \delta p_i - \sum_{j=1}^{n} \dot{q}_i \delta p_i + \sum_{j=1}^{n} \dot{p}_i \delta q_i. \tag{16}$$

It follows that

$$\dot{q}_i = \frac{\partial H}{\partial p_i}, \quad \dot{p}_i = -\frac{\partial H}{\partial q_i}, \tag{17}$$

which are again the canonical equations.

Poisson brackets. A variation in the method of expressing the canonical equations is provided by the use of the so-called *Poisson brackets*. Consider *any* two continuous and differentiable functions of the q_j, p_j, say $g(q_j, p_j)$ and $h(q_j, p_j)$. We shall write

$$(g, h) = \sum_{j=1}^{n} \left(\frac{\partial g}{\partial p_j} \frac{\partial h}{\partial q_j} - \frac{\partial g}{\partial q_j} \frac{\partial h}{\partial p_j} \right) \tag{18}$$

and call the expression the Poisson bracket for g and h. Its utility is evident. For clearly we have

$$\dot{g} = \sum_{j=1}^{n} \left(\frac{\partial g}{\partial q_j} \dot{q}_j + \frac{\partial g}{\partial p_j} \dot{p}_j \right),$$

and if we employ eq. (17) we may write

$$\dot{g} = (H, g). \tag{19}$$

The Hamiltonian equations (17) may therefore be written in the form

$$\dot{q}_j = (H, q_j), \quad \dot{p}_j = (H, p_j). \tag{20}$$

Suppose that $g = q_j$ and $h = q_k$; for any j, k (1 ... n) we have

$$(q_j, q_k) = 0. \tag{21}$$

Similarly we get

$$(p_j, p_k) = 0, \tag{22}$$

and if $j \neq k$

$$(p_j, q_k) = 0. \tag{23}$$

On the other hand, if $g = p_k$ and $h = q_k$,

$$(p_k, q_k) = 1, \quad (q_k, p_k) = -1. \tag{24}$$

Quantities (i.e., functions of the p's and q's) whose Poisson bracket vanishes are said to *commute*, while those whose Poisson bracket is equal to 1 are said to be *canonically conjugate*. From (19) we have the result that any quantity which commutes with the Hamiltonian does not change with the time. In particular H itself is constant in time since it commutes with itself. On the other hand, since

$$(H, t) = 1, \tag{25}$$

the time and the Hamiltonian are canonically conjugate. In general any function of the p's and q's whose time derivative is 1 will be canonically conjugate to H and may thus serve as a " clock function."

Part of the significance of Poisson brackets resides in the fact that they remain invariant under a canonical transformation, i.e.,

3.15 THE TRANSFORMATION THEORY OF MECHANICS

one which leaves the canonical equations unaltered (cf. Sec. 3.15). They also have a number of algebraic properties of interest. Thus the reader may verify that in general

$$(g, h) = - (h, g)$$
$$(g, c) = 0$$

if c is a constant. Also if h_1 and h_2 are any two functions of the q's and p's

$$(g_1, h_1 + h_2) = (g, h_1) + (g, h_2)$$
$$(h_1 + h_2, g) = (h_1, g) + (h_2, g)$$
$$(h_1 h_2, g) = h_2(h_1, g) + h_1(h_2, g).$$

The use of Poisson brackets might seem a merely idle curiosity were it not for the fact that they have played an important rôle in the development of quantum mechanics. By their use it has been possible to write the equations of motion of a physical system and the so-called quantum conditions (cf. Sec. 9.9) in a form in which they are closely analogous to the classical Hamiltonian equations. The reader should consult Dirac, "The Principles of Quantum Mechanics" (Oxford, 1930).

3.15. The Transformation Theory of Mechanics. It has already been emphasized that the canonical equations are usually not integrated directly. Their chief importance lies in their invariance of form with respect to certain types of transformation of coordinates. In the first place they are invariant under point transformations, that is, transformations of the form

$$Q_i = Q_i(q_1 \ldots q_n). \tag{1}$$

This, however, is not the most general kind of transformation for which the equations are invariant. Let us examine a more general one, viz.:

$$\left. \begin{array}{l} Q_i = Q_i(q_1 \ldots q_n, p_1 \ldots p_n) \\ P_i = P_i(q_1 \ldots q_n, p_1 \ldots p_n), \end{array} \right\} \tag{2}$$

or

$$\left. \begin{array}{l} q_i = q_i(Q_1 \ldots Q_n, P_1 \ldots P_n) \\ p_i = p_i(Q_1 \ldots Q_n, P_1 \ldots P_n). \end{array} \right\} \tag{2'}$$

We seek the condition that the system will be described in the new coordinates by means of canonical equations. Imagine a function $K(P_j, Q_j)$ and another function $S(q_j, Q_j, t)$, such that

$$\sum_{j=1}^{n} p_j \dot{q}_j - H = \sum_{j=1}^{n} P_j \dot{Q}_j - K + \dot{S}. \tag{3}$$

Then from (3.14–13) we have

$$\delta \int_{t_0}^{t_1} \left[\sum_{j=1}^{n} P_j \dot{Q}_j - K + \dot{S} \right] dt = 0. \tag{4}$$

But

$$\delta \int_{t_0}^{t_1} \dot{S} dt = 0 \tag{5}$$

from the terminal restrictions. Hence we are led by the method of the latter part of the preceding section to the equations

$$\dot{P}_j = -\frac{\partial K}{\partial Q_j}, \quad \dot{Q}_j = \frac{\partial K}{\partial P_j}, \tag{6}$$

which are in canonical form. It remains only to ascertain the nature of S and K. Eq. (3) sheds light on this. Multiplying through by dt gives

$$\sum_{j=1}^{n} p_j dq_j - \sum_{j=1}^{n} P_j dQ_j + (K - H) dt$$

$$= \sum_{j=1}^{n} \frac{\partial S}{\partial q_j} dq_j + \sum_{j=1}^{n} \frac{\partial S}{\partial Q_j} dQ_j + \frac{\partial S}{\partial t} dt, \tag{7}$$

whence it follows that

$$p_j = \frac{\partial S}{\partial q_j}, \quad P_j = -\frac{\partial S}{\partial Q_j}, \quad K = H + \frac{\partial S}{\partial t}. \tag{8}$$

We may therefore look upon the function $S(q_j, Q_j, t)$ as defining the transformation. It is often called the " transformation " or " substitution " function. In the special case where S does not contain the time explicitly, we have $K = H$, so that K is the new Hamiltonian resulting from the transformation.

Consider the special case of a simple harmonic oscillator with $H = p^2/2m + kq^2/2$. We introduce the transformation defined by

$$S = \frac{m\omega}{2} q^2 \cot Q, \tag{9}$$

3.15   THE TRANSFORMATION THEORY OF MECHANICS

where $\omega = \sqrt{k/m}$. Using eqs. (8) we have

$$p = m\omega q \cot Q, \quad P = \frac{m\omega}{2} q^2 \csc^2 Q.$$

The actual transformation equations in the form (2′) are therefore

$$\left. \begin{aligned} q &= \sqrt{\frac{2}{m\omega}} \cdot \sqrt{P} \sin Q, \\ p &= \sqrt{2m\omega} \cdot \sqrt{P} \cos Q. \end{aligned} \right\} \quad (10)$$

The Hamiltonian H transforms into

$$K = \omega P, \tag{11}$$

and the canonical equations (6) become

$$\dot{P} = 0, \quad \dot{Q} = \omega, \tag{12}$$

yielding

$$P = \text{constant} = \alpha, \quad Q = \omega t + \beta. \tag{13}$$

The solution of the whole problem is obtained by resubstitution into (10). Thus

$$\left. \begin{aligned} q &= \sqrt{\frac{2}{m\omega}} \cdot \sqrt{\alpha} \sin (\omega t + \beta), \\ p &= \sqrt{2m\omega} \cdot \sqrt{\alpha} \cos (\omega t + \beta). \end{aligned} \right\} \quad (14)$$

By the transformation the Hamiltonian becomes a function of only one of the canonical variables, viz., P, which is really a constant of the motion; the conjugate variable Q appears as a linear function of the time. This is a result which in certain problems proves to be of great value.

It may be noted before we proceed that it is unnecessary that S should be a function of q_i, Q_i and t. It may be a function of q_i, P_i and t, in which case we write in place of (3)

$$\sum_{j=1}^{n} p_j \dot{q}_j - H = -\sum_{j=1}^{n} Q_j \dot{P}_j - K + \dot{S}, \tag{15}$$

which yields again the Hamiltonian canonical equations (6). We now have indeed for the transformation function the equations

$$p_i = \frac{\partial S}{\partial q_i}, \quad Q_i = \frac{\partial S}{\partial P_i}. \tag{16}$$

Hamilton-Jacobi equation. The question arises: does any equation exist which may be used to determine an S function for a particular problem? The answer given by Hamilton is in the affirmative. We have seen that the Hamiltonian H is equal to the energy E of the dynamical system. Hence we have the equation

$$H(p_i, q_i) = E, \qquad (17)$$

the left-hand side being a function of the p_i and q_i while the right-hand side is the symbol representing the actual numerical energy of the system. The physical significance of this equation is not altered if we replace each p_i by the equivalent $\dfrac{\partial S}{\partial q_i}$ from eq. (16). However, its mathematical meaning is now considerably different. For

$$H\left(\frac{\partial S}{\partial q_i}, q_i\right) = E \qquad (18)$$

is a partial differential equation of the first order for the function $S(q_i)$. Clearly there are an infinity of solutions. We may, however, confine our attention to what has been called a *complete integral* of the equation, that is, a solution containing n arbitrary constants (i.e., the same as the number of the q_i) such that, when these are eliminated from the integral and the n equations obtained from it by means of differentiation with respect to the coordinates, the differential equation results.[1]

Let us agree to write a complete integral of eq. (18), which is usually known as the *Hamilton-Jacobi equation*, in the form

$$S = S(q_1 \ldots q_n, E, \alpha_1 \ldots \alpha_n), \qquad (19)$$

[1] As a simple illustration consider the equation $x^2 + y^2 + z^2 + ax + by = 0$ between a dependent variable z and the *two* independent variables x and y, where a and b are constants. Differentiation with respect to x and y, respectively, leads to

$$2x + 2z\frac{\partial z}{\partial x} + a = 0,$$

$$2y + 2z\frac{\partial z}{\partial y} + b = 0.$$

Elimination of a and b between these two differential equations and the original equation yields

$$x^2 + y^2 - z^2 + 2zx\frac{\partial z}{\partial x} + 2zy\frac{\partial z}{\partial y} = 0.$$

This is a partial differential equation of the first order of which our first equation is a complete integral.

3.15 THE TRANSFORMATION THEORY OF MECHANICS

where the necessary n constants are $\alpha_1, \alpha_2, \ldots \alpha_n$, and for the present we assume that S does not contain the time explicitly. We shall now use this as the transformation function for a transformation from the dynamical variables $q_1 \ldots q_n, p_1 \ldots p_n$ to a new set $Q_1 \ldots Q_n$, $P_1 \ldots P_n$ where we shall choose for convenience

$$Q_1 = E, \quad Q_2 = \alpha_2 \ldots Q_n = \alpha_n.$$

The one constant, α_1, left over can always be considered as an additive one. We are thus choosing the constants of integration and the energy as our new Q's. The second set of equations (8) now become

$$P_1 = -\frac{\partial S}{\partial E}, \quad P_j = -\frac{\partial S}{\partial \alpha_j} \quad (j = 2, 3, \ldots n). \tag{20}$$

But from the canonical equations, since now $K = H$,

$$\dot{P}_1 = -\frac{\partial H}{\partial E} = -1, \quad \dot{P}_j = -\frac{\partial H}{\partial \alpha_j} = 0 \quad (j = 2, 3, \ldots n)$$

or

$$P_1 = -t - \beta_1, \quad P_j = -\beta_j \quad (j = 2, 3, \ldots n), \tag{21}$$

$\beta_1 \ldots \beta_n$ being new constants of integration. The combination of (20) and (21) yields

$$\frac{\partial S}{\partial E} = t + \beta_1, \quad \frac{\partial S}{\partial \alpha_j} = \beta_j \quad (j = 2, 3, \ldots n). \tag{22}$$

The n equations (22) when they are solved yield each q_j as a function of the time. But this solves the problem of the motion of the system: eqs. (22) are precisely those which describe the path of the system in time. They contain the $2n$ arbitrary constants which of necessity enter into the motion of a dynamical system of n degrees of freedom. We thus realize the great importance of the complete integral S. To recapitulate the method: we first find a complete integral of the Hamilton-Jacobi equation (18). We differentiate this integral with respect to the energy parameter E: this equals the time plus an arbitrary constant. On differentiating the integral next with respect to the other parameters α_j, we set the result in each case equal to an arbitrary constant. Eqs. (22) are the result of this process. The beautiful simplicity of the method is particularly evident in those cases in which the complete integral of (18) is readily obtained, which happens when the variables can be separated. In this event we can split the equation into parts, whereof the first contains only q_1 and $\partial S/\partial q_1$. We then have

$$H_1\left(q_1, \frac{\partial S}{\partial q_1}, E\right) = H_2\left(q_k, \frac{\partial S}{\partial q_k}, E\right) \quad (k = 2, 3 \ldots n). \quad (23)$$

If this is to hold, each side must be equal to an arbitrary constant which we may call α_2. Now

$$H_1\left(q_1, \frac{\partial S}{\partial q_1}, E\right) = \alpha_2$$

is an ordinary differential equation. Let its solution be

$$S = S_1(q_1, E, \alpha_2).$$

This will also be a solution of $H_2 = \alpha_2$, since (23) is an identity. We can now repeat the process with H_2 and obtain eventually a set of solutions $S_1(q_1, E, \alpha_2)$, $S_2(q_2, E, \alpha_2, \alpha_3)$, $\ldots S_n(q_n, E, \alpha_2, \ldots \alpha_n)$, all of which satisfy the original equation. The sum of these will clearly be the complete integral required. For if we take

$$S = \sum_{k=1}^{n} S_k + \alpha_{n+1}, \quad (24)$$

where α_{n+1} is the last and hence additive constant, we have

$$\frac{\partial S}{\partial q_j} = \frac{\partial S_j}{\partial q_j}.$$

Moreover, S contains the requisite n constants of integration,

$$\alpha_2 \ldots \alpha_{n+1}.$$

Angle and action variables. Beautiful examples of the application of the Hamilton-Jacobi method are to be found in celestial mechanics [1] and in the classical quantum theory of atomic structure.[2] We leave the reader to investigate these for himself. We cannot refrain, however, from pursuing a little further the fundamental developments associated with the transformation theory. Let us introduce the set of quantities

$$J_k = \oint p_k dq_k, \quad (25)$$

where the integral is called a *phase* integral, the integration being conducted over a complete cycle of values of q_k. It is here to be under-

---

[1] C. V. L. Charlier, "Mechanik des Himmels." (Leipzig, Veit, 1902.)
[2] M. Born, "Mechanics of the Atom." (London, Bell, 1927.)

3.15 THE TRANSFORMATION THEORY OF MECHANICS

stood that each q_k is restricted to lie in a finite interval, i.e., q_k has a finite minimum and a finite maximum value and is allowed to assume all values between these. When we speak of a cycle of values of q_k we shall mean values running from $q_{k,\,\text{min}}$ to $q_{k,\,\text{max}}$ and back to $q_{k,\,\text{min}}$. If q_k is a variable which does not appear in the Hamiltonian, it is said to be cyclic; e.g., the longitude angle ϕ does not appear in the Hamiltonian for a free particle expressed in spherical coordinates. In the case of a cyclic coordinate the conjugate momentum is clearly constant. The integration in (25) for such a case extends by definition from 0 to 2π. Now from (16), $p_k = \partial S/\partial q_k = \partial S_k/\partial q_k$ (assuming separability of the variables) and is a function of q_k, E, and the constants $\alpha_2 \ldots \alpha_n$. Hence after conducting the integration

$$J_k = J_k(E, \alpha_2 \ldots \alpha_n). \tag{26}$$

These n equations can be solved for the α's and E in terms of the J's and the results substituted into the expression for S (24), disregarding the arbitrary additive constant. Thus S is expressed as a function of the q's and the J's, viz.

$$S = S(q_k, J_k), \tag{27}$$

This is a substitution function defining a transformation from q_j, p_j to Q_j, P_j, where we may treat the J_k as equivalent to either Q_j or P_j. If the latter choice is made and we also now choose to denote Q_j by w_j we have from eq. (16)

$$p_j = \frac{\partial S}{\partial q_j}, \quad w_j = \frac{\partial S}{\partial J_j}, \tag{28}$$

and the canonical equations become

$$\dot{w}_j = \frac{\partial K}{\partial J_j}, \quad \dot{J}_j = -\frac{\partial K}{\partial w_j}, \tag{29}$$

where K is the transformed Hamiltonian. In the new coordinates it is a function of the J_j only. Hence (29) yields

$$\dot{w}_j = \omega_j, \quad \dot{J}_j = 0, \tag{30}$$

the ω_j being constants (functions of the J_j only). Hence

$$w_j = \omega_j t + \gamma_j, \quad J_j = \text{constant}. \tag{31}$$

The new coordinates w_j therefore appear as linear functions of the time; hence the name *angle* variables. The J_j from their definition in (25) have received the name *action* variables (cf. Sec. 3.12). We have already considered an illustration of angle and action variables in connection with the simple harmonic oscillator earlier in the present section. The quantities ω_j have the dimension of reciprocal time, and one might therefore be tempted to associate them with the idea of *frequency* in the motion of the system. This view may be justified as follows. Let us consider the change in the angle variable w_j associated with a complete cycle of values of q_k. This may be written in the usual phase integral notation

$$\Delta w_j = \oint \frac{\partial w_j}{\partial q_k} dq_k.$$

But from (28) we can write this

$$\Delta w_j = \oint \frac{\partial}{\partial q_k}\left(\frac{\partial S}{\partial J_j}\right) dq_k,$$

and if S satisfies appropriate mathematical conditions, as we shall assume in physical problems, we can interchange the order of integration and write

$$\Delta w_j = \frac{\partial}{\partial J_j} \oint \frac{\partial S}{\partial q_k} dq_k. \tag{32}$$

The removal of the operator $\partial/\partial J_j$ to the outside of the integral sign need occasion no comment since the integration is taken wholly over q_k. But from (28) and (25) we write (32)

$$\left.\begin{aligned}\Delta w_j &= \frac{\partial J_k}{\partial J_j} \\ &= 0, j \neq k \\ &= 1, j = k.\end{aligned}\right\} \tag{33}$$

In other words, the change in w_j, when q_j goes through a complete cycle, is unity, while the corresponding change in w_j when any other q goes through its cycle is zero. It is clear that q_j is periodic in w_j with a period equal to 1. Hence from (30) q_j must be periodic in time with frequency equal to ω_j. Illustrations of the use of angle and action variables may again be found in celestial mechanics and atomic mechanics (cf. the references above to Charlier and Born).

3.16. Summary. We have reached the end of our brief discussion of the foundations of mechanics, which may justly be considered one of the grandest of human constructs. It is only fair to pause a moment for the sake of contemplating the subject as a whole from the standpoint of the present volume. We have seen in mechanics the

3.15 THE TRANSFORMATION THEORY OF MECHANICS

oldest example of a physical theory and in many respects the most elaborately developed one. The experimentation and thinking of many generations of philosophers, physicists, and mathematicians have gone into its making. By the middle of the nineteenth century it may be said to have reached its culmination, marked by the wellnigh universal feeling that every physical phenomenon should ultimately receive its explanation in mechanical terms. With many physicists this attitude persisted to the very close of the century. It is well expressed in the famous dictum of Lord Kelvin that he simply could not understand a physical theory which did not have a mechanical basis.[1] The contemporary physicist prides himself on a broader point of view; yet it is difficult to restrain an expression of admiration for the achievements of classical mechanics as well as its logical beauty and elegance. In Chapter V the building of mechanical theories of various physical phenomena like electric currents and thermodynamics will be touched on. We shall have there an opportunity to assess the advantages as well as the shortcomings of the mechanical viewpoint.

The historical development of mechanics affords a good illustration of the way in which primitive concepts evolve into more abstract concepts. From its close relation to everyday experience, mechanics naturally began with the use of ideas most closely associated with and suggested by such experience. The fundamental concepts—displacement, velocity, acceleration, and even mass and force—seem to possess an intuitive quality which is absent from abstractions like entropy or the intensity of an electric field. Our study in Chapter I and the present chapter has shown us that this feeling is after all more or less illusory since the precise definition of mechanical quantities demands the same idealization and abstraction that characterize the concepts of other physical theories. However, the concepts of mechanics in their long development through the ages have come to possess for us a familiarity which is not shared by more recent concepts. This will probably continue to be true as long as we use the ideas of space and time in the construction of physical theories.

Again we may note as a matter of great interest the wide variety of ways in which the principles of mechanics can be phrased. This is not surprising when we consider that in any mathematical theory emphasis may be placed on any one of a relatively large number of deductions from the initially chosen postulates. Indeed, it is possible to choose deductions from certain postulates as the fundamental assumptions and so derive the original postulates. Thus we see that on assuming the Newtonian formulation of mechanics we can deduce the principles of D'Alembert, least constraint, etc. On assuming D'Alembert's principle initially, the Newtonian principles result. The choice of a foundation is thus a matter of convenience. Yet it must again be emphasized that the greater the number of points of view from which we examine a subject the deeper should be our understanding of it.

Our examination of the foundations of mechanics has indeed been deficient in one important respect: we have not considered the modifications introduced by the theory of relativity. The form of the principles which we have discussed fails to satisfy the stringent invariance requirements of relativity. Hence our definition of mass, for example, holds only for inertial systems moving relatively to each other with velocities small compared with that of light. We shall go into this matter in detail in Chapters VII and VIII.

CHAPTER IV

PROBABILITY AND SOME OF ITS APPLICATIONS

4.1. The Meaning of Probability. Almost as fundamental as the notions of time and space, which were treated in a previous chapter, is the concept of probability, in some fields of physical analysis. Thus in our discussion in Sec. 1.5 we had occasion to comment on the way in which the notion of probability enters into the meaning of a physical law in connection with its experimental verification; we met the same concept in our discussion of boundary conditions in Sec. 1.9. It will be desirable, therefore, to treat it now in some detail. It is true that probability analysis as a mathematical technique is limited to certain special domains of physics, and it is one of the aims of the present chapter to outline clearly the conditions under which it is applicable. In Chapter V we shall make a special study of those physical theories which permit an application of probability reasoning.

Although the usefulness of the probability calculus is quite apparent in its many applications to physical problems, it is far from easy to fix its exact epistemological status or to justify its applications. In discussing the foundations of the use of probabilities in science we are entering a controversial field, a field indeed in which disputes rage hotly and opinions are nearly evenly divided. This is hardly true of any similar branch of pure science. Geometry, for instance, which stands in much the same relation to physics as does the theory of probability, contains at present fewer problematic issues than does that theory, for the voice of those who wish to deduce universally valid geometric facts from an intuition of pure reason has long ceased to be heard, and the empirical elements of geometry have been clearly recognized as such. Probability is now approximately in the state of geometry about fifty years ago, with the attempts of an *a priori* justification of its use still widely prevalent but strongly contradicted by the assertions of staunch empiricists who wish to build their theories on observations.

It is difficult for anyone to present two conflicting points of view without bias, and an attempt to do so would seem artificial. Hence we shall, in this exposition, allow ourselves the freedom of choosing that line of thought which seems logically more complete, without hesitating, however, to point out difficulties wherever they appear.

Laplace's definition. The gravest of these difficulties confronts us at the start, that is, when we seek a definition of probability. For some reason we seem to know that the probability of throwing a four with a good die is $\frac{1}{6}$, that of tossing heads with a penny is $\frac{1}{2}$, etc. Why? Laplace, to whose mathematical ingenuity we owe a great deal of the development of the probability calculus, informs us [1] that probability is to be defined as the quotient of the number of favorable cases to the total number of equally possible cases. With the die, there is one favorable case, namely, the appearance of the four, and there are 6 possible cases. At first glance this definition appears excellent indeed, for it not only defines probability, but also gives at once directions for computing it. It is essentially an *a priori* definition, for it fixes probabilities even before any die is constructed, or before any observation with it has been made. Historically it grew out of the interest in games of chance, which, up to Laplace's time, provided almost the only field to which the calculus in question was applied.

But obviously this fact is also its chief limitation. The definition can be applied to almost no other field. The statistician who is interested in the probability of death of a certain individual or class of individuals is at a loss to state the number of equally possible cases; the biometrician, in trying to determine the probability that a certain bean shall have a preassigned weight, knows nothing about the number of all possible beans; nor is the physicist interested in the total number of equally possible results of a measurement when he speaks of the probability of a given result. And even if the statistician knows how many people of the class in question are alive, how would he decide what number of them are *equally likely* to die? For nothing else than this can be the meaning of the phrase " equally possible " in this connection.

Another formal difficulty with Laplace's definition arises in all cases in which the total number of possible events is infinite. What, for instance, is the probability, according to this definition, that a point placed at random shall bisect a given line, or shall fall within a preassigned segment of it?

Another fault of every probability definition which fixes probabilities in advance of observation comes to light when we consider a bad die, or a bad penny, for here the selection of equally possible cases must definitely be made in accordance with physical observations,

[1] "Essai philosophique des probabilités" (Paris, 1814).

for instance, on the position of the center of mass of the die. To be sure, it is still possible to amend Laplace's definition and make it applicable to this anomalous situation. One might, for example, in establishing the number of equally possible cases, arbitrarily count twice the event in which a face lying closer to the center of mass appears on the bottom, while counting the others but once. But such a convention would be far fetched, and there would be no criterion for guiding our choice.

A posteriori definition. An escape from these troubles is possible if we adopt an *a posteriori* definition of probability, i.e., one based on observation. Tendencies in this direction became vocal during the nineteenth century, when Ellis, Cournot, and Fechner introduced their "frequency theories." They define probability by reference to a long-continued series of observations and thereby strip the concept of all idealistic implications. Probability becomes simply the limiting value of the ratio of the number of favorable observations to the total number of observations, when the total number becomes very large. The most systematic and complete treatment of this view has recently been given by von Mises.[1] The following remarks are in close conformity with his exposition.

The probability calculus is applied to a large number of observations, each of which will be called an *element*. The total sequence of elements is designated as a *probability aggregate*, a term chosen to be synonymous with v. Mises' "Kollektiv" (first used, but with a somewhat different connotation, by Th. Fechner).[2] It is clear that not every set of observations is of a character which justifies the application of probability arguments; in other words, a probability aggregate must satisfy very definite conditions, which will be discussed later. Every element is regarded as having one of a limited number s of mutually exclusive properties that can be observed. If n_i is the number of times which the ith property is observed and n the total number of observations or elements, the quotient n_i/n is the relative frequency of the ith property. Supposing finally that n becomes very large, i.e., that the aggregate contains an infinite number of elements, the limit of the quotient n_i/n will be defined as the probability of the ith property. It is clear that

$$\sum_{i=1}^{s} \frac{n_i}{n} = \sum_{i=1}^{s} \lim_{n\to\infty} \frac{n_i}{n} = 1.$$

[1] R. v. Mises, "Wahrscheinlichkeitsrechnung" (Leipzig, 1931).
[2] Th. Fechner, "Kollektivmasslehre" (Leipzig, 1897).

For a given aggregate, the limit of n_i/n depends only upon the index i and may be regarded as a function $w(i)$. The totality of $w(i)$'s is known as the *distribution* of the particular probability aggregate.

As an illustration consider the aggregate formed by a large number of throws of one die and the corresponding observations of the numbers which appear. Each observation subsequent to a throw is an element. There are 6 properties, the appearances of the numbers 1 . . . 6. The probabilities in this case are known to be $\frac{1}{6}$ for every property i, so that the distribution consists of 6 numbers $w(i)$ which are all equal.

In general it is not necessary, of course, that the properties be discrete and their number finite; examples, such as the determination of the coordinates of a die after its throw, where the number of properties is not finite and their distribution continuous, are easily found. It is customary to speak of the former as an *arithmetical* and of the latter as a *geometrical* distribution. No essential modification of the definitions is required when the distribution is geometrical, except that summations have to be replaced by integrations as will be indicated later. It is to be remembered, however, that in a strict sense no physical observation is made on a continuous range of properties, for every measurement is performed by means of an apparatus which yields as the value of any determinable property an integral multiple of some fundamental quantity. Even in measuring such things as the length of an object, which by its nature is felt to be continuous, the result of the measurement is stated as a multiple of the smallest scale unit of the measuring instrument. The employment of continuous distribution functions, $w(i)$, is a matter of pure convenience and must be looked upon as an idealization of experience.

The probability calculus is incompetent to provide *a priori* information regarding the distribution in a given aggregate. The fact, for instance, that in the game of dice the probability of throwing a 5 is equal to that of throwing a 2, or that either is $\frac{1}{6}$, could not have been deduced by arguments peculiar to this calculus. The distribution depends in all cases on extraneous, here physical, conditions, as is seen from the circumstance that it will change when the die is loaded. Hence the probability theorist must content himself with an experimental determination of his distribution by proceeding to observe throws for a very long time, and all concepts involved in his dealings must be based on these *a posteriori* data alone.

The only aim which the probability calculus can properly achieve is to produce the distribution of a *derived* aggregate when that of the *primary* one is known. Given, for instance, that all the $w(i)$ in our previous example of throwing one die are equal to $\frac{1}{6}$, the question, what is the probability of throwing any particular number between 2 and 12 with two dice, is one which the calculus can legitimately answer. Indeed, the derivation of new aggregates by combining primary ones, or by changing the elements or properties of a given one, is its principal business. Many such instances will occupy us later. In a systematic treatment of the problem at hand the various ways of combining or changing aggregates should be classified, and the necessary mathematical operations should be discussed. We shall deviate from this course, for the rules are simple and their application will appear natural to the reader whenever it is required in the further development of the subject.

4.2. Analytical Nature of Probability Aggregates. Before going on, let us answer the question: when is a series of observations a probability aggregate? For clearly, if there exists no definite test, the entire theory is useless. There are two criteria by which this question can be decided. The first follows immediately from previous considerations: the concepts specified by our definitions must be free from contradictions or, in a mathematical sense, must exist. Now there is no difficulty about the existence, or imaginability, of an infinitely extended sequence of observations, although it is never encountered in actual experience. Anyone who suspects trouble at this point may be reminded of the fact that geometers operate legitimately with perfect spheres although such spheres cannot be constructed. Next, the fraction n_i/n is perfectly well behaved, if, as we suppose, it is the first n observations which are being considered and n is not zero. However, the existence of $\lim_{n \to \infty} n_i/n$ is not in general guaranteed, but reflects a special property of the sequence. Hence we can cast our first condition in the form:

(I) $$\lim_{n \to \infty} \frac{n_i}{n} = w(i) \quad \text{exists.}$$

If the distribution is geometrical, and i refers to a "point" property in the usual sense, this form of the condition is useless, for $w(i)$ would be 0.

This is seen at once if we regard as properties of our aggregate the coincidences of the center of mass of a die with the various points on a line. Here the number of cases in which coincidence with a given point is exact will be zero. If, however, we consider coincidences with any of the points in a small segment of length Δx of the line their number will be finite and may be written $n_x \cdot \Delta x$. In such instances it is necessary to collect a continuous range of properties situated about x, and postulate

(I') $$\lim_{n \to \infty} \frac{n_x}{n} = w(x) \quad \text{exists.}$$

$w(x)$, thus defined, is often spoken of as a "probability density." Obviously, it must satisfy the relation

$$\int w(x)dx = 1.$$

Conditions I and I' alone do not justify the application of probability arguments to the sequence in question, for they take no account of a possible systematic interdependence of the elements. Consider, for instance, the decimal fraction for $\frac{1}{7}$, which has the period 142857. Taking each number as the result of an observation, condition I is satisfied. Nevertheless, the infinite sequence is not a probability aggregate, as is intuitively clear to everyone confronted with the task of defending such a sequence as the result of a game of chance. Qualitatively speaking, the result is too systematic to be due to "chance." We therefore require a further condition which insures a certain independence in the assignment of the various elements to the properties, and makes it conform to what we crudely denote by chance. With greater logical precision this condition may be stated as follows:

(II) Select from all the properties of the infinite sequence two mutually exclusive sets, A and B. Eliminate from the sequence all elements not belonging to A and B, and consider only the remainder, with its order unchanged. Select from the remaining elements, which now have only the (collective) properties A or B, a new sequence by means of a *preassigned* rule. If the first n' elements of this new sequence contain n_A' elements having property A, and n_B' elements having the property B, then we postulate

$$\lim_{n' \to \infty} \frac{n_A'}{n'} \quad \text{and} \quad \lim_{n' \to \infty} \frac{n_B'}{n'} \quad \text{exist;}$$

$$\frac{\lim_{n' \to \infty} \frac{n_A'}{n'}}{\lim_{n' \to \infty} \frac{n_B'}{n'}} = \frac{\lim_{n \to \infty} \frac{n_A}{n}}{\lim_{n \to \infty} \frac{n_B}{n}},$$

where n_A, n_B, n refer to the initial sequence.

4.2 ANALYTICAL NATURE OF PROBABILITY AGGREGATES   165

The word " preassigned " requires an explanation. It is intended to specify that the rule must be given *before* the property of the element to be selected or rejected is known. For instance, if the initial sequence contains the observations on an infinite number of successive throws with one die, and only the results: 1 and 2 have been retained, the remainder might be 12122121122 .... Possible proper rules would then be: select every other element; or select every element whose ordinal number is a prime; or select every element preceded by a 1, etc. An improper rule would be: select, or reject, every other 2, for this rule could evidently be applied only after the result of the throw in question is known. Conditions I and II completely characterize a probability aggregate.

Difficulties with the a posteriori definition. The foundation of the probability calculus which has now been sketched appears to have distinct advantages over the older Laplacian formulation. Its strongest points, perhaps, are its universal applicability and the closeness with which it approaches to the physicist's mode of thought. Its rejection by some physicists would therefore be surprising, were it not for an important difficulty which will now be discussed.

In conditions I and II we have postulated the existence of the limits of relative frequencies. Two objections can be, and have been, raised against this postulate, one irrelevant, the other quite serious. First, one may say that no sequence of observations, no matter how long it be continued, possesses such limits, which is true. Second, the existence of these limits is contradictory to the rules which are derived from our definitions, which, unfortunately, is also true.

To study the situation more closely, let us first discuss what mathematicians mean by a limit. Consider an infinite sequence of numbers. To find out whether or not it has a limit we proceed as follows. First we choose a positive number ε, finite but arbitrarily small. Next we strike out a finite number N of terms from our sequence and examine the remaining ones. In general the remainder will contain numbers some of which differ by amounts greater than ε. If, however, we can increase N but still leave it finite (i.e., reject more terms) and manage to be left with terms none of which differ by as much as ε, then the sequence is said to have a limit.

Sequences of observations do not possess a limit in this sense. To illustrate this, let us consider the results of tossing a penny an infinite number of times. By r_n we denote the number of heads observed in the first n tosses. If the sequence has a limit of the type here discussed, there must exist a finite number N such that, for $n > N$, all the terms r_n/n differ by less than ε. That this is not the case can be seen as follows. The infinite sequence of throws will contain every conceivable finite succession of heads and tails; among these will be runs of any finite number of heads and of any finite number of tails. Let now ε and N be given. After discarding the first N results of the sequence we proceed along the remaining elements until we come to one (whose ordinal number is N', say) where a run of s tails starts, and s satisfies the relation

$$\frac{s}{N'+s} > 2\varepsilon. \qquad (1)$$

We shall also suppose that $r_{N'}/N'$, i.e., the relative frequency of heads up to the N'th throw, is $\geq \frac{1}{2}$. There will certainly be some run for which both of these relations are satisfied. Then we need only compare the N'th and the $(N'+s)$th quotient $\frac{r_n}{n}$ to find that they differ by more than ε. For $r_{N'+s}$ is now equal to $r_{N'}$, since no heads have appeared between the N'th and the $(N'+s)$th throw. Therefore

$$\frac{r_{N'}}{N'} - \frac{r_{N'+s}}{N'+s} = \frac{r_{N'}}{N'}\left(1 - \frac{N'}{N'+s}\right) = \frac{r_{N'}}{N'}\left(\frac{s}{N'+s}\right) > \frac{1}{2} \cdot 2\varepsilon, \qquad (2)$$

on account of the relations above. Thus we are forced to admit that, if condition I is maintained with the term " limit " in its strict mathematical sense, there may be no empirical sequence of elements which satisfies the requirements of a probability aggregate. The reader is likely to feel, at first sight, that the example here chosen is not based on facts and that, in particular, inequality (1) cannot be satisfied in any sequence of observations. One supposes instinctively that there must be some correlation between s and N' which renders the inequality void. This impression, however, is as false as the supposition that the chances for the occurrence of a rare event increase the longer one waits for it.

If we still wish to retain condition I in the face of these facts we must admit that a probability aggregate is a very idealized concept, to which sequences of observation do not strictly conform. This conception is by no means so embarrassing as it might appear, for it is necessary wherever mathematical constructs are applied to observations.

Condition I receives a stronger blow, however, if we come to realize that not only experimental sequences, but even those which are derived by the very use of our probability conception, fail to conform to it. We have seen that, if the limit in question is to exist, there must never occur inconveniently long runs. So far we have avoided difficulties by saying that sequences which contain such runs simply are not probability aggregates in the sense of our definition. But now we shall find that even the ideal aggregate constructed on the basis of conditions I and II leads to the same difficulties, a fact which is indicative of some internal contradiction. The point is that the ideal aggregate, too, contains long runs. For it may be shown by the usual rules of probability, derivable from I and II (cf. von Mises, *op. cit.*), that the probability of a run of s heads is $(\frac{1}{2})^s$. Hence, according to the previous definitions, the number of such runs occurring in an infinite sequence is $\lim_{n \to \infty} n \cdot (\frac{1}{2})^s$. We cannot, therefore, simply deny away the runs without violating the definitions themselves.

For this reason many scientists have rejected altogether the foundation of probability here discussed, and reverted to the Laplacian stand. This, we feel, is a very great sacrifice which can be avoided. In fact, this logical difficulty can be overcome by introducing what amounts to an elementary uncertainty into the scheme, i.e., saying that probabilities so small that they escape experimental detection shall be regarded as 0. In that case it is necessary to fix a number δ, very small but finite, and agree to call $w(x)$ zero if $w(x) < \delta$. Then there will be a maximum of s such that runs of a greater number of heads than $s_{\max}$ do not occur, and $s_{\max}$ would be defined by the inequality:

$$(\tfrac{1}{2})^{s_{\max}} \geq \delta > (\tfrac{1}{2})^{s_{\max}+1}.$$

With this limitation placed upon s relation (1) can be violated by choosing a finite N'. This solution appears reasonable from a practical point of view.[1] We conclude, therefore, that there is no cogent reason for abandoning the *a posteriori* probability foundation in favor of an impracticable and indefinable *a priori* conception.

4.3. Some Useful Concepts. Let us now assume that a probability aggregate, together with its distribution ($w(i)$ if the properties i are discrete, $w(x)$ if they are continuous) is given. The properties are not in general numbers. They may be events, such as the drawing of a black ball, or a white ball, from an urn. To make the formulas more convenient we shall in that case assign to every property i a number x_i and allow i to designate the properties in succession. One may, for example, denote the drawing of a black ball, a white ball, and a red ball by 0, 1, 2, respectively. Thus, if there are no other colors, i would run from 1 to 3, and $x_1 = 0$, $x_2 = 1$, $x_3 = 2$. When, in the following paragraphs, we speak of a property of the probability aggre-

---

[1] Another proposal to overcome the difficulty has been made by Hohenemser, *Naturwissenschaften*, **19** (1931), 833.

gate, we shall always suppose it to be represented by a number. In accordance with our probability definition

$$\left.\begin{array}{l}\sum w(x_i) = 1 \text{ in the arithmetical case;} \\ \int w(x)dx = 1 \text{ in the geometrical case.}\end{array}\right\} \quad (1)$$

In physics one sometimes meets with distributions not satisfying these equations; they are occasionally also called probabilities.[1] To distinguish them from the true probabilities here considered we shall call the latter " mathematical probabilities," wherever distinction is necessary. Distributions not satisfying (1) have meaning only if $\sum_i w(x_i)$, or $\int w(x)dx$, exists. In that case they can be reduced to mathematical probabilities by dividing every w by $\int w(x)dx$, or $\sum_i w(x_i)$, as the case may be. This process is called " normalization."

Let $f(x)$ be a function, defined for every x_i in the arithmetical case, or for every x which has a finite probability in the geometrical case. We then denote by the " expected mean " of $f(x)$, with respect to the distribution $w(x)$, the expressions

$$\overline{f(x)} = \begin{cases} \sum_i f(x_i)w(x_i) \\ \int f(x)w(x)dx. \end{cases} \quad (2)$$

Less common, but often useful, is the definition of the " deviation " or " dispersion " of $f(x)$ with respect to the distribution $w(x)$. This is denoted by D. It is, for the arithmetical case and geometrical case, respectively,

$$D[f(x)] = \begin{cases} \sum_i [f(x_i) - \overline{f(x_i)}]^2 w(x_i) \\ \int [f(x) - \overline{f(x)}]^2 w(x)dx. \end{cases} \quad (3)$$

If we choose x itself for the function $f(x)$, (2) reduces to the " mean value $\bar{x}$ of the distribution," and (3) to its " dispersion," σ^2. Thus we have

$$\bar{x} = \begin{cases} \sum_i x_i w(x_i) \\ \int xw(x)dx, \end{cases} \quad (4)$$

and

$$\sigma^2 = \begin{cases} \sum_i (x_i - \overline{x_i})^2 w(x_i) \\ \int (x - \bar{x})^2 w(x)dx. \end{cases} \quad (5)$$

[1] Cf., for instance, "Thermodynamic Probabilities," Chapter V.

It is often conducive to clarity to think of the probability distribution as a one-dimensional distribution of mass. $\bar{x}$ is then simply the center of mass, while σ^2 is the moment of inertia about an axis through the center of mass. In general, the quantity

$$\overline{x^r} = \int x^r w(x) dx,$$

or the corresponding one for an arithmetical distribution, is known as the rth moment of the distribution.

The reader will remember a simple formula for computing the moment of inertia about any axis if that about a parallel axis through the center of mass is known. A similar theorem connects σ^2 and $\overline{x^2}$. For it is easily seen from (5) (if one remembers that $\overline{x_i}$ is a number independent of i, and $\bar{x}$ is a constant) that

$$\sigma^2 = \overline{x^2} - 2\overline{x}\overline{x} + \bar{x}^2 = \overline{x^2} - \bar{x}^2. \tag{6}$$

The dispersion is equal to the second moment minus the square of the first moment. The intuitive meaning of σ^2 becomes clear if we observe that the quantity is always positive, that it is 0 when all the mass or probability is concentrated at a point and is greater the more strongly the masses are "dispersed," i.e., recede from the center of mass. σ, the square root of the dispersion, is called the "root mean square deviation" or more commonly the "standard deviation."

The moments of a distribution are of interest because if they are known in their entirety the distribution itself can often be computed. Only when the distribution is geometrical and extends from $x = -\infty$ to $x = +\infty$ is this impossible. In that case, different distributions $w(x)$ may have identical sets of moments $\bar{x}, \overline{x^2}, \overline{x^3}$, etc.

4.4. Bernoulli's Problem. Let us now turn to an application of the notions discussed so far. The problem under consideration is an important one since it lies at the basis of many physical discussions. It was first extensively dealt with by Jacob Bernoulli; useful solutions for it were obtained by Newton, Laplace, and Poisson.

Imagine an urn containing black and white balls, different in number. Under proper conditions, successive drawings from this urn will produce a probability aggregate having two properties, the elements being the single drawings. (We shall suppose the ball to be returned to the urn after it is drawn, and the urn to be properly shaken.) Denote the appearance of a black ball by 0, that of a white ball by 1. The probabilities which have been determined by a large number of trials are found to be $w(0) = p$, $w(1) = q$. By introducing another urn containing the same number of balls with the same proportion of black and white, we can form another, somewhat more complicated, probability aggregate. For we can now regard as a single element two successive drawings, one from each urn, and as properties the four combinations 1,1; 1,0; 0,1; 0,0. By fairly obvious rules of combination which we shall not discuss here the first of these has the probability q^2, the second and third pq, the fourth p^2. If we combine three urns in a similar way and consider as a single element three drawings, one from each urn, we get a probability aggregate with eight properties 1,1,1; 1,1,0; 1,0,1; 0,1,1; 1,0,0; 0,1,0; 0,0,1; 0,0,0, and probabilities $q^3, pq^2, pq^2, pq^2, p^2q, p^2q, p^2q, p^3$. Now Bernoulli's problem is this: what is the probability $w_n(x)$ of drawing x white balls from n urns?

What we wish to find is evidently not the probability of a single property of the aggregate consisting of the repeated drawings from n urns, but the combined probability of a certain class of properties, namely, all those implying a total of x white balls. For instance, if $n = 4$ and $x = 2$, we desire to find the probability of all the following properties: 1,1,0,0; 1,0,1,0; 1,0,0,1; 0,1,1,0; 0,1,0,1; 0,0,1,1. In general, each property having x 1's and $n - x$ 0's has the same probability, namely, $q^x p^{n-x}$. If we multiply this by the total number of such properties, n_x, we have the desired result.

Now n_x can be found without much difficulty. It is the number of ways in which x 1's can be distributed among n places. The first 1 can be placed in n different ways while any given order among the remaining $(x - 1)$ 1's is maintained. For every one of these n arrangements, $(n - 1)$ places are available for the second 1, the order of the $(x - 2)$ remaining 1's being undisturbed. If there were but two 1's to distribute $(x = 2)$, then $n(n - 1)$ would be the total number of arrangements thus obtained. But it must not be forgotten that in placing the second 1 we are repeating arrangements which have already occurred, i.e., arrangements which differ only by having the two 1's interchanged. In fact, all arrangements thus obtained group themselves into pairs which are identical (unless we could distinguish between the different 1's, which, of course, is not the case). Hence, for $x = 2$, $n_x = \dfrac{n(n-1)}{2}$. The number of places available for the third 1 is $n - 2$; but it is immaterial from our point of view whether the third 1 occupies the first, second, or third of the three places filled by 1's. Hence we must divide the number $n_2(n - 2)$ by 3 in order to obtain n_3. Continuing this process, it is found that

$$n_x = \frac{n}{1} \cdot \frac{n-1}{2} \cdot \frac{n-2}{3} \cdots \frac{n-x+1}{x} = \frac{n!}{x!(n-x)!} \cdots \quad (1)$$

4.4 BERNOULLI'S PROBLEM

More generally, n_x is the number of ways in which x indistinguishable things can be distributed among n places. A symbol, due to Euler, which is commonly used for (1) is $\binom{n}{x}$. It is of interest to notice in this connection that $\binom{n}{x}$ is the general coefficient in the binomial series, a fact recognized by Newton. Thus:

$$(p+q)^n = p^n + \binom{n}{1}p^{n-1}q + \binom{n}{2}p^{n-2}q^2 + \ldots + q^n. \quad (2)$$

Collecting results, the answer of Bernoulli's problem can be stated in the form

$$w_n(x) = \binom{n}{x}p^{n-x}q^x. \quad (3)$$

Following von Mises, we shall refer to (3) as Newton's formula, and call the (arithmetical) distribution which it defines, Bernoulli's distribution. It is understood that x is an integer.

Distribution (3) satisfies (4.3-1), for if it is summed over all x up to n, there results the series (2), which is equal to 1 since $p + q = 1$.

The significance of (3) is evidently more general than the special example from which it has been derived. It represents the probability of x successes in n independent trials if the probability of success in a single trial is q.

Let us set ourselves the problem of finding the expected mean of x, that is, $\bar{x}$. In terms of our example, this quantity is the expected (loosely speaking, the average) number of white balls appearing in n drawings, if these n drawings are repeated an infinite number of times. According to (4.3-4),

$$\bar{x} = \sum_{x=0}^{x=n} x w_n(x).$$

To calculate this sum, we expand $(p + qy)^n$ by the binomial theorem (2). We have

$$\left.\begin{aligned}(p+qy)^n &= p^n + \binom{n}{1}p^{n-1}qy + \binom{n}{2}p^{n-2}(qy)^2 + \ldots + (qy)^n \\ &= \sum_{x=0}^{x=n} w_n(x) y^x,\end{aligned}\right\} \quad (4)$$

with $w_n(x)$ defined by (3). Differentiate the first and last member of this relation with respect to y:

$$n(p + qy)^{n-1}q = \sum_{x=0}^{x=n} xw_n(x)y^{x-1}.$$

If now we put $y = 1$, the right-hand side becomes $\bar{x}$, and we have

$$\bar{x} = nq. \tag{5}$$

Next let us find the dispersion, which, as pointed out before, gives a measure of the deviations of the results from the expected mean. By (4.3–5) and 4.3–6),

$$\sigma^2 = \sum_{x=0}^{n}(x - \bar{x})^2 w_n(x) = \sum_{x=0}^{n} x^2 w_n(x) - \bar{x}^2.$$

Differentiate (4) twice with respect to y and then put $y = 1$. The result is:

$$n(n-1)q^2 = \Sigma\, x(x-1)w_n(x) = \Sigma\, x^2 w_n(x) - \bar{x}.$$

Add to the first and last member, respectively,

$$nq - n^2 q^2 = px - \bar{x}^2,$$

which is true because of (5). Changing sides, the result is

$$\sigma^2 = nq - nq^2 = nq(1 - q) = npq. \tag{6}$$

The standard deviation becomes $\sigma = \sqrt{npq}$. Higher moments can be calculated by continuing this process.

4.5. The Emission of α Rays as a Probability Problem. It is of interest to apply these concepts to a physical phenomenon. According to general belief, radioactive decay proceeds according to the laws of " chance." What does this statement mean, and how is it verified?

To illustrate its meaning, consider the following experimental arrangement. A radioactive preparation emitting α rays is placed behind a set of diaphragms which screen off all but a small solid angle of the rays. The transmitted α particles fall upon a sensitive counting device, such as a Geiger-Müller tube combined with an amplifying arrangement and a counter. The number of α particles received in a given time, say one minute, is recorded. Many such one-minute counts are collected. If each such count is regarded as an element, and the number counted as its property, then the sequence of elements may, or may not, form a probability aggregate. This may be tested by conditions I and II. The statement regarding the " chance "

4.5 THE EMISSION OF α RAYS AS A PROBABILITY PROBLEM

character of radioactive decay amounts to an assertion that the sequence *is* a probability aggregate. But it tacitly implies more, for it also fixes the *distribution* of this probability aggregate. To clarify this matter let us perform a thought experiment.

Imagine the possibility of observing the given collection of radioactive atoms for a very small period of time τ, so small indeed that practically at most one α particle will be emitted. A single observation will then yield the result: no emission or emission. By repeating this observation a great number of times we can obtain a sequence of results, and we can ascertain, by the rules outlined, whether this sequence forms a probability aggregate or not. By stating that the decay process is governed by chance we assert that it, too, is a probability aggregate. This is the second assumption involved in the statement that the decay process is governed by "chance."

This matter cannot be tested directly. But it has an important influence upon the distribution of the first aggregate, discussed in the second paragraph of this section, an aggregate which is capable of experimental observation. For suppose now that we consider a great number n of such imaginary observations, and let $n\tau = 1$ minute. If q is the probability of decay, and p the probability of persistence in the interval τ, then the probability of observing x decays in n intervals, i.e., one minute, will be the same as the probability of x successes in n independent trials. But this, as we have shown in the preceding section, is given by Newton's formula (4.4-3).

One possibility of testing the conclusion is to compare the distribution of counts with this formula. The difficulty with this procedure is a technical one: Newton's formula contains factorials of large numbers which are extremely tedious to calculate. This makes the labor of direct comparison prohibitive. In the next sections we shall deduce approximate forms of Newton's formula which are more useful in this connection. At present we shall content ourselves with simpler tests.

Call the counts obtained in successive observations over one-minute intervals $x_1, x_2, \ldots, x_l$, l, the total number of observations, being large. Their experimental mean is $\frac{1}{l}\sum_i x_i$. According to Newton's formula, $\bar{x} = nq$. There is no way of comparing these two quantities, for we know neither n nor q. The dispersion of Bernoulli's distribution is, by (4.4-6), $\sigma^2 = npq$. Now it is clear that, if τ is taken small enough, q is extremely small, so that $p \sim 1$. Hence

$$\sigma^2 = \bar{x}.$$

If the experimental results conform to Bernoulli's distribution we must also have

$$\sigma^2_{\text{obs.}} = \bar{x}_{\text{obs.}},$$

that is,

$$\frac{1}{l}\sum_i (x_i - \bar{x})^2 = \frac{1}{l}\sum_i x_i,$$

at least approximately. In the following table we have listed results obtained by a method similar to the one here described. They are taken from Kohlrausch,[1] who has compiled them from the work of Rutherford and Geiger (*Phil. Mag.* (6), **20**, 698, 1910).

l	$\bar{x}_{\text{obs.}}$	$\sigma^2_{\text{obs.}}$
792	4.01	3.84
596	3.91	3.77
632	3.75	3.68
588	3.76	3.45

The data refer to four different series of observation. The agreement is seen to be satisfactory and probably supports the claim that radioactive decay is a chance phenomenon in the sense here discussed. The only troubling feature is the fact that the dispersion is in all cases too small. Subjective errors are doubtless possible in the counting of scintillations, which was the method used in the experiments discussed here. More recent careful observations by Curtiss [2] on the rate of emission of α particles from weak sources using a special double Geiger point counter have yielded results which seem to indicate beyond doubt that the distribution actually follows the laws of chance.

Perhaps it is of interest to point out that we have assumed, as is customary, that the probability of decay, q, is independent of the age of the atoms contained in the collection. This places the radioactive atom in a unique position indeed, for it is well known that all decaying organic structures have a decay constant depending on their age.

4.6. Laplace's Formula. As has been noted, the application of Newton's formula is usually highly impracticable. It is therefore necessary to find approximate formulations, valid under special conditions. One such formula has been worked out by Laplace. We shall

[1] K. W. F. Kohlrausch, *Ergebnisse der exakten Naturwissenschaften*, **5**, p. 192 (1926).

[2] L. F. Curtiss, *Bureau of Standards Journal of Research*, **8**, 339 (1932).

present a very simple derivation of it, partly sacrificing mathematical rigor.

The formula to be found is an approximation to Newton's as $n \to \infty$, n being again the total number of trials. Moreover, it represents a *continuous* distribution. As a result of the first of these facts, certain difficulties arise, and these must first be eliminated. Let us plot $w_n(x)$, as given by (4.4-3), as a function of x for four different values of n. Cf. Fig.(4.1).[1] The short vertical lines indicate the values of $\bar{x}$. q is taken as $\frac{1}{3}$. It is seen that both the mean value and the spread of the contours increase with n. In fact, both become infinite as $n \to \infty$. Analytically this is clear from eqs. (4.4-5) and (4.4-6). If we wish to prevent $w_n(x)$ from moving out and spreading indefinitely as n increases, we must suitably displace the origin of abscissas, and contract their

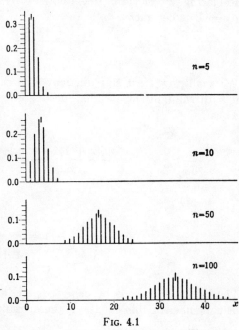

FIG. 4.1

scale. The first can evidently be done by measuring all x's from the value $\bar{x} = nq$, the second by maintaining the standard deviation finite, that is, according to (4.4-6), by choosing a quantity proportional to $\sigma/\sqrt{n}$ in place of σ as a measure of the standard deviation. All this suggests the transformation

$$u = h(x - \bar{x}), \text{ where } h = \frac{\text{constant}}{\sqrt{n}}. \tag{1}$$

Let the continuous function which $w_n(x)$ approaches as $n \to \infty$ be denoted by $w(x)$. We then have

$$\frac{d}{dx} \log w(x) = \frac{d}{du} \log w(u) \cdot \frac{du}{dx} = h \frac{d}{du} \log w(u). \tag{2}$$

[1] This figure is taken from T. C. Fry, "Probability and its Engineering Uses" (D. Van Nostrand, 1928).

On the other hand,

$$\frac{d}{dx} \log w(x) = \lim_{\Delta x \to 0} \frac{w(x + \Delta x) - w(x)}{w(x) \Delta x}.$$

Replacing all $w(x)$ in this quotient by $w_n(x)$ and observing that $\Delta x = 1$, it becomes

$$\frac{w_n(x + 1) - w_n(x)}{w_n(x)}.$$

We shall now treat this quotient as a continuous function, identify it with the limit in question, and equate it to the expression in (2). Thus

$$h \frac{d}{du} \log w(u) = \frac{w_n(x+1) - w_n(x)}{w_n(x)} = \frac{\binom{n}{x+1} p^{n-x-1} q^{x+1}}{\binom{n}{x} p^{n-x} q^x} - 1$$

$$= \frac{n-x}{x+1} \frac{q}{p} - 1 = \frac{1 - \dfrac{u}{nhp}}{1 + \dfrac{u+h}{nhq}} - 1.$$

The last step is made by using eqs. (1) and (4.4–5). Expanding the last member of this equation, there results

$$\left(1 - \frac{u}{nhp}\right)\left[1 - \frac{u+h}{nhq} + \left(\frac{u+h}{nhq}\right)^2 - \cdots\right] - 1 = -\frac{1}{nq} - \frac{u}{nhp} - \frac{u}{nhq}$$

plus terms involving higher negative powers of n which can be neglected when, as in our case, n is very large. The last two quantities on the right reduce to $-\dfrac{u}{nhpq}$, since $p + q = 1$. Even the first of the three terms may be disregarded in comparison with this one, since $h = \text{constant}/\sqrt{n}$. Therefore

$$h \frac{d}{du} \log w(u) = -\frac{u}{nhpq}.$$

Integration leads to

$$w(u) = Ce^{-\frac{u^2}{2h^2 npq}},$$

hence

$$w(x) = Ce^{-\frac{(x-nq)^2}{2npq}}.$$

To determine the constant C, we use the relation: $\int_{-\infty}^{\infty} w(x)dx = 1$.
Thus

$$C \int_{-\infty}^{\infty} e^{-\frac{(x-nq)^2}{2npq}} dx = \sqrt{2\pi npq}\, C = 1$$

whence $C = \dfrac{1}{\sqrt{2\pi npq}}$. We therefore arrive at the following result, known as Laplace's formula:

$$w(x) = \frac{1}{\sqrt{2\pi npq}} e^{-\frac{(x-nq)^2}{2npq}}. \qquad (3)$$

It was found by Laplace in 1812. In words, the result may be stated thus: For very large n, the distribution takes the form of Gauss's function, commonly known as the error function. It should always be remembered, however, that for all finite n, that is in every application, (3) is merely an approximation to the true distribution (4.4-3). It is quite good for values of x in the neighborhood of the mean, but becomes very unsatisfactory when $x - \bar{x}$ is large.

4.7. Density Fluctuations. Under suitable conditions, Laplace's formula can be applied to the density fluctuations in a gas. Although it is reasonable to suppose that variations in density should follow the so-called law of errors—and physicists sometimes content themselves with a plausibility argument of this kind—it seems well to outline the logical steps which lead to this conclusion. Our simple analysis will enable us to separate facts from assumptions, which are often intermingled and confused in this connection.

Let us first suppose that a vessel of volume V contains but a single molecule. We select a small element of volume v and make successive observations with the object of determining whether the molecule is in this element or not. The first assumption will be that this sequence of observations forms a probability aggregate. This assumption, by the way, is far from being true in a trivial sense; for it might well be that the molecule passes across the container in a definitely periodic motion, in which case condition II of Sec. 4.2 would not be satisfied. If it is true, as will now be supposed, the probability of observing the molecule in v could be determined by numerous trials. In fact, if this probability, which will be denoted by q, is independent of the particular element of volume selected for observation, it can be calculated. For then q must certainly be proportional to the size of the element v, and it must be 1 if $v = V$. Therefore

$$q = \frac{v}{V}.$$

Now let n observations be made on the volume v. The probability that, in x of them, the molecule was found in v is given by Newton's formula (4.4-3).

The next important assumption now to be made is this: n independent observations on one molecule are equivalent to one observation on n independent molecules. Instead of repeating the observations we are multiplying the object of observation. The assumption amounts to an identification of temporal succession with coexistence in time. It is very common in physical statistical theories and is usually justified by a fundamental theorem known as the ergodic hypothesis. We shall here assume its validity. We are then permitted to state that the probability of observing x of the n molecules in v is given by

$$w_n(x) = \binom{n}{x} p^{n-x} q^x.$$

From here on the work is merely analytic and involves essentially the previous steps necessary to pass from Newton's to Laplace's formula. But it must be remembered that these steps were correct only as long as n was large; not only that, but also np and nq were large, since otherwise certain terms neglected in the previous section could not have been discarded. If nq is to be large, its equivalent nv/V must be large, and this means that the "normal" number of molecules in the volume under observation must be very great. This condition may be expressed by saying that the gas must not be too rare. In that case, Laplace's formula (4.6-3) governs the density distribution. Substituting $q = v/V = 1/k$ and $p = 1 - q$ in (4.6-3), the result may be stated:

$$w(x)dx = \frac{k}{\sqrt{2\pi n(k-1)}} e^{-\frac{(kx-n)^2}{2n(k-1)}} dx, \tag{1}$$

where k is the number of small volumes v which make up the total volume V. The expected mean and dispersion of this distribution must be the same as the corresponding quantities for Bernoulli's distribution to which Laplace's formula approximates. Hence, according to (4.4-5) and (4.4-6),

$$\bar{x} = nq = \frac{n}{k}, \tag{2}$$

and

$$\sigma^2 = npq = \frac{n}{k^2}(k-1) = \bar{x}\left(1 - \frac{\bar{x}}{n}\right). \tag{3}$$

If k is sufficiently large, $\sigma^2 \sim \bar{x}$.

In physical literature it is customary to introduce a quantity δ, known as the "relative fluctuation" and defined by

$$\delta = \frac{x - \bar{x}}{\bar{x}}. \tag{4}$$

Because of (2), δ takes the form

$$\delta = \frac{kx - n}{n}.$$

If we wish to write the fluctuation law in terms of the relative fluctuation, this quantity is to be introduced in (1), which then reads:

$$\left.\begin{aligned} w(\delta)d\delta &= \sqrt{\frac{n}{2\pi(k-1)}}\, e^{-\frac{n\delta^2}{2(k-1)}}\, d\delta \\ &= \sqrt{\frac{\bar{x}}{2\pi p}}\, e^{-\frac{\bar{x}}{2p}\delta^2}\, d\delta. \end{aligned}\right\} \tag{5}$$

These formulas are well supported by experiment. As a matter of application, let it be required to find the probability that the density of molecules in a cube of edge 0.01 mm, considered as part of 1 cc of gas under normal conditions, be 0.01 per cent or more above normal. We have

$$w = \int_{10^{-4}}^{\infty} w(\delta)d\delta = \sqrt{\frac{\bar{x}}{2\pi p}}\int_{10^{-4}}^{\infty} e^{-\frac{\bar{x}}{2p}\delta^2}\, d\delta = \frac{1}{\sqrt{\pi}}\int_{\frac{10^{-4}\sqrt{\bar{x}}}{\sqrt{2p}}}^{\infty} e^{-\nu^2}dy.$$

p in this example is nearly 1, $\bar{x} = \dfrac{n \cdot 10^{-9}\,\text{cc}}{1\,\text{cc}} = 2.7 \times 10^{10}$, since the number of molecules, n, in 1 cc under standard conditions is 2.7×10^{19}. Hence the lower limit of the last integral is 11.6. The value of this integral may be found in tables. The result turns out to be

$$w \sim 10^{-60};$$

hence the occurrence of such a fluctuation is extremely improbable.

4.8. Poisson's Formula. In many cases, Laplace's formula is not a suitable approximation to Newton's, as was already mentioned, for in deriving it, terms were arranged in descending powers of n, and only the highest powers were retained. (Cf. the remarks at the end of Sec. 4.6.) In this procedure p and q were assumed to be numbers of order unity, so that $1/npq$ tends toward 0 for large n. But it happens very often that q, the probability of the elementary event in question, is very small indeed and consequently $1/nq$ may not be neglected. Whether or not this is true can be seen easily by examining the mean value, nq. In the preceding example nq was 2.7×10^{10}, i.e., certainly large enough for its reciprocal to be neglected against 1. Hence the use of Laplace's formula was proper. Considering, on the other hand, the emission of α rays (Sec. 4.5), we shall find the formula to be entirely inapplicable, for it is seen from the table in Sec. 4.5 that $\bar{x}$, or nq, is about 4. It would be equally erroneous to apply the fluctuation law derived in the preceding section to the variation in the density of colloidal particles suspended in a liquid medium if their average number is small. In all these instances we require a different approximation to Newton's formula, namely, one in which nq remains finite while $n \to \infty$. This approximation was first carried out by Poisson, and the result is known as his formula.

Let us assume, to begin with, that $\bar{x} = nq \equiv a$ is a number of order unity. Substituting in Newton's formula

$$q = \frac{a}{n}, \quad p = 1 - \frac{a}{n},$$

we find

$$w_n(x) = \binom{n}{x} q^x p^{n-x} = \frac{n(n-1)\ldots(n-x+1)}{x!} \cdot \frac{a^x}{n^x} \cdot \frac{\left(1 - \frac{a}{n}\right)^n}{\left(1 - \frac{a}{n}\right)^x}$$

$$= \left(1 - \frac{a}{n}\right)^n \cdot \frac{a^x}{x!} \cdot \frac{1\left(1 - \frac{1}{n}\right)\ldots\left(1 - \frac{x-1}{n}\right)}{\left(1 - \frac{a}{n}\right)^x}.$$

x is a finite number. Therefore, as $n \to \infty$, all factors in the numerator and denominator of the last fraction tend to 1. We also recall that

$$\lim_{n \to \infty} \left(1 + \frac{a}{n}\right)^n = e^a.$$

Consequently

$$w(x) = \lim_{n \to \infty} w_n(x) = \frac{a^x e^{-a}}{x!}. \tag{1}$$

This is Poisson's formula. There are many examples of its usefulness. To cite only one, we mention that the observations on the number of α particles emitted in equal periods of time, discussed in Sec. 4.5, conform well to (1).

4.9. Theory of Errors.

It is well known that a large number of measurements on a physical quantity, provided they are not affected by systematic errors, distribute themselves about the "true" value with frequencies given by Gauss's function, variously known as the "error function" or the "normal law." To illustrate the meaning of this remark let us suppose we are measuring a length of a units, say 50 yards, by means of a tape of length 10 yards. The numerous results may range all the way from 49 to 51 yards. We can then divide this range into 20 equal portions, extending from 49.0 to 49.1, 49.1 to 49.2, ... 50.9 to 51.0, and number them $\Delta_1, \Delta_2, \ldots, \Delta_{20}$. Let the number of observations on the length of 50 yards contained in the small range Δ_i whose midpoint is x_i be denoted by n_i. Then if we plot n_i against x_i we obtain a set of points that lie approximately on a curve of the general form

$$n(x) = \text{constant } e^{-h^2(x-a)^2}. \tag{1}$$

To be sure, in most physical experiments the "true value" a is not known. We shall later discuss this point more fully and assume at present that a has been previously determined, perhaps by the use of more accurate devices for measuring lengths. The constant h determines the width of the curve. The greater h, the narrower the distribution curve, i.e., the smaller the spread of the observations. For this reason, h is often called the "measure of precision" of Gauss's distribution. The reader will easily verify that $\sigma^2 = \dfrac{1}{2h^2}$; in other words, the dispersion is inversely proportional to h^2.

The validity of (1) is regarded as being well established experimentally; indeed, a series of measurements at variance with the law is often suspected as being peculiar and untrustworthy. But the very existence of deviations makes it impossible to classify the relation as an empirical law. And if, furthermore, we do not hesitate to use it as a criterion for the quality of observations, we are certainly confronted with the necessity of justifying it theoretically.

If no error affected the measurements in our example every one of them would give the true value a. (This statement requires no modification even in view of quantum theory and its necessary "uncertainties.") Every actual measurement is affected by very many small errors. In our example such errors arise from: lack of precision in the end points of the measured length; loss or gain of space in placing the tape; elastic stretching or contracting of the tape; thermal expansion; parallax in recording coincidences, etc. If the complex organism of the observer is included in the consideration, as it should be, the different sources of error are truly innumerable. It is safe to assume the deviation from the true value caused by each single source to be small, since large ones would undoubtedly betray their causes and be eliminated. But if the individual deviations are small, then the total deviation observed is the *sum* of all the deviations due to the single sources of error.

Let there be n independent sources of error. If, during a series of measurements, one of these sources, say the ith, were active alone, the results would not be distributed about a according to an error law, but have a distribution in $(x - a)$ which is peculiar to this source. Supposing that each source produces a probability aggregate, we can assign to the ith source a function $w_i(x - a)$ representing the distribution of errors which it alone generates. We shall now make a mathematical hypothesis which is equivalent to saying that all errors are accidental. A systematic error is usually understood to cause deviations preponderant in one direction, while an accidental one is just as likely to falsify the result in one sense as in another. On this basis one can define an accidental error as one whose w has a mean value 0, i.e., if $z = x - a$,

$$\int zw(z)dz = 0. \qquad (2)$$

All others will be regarded as systematic. Since we are excluding systematic errors, all our w_i satisfy (2). Otherwise they may differ from one another as much as we like.[1]

Given all $w_i(z)$; what is the probability $w(z)$ of making a total error between specified limits? In different phraseology: given n probabilities of individual errors, $w_1(z_1), w_2(z_2), \ldots, w_n(z_n)$, what is the probability that $z_1 + z_2 + \ldots + z_n$ shall lie, say, between $z - dz/2$ and $z + dz/2$? Clearly,

$$w_1(z_1)w_2(z_2) \ldots w_n(z_n)dz_1dz_2 \ldots dz_n$$

represents the probability of that combination of errors in which the first lies between $z_1 - dz_1/2$ and $z_1 + dz_1/2$, the second between $z_2 - dz_2/2$ and $z_2 + dz_2/2$, etc. If we integrate this expression over all those combinations of $z_1, z_2, \ldots z_n$ which are compatible with the condition

$$z - \frac{dz}{2} \leq z_1 + z_2 + \ldots + z_n \leq z + \frac{dz}{2}$$

---

[1] Of course it is assumed that $\int w_i(z)dz = 1$ for every i.

the desired result is obtained. Hence

$$w(z)dz = \int \ldots \int w_1(z_1)w_2(z_2) \ldots w_n(z_n)dz_1dz_2 \ldots dz_n, \qquad (3)$$

the integration being extended over the range in which

$$z - \frac{dz}{2} \leq \sum_j z_j \leq z + \frac{dz}{2}.$$

To evaluate this integral directly is inconvenient because of the form in which the limits are stated. But a simple artifice is useful. The definite integral

$$\frac{dz}{2\pi} \int_{-\infty}^{\infty} e^{i(z-\rho)x} dx \qquad (4)$$

has the value 1 when $z - dz/2 < \rho < z + dz/2$, and 0 when ρ is outside these limits. This formula is most easily derived from Fourier's integral theorem:

$$f(\rho) = \frac{1}{\pi} \int_{-\infty}^{\infty} dx \int_{-\infty}^{\infty} f(\lambda) \cos x(\lambda - \rho)d\lambda.$$

Let $f(\rho)$ be the function which is zero everywhere except in the range $z - dz/2 < \rho < z + dz/2$ where it takes the value 1. Then the second integration can be performed and yields $\cos x(z - \rho)dz$. Since the cosine is an even function, our result is

$$f(\rho) = \frac{dz}{2\pi} \int_{-\infty}^{\infty} dx \cos x(z - \rho)$$

$$= \frac{dz}{2\pi} \int_{-\infty}^{\infty} e^{i(z-\rho)x} dx.$$

The last step is correct because $\int_{-\infty}^{\infty} \sin ax \, dx = 0$. (The fact that the formula is not valid at the end points of the range does not disturb the use which we are making of it.) If now we identify the ρ in (4) with $\sum_j z_j$, the expression (4) takes on the value 1 precisely in the range over which we desire to integrate in (3) and is zero outside. Therefore, if we multiply the integrand in (3) by the factor (4), and then integrate over *all* values of *all* z_i, we obtain $w(z)dz$. Thus

$$w(z) = \frac{1}{2\pi} \int_{-\infty}^{\infty} \ldots \int_{-\infty}^{\infty} dz_1 dz_2 \ldots dz_n \int_{-\infty}^{\infty} dx \, e^{izx - i\sum_j z_j x} w_1(z_1) \ldots w_n(z_n).$$

The order of the different integrations is evidently immaterial. Defining

$$f_j(x) = \int_{-\infty}^{\infty} w_j(z_j) e^{-iz_j x} dz_j, \qquad (5)$$

we have

$$w(z) = \frac{1}{2\pi} \int_{-\infty}^{\infty} f_1(x) f_2(x) f_3(x) \ldots f_n(x) e^{izx} dx. \qquad (6)$$

We are now ready to make use of an interesting mathematical theorem regarding the limiting value of the product of a great number of functions, which states:

Let $f_1(x), f_2(x), \ldots, f_n(x)$ be a sequence of functions, all of which have a real maximum at the same place $x = a$, where all take the value $f_i(a) = 1$. We further assume that, while $f_i'(a) = 0, f_i''(a) = -r_i^2 < 0$, and that $f_i'''(x)$ is continuous and bounded for all i. It may then be shown that (cf., for instance, von Mises), within a certain domain about $x = a$,

$$\lim_{n \to \infty} [f_1(x) f_2(x) \ldots f_n(x)] = e^{-\frac{s^2}{2}(x-a)^2},$$

s^2 being an abbreviation for $\sum_i r_i^2$. Moreover,

$$\lim_{n \to \infty} \int_{-\infty}^{\infty} f_1(x) f_2(x) \ldots f_n(x) \psi(x) dx = \int_{-\infty}^{\infty} e^{-\frac{s^2}{2}(x-a)^2} \psi(x) dx$$

provided $|\psi(x)|$ is bounded and every $|f(x)|$ vanishes at ∞ sufficiently strongly.

Now it is easily seen that the f_j defined by (5) satisfy these conditions if the w's are probabilities, for we have

$$f_j(0) = \int_{-\infty}^{\infty} w_j(z_j) dz_j = 1,$$

$$f_j'(0) = \int_{-\infty}^{\infty} -iz_j w_j(z_j) dz_j = 0$$

because of (2), and

$$f_j''(0) = \int_{-\infty}^{\infty} -z_j^2 w_j(z_j) dz_j = -r_j^2.$$

The further necessary condition that every f_j vanishes sufficiently strongly at $|x| = \infty$ will be regarded as satisfied. For these reasons we get from (6)

$$\lim_{n \to \infty} w(z) = \frac{1}{2\pi} \int_{-\infty}^{\infty} e^{-\frac{s^2}{2} x^2} e^{izx} dx = \frac{1}{s\sqrt{2\pi}} e^{-\frac{z^2}{2s^2}}.$$

The last step is made by completing the square in the exponent of the integrand and thereafter changing suitably the variable of integration. The substitution $z = x - a$ brings us back to the form (1), where $n(x)$, the "frequency of the result x," reduces to the probability $w(x)$ if it is divided by the total number of observations. Omitting the limit sign, we can write

$$w(x) = \frac{1}{s\sqrt{2\pi}} e^{-\frac{(x-a)^2}{2s^2}}. \tag{7}$$

It will be recalled that a is the "true" value of the measurement. $s^2 = \sum_j r_j^2$, i.e., the sum of the dispersions of all the elementary error distributions. But from (7) we note that s^2 is the dispersion of the resultant Gauss function. The dispersion of the error curve is equal to the sum of all the dispersions of the contributory errors. This fact is of theoretical interest only; it is of no physical consequence since the elementary distributions, or their dispersions, are unknown. s^2 in (7) has to be regarded as a constant whose value is to be taken from experiment in every application of the error law.[1]

So far our considerations have merely satisfied the logical need of justifying our belief in Gauss's law; they have no practical bearing,

---

[1] Aside from the "measure of precision h," physicists frequently use the terms "mean deviation" and "probable error" to indicate the statistical quality of their measurements. These quantities are easily defined on the basis of (7), which, in terms of h, reads

$$w(z) = \frac{h}{\sqrt{\pi}} e^{-h^2 z^2}.$$

The "mean deviation" ϑ is the mean of the absolute values of all errors:

$$\vartheta = \frac{h}{\sqrt{\pi}} \int_{-\infty}^{\infty} |z| e^{-h^2 z^2} dz = \frac{2h}{\sqrt{\pi}} \int_0^{\infty} z e^{-h^2 z^2} = \frac{1}{h\sqrt{\pi}} = \frac{0.561}{h}.$$

The probable error ω is so defined that the range limited by $x - \omega$ and $x + \omega$ contains just half of all the observations. In other words, an error is equally likely to be greater, or to be smaller, than ω. This means

$$\frac{h}{\sqrt{\pi}} \int_{-\omega}^{\omega} e^{-h^2 z^2} dz = \frac{1}{2}; \quad \omega = \frac{0.477}{h}.$$

Integrals of this last type are evaluated by means of tables of the function (known as the Gauss integral)

$$\Phi(y) = \frac{2}{\sqrt{\pi}} \int_0^y e^{-z^2} dz.$$

for in making measurements, we are usually confronted with an entirely different problem: we do *not* know the true value of the measured quantity in advance; we make only a finite, sometimes a small, number of observations and wish to know what result we are to choose as the most probable one. To master this situation, let us propose the following question: given m observations $x_1, x_2, \ldots, x_m$ (and we shall suppose that there is no reason why any one of them should be better than the others), what is the most probable number to assume for the true value of the measured quantity?

If b were the true value, the probability that any one measurement should yield the value x_1 is given by (7) with b substituted for a and x_1 for x. Let us henceforth write this equation in the form (1) and assume h to be unknown, since we are ignorant of the value of s in (7) anyway. The probability of a set of m measurements with specified results $x_1 \ldots x_m$ will then be

$$w(b, x_1, x_2, \ldots, x_m) = Ce^{-h^2[(x_1-b)^2 + (x_2-b)^2 + \ldots + (x_m-b)^2]}. \tag{8}$$

If b had any other value than the one assumed at present, this law should also represent the probability of our particular set of x's. But we may also regard the x's as fixed and take (8) as a statement of the probability that b be the true value when the x's are given. The "most probable true value" will then be that b which makes (8) a maximum. Now

$$\frac{d}{db} w(b, x_1, \ldots, x_m)$$
$$= 2Ch^2[(x_1 - b) + (x_2 - b) + \ldots + (x_m - b)]e^{-h^2[(x_1-b)^2 + \ldots + (x_m-b)^2]},$$

and this is 0 when $b = \sum_j x_j/m$. Closer inspection shows this extremum to be a maximum. We have reached the important conclusion that *the most probable value of a set of measurements is their arithmetical mean.*

Further useful information can be obtained by finding out what the probability (8) at this maximum is. To do this, it is first necessary to determine the constant C from the relation

$$\int w(b, x_1 \ldots x_m) db = 1. \tag{9}$$

The work involved is simplified if we throw the expression $\sum_j (x_j - b)^2$ into a different form. We shall show that it is equal to $\sum_j (x_j - \beta)^2 + m(b - \beta)^2$, where β is the arithmetical mean of the x's. Expanding the last expression, one finds

$$\sum x_j^2 - 2\beta \sum x_j + m\beta^2 + mb^2 - 2mb\beta + m\beta^2.$$

Observing that $\Sigma x_j = m\beta$, this goes over into

$$\Sigma x_j^2 - 2b\Sigma x_j + mb^2 = \Sigma(x_j - b)^2.$$

Introducing this in (8), we have

$$w(b, x_1, x_2, \ldots, x_m) = C'e^{-h^2 m(b-\beta)^2}, \qquad (10)$$

where C' is written for the product of the two constants C and (9) now gives

$$C' \int e^{-h^2 m(b-\beta)^2} db = \frac{C'}{h}\sqrt{\frac{\pi}{m}} = 1; \quad C' = \frac{h}{\sqrt{\pi}} \cdot \sqrt{m}.$$

The probability of the arithmetical mean therefore becomes, according to (10),

$$w(\beta) = \frac{h}{\sqrt{\pi}} \sqrt{m}.$$

The probability that the mean of a series of observations be the true value of the measured quantity grows with the square root of the number of observations.

CHAPTER V

THE STATISTICAL POINT OF VIEW

5.1. Dynamical and Statistical Theories.[1] General dynamical principles, such as Hamilton's or that of least action, have been and can be established only by consideration of material bodies governed by mechanical laws, as has been stressed in Chapter III. Yet there are phenomena, such as those of electrodynamics and optics, which do not appear to be describable in terms of material bodies only, and there is no immediate justification for expecting dynamical laws to be applicable to them. Nevertheless a great number of attempts to apply dynamics to non-dynamical observations have been extraordinarily successful, a fact which requires explanation. It led naturally and convincingly to the supposition that *all* physical phenomena are ultimately mechanical in their nature, that all observable processes may be analyzed into elementary attractions and repulsions between small material particles obeying dynamical laws. This view is known as the mechanistic theory of the universe; it received its clearest formulation in the writings of H. von Helmholtz,[2] who characterized it by the postulate that *all forces in nature are resolvable into central forces acting between all pairs of point masses.* Simple and pleasing as this postulate may be, modern discoveries have shaken the belief in it.

The mechanistic hypothesis as we have here defined it might fail

---

[1] To minimize conflicts between symbols we use in this chapter:

  P for pressure;
  p for momentum;
  V for potential energy;
  v for velocity;
  τ for volume;
  N for number of systems in a Gibbsian ensemble;
  n for total number of degrees of freedom;
  ν for total number of constituents of a thermodynamic body;
  f for frequency.

At a few places, where ambiguities cannot arise, we have used P for a certain type of probability, and N for Avogadro's number.

[2] H. von Helmholtz, "Über die Erhaltung der Kraft," *Wiss. Abh.*, I, p. 12. See English translation, "Popular Lectures on Scientific Subjects" (London, 1873), p. 317.

to be of value in several ways. First, it is conceivable that, while nature does not in any instance directly disprove the mechanistic postulate, the analysis of macroscopic occurrences into elementary dynamical actions, or the composition of large-scale events from microscopic dynamical ones, may be a problem which present mathematical tools are hopelessly inadequate to handle. Again, it might be that such an analysis would require more observational data than the most skillful experimenter is able to supply. Under these circumstances it would be quite legitimate to regard the mechanistic theory as hypothetically valid and to uphold it as a postulate which is beyond immediate refutation. However, as a general guiding idea in the progress of science its usefulness would be impaired. Indeed it would then be necessary to formulate laws in terms foreign to classical dynamical theory. This situation has actually prevailed in physics since the middle of the last century. Thermodynamics and radioactivity refused to be completely understood in terms of the familiar dynamical concepts, and it was only the introduction of an entirely new element—that of probability—into physical reasoning, which has made these new phenomena amenable to theoretical treatment. Our present task will be to ascertain clearly in what respects these non-dynamical theories differ from the dynamical ones. But before we undertake it, let us conclude the present discussion and refer briefly to another possible instance in which mechanistic interpretation would break down. This is the obvious one in which nature presents us with *elementary* phenomena which, quite apart from difficulties of analysis or composition, contradict in themselves the dynamical laws. Physical progress during the last two decades has brought to light a number of such phenomena connected with the origin of radiation and the structure of matter. They will form the object of later discussions (Chapter IX). Here we wish merely to note their effect upon the status of dynamical laws: within their domain, ordinary dynamics, or its bolder offspring, the general mechanistic theory, not only becomes inapplicable or useless, but leads definitely to results contradictory to observation.

Uniqueness of determination vs. chance. In discussing the distinctive features of all dynamical laws and theories one must focus attention first upon their most salient characteristic: their implication of strict determinism. No chance or uncertainty is connected with the predictions of, say, the laws of motion. They take for granted the perfect knowledge of a set of facts, such as the instantaneous positions and velocities of the bodies composing a system, and then state, with a precision far greater than is experimentally obtainable, these positions and velocities at any future time.

The principles of dynamics prescribe not only the behavior of a system at every instant of time but also the precise relations between all parts of the system at any given time. Determination not only extends along the time dimension but acts in a spatial sense as well and there becomes mutual interrelation. For instance, it would be unreasonable to suppose on dynamical grounds that, of a number of atoms in close spatial proximity, each could have an isolated fate. But that is precisely what must be assumed whenever we apply the so-called "law of organic growth," so prominent in many fields of physics and chemistry. To be specific, we shall consider it in connection with radioactive phenomena. The law in question may be stated

$$dN = \pm \lambda N dt, \tag{1}$$

where N is a number of certain things, e.g., disintegrating atoms in our example; t is the time; and λ a quantity independent of N and in the present case also of t. The $+$ sign holds if the process is one of growth, the $-$ sign if it is one of decay, as in radioactivity. It says that, for time intervals of equal length dt, the number of atoms disintegrating is proportional to the number of atoms present. In other words, how many atoms and which atoms will disrupt during the next small fraction of a second is entirely independent of the location and number of atoms disintegrating now. It depends on the total number present. It means that on the *average* each atom is transformed after it has existed for a perfectly definite length of time, e.g., many millions of years in the case of uranium; it means also that the explosion of one atom, which occurs with a violence having no analogue on a macroscopic scale, leaves its neighbor undisturbed until perhaps after millions of years the neighbor succumbs to the same fate. The life of the atoms appears to be regulated only by chance.

Eq. (1) is even outwardly distinct from dynamical laws which carry with them deterministic certainty. It contains no dynamical quantities, such as coordinates, velocities, momenta, etc., either in the direct or in the generalized sense. (Cf. Sec. 3.13.) But it displays very characteristically an integral *number* N as a mark of its *statistical* origin. Like all statistical laws, it has meaning only if N (as well as dN) is large, and may not be applied to elementary phenomena. This is not a defect in any sense of the word, since the lack of applicability arises from the very nature of the law and cannot lead to errors. Again, as will appear more clearly later, all statistical laws or theories refer to large numbers, although these may not

5.1 DYNAMICAL AND STATISTICAL THEORIES

always appear explicitly in equations. Eq. (1), when integrated, yields the well-known law of radioactive decay,

$$N = N_0 e^{-\lambda t}, \qquad (2)$$

λ appearing as the "radioactive constant" of the decaying product. To be sure, the preceding considerations are not a final argument against the mechanistic hypothesis. It may indeed be that dynamical laws exist, but that under their action an aggregate consisting of many such elements, in our case disintegrating atoms, does not reveal their exact nature and can be described more adequately by obliterating every reference to these laws. Whatever the interpretation, a statement of the regularities in the behavior of the aggregate in question can be made more precisely and simply in terms of *probabilities*; whether these probabilities point ultimately to the action of unknown dynamical laws is obscure.

It has been emphasized by Poincaré [1] and others that every phenomenon in nature is capable of a mechanistic explanation, whence dynamical laws alone are sufficient to account for every possible observation. Why, then, is it necessary to introduce statistical laws? The answer is: in order to save the simplicity of physical theories. No one will doubt that an arbitrary occurrence may be produced by purely mechanical elements of extreme smallness if it is permissible to make use of an unlimited number of indetectable mechanical devices, nor will anyone fail to see the undesirable complexity that such a procedure introduces. In our example of radioactive decay; for instance, one might crudely propose that every atom is surrounded by a very high potential "wall" screening from it the influence of others, and that it contains internal motions and a mechanism subject to "wear." To this mechanism may then be assigned a certain lifetime in the same manner as every type of automobile has a fairly definite average lifetime. Imagine further that, at a certain stage during its progressive deterioration, the mechanism breaks down and causes the atom to explode—and we have a mechanistic picture of radioactive decay. To complete it we require, of course, a random distribution of the birth dates of all the atoms, but this again may be produced by mechanistic assumptions. The design of the mechanism within the atom would be an engineering problem capable of solution. It is at once apparent, however, that this fantastic model, regardless of its unreasonableness, calls for a very

[1] "Cf. Foundations of Science" (New York, 1913), pp. 144, 177.

great number of independent hypothetical features, intrinsically without coherence, and joined only by our effort to explain one isolated phenomenon. For one thing, we have no assurance that the properties of the atom here imagined will manifest themselves in other fields of observation. We recognize that the strict adherence to dynamical laws may be wasteful of hypotheses, and contradictory to one of the most powerful principles guiding us in our physical reasoning: that of simplicity.

The element of chance which, as we saw, is involved in the formulation of statistical laws, naturally also affects their operation. Consider, for instance, the second law of thermodynamics, expressing the well-known fact that heat cannot be transferred from a cold to a hot body without expenditure of external work. We are not interested at present in its precise formulation and postpone its analysis to Sec. 5.4. Such an analysis shows that this far-reaching principle, which among statistical pronouncements assumes a rank of the same importance as that of the principle of conservation of energy among dynamical principles, is true only in the overwhelming majority of cases, but not in an absolute sense. It may be shown that the probability of its violation is finite, though very small. Numerical computations indicate that a case in which heat is spontaneously transferred from a colder to a hot body in contradiction to the second law should occur once during an age vastly greater than the estimated age of the universe. A similar statement would apply to violations of the law of radioactive decay and to all other statistical laws. Hence a practically minded critic might object that the distinction we have established between dynamical and statistical principles from the point of view of determinism is an insignificant and artificial one. If, in spite of the indeterministic theoretical basis of a statistical law we may expect with perfect confidence that it will predict the behavior of the universe for an immeasurable time, why then contrast it to the dynamical ones which were characterized as deterministic? The answer is that we are, and should be, concerned not only with the practical consequences of physical theories but also with their metaphysical implications. Dynamical laws imply uniqueness in the succession of phenomena. Any possible deviation from a rigid and fixed course of events, however small, destroys uniqueness *in toto*. It is therefore clear that the distinction here drawn is an essential one, although it becomes diffuse if viewed from a practical standpoint.

There is some danger of confusion at this point; by uniqueness of succession is *not* meant necessity of enforcement. The last concept is of an entirely different order and plays no rôle in physics what-

5.1 DYNAMICAL AND STATISTICAL THEORIES

ever. It implies metaphysical elements which properly belong to the field of ethics and require no discussion here.[1] Furthermore, by uniqueness we do not mean perfect accuracy of prediction. As was discussed in Chapter I, every law is but an approximation. This, however, does not render the foregoing distinction illusory. From this angle, the difference appears in the circumstance that a dynamical law, if found to be in error, is either abandoned as false, or modified to meet the situation more precisely, whereas a statistical law requires in principle no alteration if one of its rare violations should be encountered.

The workings of chance are visually manifest in a phenomenon much cited in this connection, known as the Brownian movement. If minute drops of an opaque liquid are suspended in a transparent one and a portion of the liquid is then examined under a microscope, the small visible drops or particles are seen to execute swift and erratic movements which can be studied in considerable detail. The phenomenon, predicted by the French physicist Gouy, is named after its discoverer Brown, an English botanist, who observed it in 1827. It results from the impacts of the relatively heavy drops with the molecules of the liquid which are in thermal motion. An attempt to calculate the details of the Brownian movement by means of dynamical laws would at once appear hopeless. But the motion possesses regular physical features. There is, for example, a definite dependence of the average speed of the drops upon the temperature and upon their own size; the number of drops having a given velocity is, on the average, a function of that velocity; one also finds a definite average range of the motion, the mean free path, depending on the size of the drops and the temperature of the liquids, and finally one notes a regularity even in the deviations from this mean free path. All these details were worked out by Einstein and Smoluchowski (1905) on the basis of a statistical theory using theorems of probability, and their theory received a striking confirmation through the quantitative experimental work of the French physicist Jean Perrin (1912).

---

[1] The term "necessity" sometimes creeps into physics under the guise of the limit of a probability; i.e., a phenomenon is said to be necessary if its probability of occurrence is unity. This is to be regretted, for although it is true that the only *operational* meaning of the term necessity is identical with certainty (and this is the meaning of "unit probability") it is unwise to deprive the term necessity of its metaphysical implications in fields other than physics. The positivist, who denies such implications, is at liberty to refrain from using the term, as we are doing for the most part in this volume. The open-minded reader will observe that the physicist who speaks of necessity in the physical world is just as careless in his terminology as the moralist who speaks of the force of motivation; both phrases are at best metaphorical.

The statistical nature of some physical propositions is not apparent from the form in which they are occasionally stated. It is instructive to discuss briefly the following three statements, every one of which expresses a recurrent regularity or routine in nature, though it is not a law in the more precise sense of Chapter I. (1) bodies fall from higher to lower levels; (2) heat passes from higher to lower temperatures; (3) electric currents flow from higher to lower potentials. The three statements are so similar that one is frequently inclined to subsume them under one more general principle which assigns an intensity or "level" to every form of energy, which then is supposed to direct its flow. From this point of view, level, temperature, and electric potential would appear to be analogous quantities; the statements would become equivalent except in so far as they refer to different forms of energy

But this view is entirely misleading. The first proposition is evidently a dynamical one; the second and third are statistical in character. The difference is apparent on closer inspection. Bodies *in vacuo* will *always* fall from high to low levels; the statement carries with it a claim to universality and uniqueness which is absent in the other two. In an extremely long time there will presumably occur a case in which heat will pass from lower to higher temperatures, and the electrons in a metal drifting down under the influence of an electric field will, by some very improbable but possible combination of circumstances, derive enough of an impulse in the direction against the field from their impacts with the metallic atoms (or ions) as to be able to progress, on the average, a small distance in opposition to the field.

Another interesting point is noticed when we attempt to resolve the processes described above into their microcosmic constituents. Every part of the falling body is subject to statement (1). Heat, on the other hand, is to be interpreted as the rapid motions of the molecules, and temperature corresponds to the mean kinetic energy of the molecules. If heat flows from high to low temperatures kinetic energy must be transmitted through impacts between the molecules from the swifter to the slower ones. However, although this occurs as far as the whole aggregate of molecules is concerned, it is not true in general for a single impact; it may well be that a swift molecule receives part of the energy of a more slowly moving one. Imagine, for instance, a molecule moving rapidly along the x axis to be struck by another, slower one whose velocity lies entirely along the y axis. Let the centers of the two molecules, at the instant of contact, lie in a line parallel to the y axis. Then, if we treat the molecules as smooth, perfectly elastic spheres and assume, for simplicity, that they have equal masses, conservation of momentum and of energy requires that the slower molecule lose all its kinetic energy while the faster one gains it. Here, every single microcosmic process is not the exact analogue of the macrocosmic one: the latter may be said to represent the *average* result of all elementary impacts.

This is equally true of the metallic conduction of electricity. The motion of an electron is one of thermal agitation with a relatively slow drift in the direction of the field, i.e., from high to low potential, superposed. At a given instant its motion may very well be opposed to the field; the existence of a current results from a slight preponderance of motion along the field; hence it is again an average effect.

Reversibility. Next we come to another important distinction between dynamical and statistical theories. The former entail reversibility; the latter do not. Let us see what this means and how it comes about.

The most general energetical principle governing all dynamical motions is Hamilton's principle (Sec. 3.12). It endows all phenomena which it regulates with the property of reversibility, as will now be shown. Hamilton's principle asserts that, of all hypothetically possible motions, nature realizes the one for which the integral

$$\int_{t_1}^{t_2} (T - V) dt \tag{3}$$

has a stationary value (usually a minimum). The symbols have the significance attached to them in Sec. 3.12. Both T and V are functions of the time implicitly, and V may even depend explicitly on t without disturbing the validity of the principle. By changing the independent variable from t to $-t$ and making the corresponding change in the limits we obtain the identity

$$\int_{t_1}^{t_2} [T(t) - V(t)] dt = \int_{-t_1}^{-t_2} [T(-t) - V(-t)](-dt)$$
$$= \int_{-t_2}^{-t_1} [T(-t) - V(-t)] dt. \tag{4}$$

We shall suppose the motion for which eq. (3) is a minimum to be known and refer to it as the original motion. The integration on the right of eq. (4) yields the same numerical result as eq. (3) but calls for a different interpretation. The motion which it represents will be a dynamically possible one if the original motion was dynamically possible. Interpreting the right-hand side of eq. (4) we find first that it defines a motion lasting from time $-t_2$ to $-t_1$, that is, extending in the proper time direction, for if $t_2 > t_1$, then $-t_1 > -t_2$. The appearance of $-t$ as the argument of T, which is a function of the velocities, reverses the direction of all velocities. But since T is a quadratic function of the velocities, its sign remains unchanged. Finally we notice, most easily by substituting in the integrand all values of the argument from $-t_2$ to $-t_1$ successively, that the motion under consideration starts with the final T and the final V of the original motion, passes through all its intermediate stages in the reverse order, ending with the initial T and V of the original motion. After the completion of the process the system involved is in all respects in the same state as before the beginning of the original motion. By the system is meant, of course, the totality of bodies undergoing changes, so that one might well say that the second motion has restored the initial state of the universe. *All processes, the effects of which can be completely annulled, are called reversible.* Irreversible processes are such that, with the use of all possible physical means, a complete restoration of the initial state *everywhere* cannot be achieved. From the preceding remarks it follows that dynamical laws describe reversible processes, at least in so far as they obey Hamilton's principle. It is thus proper to consider reversibility as a characteristic consequence of dynamical laws.

If the integrand in eq. (3) depended on any odd power of the velocity the argument in the last paragraph would not be applicable. Suppose, for instance, that V contains a term proportional to ds/dt; this term would, after the transformation (4), appear with its sign reversed, and we could no longer correlate the resultant motion with the original one in the manner discussed. Forces proportional to velocities usually occur when the process involves friction, and if it were permissible at all to associate a potential energy with frictional forces these, too, would be proportional to the velocities. This indicates the irreversibility of frictional action.

Strictly speaking, therefore, processes involving friction do not come under the jurisdiction of dynamical laws. Nevertheless, dynamical principles have been formulated which include, or are applicable to, frictional processes. The usual method in dynamics is to introduce into Lagrange's equations a dissipation function, which takes proper care of dissipative resistances such as friction (recall Sec. 3.11). But this should be regarded merely as a *formal* generalization of the dynamical equations, and not as a fundamental extension of dynamical laws to irreversible phenomena.

5.1 DYNAMICAL AND STATISTICAL THEORIES

As examples of reversible processes may be cited all movements of perfectly rigid bodies and of incompressible fluids so far as friction is absent, most notably all periodic motions, e.g., the oscillations of a simple pendulum, or the revolutions of the planets. Returning in this connection to our previous discussion we observe that falling bodies, for instance a liquid in a U-tube, are capable of executing periodic motions, while heat in passing from higher temperatures to lower temperatures never oscillates. Again, an electric current in a resisting medium (without capacity) does not produce oscillations, and is analogous to the passage of heat.

A typical irreversible process is that of heat conduction. This is at once seen to be the case from the second law of thermodynamics which may be considered as established experimentally. For suppose that conduction of heat from one part A of a body initially at temperature T_1 to another part B initially at temperature T_2 has taken place, T_1 being greater than T_2, and that we attempt to reverse this change. As long as A and B are at the same temperature we can transfer heat back from B to A without engaging external agencies, but as soon as an infinitesimal quantity of heat has passed to A the temperature of A will immediately be greater than that of B, and the second law becomes operative. Any further transfer of heat will then necessitate the expenditure of external work, and when A is again at the temperature T_1 and B at T_2, the body itself will be in its initial state but external bodies will have undergone changes. Hence the process in question is irreversible.

Other distinctly irreversible phenomena are the generation of heat by friction, the free expansion of a gas, diffusion, the freezing of an undercooled liquid, the condensation of a supersaturated vapor, and all explosions.

It was noted in our discussion of the probability elements of statistical laws that, if human inventiveness is allowed free play, it is possible to base the uncertainty of these laws upon the action of definite dynamical principles. It is equally possible in this manner to eliminate by hypothesis the irreversibility manifest in the instances cited above. For example, we may properly think of thermal diffusion or equalization of temperatures within a gas as the mixing of molecules having different kinetic energies, retaining the assumption that the motion of every individual molecule obeys purely dynamical laws. Then irreversibility attends only the behavior of the gas as a whole and results from our inability to separate the individuals. A microcosmic being endowed with powers of handling single molecules, such as Maxwell's well-known demon, would be perfectly capable of performing the separation, even without the expenditure of work. But the assignment of dynamical laws together with detailed statements of the individual fates of all the molecules would introduce irrelevant features into the theory and would, moreover, be arbitrary within certain limits; it would be out of harmony with the simplicity of the laws observed to govern the changes of the gas as a whole, which imply irreversibility. We conclude, therefore, that even though there is no logical necessity for the distinction between reversible and irreversible phenomena, in order to maintain simplicity of description we are forced to adhere to it.

But there is another objection to the distinction we have drawn. Perfectly reversible phenomena do not exist in nature. It is an idealization when we speak, for instance, of dynamical processes without friction. Transformation of kinetic energy into heat is always present. Even the most nearly periodic motion we know of, the revolution of the planets in their orbits, is accompanied by the dissipative phenomenon of the tides. Should we not, in view of this situation, wipe out the line we have just drawn and consider the existence of irreversible processes only? If we did, we should fail to take cognizance of another essential difference between dynamical and statistical occurrences. Let us go back to consider the two statements: (1) bodies fall from higher to lower levels; (2) heat passes from higher to lower temperatures, the first of which was characterized as dynamical, the second as statistical. Drop a body from a given height to the ground; on impact it will lose part of its kinetic energy, and if it rebounds, it will not reach the level from which it is dropped. The process is irreversible. Heat conduction is always irreversible, as was shown. Now select another, more elastic body and repeat the first procedure, maintaining the same level. The process will be seen to be more nearly reversible. If, however, different bodies are used in connection with the conduction of heat there is, as long as the temperature difference remains the same, no change in the degree to which the phenomenon is irreversible; there is involved a natural minimum amount of unavailable energy which cannot be eliminated. Hence in processes governed by dynamical laws one may attain reversibility progressively as a limit by a suitable selection of circumstances, whereas in statistical phenomena we encounter a natural limitation in the attempt of making them reversible. This consideration justifies the distinction.

There is indeed an important artifice, much used in the theory of thermodynamics, by means of which many statistical phenomena may be conceived as reversible ones. As was pointed out, dissipative

5.1 DYNAMICAL AND STATISTICAL THEORIES

action is closely connected with velocities; it clearly vanishes for zero velocities. Consequently one would expect that if any mechanical process can proceed with infinite slowness it will be reversible. Now mechanical processes always involve, by their very nature, a transformation of kinetic energy into potential energy or the converse, which is impossible if one assigns to the velocities an arbitrary lower limit. However, though the device is inapplicable there, it is quite useful in discussing thermodynamic changes, which may be imagined to occur as slowly as desired.

Consider the free expansion of a gas which takes place in a finite time and is irreversible. Initially the gas has a pressure P_1 and volume τ_1, finally a pressure P_2 and volume τ_2; $P_1 > P_2$, and $\tau_1 < \tau_2$. The same change may be brought about reversibly in an infinite variety of ways, of which the following is one. Let the gas be enclosed in a cylinder with a movable piston of area A, and suppose the whole enclosure to be impermeable to heat. The gas will exert a force $P_1 A$ upon the piston, and if a force of equal magnitude acts upon the other face of the piston no change will occur. Now let this latter force be diminished by an infinitesimal amount. The result will be an infinitely slow expansion of the gas and a similar outward motion of the piston until the pressure in the cylinder has decreased sufficiently to be balanced by the outside force. Next, reduce this force again by an infinitesimal decrement and permit the process of extremely slow expansion to continue. Repeat this procedure an infinite number of times until the final volume τ_2 is reached. This process of expansion has been reversible, for an infinitely small increase in the outside force at every instant would have caused the opposite change. More explicitly, if the gas were to be restored to its initial state the outside force would have to do the work

$$\int (F + \Delta F) dx,$$

where $F = PA$ at every instant and ΔF is infinitesimal and a constant, the integration extending from the largest to the smallest displacement of the piston. But during the expansion the external agency doing this work has received and may be conceived to have stored up an amount of energy $\int (F - \Delta F) dx$, which differs from the energy spent during the restoration of the initial state by the amount

$$2 \int \Delta F dx = \Delta F \cdot d,$$

where d is simply the total displacement of the piston which is finite, so that the difference can be made as small as is desired by making ΔF

sufficiently small. This means, however, that the final state of all bodies is, except for infinitesimal changes, the same as the initial state.

The condition of the gas at this stage is not the one in which it was found at the end of the irreversible process of free expansion. When the volume is τ_2, the average kinetic energy of the molecules, and hence the temperature, must be lower, for the gas has done external work. To arrive at the state of the gas, prevailing after the *irreversible* free expansion which we set out to bring about by reversible changes, it is necessary to add heat to the gas. This can also be done reversibly. If the temperature is at present T, one may bring the gas in thermal contact with a body at temperature $T + \Delta T$, perhaps by removing part of the impermeable cylinder wall. Then, if ΔT is infinitesimal, a very slow transfer of heat, reversed on changing ΔT into $-\Delta T$, will take place. After equalization of temperatures, contact may be made with a slightly warmer body and so on until the temperature corresponding to P_2 at the volume τ_2, which during this last procedure is maintained constant, is attained. This, to be sure, will again require an infinite time but is, in principle, a reversible transformation. Thus we have linked the states $(P_1\tau_1)$ and $(P_2\tau_2)$, which in the process of free expansion were connected irreversibly, by means of a hypothetical reversible transformation.

A body which has no tendency to change its thermal state is said to be in thermal equilibrium. This obtains in general if a certain relation between p, v, and T, known as the equation of state, is satisfied. The infinitely slow reversible process just outlined is essentially a succession of equilibrium states. This is true of all reversible thermodynamic processes and shows at once the necessity of infinite slowness. During free expansion, on the other hand, the gas passes through a sequence of states in which there is no equilibrium and which do not obey the equation of state. A similar statement holds for all other irreversible thermodynamic phenomena.

In theoretical considerations it is frequently immaterial whether a change occurs irreversibly in a finite time, as it always does in nature, or infinitely slowly in the manner just discussed, provided that the initial and the final states are identical. In these cases substitution of the ideal reversible for the real irreversible process is very useful. But there are changes which do not yield to this device and which may therefore be termed intrinsically irreversible; they include heat conduction or radiation at a finite temperature difference, generation of heat by friction, and diffusion.

Reversible changes are often defined as changes which can be undone, at every state, by an infinitesimal increment or decrement in

5.2   THE DYNAMICAL THEORY OF ELECTRIC CURRENTS

one parameter (ΔF or ΔT in our example of reversible expansion), and which therefore proceed with infinite slowness. It is clear that this definition does not cover the most important class of reversible phenomena, namely dynamical ones, and that it is a criterion applicable only to thermodynamics.

Reviewing the situation as a whole we are led to two conclusions regarding the nature of dynamical and statistical laws, or theories: (1) Dynamical laws though not necessarily restricted to the field of mechanics are strictly deterministic and operate with absolute certainty; statistical laws are capable of establishing probabilities only, they hold for a great number of individuals and lose their meaning if applied to a small number of them. (2) Dynamical laws describe reversible processes; statistical laws deal with irreversible phenomena, except so far as they can be theoretically converted into reversible ones by the artifice of allowing them an infinite time for their completion.

After these general considerations it seems well to illustrate the detailed properties of the two antithetical types of physical theories by discussing a few of them. Dynamical principles have already been derived (Chapter III), and applications to the macroscopic motions of bodies have been considered. Let us now extend them to fields other than mechanics.

5.2. The Dynamical Theory of Electric Currents. In presenting a dynamical theory of electric currents we are not proposing to explain their nature in mechanistic terms. We shall merely show that with due generalization of dynamical terms the formal laws which imply determinism and reversibility describe certain features of electric currents.

It is most convenient to use as a starting point Lagrange's equations in the general form for a holonomic system, namely,

$$\frac{d}{dt}\frac{\partial L}{\partial \dot{q}_r} - \frac{\partial L}{\partial q_r} = Q_r \tag{1}$$

already derived in Sec. 3.13 (eq. 3.13–27). The number of equations equals the number of generalized coordinates q_r. We recall that L is the so-called Lagrangian function and stands for $T - V$, T being the kinetic and V the potential energy. Q_r is the generalized *external* force associated with the coordinate q_r; it can always be identified by its physical dimensions: energy/q_r. Eq. (1) will be the basis for all considerations of this section. Hence, if it is possible to assign to a system a definite Lagrangian function depending upon generalized coordinates and velocities, its behavior may be described in dynamical terms. Now whereas the total energy $W = T + V$ may be experimentally determined in terms of dynamical variables, one is able to form L only if it is known which part of the energy is kinetic and which part is potential. Furthermore, there may be an ambiguity with regard to the interpretation of the dynamical variables, that is, it may not be evident which of them are to be regarded as generalized velocities and which as generalized coordinates. It happens, however, that in our simple examples they can be easily identified and L can be determined.

Faraday's law of induction. We begin by deducing Faraday's law of electromagnetic induction. Let us consider any number of closed circuits numbered consecutively, carrying currents of magnitudes i_1, i_2, etc. It will be recalled from the elementary theory of electromagnetism that the flux or the total number of tubes of induction through any one circuit at a given time depends linearly on all the currents, so that, for instance, the flux through circuit 1 is

$$N_1 = M_{11}i_1 + M_{12}i_2 + M_{13}i_3 + \ldots$$

The quantities M_{11}, etc., are known as coefficients of induction, M_{1i}, for example, being the flux through circuit 1 produced by current of unit magnitude in the ith circuit. M's with different indices are clearly coefficients of *mutual induction*; those having equal indices are known as coefficients of *self-induction*. Each M_{ij} will in general be a function of the position coordinates of the ith and of the jth circuit; M_{ii} will depend on the coordinates of the various parts of the ith circuit only and will be a constant if this circuit is rigid. Moreover

$$M_{ij} = M_{ji}. \tag{2}$$

The dependence of M on the geometrical coordinates may be indicated by writing it as $M(x)$, where by x we mean any or all position coordinates. We thus have in general

$$N_s = \sum_r M_{sr}(x)i_r. \tag{3}$$

Now it may be shown that the electrical energy of a system of currents is given by [1]

$$W_e = \tfrac{1}{2}(N_1 i_1 + N_2 i_2 + \ldots) = \tfrac{1}{2} \sum_s N_s i_s. \tag{4}$$

By substitution of eq. (3) this becomes

$$W_e = \tfrac{1}{2} \sum_{r,s} M_{sr}(x) i_r i_s. \tag{5}$$

[1] See, for example, Page and Adams, "Principles of Electricity" (D. Van Nostrand Co., N. Y., 1931), p. 366.

5.2 THE DYNAMICAL THEORY OF ELECTRIC CURRENTS

If the conductors are allowed to move freely there will be an additional energy of a mechanical nature which we may write symbolically with the previous conventions

$$W_m = \tfrac{1}{2} \sum_{r,s} a_{rs}(x)\dot{x}_r \dot{x}_s, \qquad (6)$$

the $\dot{x}$'s being time rates of coordinates (distances or angles) and the a's quantities depending on the coordinates. The total energy will be the sum of eqs. (5) and (6). In order to form L we must know whether to interpret W_e as potential or as kinetic energy; W_m is certainly kinetic. The decision with regard to W_e will depend on the interpretation of the current i. To be sure, the current may be a generalized coordinate. On the other hand, it is the time derivative of the charge q and hence may be taken as a generalized velocity. Then the coordinate q would not appear in W and would be a cyclic one (recall Sec. 3.13). We note another point in favor of this latter choice: W_e has the characteristic quadratic form of the kinetic energy expression. Our expectation is strengthened by observing the following circumstance: a coordinate may change its value in the absence of external forces; a velocity must remain constant. But a current cannot change as long as no electromotive force is impressed upon the circuit, hence it is more naturally regarded as a *velocity*. It is important to remember that we are not including dissipative effects in our discussion as long as we are dealing with strictly dynamical laws, and hence are at present confining our attention to what may be called perfect conductors.

We assume therefore that both W_e and W_m are kinetic energies, so that $V = 0$. Hence

$$L = T = \tfrac{1}{2} \sum_{r,s} M_{rs}(x) \dot{q}_r \dot{q}_s + \tfrac{1}{2} \sum_{r,s} a_{rs}(x) \dot{x}_r \dot{x}_s \qquad (7)$$

in our present example. Writing down Lagrange's equations (1) for the charge coordinate q_i we find

$$\frac{d}{dt}\left(\frac{\partial T}{\partial \dot{q}_i}\right) - \frac{\partial T}{\partial q_i} = E_i, \qquad (8)$$

where we have substituted E_i for the generalized force Q_i because the quotient energy/charge has the dimensions of electromotive force. Differentiating eq. (7) and remembering eq. (2) we obtain

$$\frac{\partial T}{\partial \dot{q}_i} = \frac{1}{2} \sum_r M_{ri}(x) \dot{q}_r + \frac{1}{2} \sum_s M_{is}(x) \dot{q}_s = \sum_s M_{is}(x) \dot{q}_s, \qquad (9)$$

which, by eq. (3) is just N_i, the total flux through the ith circuit.

According to the general definition of Sec. 3.13, $\partial T/\partial \dot{q}_i$ is the generalized momentum associated with the coordinate q_i. For this reason Maxwell called the right-hand side of eq. (9), or N_i, the *electrokinetic momentum* of the ith circuit. The term $\partial T/\partial q_i$ is evidently zero, so that eq. (8) reduces simply to

$$\frac{d}{dt}(N_i) = E_i, \qquad (10)$$

which is *Faraday's law for a perfect conductor*. Its more usual form

$$iR = E - \frac{dN}{dt} \qquad (11)$$

results if we suppose that not all the external electromotive force E is available for balancing the inductive effect, but only the part $E - iR$ which is not spent in overcoming the resistance R, so that this difference is to be substituted in eq. (10) in place of E. Since the result is true for every circuit it is permissible to omit the index i in eq. (11).

Mechanical force acting on a circuit. To obtain the external mechanical force acting upon the ith circuit and tending to change the coordinate x_i (there may be a number of x_i's, since more than one coordinate is usually required to fix the position of the ith circuit, but we are here choosing a particular one) we write down Lagrange's equations for x_i. The Q of eq. (1) will now be a force in the proper sense if x_i is a distance and a torque if it is an angle. In either case we shall denote it by F. Eq. (1) now reads:

$$\frac{d}{dt}\left\{\frac{\partial}{\partial \dot{x}_i}\left(\frac{1}{2}\sum_{rs} M_{rs}(x)\dot{q}_r\dot{q}_s\right)\right\} + \frac{d}{dt}\left\{\frac{\partial}{\partial \dot{x}_i}\left(\frac{1}{2}\sum_{rs} a_{rs}(x)\dot{x}_r\dot{x}_s\right)\right\}$$

$$- \frac{1}{2}\frac{\partial}{\partial x_i}\left(\frac{1}{2}\sum_{rs} M_{rs}(x)\dot{q}_r\dot{q}_s\right) - \frac{1}{2}\frac{\partial}{\partial x_i}\left(\frac{1}{2}\sum_{rs} a_{rs}(x)\dot{x}_r\dot{x}_s\right) = F_i. \qquad (12)$$

The first term on the left of this equation vanishes since none of the quantities in the parenthesis is a function of $\dot{x}_i$; the second term represents the rate of change of the mechanical momentum corresponding to x_i. We now assume that all circuits are at rest and have no acceleration. In this case, the second term on the left as well as the fourth is zero, because the result of the differentiation will still contain the velocities which are zero. F_i is now the external force required to

5.2 THE DYNAMICAL THEORY OF ELECTRIC CURRENTS

maintain the state of rest and is equal and opposite to the force due to the currents themselves. Thus finally

$$-F_i = \frac{1}{2} \sum_{rs} \dot{q}_r \dot{q}_s \frac{\partial}{\partial x_i} M_{rs}(x). \tag{13}$$

Let us illustrate this formula by applying it to the interaction of two non-deformable conductors. Since

$$M_{11} = \text{const.}, \quad M_{22} = \text{const.}, \quad M_{12} = M_{21},$$

eq. (13) reduces to

$$-F_1 = \frac{1}{2}\left(\frac{\partial M_{12}}{\partial x_1}\dot{q}_1\dot{q}_2 + \frac{\partial M_{21}}{\partial x_1}\dot{q}_2\dot{q}_1\right) = \frac{\partial M_{12}}{\partial x_1}\dot{q}_1\dot{q}_2, \tag{14}$$

and a similar relation for $-F_2$. To specialize even further, we can compute the force between two parallel long straight wires each carrying a current i, and having its return circuit far enough away to be neglected. Placing the first wire in the y axis and the second one a distance x from it, the force on the second is, according to eq. (14),

$$\frac{\partial M_{12}}{\partial x} i_1 i_2.$$

M_{12} is, by definition, the flux through the second circuit per unit current in the first. But unit current sets up a field H, i.e., a flux per unit area (in air), of approximate magnitude $2/r$, r being the distance of the point in question from the current; hence

$$M_{12} = \int_x^\infty H \cdot l dr = 2l \int_x^\infty \frac{dr}{r},$$

where l is the length of each wire. Therefore

$$\frac{\partial M_{12}}{\partial x} = -\frac{2l}{x},$$

and the external force becomes

$$F = \frac{2l i_1 i_2}{x},$$

which is the elementary formula for the force between two parallel straight conductors. The force is an attraction or a repulsion according as i_1 and i_2 are of the same or opposite sign.

Electric oscillations. In the preceding examples the conductors have not been assumed to possess electrical capacity. Let us now apply our dynamical theory to a single circuit with a capacity C and a coefficient of self-induction M. In addition to the kinetic energy $\frac{1}{2}M\dot{q}^2$ there will then be the electrostatic energy arising from an accumulation of charges. This is known to be $\frac{1}{2}q^2/C$. The total energy is therefore

$$W = \frac{1}{2} M\dot{q}^2 + \frac{1}{2}\frac{q^2}{C}. \tag{15}$$

Having established that $\dot{q}$ is to be treated as a generalized velocity, and consequently the first term as T, it is obvious that q must be a coordinate and the second term, therefore, V. This makes

$$L = \frac{1}{2} M\dot{q}^2 - \frac{1}{2}\frac{q^2}{C},$$

and Lagrange's equation becomes

$$M\ddot{q} + \frac{q}{C} = 0, \tag{16}$$

in the absence of an external electromotive force. As is well known, the solution of this equation represents oscillations of period $2\pi\sqrt{MC}$.

Many other interesting results can be obtained by the use of Lagrange's equations in electrodynamic theory; however, since our chief concern is a discussion and illustration of the dynamical method we conclude our account of electric phenomena somewhat prematurely and consider now in what manner, and to what extent, dynamical theory may be applied to processes that are even further removed from dynamics proper than the examples of this section.

5.3. Application of Dynamical Theory of Thermodynamics. Thermodynamics describes a group of physical observations which require almost exclusively statistical explanations. We should expect, therefore, that the dynamical method of working with a Lagrangian function in the ordinary manner must fail. Consider a system that is typical for thermodynamic investigations, namely, a given volume filled with the molecules of a gas. It is quite possible to write down an expression for the total kinetic energy of this system in terms of dynamical variables, that is, in terms of the speeds of all individual molecules. An accurate expression for the potential energy would be more difficult to obtain if we did not limit the consideration to *ideal* gases; for the molecule of a *real* gas is known to be surrounded by a rather complex field of force. The potential energy of every pair

5.3   DYNAMICAL THEORY OF THERMODYNAMICS

of molecules of an ideal gas, on the other hand, is zero when the pair is separated and becomes infinite suddenly when the molecules come in contact. It is seen, then, that nothing prevents us from obtaining the T and V appearing in the Lagrangian function at least for an ideal gas, and approximately also for a real one.

However, a difficulty is immediately encountered when, after having constructed L, we continue to apply strictly the dynamical method. For, if L is a function of the microscopic properties of the molecules composing the gas, taking the variation of L is physically meaningless, since we are unable to manipulate single molecules. In other words, L is expressed in terms of *uncontrollable* coordinates. As a result, Lagrange's equations obtained by such a method would lack physical verification and would be useless.

Another more formal objection may be raised against this procedure. The L we have formed depends on the number of molecules in the gas, since every added molecule introduces at least three new positional coordinates and three corresponding momenta. But we know that the laws of thermodynamics do not depend on the number of molecules in an essential manner. Hence it is clear that the procedure in question places emphasis on the physically insignificant features of thermodynamic problems and introduces irrelevant elements into the calculations.

The average Lagrangian function. If we are to make any progress at all by applying dynamical methods we must avoid the use of an L containing uncontrollable coordinates and attempt to use instead a quantity which is a function of measurable physical coordinates, such as temperature, volume, pressure, etc. This is indeed possible, as will now be shown. The quantity

$$\int_{t_1}^{t_2} L\,dt,$$

as we have already seen (Sec. 3.12), is the *time average* of L multiplied by $t_2 - t_1$. Let us denote this time average by $\bar{L}$. This quantity will not depend on the coordinates and velocities of the individual molecules if the average is taken over a time that is large compared with the time characteristics of the motion of the molecules. But the molecules are known to move very swiftly, and the average time between their collisions is of the order of 3×10^{-10} sec (in air under normal conditions). Hence, if the state of the gas is steady and we consider averages over ordinary measurable intervals of time, $\bar{L}$ may be expected to be a function of measurable quantities characterizing the state of the gas. Using the principle of varying action we may now write (eq. 3.13–29)

$$\delta \overline{L} = \frac{\left[\sum_r \frac{\partial T}{\partial \dot{q}_r} \delta q_r\right]_{t_1}^{t_2}}{t_2 - t_1}. \tag{1}$$

Clearly, provided that the numerator on the right of eq. (1) does not increase beyond bounds as the interval of time $t_2 - t_1$ is made longer, the right-hand side can be made as small as is desired if the time interval is chosen sufficiently long. As a matter of fact, our premise holds, and the behavior of the gas obeys the simple equation

$$\delta \overline{L} = 0. \tag{2}$$

The existence of a function $\overline{L}$ whose variation vanishes for every small change from equilibrium was shown in a more detailed and rigorous manner by J. J. Thomson.[1]

The next step will be to construct $\overline{L}$. It is evident that the direct procedure of setting up T and V and then averaging will be complicated and of no avail, so that an indirect and more empirical method must be used. By Lagrange's equations, $\partial L / \partial q$ is the force corresponding to the generalized coordinate q produced by the system. Hence, when the system is in a steady state and q is an external coordinate, $\partial \overline{L} / \partial q$ must be the average force exerted by the system and tending to effect a change in q. There may, in general, be a number of external coordinates q_i satisfying such a relation, so that

$$\frac{\partial \overline{L}}{\partial q_i} = \overline{Q}_i \ldots (i = 1, 2, 3 \ldots), \tag{3}$$

Q_i being the generalized force *exerted by the system* (the negative of the force required to maintain equilibrium) and corresponding to q_i. The average Lagrangian function is then given by

$$\overline{L} = \sum_i \int \overline{Q}_i dq_i - f. \tag{4}$$

It is necessary, of course, to introduce a term $-f$, which does not depend on the external coordinates q_i but may result from the internal structure of the molecules. f is of the nature of a potential energy and must therefore be subtracted.

Let us illustrate this reasoning by applying it to the simplest

[1] J. J. Thomson, "Applications of Dynamics to Physics and Chemistry" (Macmillan and Co., 1888), p. 140.

possible case, that of an ideal gas, satisfying Boyle's law, i.e., with equation of state

$$P\tau = NkT, \tag{5}$$

where P is the pressure, τ the volume, k Boltzmann's constant (1.37×10^{-16} erg/°C) and T the absolute temperature. If we let τ be the volume of one mol of the gas, then N is Avogadro's number, 6.064×10^{23}.

Suppose the gas is confined in a cylinder closed by means of a movable piston, and let the distance from the base of the cylinder to the piston be x. If A is the area of the piston, there exists an average force tending to increase x of magnitude AP. Consequently by eq. (3)

$$\frac{\partial \overline{L}}{\partial x} = AP.$$

But since $A = \dfrac{d\tau}{dx}$ and P can be expressed in terms of τ and T by means of (5), there results on substitution

$$\frac{\partial \overline{L}}{\partial x} = \frac{NkT}{\tau} \frac{\partial \tau}{\partial x}, \tag{6}$$

an equation which if integrated gives the part of $\overline{L}$ that depends on the external coordinate τ. To this part we must add, according to (4), parts depending on other external quantities with which, however, we shall not be concerned, and finally subtract f. Eq. (6) integrates into

$$\overline{L} = NkT \log \frac{\tau}{\tau_0} + F(T),$$

where τ_0 is an arbitrary constant and $F(T)$ an unknown function of the temperature. Some interesting information may be obtained even if we do not determine the way in which $\overline{L}$ depends on T other than through the first term of this last expression.

Every molecule contains a certain amount of internal energy. Suppose, for instance, that it is made up of 2 atoms and that the dissociation energy is ε. This means that an amount of work equal to ε ergs must be done upon the molecule to separate its constituent 2 atoms. We may arbitrarily assume the state in which all molecules are dissociated as the state of zero potential energy. Then the state in which $2N$ atoms form N molecules, as in our example, will represent a potential energy $-N\varepsilon$ ergs. Consequently $f = -N\varepsilon$, and eq. (4) becomes

$$\overline{L} = NkT \log \tau/\tau_0 + F(T) + N\varepsilon.$$

If we consider m mols of gas instead of one, we have

$$\overline{L} = mNkT \log \tau/\tau_0 + mF(T) + mN\varepsilon. \qquad (7)$$

Reviewing the path which led to eq. (7) one cannot fail to observe that the dynamical method, so forceful and elegant when applied to mechanical and some electrical problems, has become somewhat indirect and inadequate. It is no longer possible to construct a Lagrangian function by very general and independent considerations; we are compelled to appeal to Boyle's law in order to make progress. The most interesting feature of the analysis, however, is the appearance of a bar over the L. We are definitely dealing with averages and are not referring to the instantaneous state of the system. Our method is beginning to exhibit a statistical character, and the epithet dynamical, usually applied to it, should not be understood too literally. It merely emphasizes, perhaps excessively, the formal similarity between the procedure here pursued and the rigorously dynamical method previously outlined.

Eq. (2) represents an interesting analogy to the well-known principle that the thermodynamical quantity called *entropy* is a maximum for a steady state. This may be expressed by writing

$$\delta S = 0, \qquad (8)$$

if S denotes entropy. The meaning of S will become clearer in the subsequent discussion of statistical mechanics. In this connection it is important to note that an extremum principle of the type (8) exists. Now anyone familiar with thermodynamic calculations will observe a close resemblance between the average Lagrangian function (7) and the usual expression for the entropy (cf. 5.4) of an ideal gas. Neglecting the last term in eq. (7) this latter expression is simply

$$S = \frac{\overline{L}}{T}. \qquad (9)$$

But as long as a gas is in a steady state the temperature T is not subject to variation; hence eqs. (2) and (8) are equivalent. This equivalence may be looked upon as establishing a justification *a posteriori* for the use of eq. (2) at least in problems relating to ideal gases. Although it is very useful and suggestive to extend the analogy which

5.3 DYNAMICAL THEORY OF THERMODYNAMICS

we have obtained for a special case, its universality appears dubious, particularly in view of the limitations peculiar to the so-called dynamical method. There have in fact been far-reaching attempts to show that the second law of thermodynamics is a consequence of the principle of least action; but these proofs seem hardly satisfactory. The dominating attitude of physicists at present is to regard the second law as a primary principle verified by innumerable observations without appeal to dynamical theories, and to concede to the concept of entropy a statistical meaning independent of the Lagrangian function.

Dissociation. Let us now, by way of further illustration, use eq. (2) to gain information concerning the phenomenon of dissociation, i.e., the splitting up of molecules into atoms. To be specific, we will consider iodine vapor at a given absolute temperature T. Part of the constituents of the vapor will be in the molecular form I_2, part will have dissociated and exist in the atomic state I. We require to know what fraction of the total number of the molecules is dissociated in the steady state.

Let the vapor be contained in a closed vessel of volume τ. We assume that the vapor is sufficiently rarefied so that it may be supposed to behave like an ideal gas. This means that τ must be large. For the sake of convenience we shall make some minor changes in eq. (7). The τ appearing there was *specific* volume, i.e., volume per mol, and is equivalent to τ/m in our present notation. Furthermore, if N is the number of molecules per mol (= Avogadro's number), $mN = \nu$, the total number of molecules. Finally, we write in place of $F(T)$ in eq. (7) $NF'(T)$.

The average Lagrangian of our composite assembly will be made up of the contribution $\bar{L}_1$ due to ν_1 I_2 molecules and $\bar{L}_2$ due to ν_2 I atoms. Thus

$$\left.\begin{aligned}\bar{L} &= \bar{L}_1 + \bar{L}_2, \\ \bar{L}_1 &= \nu_1 kT \log \frac{\tau N}{\nu_1 \tau_0} + \nu_1 F_1'(T) + \nu_1 \varepsilon, \\ \bar{L}_2 &= \nu_2 kT \log \frac{\tau N}{\nu_2 \tau'_0} + \nu_2 F_2'(T).\end{aligned}\right\} \quad (10)$$

As was pointed out before, ε is the dissociation energy of the I_2 molecule. The corresponding term in $\bar{L}_2$ is missing because the constituents in this case are atoms which do not dissociate under conditions which we are considering. τ_0 and τ'_0 are unknown, but in general different, constants.

Let us now vary $\bar{L}$ by changing the numbers ν_1 and ν_2, maintaining the same volume and the same temperature during the variation, noting that ν_1 and ν_2 cannot be changed independently; if ν is the constant total number of I atoms either in combination or free, then

$$2\nu_1 + \nu_2 = \nu,$$

so that

$$2\delta\nu_1 + \delta\nu_2 = 0. \quad (11)$$

But

$$\delta\bar{L} = \frac{\partial \bar{L}}{\partial \nu_1} \delta\nu_1 + \frac{\partial \bar{L}}{\partial \nu_2} \delta\nu_2 = \left(\frac{\partial \bar{L}_1}{\partial \nu_1} - 2\frac{\partial \bar{L}_2}{\partial \nu_2}\right) \delta\nu_1,$$

according to eq. (11). Now eq. (2) requires this to be zero; therefore, since $\delta\nu_1$ is arbitrary,

$$\frac{\partial \bar{L}_1}{\partial \nu_1} - 2\frac{\partial \bar{L}_2}{\partial \nu_2} = 0.$$

Substituting eq. (10) and writing $\frac{N}{\tau_0} = \alpha$, $\frac{N}{\tau'_0} = \alpha'$, $\frac{\alpha}{\alpha'^2} = \beta$, we obt

$$kT \log \frac{\alpha\tau}{\nu_1} - kT + F_1'(T) + \varepsilon - 2kT \log \frac{\alpha'\tau}{\nu_2} + 2kT - F_2'(T) = 0,$$

$$\log \frac{\beta\nu_2^2}{\tau\nu_1} + 1 + \frac{F_1'(T) - F_2'(T)}{kT} = -\frac{\varepsilon}{kT},$$

and this becomes

$$\frac{\nu_2^2}{\tau\nu_1} = \phi(T)e^{-\frac{\varepsilon}{kT}}, \qquad (12)$$

where $\phi(T)$ is an undetermined function of the temperature. If we desire to express the left-hand side in terms of numbers per unit volume, c, instead of total numbers of molecules, ν, it reduces to c_2^2/c_1; for $\nu = c\tau$. Eq. (12) is known to chemists as the *reaction isochore*. It is of some interest even in its present indefinite form, which does not state the exact manner in which dissociation depends on the temperature. Evidently, the greater the right-hand side the greater the degree of dissociation. Hence we see at once that of two substances at the same temperature the one having the smaller dissociation energy will be more highly dissociated. It is possible to determine the form of $\phi(T)$ by other considerations; $\phi(T)$ turns out to be proportional to some small finite power of T. Hence it varies with the temperature much more slowly than the factor $e^{-\varepsilon/kT}$. But $e^{-\varepsilon/kT}$ is very small for values of $T \ll \frac{\varepsilon}{k}$ and becomes appreciable when $T \sim \frac{\varepsilon}{k}$. This shows that in general dissociation sets in to an appreciable degree at a temperature comparable to ε/k. All these conditions have been verified experimentally in a more precise quantitative way.

5.4. The Fundamental Facts of Thermodynamics. In the last section we have shown how it is possible to understand formally some of the facts of thermodynamics by extending the principles of dynamics. The success of such an extension hardly appears very substantial, and if it were continued very much farther, the analysis would become very artificial and finally fail. The purpose of the remainder of this chapter is to demonstrate how naturally statistical reasoning lends itself to an elucidation of thermodynamical phenomena and to outline the most successful theories of statistical mechanics. But before embarking on this project, let us pause to review the principal data to be explained.

Thermodynamics is predominantly an empirical science based upon axioms of experimentally established validity. It deals with the

5.4 THE FUNDAMENTAL FACTS OF THERMODYNAMICS

behavior of complex bodies, termed thermodynamic systems, in particular with their reactions to heat. The dynamical specification of the state of such a system would require the knowledge of instantaneous values of an enormous number of variables fixing the state of all its constituents. Empirically, however, it is found that *two* quantities suffice to determine the state of a homogeneous body (to which our considerations will be confined) in all respects which are of interest in thermodynamics. These two are pressure, P, and volume, τ. For this reason, they are called variables of state or thermodynamic variables.

The foregoing remarks require one important qualification. It is certainly possible for a system, such as a gas, to exist under identical conditions of pressure and volume without exhibiting the same properties. For instance, the gas may be in the process of expanding into parts of the total volume which are not yet occupied. But such a state would be only a transitory one, in which the gas is said to be not in equilibrium. Equilibrium states are the only ones that are capable of explicit analysis in thermodynamics, and our attention will be confined mainly to them. They are defined as states in which the system persists for a long time when all external influences are removed.

It is possible, in terms of empirical operations, to define the temperature, T, of a system. It is known that, for every body in equilibrium, there exists a characteristic function f of P and τ which has the same numerical value at a given temperature. Hence one may write

$$f_1(P_1, \tau_1) = f_2(P_2, \tau_2) = \ldots = F(T), \tag{1}$$

where the subscripts refer to different bodies. Each of these relations is known as an *equation of state*. The evidence for its existence is largely empirical, although it has not been possible to formulate the equation in finite or closed form for most substances. Its existence will henceforth be taken for granted. Eq. (1) means that, in addition to P and τ, we have found another thermodynamic variable, T, in terms of which the state of a system may be expressed. Only two of the three variables P, τ, and T, are independent, however.

There are quantities of interest in thermodynamics which are not variables of state, which means that they do not always assume the same value when the state of the system is the same. This is easily seen by considering a Carnot cycle. Starting with a substance in a state P, τ, T, we may allow it to pass through a cycle in which heat Q is absorbed and external work W done by the substance. The final state is again characterized by P, τ, T, but it is clearly not true that the variables Q and W return to their initial values. Hence Q and W are not variables of state.

The condition that a thermodynamic quantity shall be a variable of state is simply that it be completely expressible as a function of any two variables of state. Another way of expressing this is to say that the differential of any variable of state must be perfect if two other variables of state are regarded as independent ones. After this preliminary survey we are able to state the two fundamental laws of thermodynamics, from which all other relations of interest in this science may be inferred by deduction.

First law of thermodynamics. A complete statement of the first law comprises two assertions:

(a) Heat is a form of energy.
(b) Energy is conserved.

Neither of these requires explanatory comment; both are well substantiated by experience. If all bodies are regarded as being composed of particles subject to mechanical principles, then (b) can be proved. It is not irrelevant to remark, however, that even if mechanical laws do not govern completely the behavior of elementary particles, the statistical energy balance for the whole system may still be true within the accuracy of measurement. The first law permits us to define a new thermodynamic variable: the *total internal energy*, E, of a system. For it follows at once that, for any transformation in which the changes ΔE, ΔQ, and ΔW occur,

$$\Delta E = \Delta Q - \Delta W, \qquad (2)$$

where ΔQ is the heat *absorbed* by the substance and ΔW the external work done *by* it. It does not follow from eq. (2) that E is a variable of state, but this may be proved as follows. Starting with a given state P, τ, T, we allow the system to pass through a series of different transformations, each of which finally restores the state P, τ, T. We shall in general find different values of ΔQ and ΔW for the different transformations. Now it is conceivable and consistent with the first law that the differences $\Delta Q - \Delta W$ might also be different for all transformations. In that case we should conclude that E is not a variable of state, for, although the system has returned to its initial state, it has undergone a finite change in its internal energy as seen from eq. (2). But experiment shows that all the differences $\Delta Q - \Delta W$ are equal, which establishes the character of E as a variable of state. The absolute value of E cannot be determined, of course, except by arbitrary fixation.

5.4 THE FUNDAMENTAL FACTS OF THERMODYNAMICS

Second law of thermodynamics. The first law is not sufficiently restrictive to serve as the basis of thermodynamics without an essential supplement. This circumstance is not surprising, for we have already emphasized that mechanics, too, would be a very incomplete structure if it had no other guiding principle than the law of conservation of energy (see Sec. 3.11). The fall and rise of a weight under gravity are both compatible with this law; by itself alone it provides no means for discerning which will happen under given conditions. Similarly, when heat is transferred between two systems at different temperatures, the first law can be satisfied by a transfer in either direction. But we know that, as long as the two systems are in mutual contact and otherwise isolated, heat does not pass from lower to higher temperatures. Numerous examples of this type, where the first law is indifferent with regard to an actual selection made by nature, can be cited. They require that we introduce a further criterion by means of which processes compatible with the energy principle but unrealized in nature may be rejected.

It is natural, for the sake of precision, to look for a thermodynamical quantity different from energy which has the property of changing in one direction only during all possible transformations. This quantity, if it exists, should be a variable of state, for otherwise it could not be uniquely specified with reference to a given state. To be sure, variables of state have meaning only if they define an equilibrium state. Hence the quantity we are seeking will be meaningless unless it refers to equilibrium states. But this is no serious shortcoming. Variables of state, by their definition, are independent of the processes leading to the state in question. If now we are to decide whether a certain transformation leading from state I to state II is a possible one, we merely determine the value of the quantity under discussion for state I and then for state II, treating both as equilibrium states, and inquire whether or not its change is in the proper direction.

If this quantity, to be denoted by S, is chosen as

$$S = S_0 + \int_{S_0}^{S} \frac{dQ}{T}, \qquad (3)$$

where S_0 refers to any arbitrary but fixed normal state with respect to which all values of S are calculated, and the integration is performed along any path consisting only of equilibrium states, then all experiments indicate that S never decreases. One can also show by elementary reasoning that S is a variable of state. S is known as the *entropy* (a name first used by Clausius). To find its absolute value requires an explicit evaluation or fixation of S_0, but this is not of interest at present. We shall regard eq. (3) as the thermodynamic *definition* of entropy. All efforts to deduce it from a more general principle, though they have been numerous, have failed.

In asserting that S never decreases care must be taken to refer S to a closed system, i.e., one that has no interactions of any sort with its surroundings, which amounts to saying that its energy is constant in time. The question as to the existence of closed systems will be considered later (Chapter X). It is sufficient at present to assume that they are realized in nature to a degree which allows the increasing property of S to be verified. If any one part of a closed system (i.e., an open system) is considered it may well be that the entropy of this part decreases.

A study of eq. (3) in connection with various thermodynamic transformations leads to this important result: if again I is the initial and II the final state, $S_{II} - S_I > 0$ whenever the process leading from I to II is an irreversible one in the sense previously specified (Sec. 5.1); $S_{II} - S_I = 0$ whenever the process is reversible, $S_{II} - S_I$ referring in both instances to the change in entropy of a closed system. For an open system these relations do not hold.

We combine all these remarks in the following statement of the second law:

(*a*) The entropy, defined by eq. (3), is a variable of state.

(*b*) Its value, for a closed system, can never decrease.

An application of the principle in this form to various thermodynamical transformations, which we shall forego here, would show that it excludes all those processes in which heat is transferred from a body of lower to one of higher temperature by means of a continually self-acting engine. Another consequence is the impossibility of a periodic engine which produces nothing but the lifting of a weight and the cooling of a heat reservoir. An engine of this latter type has been called by Ostwald a " perpetual motion machine of the second kind." Verbal statements like the preceding are often taken as definitions of the second law. They are necessary consequences of our more formal statement and logically equivalent to it.

Eqs. (2) and (3) may be conveniently combined in the form

$$dE = TdS - dW = TdS - \sum_i F_i d\xi_i, \qquad (4)$$

where the F_i are generalized external forces exerted by the system upon bodies surrounding it and the ξ_i are external coordinates. It

5.4 THE FUNDAMENTAL FACTS OF THERMODYNAMICS

must not be understood, however, that eq. (4) is a combination of the first and second laws, for the latter requires in addition to eq. (4) that S be a variable of state, and that the value of S for a closed system may never decrease. Eq. (4) does not refer to a closed system, hence the value of dS in it is not of necessity positive (or 0).

It seems worth while to remark briefly on the entropy of systems which are not in equilibrium. Non-equilibrium conditions cannot be specified by variables of state, and their entropy cannot be computed. Eq. (3) breaks down since such conditions cannot be reached from the normal state by a path of integration that consists only of equilibrium states. Nevertheless, if an estimate is to be made, the entropy of a system not in equilibrium must be smaller than that of the resulting equilibrium state, as the two conditions are always connected by an irreversible process, i.e., one in which S increases. Consequently it is proper to say, in this sense, that the condition of equilibrium is one of maximum entropy.

From the basic set of variables of state discussed so far (P, τ, T, E, S) one may deduce others *ad libitum* by combining them into functions. One such combination, known as Gibbs's ψ function, is particularly convenient, as we shall see. Its definition is

$$\psi = E - TS. \tag{5}$$

Hence
$$d\psi = dE - TdS - SdT.$$

But $dE - TdS = -\sum_i F_i d\xi_i$ on account of eq. (4), so that

$$d\psi = -SdT - \sum_i F_i d\xi_i. \tag{6}$$

This implies that during an isothermal (reversible) transformation (i.e., $dT = 0$) the decrease in ψ is equal to the external work done by the system. For this reason ψ is often called the "free energy" of the system. It will be shown that the ψ function is very fundamental in statistical mechanics. It seems worth while to point out one important application of eq. (6) as an example of its usefulness. If the only way in which the system can react with its surroundings is to change its volume, the term $-\sum_i F_i d\xi_i$ is equivalent to $-Pd\tau$, so that, unless T changes,

$$P = -\frac{\partial \psi}{\partial \tau}. \tag{7}$$

Now it is often possible, as will appear later, to express ψ as a function of τ and T. In that case eq. (7) is identical with the equation of state and may be used for its explicit calculation.

All the fundamental concepts with which thermodynamics operates have now been introduced. We have emphasized that although their structure is free from internal contradictions, the relations between them are not always based upon logical constraints but more frequently upon experimental facts. To satisfy our desire for scientific explanation it is necessary to design a logical structure that will endow these relations with a greater degree of internal coherence and necessity than experiment alone can provide. A theory accomplishes this aim if it is competent to derive a set of quantities analogous to our variables of state and a set of relations among them by starting from very general assumptions. The relations among these quantities must be identical with eqs. (1) to (6). It must further show that the quantity corresponding to entropy never decreases, and that it is a maximum for an equilibrium state. This result is achieved, within limits, by the theory of statistical mechanics.

5.5. Statistical Mechanics. The Method of Gibbs. Among the various methods of building up the science of statistical mechanics two stand out for their logical consistency and mathematical rigor. One is the treatment of Gibbs, the other that of Darwin and Fowler. In presenting only these, we are fully aware of the great importance of the work of other investigators, particularly Boltzmann, Maxwell, and Planck. Nevertheless, the criticism which has been directed (notably by Fowler) against all methods working with " thermodynamic probabilities " and " most probable states " is indeed incisive and should be respected whenever methods are available which are not subject to it. It is true that " thermodynamic probabilities "— though they are not probabilities at all and lead to difficulties if converted into true mathematical probabilities—appeal very directly to physical intuition, but it seems unwise to sacrifice rigor in order to avoid abstractness. We derive further justification for our choice of the two first-named methods by considering more closely the aim of our discussion. We are interested in statistical mechanics as a general basis of thermodynamics, but not in those more detailed properties of thermodynamic systems which find their explanation in the kinetic theory, a science not to be confused with statistical mechanics. The latter proceeds in a very general way and produces theorems of wide validity, while the kinetic theory makes frequent reference to particular properties of bodies; for instance, it discusses collisions, laws of force between molecules, and the like. The work of Maxwell and Boltzmann is largely in the field of kinetic theory, while the methods of Gibbs and of Fowler are more directly adaptable to our needs. Each of these, too, has its own peculiar difficulties which we shall not fail to point out.

5.5   STATISTICAL MECHANICS. THE METHOD OF GIBBS

Terminology. The term *system*, as used by Gibbs,[1] is synonymous with the thermodynamical system previously discussed. Such a system may be any solid body of ordinary dimensions, or any finite volume of liquid or gas. It is composed of a very great number of constituents each of which is capable of motion. The motion of each element of the system may or may not be independent of that of the others. Complete independence exists only in the case of an ideal gas. The constituents of real bodies do not move independently, but enjoy at least sufficient freedom to be distinguishable as separate identities (groups of molecules, molecules, atoms, ions, etc.). Their number ν within one system is in general very great. Since the number of degrees of freedom of the entire system is ν times the number of degrees of freedom of one constituent, it is also very great. Let this number be n.

Dynamically, the state of the system at a given instant is to be specified by $2n$ numbers, the n *coordinates*, fixing the positions of all constituents, and their n corresponding *momenta*. These $2n$ numbers define one point in a space of $2n$ dimensions, called the *phase space* of the system. Each system, i.e., each body under consideration, has its own phase space, which implies that two different systems, such as two portions of the same gas, cannot be compared in the same phase space. As the dynamical state of the given system changes, some or all of the $2n$ numbers take on different values and the representative point moves in phase space.

We shall denote the n generalized coordinates by $q_1 \ldots q_n$, the n generalized momenta by $p_1 \ldots p_n$. If all p's and q's and their derivatives were independent, our point would be free to move with any velocity in any path. It is assumed that the constituents of the system obey the general dynamical laws. These may be written in Hamilton's canonical form:

$$\dot{p}_s = -\frac{\partial H}{\partial q_s}; \quad \dot{q}_s = \frac{\partial H}{\partial p_s}, \quad (s = 1 \ldots n) \tag{1}$$

(cf. Sec. 3.14). Hence our representative point is restricted in its motion in such a way that, while it is permitted to pass through every point of phase space, the direction of its passage is fixed for every point. As a result, it can never cross its previous path. This is all that can be said about the motion of a single representative point without introducing a further hypothesis. There is no consideration that would show, for instance, that the point will not traverse a

---

[1] J. W. Gibbs, "Elementary Principles in Statistical Mechanics" (C. Scribner's Sons, 1902).

periodic path, or that its motion is not confined to certain regions of phase space. Such a behavior would be disastrous to some statistical theories and is usually eliminated by a special postulate, known as the ergodic hypothesis. It asserts that the representative point, during its motion, passes through the immediate vicinity of every point of phase space, so that it cannot be confined to a certain orbit or a fractional part of phase space.

In Gibbs's terminology, the *phase* of a system is its complete dynamical state, as specified by the $2n$ numbers. If these are taken to be the Cartesian coordinates of the representative point, then the product $dq_1 dq_2 \ldots dq_n dp_1 dp_2 \ldots dp_n$ defines an element of phase space. When we do not wish to direct attention to a particular differential of this group, we abbreviate this element of phase space into $d\phi$.

The ensemble. Thus far no statistical concepts have been introduced. To provide a basis for statistical reasoning we construct an *ensemble* consisting of a very great number of independent systems, all identical in nature with the first but differing in phase. In other words, we imagine the first system to be repeated many times, each time with a different arrangement and with different momenta of its constituents. External bodies are located with respect to every system of the ensemble as they are with respect to the first. It is to be understood that the construction of such an ensemble is a purely creative act of reason, and that there is nothing in nature which corresponds to it. Nothing is said at present regarding the exact distribution of the various systems in phase, nor do we suppose that all have the same total energy. If the original system is a closed one, all others are necessarily closed and the energy of each is constant in time, but this constant will in general differ as we pass from one system of the ensemble to the next.

Let N be the number of systems constituting the ensemble. Each may be represented by one point in phase space, so that the ensemble amounts to a certain distribution of N points in phase space. If N is very large, as we shall assume, the distribution may be thought of as nearly continuous, and it is proper to speak of the density of representative points at every point in phase space. D, the density at the point $p_1, \ldots p_n, q_1, \ldots q_n$, will in general be a function of $p_1, \ldots p_n, q_1, \ldots q_n$. Moreover, it may depend on the time t explicitly, for, while we are at liberty to fix the distribution at any given time t_0, we have no assurance that this distribution will be permanent. If it were, the particular distribution would be one of equilibrium, as will be discussed later. Hence, in general, $D = D(p_1, \ldots p_n, q_1, \ldots q_n, t)$.

By the definition of density, the number of systems in the element

5.5 STATISTICAL MECHANICS. THE METHOD OF GIBBS

of phase $d\phi$ surrounding the phase point $p_1, \ldots p_n, q_1, \ldots q_n$ at the instant t is

$$D(p_1 \ldots p_n, q_1 \ldots q_n, t)d\phi,$$

and it is equally evident that at every instant t

$$\int D d\phi = N. \tag{2}$$

If a system were selected at random from the ensemble, the probability of choosing one whose phase lies in the element enclosing the point $p_1, \ldots p_n, q_1, \ldots 'q_n$ is simply $P = D/N$. This follows from the definition of mathematical probability (cf. Chapter IV) and requires no further hypothesis. As a result of eq. (2), at every instant

$$\int P d\phi = 1. \tag{3}$$

Liouville's theorem. The number of systems contained within the domain of phase extending between two phases $p_1^{(1)}, \ldots p_n^{(1)}, q_1^{(1)}, \ldots q_n^{(1)}$ and $p_1^{(2)}, \ldots p_n^{(2)}, q_1^{(2)}, \ldots q_n^{(2)}$, which will be abbreviated as ϕ_1 and ϕ_2, is given by

$$N_{12} = \int_{\phi_1}^{\phi_2} D d\phi. \tag{4}$$

In the course of time, this particular group of points will move, each point according to the dynamical laws of motion, and in moving the group may spread or contract. Their number, however, will remain unchanged, since no systems are destroyed or created. Hence if we consider the total time derivative of N_{12}, taking account of the change in the limits ϕ_1 and ϕ_2 in time, we find that it must vanish.

$$\frac{dN_{12}}{dt} = 0. \tag{5}$$

To perform the differentiation N_{12} must be written in unabbreviated form. The integration is $2n$ fold, and the limits as well as the integrand are functions of t. The method of differentiating such an integral is best explained by discussing a simple case involving but a single integration. Consider the function

$$F(t) = \int_{x_1(t)}^{x_2(t)} f(x, t) dx,$$

and let it be required to find dF/dt. By definition

$$\frac{dF}{dt} = \lim_{\Delta t \to 0} \frac{\int_{x_1(t+\Delta t)}^{x_2(t+\Delta t)} f(x, t + \Delta t)dx - \int_{x_1(t)}^{x_2(t)} f(x, t)dx}{\Delta t}. \quad (6)$$

Expand in a Taylor series, denoting partial time derivatives by dotted symbols, as follows:

$$f(x, t + \Delta t) = f(x, t) + \dot{f}\Delta t + \ldots \, ; \; x_2(t + \Delta t) = x_2(t) + \dot{x}_2\Delta t + \ldots \, ;$$

$$x_1(t + \Delta t) = x_1(t) + \dot{x}_1\Delta t + \ldots \, .$$

We may also split the range of integration of the first term in the numerator of eq. (6), thus:

$$\int_{x_1(t+\Delta t)}^{x_2(t+\Delta t)} = \int_{x_1(t)}^{x_2(t)} + \int_{x_2(t)}^{x_2(t)+\dot{x}_2(t)\Delta t + \ldots} - \int_{x_1(t)}^{x_1(t)+\dot{x}_1(t)\Delta t + \ldots}$$

The numerator of eq. (6) then becomes, before the limit is taken,

$$\int_{x_1(t)}^{x_2(t)} [f(x, t) + \dot{f}(x, t)\Delta t + \ldots]dx$$

$$+ \int_{x_2(t)}^{x_2(t)+\dot{x}_2(t)\Delta t + \ldots} [f(x, t) + \dot{f}(x, t)\Delta t + \ldots]dx$$

$$- \int_{x_1(t)}^{x_1(t)+\dot{x}_1(t)\Delta t + \ldots} [f(x, t) + \dot{f}(x, t)\Delta t + \ldots]dx - \int_{x_1(t)}^{x_2(t)} f(x, t)dx.$$

Of the 7 integrations here explicitly indicated, the first cancels the last. The fourth and sixth yield a result of the second order in Δt, since the ranges of integration are very small (proportional to Δt) and the integrands contain Δt as a factor. Hence they vanish when we proceed to the limit $\Delta t \to 0$. The third and fifth integrations give simply

$$f(x_2, t)\dot{x}_2(t)\Delta t - f(x_1, t)\dot{x}_1(t)\Delta t = \Delta t \cdot \int_{x_1(t)}^{x_2(t)} \frac{\partial}{\partial x}[f(x, t)\dot{x}]dx.$$

Hence we have found

$$\frac{dF}{dt} = \int_{x_1(t)}^{x_2(t)} \dot{f}(x, t)dx + \int_{x_1(t)}^{x_2(t)} \frac{\partial}{\partial x}[f(x, t)\dot{x}]dx. \quad (7)$$

$\frac{dN_{12}}{dt}$ is computed in exactly the same way, by applying this process to every one of the $2n$ integrations and retaining only terms linear in Δt.

5.5 STATISTICAL MECHANICS. THE METHOD OF GIBBS 223

Instead of the one variable x, there are now n p's and n q's, and the result is

$$\frac{dN_{12}}{dt} = \int_{\phi_1}^{\phi_2} \dot{D} d\phi + \sum_{i=1}^{n} \int_{\phi_1}^{\phi_2} \frac{\partial(D\dot{p}_i)}{\partial p_i} d\phi + \sum_{i=1}^{n} \int_{\phi_1}^{\phi_2} \frac{\partial(D\dot{q}_i)}{\partial q_i} d\phi.$$

By eq. (5) this is zero. But the range of integration $\phi_1 \to \phi_2$ is entirely arbitrary. We may, if we desire, contract it to such an extent that D, as well as all the p's and q's, are constant within it. Then the integral signs may be omitted, and it is shown that

$$\dot{D} + \sum_{i=1}^{n} \frac{\partial(D\dot{p}_i)}{\partial p_i} + \sum_{i=1}^{n} \frac{\partial(D\dot{q}_i)}{\partial q_i} = 0. \tag{8}$$

In performing the differentiations in eq. (8) there appear, besides two other summations, the terms

$$D\left[\sum_{i=1}^{n} \frac{\partial \dot{p}_i}{\partial p_i} + \sum_{i=1}^{n} \frac{\partial \dot{q}_i}{\partial q_i}\right].$$

The bracket contains purely dynamical quantities which can be evaluated by means of the equations of motion in the form (1). It is seen at once that they vanish, since $\frac{\partial^2 H}{\partial q_i \partial p_i} = \frac{\partial^2 H}{\partial p_i \partial q_i}$. Therefore, eq. (8) takes the final form

$$\dot{D} + \sum_{i=1}^{n} \frac{\partial D}{\partial p_i} \dot{p}_i + \sum_{i=1}^{n} \frac{\partial D}{\partial q_i} \dot{q}_i = 0. \tag{9}$$

This equation represents a theorem of great importance, known as *Liouville's theorem*, or as the *principle of conservation of density-in-phase* (Gibbs). As D is a function of the p's, the q's, and t, the left-hand side of eq. (9) is merely the total rate of change of D as all these quantities vary. If the equation referred to three dimensions instead of $2n$ dimensions, we should interpret it properly by saying that the density of any group of points, as we follow them in their motion, must remain constant. This type of motion is characteristic of an incompressible fluid. Generalizing these terms to apply to the case of very many dimensions, Liouville's theorem may therefore be expressed verbally as follows: the points representing the systems which constitute an ensemble move in phase space as though they were an incompressible fluid; or any group of representative points filling a certain region of phase space can neither contract nor expand during its motion; it can only alter its shape. The last two statements are clearly equivalent.

Meaning of equilibrium. Let us again consider the number of systems within the domain of phase between ϕ_1 and ϕ_2. But now assume ϕ_1 and ϕ_2 to be fixed, not changing according to the laws of motion, as was supposed before. This number, N'_{12}, will not in general be a constant. However, if it is, the ratio N'_{12}/N is independent of the time, which means that the probability of selecting at random a system whose phase lies within the given range is the same at all times. *Equilibrium* exists when this is true for every range $\phi_1 - \phi_2$.

Analytically, then, the condition of equilibrium is found by putting $dN'_{12}/dt = 0$. N'_{12} is of course defined by eq. (4), where ϕ_1 and ϕ_2 are constants. Hence

$$\frac{dN'_{12}}{dt} = \int_{\phi_1}^{\phi_2} \dot{D} d\phi = 0,$$

which amounts to $\dot{D} = 0$, is the condition of equilibrium. In view of eq. (9), which holds whether there is equilibrium or not, this condition may be stated

$$\sum_{i=1}^{n} \frac{\partial D}{\partial p_i} \dot{p}_i + \sum_{i=1}^{n} \frac{\partial D}{\partial q_i} \dot{q}_i = 0. \qquad (10)$$

Basic postulate. The preceding considerations regarding ensembles bear as yet no relation to experience; they do not permit any prediction as to the behavior of material bodies. But we are able, now, to construct an ensemble when a real system is given; in fact, we are in possession of certain theorems characterizing the properties of such an ensemble. To make our knowledge useful it is necessary to connect the ensemble in some manner with the real system by setting up a general correspondence between the two. This can be done only by means of a postulate, which will be phrased as follows:

The probability that, at any instant t, a given real system be found in the dynamical state characterized by the coordinates $q_1, \ldots q_n$ and the momenta $p_1, \ldots p_n$, is equal to the probability $P(p_1, \ldots p_n, q_1, \ldots q_n, t)$ that a system selected at random from the corresponding ensemble at the same instant t shall have the phase $p_1, \ldots p_n, q_1, \ldots q_n$. P is the quantity previously defined as

$$P(p_1, \ldots p_n, q_1, \ldots q_n, t) = \frac{D(p_1, \ldots p_n, q_1, \ldots q_n, t)}{N}. \qquad (11)$$

Strictly, in stating the postulate we should have specified both the

5.5   STATISTICAL MECHANICS. THE METHOD OF GIBBS

dynamical state and the probability P in terms of ranges p_1 to $p_1 + \Delta p_1, \ldots q_n$ to $q_n + \Delta q_n$. For the sake of brevity the inexact statement was preferred.

So far, even this postulate does not allow us to construct a bridge between our theory of ensembles and real systems. We should be able to make predictions regarding the latter if we knew the initial distribution within the ensemble, but this has been left arbitrary. At this point, specific directions as to the initial distribution in phase space of the systems constituting the ensemble evidently become necessary.

The canonical ensemble. There is nothing to guide us in making the correct choice of the initial distribution except the endeavor to produce agreement with thermodynamic experience. Every choice of our initial distribution will, by way of the postulate, describe a particular process occurring in the real system. Because of the confusing variety of different possible processes the task of assigning a different distribution to each would appear hopeless at the very outset. Fortunately, however, thermodynamics is interested primarily in equilibrium states and in reversible processes which consist of a series of equilibrium states. For these states, it is possible to find a distribution that will always lead to agreement with experiment, a fact that will lead to the thermodynamic equations which we desire ultimately to derive.

The condition expressed by eq. (10) is our starting point. Since it is equivalent to $\dot{D} = 0$, we have the assurance that the distribution sought will be permanent once it is realized. Hence we need no longer speak of the initial distribution, but may refer to the equilibrium distribution without further qualification. From now on the discussion will be restricted to conservative systems, i.e., systems each having total energy H which is constant in time. If then we put D equal to any function of the energy, $D(H)$, eq. (10) is identically satisfied, for it may be written

$$\frac{\partial D}{\partial H}\left[\sum_i \frac{\partial H}{\partial p_i}\dot{p}_i + \sum_i \frac{\partial H}{\partial q_i}\dot{q}_i\right] = 0,$$

and the quantity in brackets is the time rate of change of H, which vanishes by hypothesis. Eq. (10) provides no further help in determining the accurate form of $D(H)$. But there are other conditions which D must satisfy. First, it is to be single-valued and never negative; second, it must satisfy eq. (2); finally, we desire for convenience that it shall be a continuous function with a continuous derivative. These facts eliminate from our choice a large variety of functions.

$H(p_1 \ldots p_n, q_1 \ldots q_n)$ will be a different function for every kind of system, but $D(H)$ must satisfy the conditions just mentioned for every possible H. Therefore, if a certain function $D(H)$, with H referring to any one system chosen at will, is shown to violate these conditions, that D is to be discarded. For convenience, let us consider an ideal gas consisting of ν free mass points each of mass m. For this system, $n = 3\nu$, and

$$H = \sum_{i=1}^{n} \frac{1}{2} m \dot{q}_i^2 = \sum_{i=1}^{n} \frac{p_i^2}{2m}, \qquad (12)$$

where $p_1 = mv_{x1}$, $p_2 = mv_{y1}$, $p_3 = mv_{z1}$, $p_4 = mvx_2$, etc., and v_{yi} is the y component of the velocity of the ith mass point, etc. Eq. (2) then takes the form

$$\int \ldots \int D\left(\sum_i \frac{p_i^2}{2m}\right) dq_1 \ldots dq_n \, dp_1 \ldots dp_n = N. \qquad (13)$$

The integration is $2n$ fold. Since H does not depend on the q's, the integration over the coordinates gives simply τ^ν, τ being the volume of the gas. Let us now take $D(H)$ to be any finite power, say λ, of H. The integrand of eq. (13) will then be a sum of terms each of which is a product of various powers, up to and including the (2λ)th, of the various p_i. All integrations over the momenta can be carried out independently between the fixed limits $-\infty$ and $+\infty$. But the integral of any power $(p_i)^\mu$ between these limits diverges since μ is even. Similarly, the total integral will diverge so that eq. (13) can never be satisfied by this choice of D. The range of functions from which D may be selected is now strongly limited: no finite power of H, and hence no finite polynomial, is eligible. This turns our attention immediately to functions of the type: constant $e^{f(H)}$. Clearly if eq. (13) is to be satisfied, $f(H)$ must not be $+\infty$ for any value of the p's. Any polynomial in H whose highest power has a negative coefficient is compatible with this requirement. Of all these we select, merely for convenience at present, a polynomial of the first degree and write $f(H) = \dfrac{-H + \psi}{\theta}$, where ψ and θ are constants whose values and significance will be determined later. This choice will be considered correct if it leads to the thermodynamical equations. $P = D/N$ now takes the form

$$P = e^{\frac{\psi - H}{\theta}}; \qquad (14)$$

5.5 STATISTICAL MECHANICS. THE METHOD OF GIBBS

the constant multiplier which is omitted can evidently be absorbed in ψ. The distribution given by eq. (14) is called by Gibbs the *canonical* distribution-in-phase, and an ensemble having this distribution is called a canonical ensemble. It is of peculiar interest because it permits a very simple interpretation of thermodynamic phenomena, as will be shown. We have seen that it describes a state of equilibrium. The distribution given by eq. (14) is not completely fixed, for θ may assume different values. θ is frequently called the *modulus* of the canonical distribution; it must of course be a positive parameter.

ψ and θ must be so chosen that P satisfies eq. (3). In attempting to do this one encounters a difficulty regarding dimensions. P is a pure number; the element of phase $d\phi$ has the dimensions $[L^2MT^{-1}]^n$; hence the product cannot be a number as eq. (3) requires. To set matters straight, $d\phi$ has to be divided by a constant of the same dimensions as $d\phi$, a constant which may be regarded as the unit volume of phase space. As long as we are attempting to derive formulas of classical thermodynamics—as distinct from the quantum theory—the size of this unit is immaterial as one may verify by carrying it through all the analysis. For this reason it will not be stated explicitly. It seems well to point out at this stage, however, that Planck's constant h, if raised to the nth power, has the proper dimensions of such a divisor. Indeed, quantum phenomena receive a very natural explanation if h is chosen as the unit for every product $dp_i dq_i$.

As a result of eq. (3)

$$e^{-\frac{\psi}{\theta}} = \int e^{-\frac{H}{\theta}} d\phi, \tag{15}$$

an equation that allows the calculation of ψ.

Maxwell-Boltzmann law. Let us again consider the case of a gas of ν mass points without mutual interactions, but with each mass subject to a potential energy V which is a function of position only (e.g., an ideal gas in a static field). The energy is given by [1]

$$H = \sum_{i=1}^{n} \frac{p_i^2}{2m} + \sum_{j=1}^{\nu} V(x_j, y_j, z_j). \tag{16}$$

While it is experimentally impossible to determine the position and momentum of any one mass point, our theory in its present form permits us to calculate the probability of finding a single mass of the system in a specified dynamical state. In eq. (16), the indices

---

[1] The summation over the potential energies goes to $i = \nu$ only; it is written in this form because V is in general a function of all three coordinates of each particle.

$i = 1, 2, 3$ and $j = 1$ refer to the first mass point. Suppose that we give it fixed momenta and fixed coordinates $p_1, p_2, p_3, x_1 y_1 z_1$. If we are interested in finding the probability that the first mass point be in this fixed state while all others may have any coordinates and momenta, we insert eq. (16) in eq. (14) and integrate over all p's and q's except $p_1, p_2, p_3, x_1 y_1 z_1$. Thus

$$P_1 = e^{\frac{\psi}{\theta}} e^{-\frac{1}{2m\theta}(p_1^2 + p_2^2 + p_3^2) - \frac{V(x_1, y_1, z_1)}{\theta}} dp_1 dp_2 dp_3 dx_1 dy_1 dz_1.$$

$$\times \int \ldots \int e^{-\frac{1}{2m\theta}(p_4^2 + \ldots + p_n^2) - \frac{1}{\theta}[V(x_2, y_2, z_2) + \ldots + V(x_\nu, y_\nu, z_\nu)]}$$

$$dp_4 \ldots dp_n dx_2 dy_2 dz_2 \ldots dx_\nu dy_\nu dz_\nu.$$

The integration on the right is $(2n - 6)$ fold, it extends over $\nu - 1$ mass points. It decomposes into $\nu - 1$ independent sixfold integrations, each of which gives the same result:[1]

$$R = \int (6) \int e^{-\frac{1}{2m\theta}(p_x^2 + p_y^2 + p_z^2) - \frac{1}{\theta} V(x, y, z)} dp_x dp_y dp_z dx dy dz. \tag{17}$$

Hence

$$P_1 = e^{\frac{\psi}{\theta}} e^{-\frac{1}{2m\theta}(p_1^2 + p_2^2 + p_3^2) - \frac{1}{\theta} V(x_1, y_1, z_1)} R^{\nu-1} dp_1 dp_2 dp_3 dx_1 dy_1 dz_1.$$

But by eq. (15)

$$e^{\frac{\psi}{\theta}} = R^{-\nu},$$

so that

$$P_1 = \frac{1}{R} e^{-\frac{1}{2m\theta}(p_1^2 + p_2^2 + p_3^2) - \frac{1}{\theta} V(x_1, y_1, z_1)} dp_1 dp_2 dp_3 dx_1 dy_1 dz_1. \tag{18}$$

R is a constant which can be evaluated by eq. (17) when $V(x, y, z)$ is known. If any other mass point had been selected in place of the first, the probability of finding it in the state with the same momenta and coordinates would have been the same as eq. (18) since it is quite immaterial in what order the mass points are numbered. Eq. (18) therefore, represents the probability that any one mass point picked out at random will have the coordinates and momenta whose numerical values appear in the exponent. But this can be interpreted in no other way than by supposing that, on the average, the number of mass points with coordinates x, y, z and momenta p_1, p_2, p_3 is equal to νP_1. (This number is not to be confused with the number of *systems in the*

---

[1] We shall often use the symbol $\int \overset{(n)}{\ldots} \int$ to stand for n-fold integration.

5.5 STATISTICAL MECHANICS. THE METHOD OF GIBBS

ensemble in which a given mass point has these particular coordinates and momenta. The latter number would be NP_1). Later it will be shown that $\theta = kT$, where k = Boltzmann's constant and T the absolute temperature. With this substitution for θ, eq. (18) is the well-known Maxwell-Boltzmann distribution law. It is seen to follow at once from Gibbs's definition of the canonical ensemble.

When $V = 0$, the term $\frac{1}{\theta}V(x,y,z)$ disappears from eq. (18). Hence, for the molecules of an ideal gas not subject to external forces, *any* position is as likely as any other. This directs our interest to the distribution in momenta alone, regardless of position. But to find the momentum distribution when no consideration is given to position, we must integrate eq. (18) over $dx_1 dy_1 dz_1$. This merely introduces the factor τ, the total volume of the gas. In this case R also takes a simple form, viz.,

$$R = \tau \int\int\int e^{-\frac{1}{2m\theta}(p_x^2 + p_y^2 + p_z^2)} dp_x dp_y dp_z$$

Making the substitution $\sqrt{\frac{1}{2m\theta}}\, p = x$,

$$R = \tau(2m\theta)^{3/2}\left[\int_{-\infty}^{\infty} e^{-x^2}dx\right]^3 = \tau(2m\theta)^{3/2}\cdot \pi^{3/2}.$$

Therefore eq. (18) becomes

$$P_1' = (2\pi m\theta)^{-3/2} e^{-\frac{1}{2m\theta}(p_1^2 + p_2^2 + p_3^2)} dp_1 dp_2 dp_3.$$

This may be written in terms of velocities if we remember that $p_1 = mv_x$, $p_2 = mv_y$, $p_3 = mv_z$. Then

$$P_1' = \left(\frac{m}{2\pi kT}\right)^{3/2} e^{-\frac{\frac{1}{2}m(v_x^2 + v_y^2 + v_z^2)}{kT}} dv_x dv_y dv_z, \tag{18a}$$

where kT has been substituted for θ. Eq. (18a) is a special form of (18) and is often referred to as Maxwell's law of distribution of velocities.

The validity of the Maxwell-Boltzmann law cannot be tested by thermodynamic experiments, but other, more direct methods are available.[1] The intensity of spectral absorption lines is proportional to the number of absorbing atoms in the initial quantum state. Thus,

---

[1] For a detailed discussion see K. F. Herzfeld, "Kinetische Theorie der Wärme" (third volume of Mueller-Pouillet's "Lehrbuch der Physik"), p. 36.

by comparing the intensities of lines originating in different quantum states, one can subject the distribution law to direct verification, and it has been found to be true. In other experiments, the velocities of the molecules in a beam of molecular rays is directly analyzed and the relative numbers of constituents with different velocities are measured. Finally, the empirical density variation of the atmosphere with height of ascent leads to conclusions equivalent to the Maxwell-Boltzmann law.

Average values. From now on, attention will be given no longer to the behavior of individual constituents of the system, but to the properties of the system as a whole, which Gibbs's theory is more naturally fitted to describe. They are the only properties of interest from the point of view of thermodynamics; our derivation of the Maxwell-Boltzmann law has really been a digression into kinetic theory and should be regarded as subsidiary to the principal aim of this section. In order to determine an average property of the real system it is necessary, in the present scheme, to calculate the average of this property for all the systems of the ensemble. It will be noticed that this follows at once from the postulate introduced on p. 224. Among the dynamical averages that of the total energy is of particular interest.

In general, the average of any quantity u depending on the various p's and q's (i.e., on the phase) of the system is given by

$$\bar{u} = \int u e^{\frac{\psi - H}{\theta}} d\phi. \tag{19}$$

The average total energy is

$$\bar{H} = \int H e^{\frac{\psi - H}{\theta}} d\phi. \tag{20}$$

Since ψ and θ are independent of the phase, this may be transformed as follows:

$$\bar{H} = e^{\frac{\psi}{\theta}} \int H e^{-\frac{H}{\theta}} d\phi = - e^{\frac{\psi}{\theta}} \frac{\partial}{\partial \left(\frac{1}{\theta}\right)} \int e^{-\frac{H}{\theta}} d\phi.$$

The integral $\int e^{-\frac{H}{\theta}} d\phi$ occurs often in this work; it is of considerable importance as we shall see. Let us denote it by I. According to eq. (15)

$$e^{\frac{\psi}{\theta}} = \frac{1}{I}. \tag{21}$$

Hence

$$\overline{H} = -\frac{1}{I}\frac{\partial}{\partial\left(\frac{1}{\theta}\right)} I = -\frac{\partial}{\partial\left(\frac{1}{\theta}\right)} \log I, \qquad (22)$$

as a general result. If, for any sytem, I can be calculated, the average energy is easily computed by this relation. This is feasible when the exact functional form of H is known.

The average value of the *kinetic* energy can be calculated explicitly for any system, since the kinetic energy can always be written in the form

$$E_{\text{Kin}} = \sum_{i=1}^{n} u_i p_i^2, \qquad (23)$$

where the u_i may be different functions of the generalized coordinates. If the particular coordinates chosen in a given case do not yield an expression for E_{Kin} which, like eq. (23), is a sum of squares, then it is always possible to make a linear transformation that will throw the result into this form. When such a transformation is necessary, it is no longer permissible to write $dp_1 \ldots dp_n\, dq_1 \ldots dq_n$ for the element of extension in phase space. This must then be multiplied by the Jacobian determinant connecting the old p's with the new. This determinant, however, will be a function of the coordinates only, since the transformation is linear in the p's. As such it may be regarded constant in integrations over the p's and will, if carried in the formulae, finally drop out, as the reader will easily verify. Hence we shall omit it from the beginning. In accordance with (23)

$$\overline{E}_{\text{Kin}} = \int \sum_{i=1}^{n} u_i p_i^2 e^{\frac{\psi - \Sigma u_i p_i^2 - V}{\theta}} d\phi.$$

We have written $\sum_{i=1}^{n} u_i p_i^2 + V$ for H, where V is a function of the coordinates alone. The integration extends over the n p's and the n q's, $d\phi$ being a product of $2n$ differentials. The first summation sign on the right of this equation may be placed in front of the symbol of integration. Of the n terms, let us select the jth and denote its contribution to $\overline{E}_{\text{Kin}}$ by $\overline{E}_j$.

$$\overline{E}_j = \int u_j p_j^2 e^{-\frac{u_j p_j^2}{\theta}} dp_j \cdot \int (n-1) \int e^{\frac{\psi - \sum_{i \neq j} u_i p_i^2 - V}{\theta}} dp_1 \ldots dp_{j-1} dp_{j+1} \ldots$$
$$dp_n dq_1 \ldots dq_n.$$

Put

$$\int (n) \int e^{-\frac{V}{\theta}} dq_1 \ldots dq_n = X; \quad \int e^{-\frac{u_i p_i^2}{\theta}} dp_i = Y_i.$$

As a result of eq. (15)

$$e^{\frac{\psi}{\theta}} = \left(X \prod_{k=1}^{n} Y_k \right)^{-1}$$

Then

$$\overline{E}_j = \frac{X \prod_{i \neq j} Y_i \int u_j p_j^2 e^{-\frac{u_j p_j^2}{\theta}} dp_j}{X \prod_{k=1}^{n} Y_k} = \frac{\int u_j p_j^2 e^{-\frac{u_j p_j^2}{\theta}} dp_j}{\int e^{-\frac{u_j p_j^2}{\theta}} dp_j}.$$

On making the substitution $\sqrt{u_j/\theta}\, p_j = x$, this becomes

$$\overline{E}_j = \frac{\theta^{\frac{3}{2}} \int_{-\infty}^{\infty} x^2 e^{-x^2} dx}{\theta^{\frac{1}{2}} \int_{-\infty}^{\infty} e^{-x^2} dx} = \frac{\theta^{\frac{3}{2}} \cdot \frac{1}{2}\sqrt{\pi}}{\theta^{\frac{1}{2}} \cdot \sqrt{\pi}} = \frac{1}{2}\theta.$$

Quantities with an index j disappear on integration. Consequently, every term of $\overline{E}_{\text{Kin}}$ has the same value, and the result is simply

$$\overline{E}_{\text{Kin}} = n\overline{E}_j = \frac{n}{2}\theta. \tag{24}$$

Anticipating again the later result that $\theta = kT$, eq. (24) expresses the so-called *equipartition theorem*. In words, it states that the average kinetic energy associated with each of the n degrees of freedom is equal to $\frac{1}{2}kT$. It also shows that θ is proportional to the mean kinetic energy regardless of the nature of the system.

The derivation of eq. (24) has involved a tacit assumption which must be exposed if the validity of the equipartition theorem is to be understood. Gibbs's method arrives at observable properties of a real system by averaging over all systems of an ensemble. These systems are continuously distributed in phase, as shown. The integrations in the expression for $\overline{E}_{\text{Kin}}$, for instance, extend over all real positive values of the energy. But quantum theory shows that the possible energy values often form a discrete set, excluding large domains of phase space. Hence, whenever quantization is present, the integrations involved in Gibbs's method are improper, and the results to which they lead are wrong. The equipartition theorem, too, is therefore correct only as long as quantization does not occur

5.5 STATISTICAL MECHANICS. THE METHOD OF GIBBS

or can be neglected. Fowler's method, which will be described in the next section, is beautifully adapted to handling the quantized states. At present, however, we shall confine ourselves to classical statistical mechanics.

The calculation of the average potential energy $\overline{V}$ is usually more difficult. In the first place, V has a different functional form for all real systems so that no general calculation of $\overline{V}$ is possible. Second, V depends on the coordinates in a manner essentially different from that in which E_{Kin} depends on momenta. V cannot be split up into a sum of terms each of which is a function of only one coordinate, but involves at once all coordinates in an inseparable manner. For these reasons, the calculation of $\overline{V}$ presents insuperable difficulties and has been carried out only by means of approximations.

For an ideal gas V is zero. Hence the average total energy of such a gas is equal to $\dfrac{n}{2} \theta$. Some of the following considerations are best understood if we refer to an ideal gas as the sample system whenever detailed calculations are to be made. Their validity is not restricted to ideal gases, however.

Deviation of properties from their averages. Eq. (19) represents the average of the property u taken over all systems of the ensemble, but not its unique value for every system. In general, there will be considerable deviations from this average. When we realize this in connection with the total energy we are confronted with a situation which seems at first alarming, for it means that there is a finite probability of finding systems in the ensemble with energies differing from $\overline{H}$, and hence a finite probability that the real system be observed to have an energy different from $\overline{H}$. But we know that the real system conserves its energy; thermodynamic experiments show the intrinsic energy to be invariable. It would appear, therefore, that Gibbs's statistics is fundamentally incompatible with the facts. This difficulty will disappear when we attempt to reason about the matter in a more quantitative way. Let us calculate the actual deviations from $\overline{H}$ which we may expect to find among the systems constituting the ensemble.

The natural measure of these deviations would be the average difference between the individual energies and their mean: $\overline{H - \overline{H}}$, averages being defined by eq. (19). But this quantity clearly vanishes, since $\overline{H}$ is a constant, and the first H becomes $\overline{H}$ after integration. We then turn to the standard deviation (cf. Sec. 4.3)

$$\left[\overline{(H - \overline{H})^2}\right]^{1/2}.$$

Since we are not interested in the absolute value of the deviations, we choose as their measure

$$\Delta = \left[\frac{\overline{(H-\overline{H})^2}}{\overline{H}^2}\right]^{1/2},$$

the standard deviation divided by the average. This we compute for the case of an ideal gas as a typical example. Then $\overline{H} = \frac{1}{2}n\theta$. $\overline{(H-\overline{H})^2} = \overline{(H^2 - 2H\overline{H} + \overline{H}^2)} = \overline{H^2} - \overline{H}^2$. This difference is easily obtained. First, differentiate the identity

$$\overline{H} = \int H e^{\frac{\psi-H}{\theta}} d\phi$$

with respect to θ. The result is

$$\frac{\partial \overline{H}}{\partial \theta} = \int H e^{\frac{\psi-H}{\theta}} \left[\frac{H-\psi}{\theta^2} + \frac{1}{\theta}\frac{\partial \psi}{\partial \theta}\right] d\phi,$$

since ψ, by eq. (15), is a function of θ. This can be written

$$\frac{\partial \overline{H}}{\partial \theta} = \frac{\overline{H^2}}{\theta^2} - \overline{H}\left(\frac{\psi}{\theta^2} - \frac{1}{\theta}\frac{\partial \psi}{\partial \theta}\right). \tag{25}$$

Next, differentiate eq. (15) with respect to θ. Thus

$$e^{-\frac{\psi}{\theta}}\left(\frac{\psi}{\theta^2} - \frac{1}{\theta}\frac{\partial \psi}{\partial \theta}\right) = \int e^{-\frac{H}{\theta}} \cdot \frac{H}{\theta^2} d\phi = \frac{e^{-\frac{\psi}{\theta}}}{\theta^2} \int H e^{\frac{\psi-H}{\theta}} d\phi = e^{-\frac{\psi}{\theta}} \frac{\overline{H}}{\theta^2}.$$

Therefore, the quantity in the parenthesis of eq. (25) is simply $\overline{H}/\theta^2$. Thus we have

$$\frac{\partial \overline{H}}{\partial \theta} = \frac{\overline{H^2} - \overline{H}^2}{\theta^2} = \frac{n}{2}.$$

Solving for Δ, we get

$$\Delta = \sqrt{\frac{2}{n}}. \tag{26}$$

This result is very significant inasmuch as it shows the spread of the energies within the ensemble to be extremely small, for n, the total number of degrees of freedom, is three times the number of mass points, which for 1 cc of gas under normal conditions is 2.7×10^{19}. One may say, therefore, that the chance of observing a system in a momentary state in which its energy differs appreciably from its average, though finite, is imperceptibly small. This circumstance,

5.5 STATISTICAL MECHANICS. THE METHOD OF GIBBS

which is essential to the consistency of Gibbs's method, restores our confidence in the theory.

The importance of this point makes it appear worth while to look upon the problem from a somewhat different point of view, and actually to work out the number of systems in a Gibbsian ensemble which have an energy within specified limits. If these limits are $E - \dfrac{\Delta E}{2}$ and $E + \dfrac{\Delta E}{2}$, then the number in question is

$$\Delta N = N \int_{H=E-\frac{\Delta E}{2}}^{H=E+\frac{\Delta E}{2}} e^{\frac{\psi - H}{\theta}} d\phi. \tag{27}$$

The limits appearing here are very inconvenient; it is desirable to eliminate them by the use of a Dirichlet factor, as will now be explained. The expression

$$\frac{1}{\pi} \int_{-\infty}^{\infty} \frac{\sin\left(\dfrac{x}{2} \Delta E\right)}{x} e^{i(E-H)x} dx$$

equals 1 if $E - \dfrac{\Delta E}{2} < H < E + \dfrac{\Delta E}{2}$, but is otherwise 0. Therefore, by multiplying the integrand in eq. (27) by this factor and then integrating between infinite limits for the energy, the correct value of ΔN will be obtained. But if the energy range ΔE is made small enough, it is permissible to substitute $\dfrac{\Delta E}{2}$ for $\dfrac{\sin\left(\dfrac{x}{2}\Delta E\right)}{x}$. Hence

$$\Delta N = \frac{N}{2\pi} \Delta E \int \int e^{\frac{\psi}{\theta} + iEx - H\left(ix + \frac{1}{\theta}\right)} d\phi dx. \tag{28}$$

The integration over x is a single one, while that over the phase, although indicated by a single integration sign, is really $2n$ fold. $e^{\psi/\theta}$ can be expressed from the relation

$$\begin{aligned}
e^{-\frac{\psi}{\theta}} &= \int e^{-\frac{H}{\theta}} d\phi = \int (2n) \int e^{-\frac{1}{2m\theta} \sum_{i=1}^{n} p_i^2} dp_1 \ldots dp_n \, dq_1 \ldots dq_n \\
&= \tau^\nu \int (n) \int e^{-\frac{1}{2m\theta}\sum p_i^2} dp_1 \ldots dp_n = \tau^\nu \left[\int_{-\infty}^{\infty} e^{-\frac{p^2}{2m\theta}} dp\right]^n \\
&= \tau^\nu (\sqrt{2\pi m\theta})^n,
\end{aligned} \tag{29}$$

where τ is written for the volume of the gas; τ' appears as a result of the integration over the n q's, ν being the number of mass points, as before. Now substitute eq. (29) into eq. (28), expand H, and put

$$\frac{ix}{2m} + \frac{1}{2m\theta} = \lambda.$$

Then

$$\Delta N = \frac{N\Delta E}{2\pi}(2\pi m\theta)^{-n/2} \tau^{-\nu} \int_{-\infty}^{\infty} e^{iEx} dx \int \underset{\cdots}{(2n)} \int e^{-\lambda \Sigma p_i^2} dp_1 \ldots dp_n \\ dq_1 \ldots dq_n.$$

The integration over the q's gives τ' again, that over the p's leads to

$$\left[\int_{-\infty}^{\infty} e^{-\lambda p^2} dp\right]^n = \left(\frac{\pi}{\lambda}\right)^{n/2}$$

Therefore

$$\Delta N = \frac{N\Delta E}{2\pi} \cdot (2m\theta)^{-n/2} \int_{-\infty}^{\infty} \frac{e^{iEx}}{\lambda^{n/2}} dx = \frac{N\Delta E}{2\pi}\left(\frac{E}{\theta}\right)^{n/2} \int_{-\infty}^{+\infty} \frac{e^{iEx}}{\left(iEx + \frac{\theta}{E}\right)^{n/2}} dx.$$

The last integral is most easily evaluated by complex integration; it has the value

$$\frac{2\pi}{E} \frac{e^{-E/\theta}}{\left(\frac{n}{2} - 1\right)!}.$$

The result is, therefore,

$$\Delta N = N\left(\frac{E}{\theta}\right)^{n/2-1} \frac{e^{-E/\theta}}{\left(\frac{n}{2} - 1\right)!} \frac{\Delta E}{\theta}. \qquad (30)$$

The right-hand side of eq. (30), if summed over all energies, should give the value N. To verify this we observe that

$$\int_0^{\infty} x^K e^{-x} dx = K!$$

Substituting x for E/θ in eq. (30), dx for $\Delta E/\theta$, and integrating, the result on the right is N.

Let us now inquire what is "the most popular value" of the energy, i.e., for what E the corresponding ΔN is a maximum. Differ-

5.5 STATISTICAL MECHANICS. THE METHOD OF GIBBS

entiating eq. (30) with respect to E and putting the result equal to zero, we find

$$E' = \left(\frac{n}{2} - 1\right)\theta \tag{31}$$

for the most popular value. Since n is extremely large, $n/2 - 1$ is indistinguishable from $n/2$, and eq. (31) is seen to be practically identical with eq. (24), which represents the average total energy of the ideal gas. Consequently, we may say that the average energy is at the same time the most popular energy of the ensemble.

Next we wish to answer the question as to the number of systems having an energy differing slightly from E' relative to those whose energy is E'. Let us put $E = 1.01 \, E' = 1.01(n/2 - 1)\theta$ and calculate $\dfrac{\Delta N(E)}{\Delta N(E')}$. By eq. (30)

$$\frac{\Delta N(E)}{\Delta N(E')} = \frac{(1.01\alpha)^\alpha e^{-1.01\alpha}}{\alpha^\alpha e^{-\alpha}} = (1.01)^\alpha e^{-.01\alpha},$$

where $\alpha = n/2 - 1$. Then

$$\log_e \frac{\Delta N(E)}{\Delta N(E')} = \alpha \log_e 1.01 - 0.01\alpha$$

$$= \alpha\left(0.01 - \frac{(0.01)^2}{2} + \frac{(0.01)^3}{3} - \ldots\right) - 0.01\alpha$$

$$= -\tfrac{1}{2}(0.01)^2\alpha.$$

As mentioned before, for 1 cc of gas under standard conditions, $\nu = 2.7 \times 10^{19}$, so that $\alpha \approx 4 \times 10^{19}$. Therefore

$$\frac{\Delta N(E)}{\Delta N(E')} = e^{-\frac{1}{2}(0.01)^2\alpha} \approx e^{-10^{15}}.$$

Thus we have shown that the number of systems in the ensemble having an energy differing by as little as 1 per cent from the average, or the most popular, value, is extremely small, so small indeed that we may well say that the real system has no chance of occurring in states in which its energy is appreciably different from $\overline{E}$. It is possible to show that the standard deviation of the energies is in general proportional to $\sqrt{1/n}$ for an ensemble of systems the constituents of which exert forces upon one another; moreover, this property is not restricted to the energy alone: the standard deviation of every quantity of thermodynamic interest is very small when the system has a large number of degrees of freedom. But we shall not stop to prove this here.

The result at which we have now arrived may be stated: while the canonical ensemble, strictly, contains systems with all possible energies, the "spread" of these energies is so small that, for purposes of observation, all systems may be thought of as having the average energy. If this were not the case the choice $D = Ne^{(\psi-H)/\theta}$, characterizing the canonical ensemble, would have been contradictory to experience. Then it would have been necessary to limit the energy of all systems artificially to the range between E and $E + dE$ by selecting a function

$$D = \begin{cases} \text{constant when } E \leq H \leq E + dE \\ 0 \text{ otherwise.} \end{cases} \quad (32)$$

This function would also satisfy the necessary condition of equilibrium (10) since it depends on H alone. D, as given by eq. (32), is known as the *microcanonical* distribution, and the ensemble possessing this distribution is called a *microcanonical ensemble*. Its properties have been investigated by Gibbs. It turns out that it also leads to correct thermodynamical equations. It is less satisfactory in other respects, however, a main disadvantage being the difficulty of handling it analytically. We shall not consider it in detail, but proceed now to show that the canonical ensemble is governed by relations analogous to those of thermodynamics.

Thermodynamic relations. If there is to be a definite correspondence between the properties of a canonical ensemble on the one hand and thermodynamic quantities on the other, the foregoing analysis compels us to associate $\bar{H}$ with E. Consider the equation

$$\int e^{\frac{\psi-H}{\theta}} d\phi = 1, \quad (33)$$

which is true for every canonical ensemble. As was pointed out before, ψ is a function of the modulus of distribution θ, but is independent of the phase. H contains in general not only the p's and q's of the internal constituents of the system, but also the parameters of external bodies which react with the system.[1] Indeed, H is such that

$$\frac{\partial H}{\partial \xi_i} = -F_i, \quad (34)$$

[1] When the system is a fluid confined in a given volume, the reaction with the walls of the container must appear in the form of H. Analytically this is done most easily by introducing a fictitious potential energy which is constant throughout the interior of the container and rises to $+\infty$ suddenly near the walls. This procedure will be illustrated more clearly in the next section.

5.5 STATISTICAL MECHANICS. THE METHOD OF GIBBS   239

F_i being the force which the system in its totality exerts upon external bodies when the external coordinate ξ_i is slightly varied. Suppose now that the thermodynamic system has its external parameters ξ_i changed by an infinitesimal amount and that it undergoes at the same time a transformation equivalent to an infinitesimal change in the modulus of its ensemble (which will be interpreted thermodynamically later). Throughout this process the system is assumed to be in equilibrium so that eq. (33) holds for the distribution within the corresponding ensemble at every instant during the infinitesimal change. Hence

$$d\int e^{\frac{\psi-H}{\theta}}d\phi = \int e^{\frac{\psi-H}{\theta}}\left[\frac{d\psi}{\theta} - \frac{\psi-H}{\theta^2}d\theta - \frac{1}{\theta}\sum_i \frac{\partial H}{\partial \xi_i}d\xi_i\right]d\phi = 0. \qquad (35)$$

Following Gibbs, we now introduce a quantity η defined by the relation

$$\eta = \log P = \frac{\psi - H}{\theta}, \qquad (36)$$

and call it the *index of probability* of the phase. η, through H, is a function of the phase. In view of eqs. (34, 36) and the relation defining averages eq. (19), eq. (35) takes the form

$$\frac{d\psi}{\theta} - \frac{\bar{\eta}d\theta}{\theta} + \frac{1}{\theta}\sum_i \bar{F}_i d\xi_i = 0. \qquad (37)$$

But

$$\bar{\eta} = \frac{\psi - \bar{H}}{\theta},$$

so that

$$d\psi = d\bar{H} + \bar{\eta}d\theta + \theta d\bar{\eta}.$$

Substituting this in (37), we obtain

$$d\bar{H} = -\theta d\bar{\eta} - \sum_i \bar{F}_i d\xi_i. \qquad (38)$$

Let us compare this with eq. (5.4-4), the fundamental equation of thermodynamics. Complete correspondence is at once established if H is interpreted as E, θ as the temperature, $-\bar{\eta}$ as entropy, and if finally the average of the external forces taken over the ensemble is identified with the force exerted by the thermodynamic system. Eq. (38) will not guarantee that, if these correlations are made, θ and $-\bar{\eta}$ will agree with the empirical temperature and entropy as measured on their usual scales; they may indeed differ by a constant factor.

But if $\theta = cT$, then $-\bar{\eta} = S/c$, necessarily. The constant c will be determined later.

Relation between θ and temperature. The formal similarity between eqs. (38) and (5.4–4) is hardly sufficient to permit us to regard the Gibbsian ensemble as an abstract model from which all thermodynamic properties of real systems can be deduced. Further evidence for the interpretations to which eq. (38) has led seems desirable. We shall first strengthen our confidence by proving that θ is analogous to T in other important respects.

If two thermodynamic systems, both in equilibrium and at the same temperature, are brought into thermal contact, the result is an enlarged system in equilibrium at the same temperature. The ensemble corresponding to the enlarged system, C, will therefore have a probability distribution in phase

$$P_C = e^{\psi_C - H_C/\theta}.$$

If A and B denote the two initial systems

$$H_C = H_A + H_B + H_{AB},$$

where H_{AB} is the energy of interaction between A and B and is analytically the sum of a great number of terms each representing the energy of attraction or repulsion between one constituent of A and one of B. Now unless H_{AB} is small compared with $H_A + H_B$ there will be no resultant state of equilibrium at the same temperature. For instance, if A and B can combine chemically, the resultant temperature is known to be different from that of A and B. Hence, if we wish to describe the process of thermal contact as it is usually understood, we must consider only cases for which H_{AB} is negligibly small. Then

$$P_C = e^{(\psi_C - H_A - H_B)/\theta}. \tag{39}$$

But this probability of phase is precisely the one which results if each system in the ensemble corresponding to A is combined with each system in the ensemble corresponding to B. Let A have m, and B, n degrees of freedom. The probability of selecting a system A having the phase $p_1, \ldots p_m, q_1, \ldots q_m$ from its own ensemble is

$$P_A d\phi_A = e^{(\psi_A - H_A)/\theta} dp_1 \ldots dp_m \, dq_1 \ldots dq_m. \tag{40}$$

Numbering the degrees of freedom of system B from $m + 1$ to $m + n$,

5.5 STATISTICAL MECHANICS. THE METHOD OF GIBBS

the probability of selecting such a system having the phase $p_{m+1}, \ldots p_{m+n}, q_{m+1}, \ldots q_{m+n}$ from its ensemble is

$$P_B \, d\phi_B = e^{(\psi_B - H_B)/\theta} dp_{m+1} \ldots dp_{m+n} \, dq_{m+1} \ldots dq_{m+n}. \quad (41)$$

The combined ensemble has $m + n$ degrees of freedom. The probability of finding a system in this ensemble whose A component has the phase $p_1 \ldots q_m$ while its B component has phase $p_{m+1} \ldots q_{m+n}$ is the product of eqs. (40) and (41), i.e.,

$$P_A \, d\phi_A \, P_B \, d\phi_B = e^{(\psi_A + \psi_B - H_A - H_B)/\theta} dp_1 \ldots dp_{m+n} \, dq_1 \ldots dq_{m+n}.$$

This agrees with eq. (39), for $\psi_A + \psi_B = \psi_C$, as is seen from the fact that both the last expression and eq. (39) are subject to the normalizing condition $\int P d\phi = 1$.

A similar result would not have been obtained if $\theta_A \neq \theta_B$, for then the probability coefficient of the combined ensemble is

$$e^{\frac{\psi_A - H_A}{\theta_A} + \frac{\psi_B - H_B}{\theta_B}},$$

a quantity not identical with (39).

The foregoing considerations are important in two respects: first, they establish another point of correspondence between θ and the temperature; second, they illustrate what is meant by thermodynamic contact in terms of ensembles, showing that, when bodies are placed in contact without appreciable interaction, the corresponding systems in the ensembles are to be combined in phase space, one by one.

If two substances at different temperatures T_A and T_B are brought in contact and equilibrium is established, the resultant temperature T_C is given by the law of mixtures

$$M_A \int_{T_C}^{T_A} s_A \, dT = M_B \int_{T_B}^{T_C} s_B \, dT,$$

where M_A and M_B are the total masses of substances A and B, respectively, and s_A, s_B are their specific heats. If A and B are both ideal gases, their specific heats (at constant volume) are $s = \dfrac{3\nu}{2M} k$, ν being the number of mass points and k Boltzmann's constant, so that the law reduces to

$$\nu_A T_A + \nu_B T_B = (\nu_A + \nu_B) T_C. \quad (42)$$

Describing this process in terms of ensembles we start with two canonical distributions having moduli θ_A and θ_B. The new distribution formed on combining them, which we are supposing to be canonical, has a modulus θ_C. Its average energy will be $\overline{H}_A + \overline{H}_B$. But according to eq. (24) the average energy in the ensemble of an ideal gas is $\frac{3}{2}\nu\theta$. The statement of conservation of energy therefore amounts to:

$$\tfrac{3}{2}\nu_A\theta_A + \tfrac{3}{2}\nu_B\theta_B = \tfrac{3}{2}\nu_C\theta_C = \tfrac{3}{2}(\nu_A + \nu_B)\theta_C,$$

an equation analogous to (42), if θ is interpreted as the temperature. θ_C will always lie between θ_A and θ_B unless these are equal, which implies that the modulus of the warmer body decreases while that of the colder body increases. In this respect, too, the correspondence to temperature is complete.

It will now be desirable to fix the relation between θ and T more precisely. As was pointed out, the constant multiplier in $\theta = cT$ is not fixed by any of the analogies so far considered. The value of c may be determined, however, by noting the empirical relation for an ideal gas

$$\frac{\partial E}{\partial T} = \frac{3}{2}\nu k,$$

T being the absolute temperature. We obtain from eq. (24)

$$\frac{\partial \overline{H}}{\partial \theta} = \frac{3}{2}\nu.$$

But we have already been compelled to identify $\overline{H}$ with E. Consequently

$$c = \frac{\partial \theta}{\partial T} = k.$$

θ will henceforth be interpreted as kT, which in turn leads to the relation $-\overline{\eta} = S/k$, where S is entropy. The consistency of this procedure is at once evident when the thermodynamic definition of absolute temperature

$$\frac{\partial S}{\partial E} = \frac{1}{T}$$

is considered. Substituting $-\overline{\eta}k$ for S, $\overline{H}$ for E, and θ/k for T, and determining $\dfrac{\partial \overline{\eta}}{\partial \overline{H}}$ from eq. (38), this relation is seen to be satisfied.

5.5 STATISTICAL MECHANICS. THE METHOD OF GIBBS

Relation between $\bar{\eta}$ and entropy. There are further reasons for identifying the quantity $-\bar{\eta}$ with the entropy of thermodynamics. (The constant $1/k$ will be omitted when attention is not given to the units in which entropy is expressed.) First, it can be shown that, if an ensemble of systems is canonically distributed in phase, $-\bar{\eta}$ is larger algebraically than it would be in any other distribution of the ensemble having the same average energy. This theorem is equivalent to the thermodynamic proposition (Sec. 5.4) characterizing the condition of equilibrium as one of maximum entropy. To prove it we observe again that the index of probability η for the canonical distribution is $(\psi - H)/\theta$. Let the index η' for the non-canonical distribution be $\dfrac{\psi - H}{\theta} + \Delta\eta$, where $\Delta\eta$ is an arbitrary function of the phase (the p's and q's). Since the number of systems as well as the average energy is the same in both distributions,

$$\int e^{(\psi-H)/\theta + \Delta\eta} d\phi = \int e^{(\psi-H)/\theta} d\phi = 1 \tag{42}$$

and

$$\int H e^{(\psi-H)/\theta + \Delta\eta} d\phi = \int H e^{(\psi-H)/\theta} d\phi. \tag{43}$$

We shall prove that $\overline{\eta'} > \bar{\eta}$, that is,

$$\int \left(\frac{\psi - H}{\theta} + \Delta\eta\right) e^{(\psi-H)/\theta + \Delta\eta} d\phi > \int \frac{\psi - H}{\theta} e^{(\psi-H)/\theta} d\phi. \tag{44}$$

All integrations extend, of course, over the n p's and the n q's. Since ψ and θ do not depend on the p's and q's, the term ψ/θ may be dropped from the integrands of (44) in view of eq. (42), while eq. (43) permits us to omit H/θ on both sides of the inequality. Hence (44) is equivalent to

$$\int \Delta\eta \, e^{(\psi-H)/\theta + \Delta\eta} d\phi > 0.$$

If the second member of eq. (42) is added to, and the first is subtracted from, the left side of this inequality, there results

$$\int (\Delta\eta \, e^{\Delta\eta} + 1 - e^{\Delta\eta}) e^{(\psi-H)/\theta} d\phi > 0. \tag{45}$$

The factor $e^{(\psi-H)/\theta}$ is always positive; the function $xe^x + 1 - e^x$ has only one minimum for all real values of x, and this occurs at $x = 0$. Therefore the integrand of (45) is positive for all values of $\Delta\eta$ except 0. Hence, unless $\Delta\eta$ is zero everywhere in the range of integration—in which case the two distributions would be identical and both canonical—the left-hand side of (45) must be positive. This proves the proposition, for if $\overline{\eta'} > \overline{\eta}$, then $-\overline{\eta} > \overline{\eta'}$.

The most important property of the entropy is its inability to decrease if the system under consideration is isolated from its surroundings (closed). Hence, in order to be sure of having interpreted $\overline{\eta}$ correctly, it is necessary to show that, in an ensemble not reacting with external bodies, $-\overline{\eta}$ tends to become larger and larger. To be sure, if the initial distribution is canonical and as such constant in time, $-\overline{\eta}$ will remain the same. Correspondingly, the entropy of a closed system in equilibrium is constant in time. But now suppose that the distribution of systems in the ensemble is not canonical initially; then the distribution will vary and $-\overline{\eta}$ will change in time. To investigate this change, we must start with an ensemble not canonically distributed in phase.

Our analysis will be greatly facilitated if we associate with $-\overline{\eta}$ a notion less abstract than that of "the average index of probability with its sign reversed," as it is described by Gibbs. We shall think of it as a *measure of the uniformity* of a distribution. Having satisfied ourselves as to the legitimacy of this interpretation we may then proceed to inquire whether or not distributions of the type occurring in Gibbsian ensembles tend toward greater uniformity. This proposed method of reasoning will not appear to be straightforward and is slightly lacking in rigor, but it has the advantages of clarity and intuitive directness.

The first step is to show that $-\overline{\eta}$ is a measure of uniformity. To do this, we prove that for a uniform distribution, i.e., one in which P is independent of the phase, $\overline{\eta}$ is a minimum, or $-\overline{\eta}$ a maximum. More accurately: for a uniform distribution of a given number of systems within given limits of phase $\overline{\eta}$ is smaller than for any other distribution.

The constant uniform distribution can be represented by $P = e^{\eta}$, where η is constant, and any other distribution by $P' = e^{\eta + \Delta \eta}$, $\Delta \eta$ being an arbitrary function of the phase. The number of systems is the same in both distributions, hence

$$\int e^{\eta + \Delta \eta} d\phi = \int e^{\eta} d\phi,$$

or, since η is constant,

$$\int (e^{\Delta \eta} - 1) d\phi = 0. \tag{46}$$

The integrations are to be extended between definite finite limits of phase, since the integrals must converge. A constant distribution in phase with infinite limits of phase would of course require an

5.5 STATISTICAL MECHANICS. THE METHOD OF GIBBS

infinite number of systems, which is contrary to the initial assumptions. The theorem to be proved may be written

$$\int (\eta + \Delta\eta) e^{\eta + \Delta\eta} d\phi > \int \eta e^{\eta} d\phi,$$

or, canceling e^{η} again, and transposing,

$$\int [(\eta + \Delta\eta) e^{\Delta\eta} - \eta] d\phi > 0. \tag{47}$$

Multiply eq. (46) by the constant $\eta + 1$ and subtract it from the left-hand side of eq. (47). The result is

$$\int (\Delta\eta e^{\Delta\eta} + 1 - e^{\Delta\eta}) d\phi > 0.$$

The integrand is identical with the first factor in the integrand of (45), which was shown to be always positive except for $\Delta\eta = 0$. Hence the last inequality is true, which proves the theorem.

Since $-\bar{\eta}$ is an absolute maximum for a perfectly uniform distribution we feel justified in assuming $-\bar{\eta}$ to be in general a measure of the degree of uniformity of any distribution. Logically, this is to be regarded as a definition of "degree of uniformity." In view of this fact, the result expressed by (44) amounts to the statement that the canonical distribution of systems in an ensemble is the most uniform one *consistent with a preassigned value of the average energy*. Thus it is proper to say that a conservative thermodynamic system, in seeking an equilibrium state described by a canonical ensemble, seeks the state of maximum uniformity of distribution.

The problem now is to show that an ensemble tends toward greater uniformity of distribution as time goes on. If we can show that, without any restrictions, the distribution after a long time is perfectly uniform regardless of the initial distribution, then our problem is solved. Now the behavior of the points in $2n$-dimensional phase space representing the ensemble has been proved to be precisely analogous to the motion of an incompressible fluid, or of incompressible mass points. Therefore the proof which we are going to conduct is essentially identical with the verification of the proposition: a great number of mass points, distributed initially in any manner, will, after a long time, be found to be uniformly distributed if they are subject only to their laws of motion.

Probability predictions. In this simplified form the problem is a very basic one, for it is at the bottom of every application of statistics to mechanics. Its philosophical importance is more readily understood by analyzing it further. Fundamentally, it amounts to this. We are given a group of bodies, and we know their laws of motion. We know nothing, however, about their initial state. What can we say about their state at a later time? Philosophers have often been baffled by the fact that physicists find it possible to make any predictions at all in view of the total ignorance regarding the initial conditions. It is, indeed, one of the strangest things for a thoughtful observer to encounter; yet it is no miracle. In order to exhibit as clearly as possible all the logical machinery which produces such remarkable results we shall abstain from the generalities that obscure the usual proofs, and set forth the reasoning in connection with a simple concrete example. Let us select a problem, discussed by Poincaré, regarding the distribution of the minor planets on the zodiac.

We suppose these planets to lie all in one plane. Their laws of motion are Kepler's laws. The problem may be simplified by supposing that the planets move in circular orbits. Each has a different angular velocity ω_i. The angular position of the ith planet with respect to a given reference line at the time t is

$$\theta_i = \omega_i t + \alpha_i, \tag{48}$$

where α_i is the angular position of that planet at $t = 0$. Nothing is known about the distribution of the ω's, or that of the α's as long as a detailed theory of the origin of the solar system is not available. Nevertheless, astronomers claim that the angular distribution of the minor planets is very nearly uniform, and observations bear out this assertion. How can we explain this result?

Let us first examine how the condition of uniformity can be expressed in this case. Suppose that the angular distribution is not uniform, but that there is a condensation of planets for values of θ somewhere in the first half of the range from 0 to 2π. Then one can see at once that the sum, taken over all the planets,

$$\sum_i \sin \theta_i$$

is not zero. If, however, there is a similar cluster at a point on the circle diametrically opposite to the first cluster, this sum will vanish. Hence the vanishing of $\sum_i \sin \theta_i$ is not a general criterion for uniformity of distribution. But in the case of two symmetrical clusters it is possible to select an integer λ (e.g., 2), such that $\sum_i \sin \lambda \theta_i$ is not zero. A moment's reflection will show that the planets are distributed with angular uniformity if

$$\sum_i \sin \lambda \theta_i = 0 \tag{49}$$

for every integer λ.

5.5 STATISTICAL MECHANICS. THE METHOD OF GIBBS

Now let $n(\omega, \alpha)\Delta\omega\Delta\alpha$ denote the number of planets with angular velocity in the range between ω and $\omega + \Delta\omega$ and with an initial angular position between α and $\alpha + \Delta\alpha$, and make the following assumptions:

1. The total number of planets is so large that $n(\omega, \alpha)$ may be regarded as a continuous function of ω and α.
2. n and its first derivatives are finite for all occurring values of ω and α.

Let ω_0 be the smallest and ω_1 the largest value of ω respectively, α_0 and α_1 the smallest and largest values of α. Then the condition for uniform distribution eq. (49) if combined with eq. (48) reads

$$\int_{\alpha_0}^{\alpha_1} \int_{\omega_0}^{\omega_1} n(\omega, \alpha) \sin[\lambda(\omega t + \alpha)] d\omega d\alpha = 0.$$

If we write

$$I(\alpha) = \int_{\omega_0}^{\omega_1} n(\omega, \alpha) \sin[\lambda(\omega t + \alpha)] d\omega,$$

the condition amounts to

$$\int_{\alpha_0}^{\alpha_1} I(\alpha) d\alpha = 0,$$

which is certainly true if $I(\alpha) = 0$. This we wish to show. Integrating I by parts,

$$I = -\frac{1}{\lambda t}\left\{n(\omega, \alpha) \cos[\lambda(\omega t + \alpha)]\right\}_{\omega_0}^{\omega_1} + \frac{1}{\lambda t}\int_{\omega_0}^{\omega_1} \frac{\partial n}{\partial \omega} \cos[\lambda(\omega t + \alpha)] d\omega.$$

As t increases, both terms tend to zero as a result of assumptions 1 and 2. Hence, as time goes on, the distribution becomes more and more uniform.

This method of reasoning is typical, and its features can be generalized to apply to Gibbsian ensembles. Its limitations are apparent. Suppose, for instance, that all the ω_i are equal. Then it is evident that uniformity will never occur, for the planets will preserve their initial distributions. Analytically, this means that assumption 2 is not valid, since $n(\omega, \alpha)$ is infinite at $\omega = \omega_i$. A condition of this kind is by no means impossible, though quite improbable. Correspondingly peculiar conditions, which do not lead to uniformity, may occur in Gibbsian ensembles, but there, too, they would appeal to physical intuition as very improbable, even more improbable than in the simple case discussed here because of the much greater number of degrees of freedom. At any rate, it is apparent that a rigorous logical demonstration of the increasing property of entropy requires additional hypotheses barring certain initial distributions that are troublesome.

Simpler empirical evidence for the tendency of ensembles· to establish a uniform distribution may be obtained by considering a 3-dimensional system with properties analogous to those of a $2n$-dimensional ensemble. As was pointed out, an incompressible liquid is such a system. Imagine a drop of ink to be inserted in a stationary body of water and suppose that there is no diffusion. If the water is stirred, the particles of ink will move with it, and their paths in this ordinary space are exact three-dimensional projections of the points of an ensemble in phase space. Now it is common experience that after the mixture has been stirred for some time the ink will be uniformly distributed throughout the water. But here again, if the body of water rotated as a whole, or if it were in a cylindrical vessel and the stirring produced an angular rotation about the axis with an angular velocity equal to a function of the distance of the liquid particles from the axis, perfect uniformity would never occur.

What we have proved may well be expressed by saying that in Gibbsian ensembles the entropy *practically* never decreases. The necessity for this more cautious statement arises, not from accidental flaws of our somewhat indirect proof, but from an essential difficulty. Another way of expressing this difficulty is this: after a long time the ensemble will certainly be in a stationary distribution *if such a distribution can be reached at all*. The distribution with maximum entropy, i.e., the canonical distribution (since the average energy must be conserved), is a stationary one, as we have shown. But it is not the only one! Before the ensemble reaches its canonical distribution it may be entrapped in another, as it were, "metastable" one from which it cannot escape.

The example of the distribution of minor planets presents a very curious feature on closer inspection. Let us assume for the moment that they are irregularly distributed at present and inquire what their distribution was a long time ago. Eq. (48) holds for any time t. Furthermore, there is no reason why we should not designate by $t = 0$ the present moment. Then the α_i will represent the distribution at present. To find the condition in the past it is only necessary to repeat the argument with t replaced by $-t$. But this does not modify the result; we find that the average of $\sin \lambda\theta$ at the time $t \to -\infty$ to be 0, and the distribution to be uniform. This peculiar symmetry with respect to the moment at which an irregularity exists may be surprising, but is by no means illogical. For we are considering the system of planets to be governed entirely by its laws of motion. Our

5.5 STATISTICAL MECHANICS. THE METHOD OF GIBBS

analysis, if correctly interpreted, shows that the probability of finding the small planets uniformly distributed is vastly greater than the probability for any other distribution. The condition of irregularity is a very improbable one, and, if it is realized at any moment, must be of short duration, which means that the states preceding is as well as those following it are more probable ones, i.e., have a more uniform distribution. We are not entitled, however, to the following conclusion: at present, the distribution of the planets is known to be uniform, hence $n(\omega\alpha)$ is a constant; the calculation following eq. (49) leads to $\sum_i \sin \lambda \theta_i = 0$ at $t \to -\infty$, therefore the distribution must have been uniform to begin with. The absurdity of this argument is evident. Analytically it fails because the production of the initial irregular state was not a result of the laws of motion in the simple form here chosen, but depends on the agencies to which the solar system owes its origin.

The same considerations apply almost literally to Gibbsian ensembles. If at any time t the quantity $-\bar{\eta}$ is not a maximum, then any method of calculation will lead to the result that, at $t - \infty$ or at $t + \infty$, $-\bar{\eta}$ is a maximum. The meaning of this will now be clear.

Having satisfied ourselves that the quantity $-\bar{\eta}$ "practically" never decreases and is therefore analogous to the entropy of thermodynamic systems, we complete the identification by showing that $-\bar{\eta}$ is a variable of state. As previously discussed, such a variable is one that depends only on two quantities like P, τ, T, E, for any given system. Now

$$\bar{\eta} = \int \frac{\psi - H}{\theta} e^{(\psi - H)/\theta} d\phi.$$

To carry out this integration it is necessary to know: (1) the form of H, which amounts to a knowledge of the characteristics of the system under consideration and is always supposed given; (2) θ, a quantity shown to be identical with the temperature; (3) the limits of integration. The momenta always range from $-\infty$ to $+\infty$; each q_i varies over the volume of the system. Hence requisite (3) amounts to a knowledge of the volume. Consequently, for a system with given characteristics, $-\bar{\eta}$ is a function only of the temperature and the volume.

Conclusion. We have now sufficient evidence to permit identification of $-\bar{\eta}$ with S/k, and of θ with kT, S, T, and k being, respectively, the entropy, the absolute temperature, and Boltzmann's constant in their usual thermodynamic sense. But what is the meaning of Gibbs's ψ function? It happens to be identical with the free energy, ψ, appearing in eqs. (5.4–5) and (5.4–6), as is easily seen. On multiplying eq. (37) by θ, we have

$$d\psi = \bar{\eta}d\theta - \sum_i \bar{F}_i d\xi_i, \qquad (50)$$

and this agrees with eq. (5.4–6) if the proper identifications are made.

This fact is of considerable practical significance. We have already pointed out (cf. eq. 5.4–6, et seq.) that, if ψ is known as a function of T and τ, important results can be computed. Thermodynamics alone provides no method of calculating ψ except from empirical data; Gibbs's theory, however, does. For it follows at once from eq. (15) that

$$\psi = -\theta \log I,$$

where the quantity $I = \int e^{-\frac{H}{\theta}} d\phi$ is known as Gibbs's phase integral. Thus the calculation of ψ, and hence of equations of state and many properties of thermodynamic substances, reduce essentially to an evaluation of the phase integral, a quantity of central significance in the mathematical structure of Gibbs's theory. We shall have occasion later to return to it (cf. 5.7).

This concludes our discussion of the major points of Gibbs's statistics. We have seen that it leads to a set of relations formally identical with the thermodyamic equations, and that all quantities occurring in these relations possess the properties characteristic of thermodynamic variables. Therefore, this entire scheme may be regarded as an adequate explanation of the empirical data of thermodynamics.

5.6. Survey and Critique of Gibbs's Method. In looking back over the essential features of Gibbs's method it is difficult to suppress a feeling of esthetic satisfaction. To those who enjoy abstract reasoning the generality and loftiness of the scheme are particularly impressive, while others admire the simplicity with which it explains the thermodynamic relations. There are very few notions in physics so abstract and so far removed from experience as the Gibbsian ensemble. In introducing this concept no direct relation is sought to reality, for, as we have shown, the ensemble is no physical thing. Gibbs does not set himself the problem of finding a mechanism which, if identified with the object of observation, will give rise to the phenomena observed. He merely and modestly designs a logical apparatus, regulated by postulates, that produces relations between well-defined quantities, relations which can be interpreted as analogues to the thermodynamical relations. If an explanation amounts to

5.6 SURVEY AND CRITIQUE OF GIBBS'S METHOD 251

analytic decomposition of a phenomenon into the behavior of its various parts, Gibbs's theory is not a physical explanation of thermodynamics. In another, more adequate sense it is a very successful one.

It is especially noteworthy that in this theory no attempt is made to derive distribution laws (such as the Maxwell-Boltzmann law) from other experimental facts. They are all introduced in one grand sweep by the postulation of the canonical ensemble. No question as to the justification of this concept can arise, for it is a definition which need not satisfy any experience but only the requirement of simplicity. Indeed, it is widely divergent from anything that observation provides. We have shown that a canonical ensemble of conservative systems is not conservative in the sense that all systems in the ensemble have the same energy. As a consequence of this fact the energy of a real conservative system, if calculated from the ensemble, is subject to fluctuations which are not observed. Agreement with experience is restored by noting that these fluctuations are imperceptibly small. Had it been Gibbs's chief aim to construct a model of so-called reality he would have chosen the microcanonical ensemble as the basis for his analysis, for this type of ensemble corresponds more closely to the physical notion of a conservative system. Thermodynamic relations can indeed be derived from it, but less simply. Because of its closer relation to experience, writers on statistical mechanics occasionally prefer the microcanonical ensemble to the canonical one as a starting point of their investigations. Logical simplicity appears to favor the latter.

Although the canonical ensemble is to be regarded initially as a purely logical construct it is possible to find a physical correlate for it if this is desired. This correlate, however, reflects by no means the power and generality of the concept on the logical side and should be considered merely as an illustration. A thermometer of small heat capacity, if placed in a large bath at constant temperature, is subject to precisely the same fluctuations in its internal energy as are present among the systems of the canonical ensemble. This observation is no doubt interesting but hardly of basic significance.

A considerable advantage of Gibbs's method is the elegance with which it escapes all difficulties regarding statistical weights, which, as we shall see, emerge very annoyingly in connection with other methods. Whenever we attempt to find "most probable" or "average" states of a *real* system without reference to postulated ensembles we must know the statistical weight of each state of the particular system in question. These statistical weights (measures of probability of a real state) are essentially experimental quantities, physical magnitudes concerning which no arbitrary postulation is permissible. But experiment is not detailed enough to yield information regarding the weight of individual dynamical states. Hence, in dealing with physical systems without having recourse to fictitious ensembles, one must follow the cumbersome method of assuming weights to start with and seeing if they lead to correct results. This procedure is to be applied until agreement with experiment is achieved—and even then there lingers the embarassing question as to whether or not the correct assignment of weights is unique.

Gibbs does not trouble himself with averages over real systems. He averages over ensembles and shows that this leads to formulas identical with the empirical ones. Now an ensemble has certain convenient properties expressed by the theorem of conservation of phase. This implies that there is no tendency for the representative points to crowd into special regions of phase space, in other words, that there is no preferential distribution of points among the various parts of phase space. Hence there can only be uniform weighting of all states of the ensemble.

It is to be observed, however, that this manner of proceeding eliminates the difficulties by cutting the Gordian knot. It introduces all significant features into the analysis by means of one far-reaching postulate (p. 224) and the concomitant definition of the canonical ensemble. Thus, if any part of Gibbs's theory is found defective the whole structure is likely to fall. Its elegance and simplicity refuse to be disfigured by being patched up. Now, unfortunately, recent developments have shown the results of Gibbs's method to be definitely in error. Quantum theory in particular indicates that something is wrong with the notion of an ensemble, for some of its states (over which Gibbs extends the averaging process) are dynamically impossible. The motion of representative points would have to be thought of as partly discontinuous. Experimentally, the first infelicities in connection with Gibbs's statistics were encountered in observations on specific heats which did not agree with those calculated. Thus we are forced to regard the method of Gibbs, which is eminently successful in classical physics, as definitely incomplete, and to look for a system of statistical mechanics that will stand in the face of quantum mechanics. A completely successful system of this type has, in our opinion, not yet been designed. But the one due to Darwin and Fowler appears to be the nearest approach to it.

5.7. The Method of Darwin and Fowler.[1] *Basic concepts.* As in

[1] *Phil. Mag.*, **44**, 1922, pp. 450, 823; **45**, 1923, pp. 1, 497. *Proc. Cam. Phil. Soc.*, **21**, 1922, p. 262; 1923, pp. 392, 730. For a general exposition, see R. H. Fowler "Statistical Mechanics" (Cambridge University Press, 1929).

5.7 THE METHOD OF DARWIN AND FOWLER

the last section, statistical analysis will be applied to a material system of many degrees of freedom. Its number of constituents (atoms, molecules, etc.) will again be called ν. But we shall now confine our attention at once to conservative systems. No longer following Gibbs, we shall not construct an ensemble of a great number of systems similar to the one under consideration, but consider it to occur *only once*. Average values are therefore not taken over a great number of such systems, but over the large number of possible *states* of one. It is of advantage to change our terminology at this point. In the present analysis, the constituents of the material body play a more prominent rôle, and frequent reference to them will be necessary. For the sake of brevity and clearness, let us now call them *systems*, and refer to the body under consideration, to which the term system was previously applied, as an *assembly* of systems. This change not only is convenient, but also makes our terminology agree with that used by Fowler and Darwin.

Procedure. Each individual system contained in the assembly is in general capable of existing in a number of different states.[1] By specifying the states of all individual systems the composite state of the assembly can be fixed. All such composite states would be equally likely, if the individual states of a system were equally likely, but this is not always the case. As was pointed out previously, these states may have different weights which have to be considered in determining the probability, or the weight, of the composite state of the assembly.

It is now necessary to investigate more closely what is meant by a composite state, defined as the totality of individual states of all the systems. We could represent it most easily by numbering the systems as well as the possible states of one system (the number of states may be infinite!) and then writing below the number of each system that of the state in which it exists. This manner of description is very detailed, more detailed indeed than experiment allows it to be made, for there is no way in which individual systems can be distinguished; we can at most state *how many* systems are in a given state, not precisely which ones. The former detailed mode of specification which assumes the possibility of distinguishing between the systems as individuals is called the microscopic one; it is said to define a microscopic state. The latter, which fails to recognize systems as individuals and assigns merely a number of systems to every possible

---

[1] We require a criterion, of course, by which we classify states. Thus we might think of states of different velocity, or states of different position, or states of different energy, etc. In general, any principle of division can be used as the basis of statistical reasoning. Later we shall restrict ourselves to "energy states," however.

state, specifies what will be called a macroscopic state. Clearly, to every macroscopic state there correspond in general numerous microscopic ones. Cf. Fig. 5.1. The weight of a microscopic state of the assembly is determined only by the various elementary weights of the states of the individual systems of which it is composed. If these elementary weights are all equal, all microscopic states can plausibly

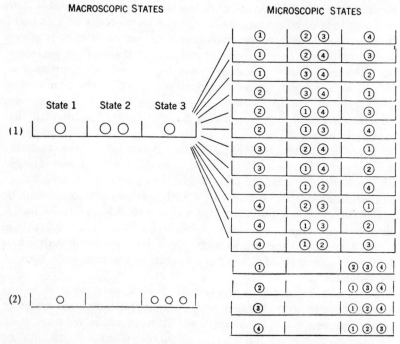

FIG. 5.1. Each section of a horizontal line represents a state of a single system. Circles symbolize individual systems, in this case four in number. Macroscopic state (1) is the composite state which has 1 system in state 1, 2 systems in state 2, 1 system in state 3. It corresponds to the 12 microscopic states on the right. Only 4 microscopic states correspond to macroscopic state (2), as is seen.

be assumed to have the same weight. At any rate this will be assumed at present. (We are by no means logically compelled to do this, as will be discussed in a later chapter. By substituting different postulates for the one here under discussion one arrives at "non-classical" statistics, such as the Einstein-Bose or the Fermi-Dirac forms.) The weight, or probability, of a macroscopic state is then proportional to the number of microscopic states of which it may be said to "consist."

5.7 THE METHOD OF DARWIN AND FOWLER

Every macroscopic state will be characterized by definite properties, such as the number of systems with a given energy or a given momentum, or the total electric moment of the assembly. Let us think for the moment of the first mentioned, the number of systems with a given energy, say between 1 and 2 ergs. If we number all possible macroscopic states $1, 2, \ldots r, \ldots$ and assign to each its correct weight $(w_1, w_2, \ldots, w_r, \ldots)$, then we may find that in state 1 there are N_1 systems having an energy between 1 and 2 ergs; in state 2 there are N_2; in state r, N_r; etc. Hence the *average* number of systems within this energy range is given by

$$\frac{\sum_i N_i w_i}{\sum_i w_i}.$$

This simple reasoning underlies all the analysis which we are now presenting. In order to find an average property of the assembly we compute (1) the weights of all macroscopic states, (2) the value of the property in question for every macroscopic state; finally we average in the manner just indicated.

This brings us face to face with an important question: what do these averages mean in terms of experience? Are they the values of the properties exhibited by the assembly when it is in equilibrium? This is generally assumed to be the case. Let us call the value of a certain property (say the number of systems in a given energy state) which the assembly is experimentally found to possess (in equilibrium) the *normal value* of this property. A normal value then is one which the assembly generally exhibits, i.e., it is the most frequently occurring one. Hence, if it could be shown that the values obtained by averaging over the various macroscopic states of an assembly, in the manner here discussed, are identical with those realized for the greatest length of time during the total life of the assembly, the situation would be quite satisfactory. As a matter of fact, however, this proof has not been conducted as far as we are aware; the agreement between normal values and average (or most probable) values is still an article of faith. On the logical side, therefore, the present method is no better than that of Gibbs, who found it necessary to bridge the gap between ensembles and experience by an (implicit) postulate.

Hypothesis regarding elementary weights. We can now proceed to the mathematical details. The first step is to fix the weights of the elementary states of one system. This requires the introduction of a few simple quantum mechanical notions. For a more extensive treatment of quantum mechanics the reader is referred to Chapter IX;

the following remarks suffice for the purposes at hand. We shall consider a very simple dynamical system, the harmonic oscillator. A crystal, for instance, may be thought of as composed of a multitude of such systems, because the atoms, ions, or molecules composing it are bound to their positions of equilibrium by forces which are approximately simple harmonic (for small displacements).

In classical mechanics every harmonic oscillator has a definite fundamental frequency f which is determined by its stiffness and its mass. But it may vibrate with any amplitude, and hence its energy of vibration is unrestricted (except for the trivial absence of negative kinetic energies). In quantum mechanics, this is not found to be the case. To state accurately what happens, let us distinguish three different types of simple harmonic oscillators: (1) the linear oscillator, which is constrained to vibrate along a line; (2) the isotropic two-dimensional oscillator, constrained to vibrate in a plane; (3) the isotropic three-dimensional oscillator, free to vibrate in any direction in space. Quantum mechanics shows that there are (h is Planck's constant):

$$
\begin{array}{lll}
\textit{in Case 1} & \textit{in Case 2} & \textit{in Case 3} \\
\text{1 state of energy } \tfrac{1}{2} hf, & \text{1 state of energy } hf, & \text{1 state of energy } \tfrac{3}{2} hf, \\
\text{1 " " " } \tfrac{3}{2} hf, & \text{2 states " " } 2hf, & \text{3 states " " } \tfrac{5}{2} hf, \\
\text{1 " " " } \tfrac{5}{2} hf, & \text{3 " " " } 3hf, & \text{6 " " " } \tfrac{7}{2} hf, \\
\cdots & \cdots & \cdots \\
\text{1 " " " } & (n+1) \text{ states of energy} & \tfrac{1}{2}(n+1)(n+2) \text{ states} \\
(n+\tfrac{1}{2})hf, & (n+1)hf, & \text{of energy } (n+\tfrac{3}{2})hf
\end{array} \quad (1)
$$

By "state" we mean here an elementary state of motion characterized by a certain function (ψ function) in a sense to be discussed more thoroughly later. There is no reason why these elementary states should not be of equal weights; hence we assign the weight unity to each. Now it is customary in statistical mechanics to refer to states as *energy states*. For instance, one would specify the state of a three-dimensional oscillator by specifying its energy, say $\tfrac{5}{2}h\nu$. This energy state would then clearly have a weight of 3, since three elementary states, each of weight unity, are contained in it. The occurrence of discrete values, such as the energy is seen to have in this example, is termed *quantization*. Now the energy is a very important concept; its values can be very easily found in most problems. Hence it has become customary in quantum mechanics to enumerate all possible states according to the energies which they possess, and to call different elementary states having the same energy " degenerate " states as opposed to single or non-degenerate ones. The number of different elementary states coalescing to a given energy state is known as the degree of degeneracy. If, from now on, we agree to mean by a state of a system its energy state, our example shows us that the weight of each state is equal to its degree of degeneracy.

We have outlined how quantum mechanical systems are to be treated with regard to elementary weighting. Strictly speaking, all possible systems are subject to the laws of quantum mechanics, and we require no further hypotheses. But it is very convenient to introduce a special statement regarding systems, such as freely moving particles, which are commonly considered as non-quantized. The reason for this assumption is that a free mass point moving in a volume of ordinary dimensions, while having definitely quantized energy states, has these states lying so close together that the practical effect of their discreteness is negligible. If, in particular, we allow the volume to increase beyond bounds, the discrete energy values of a free mass point merge to form a continuous distribution (cf. Sec. 9.7). For a finite volume, we can count the number of states falling within a given range of energies—or, if we desire, a given range of momenta. We thus find, for the simple case of a free mass point moving in one dimension only, that the number of states of motion falling within any range of position Δq, and at the same time within any range of momentum Δp, is equal to $\Delta q \Delta p / h$. For a group of mass points of n degrees of freedom the number of elementary states of motion falling within $dq_1 \ldots dq_n\, dp_1 \ldots dp_n$ is found to be

$$\frac{dq_1 \ldots dq_n\, dp_1 \ldots dp_n}{h^n}.$$

This, then, must be taken as the weight of the classical element of extension in phase $dq_1 \ldots dq_n\, dp_1 \ldots dp_n$, i.e., the quantity called $d\phi$ in Sec. 5.5. Hence we see: (1) that equal elements of extension in phase of classical systems have equal weights; (2) that the weight unity corresponds to any portion of phase space of size h^n. Result (1) was implied in Gibbs's statistics; (2) was found to be needed to produce correct physical dimensions in eq. (5.5–3).

To summarize our hypothesis regarding elementary statistical weights:

(*a*) Each energy state of a quantum mechanical system has a weight equal to its degree of degeneracy.

(*b*) Each element of phase space $dq_1 \ldots dq_n\, dp_1 \ldots dp_n$ of a classical system of n degrees of freedom has a weight equal to

$$\frac{dq_1 \ldots dq_n\, dp_1 \ldots dp_n}{h^n}.$$

We have indicated that (b) is not a separate hypothesis but merely a result of (a). Nevertheless, it is useful to state it separately.

Application to an assembly of quantized systems. Let us now apply these considerations to find the "distribution law" for a quantized assembly. We suppose the assembly to consist of ν identical systems without appreciable interactions,[1] so that the energy of each system can be stated without reference to any other. The energy states of one system will be denoted by $\varepsilon_0, \varepsilon_1, \varepsilon_2 \ldots \varepsilon_r \ldots$, and their weights, i.e., their respective degrees of degeneracy, by $g_0, g_1 \ldots g_r \ldots$. Our first aim will be to determine the weight of a macroscopic state (sometimes called a statistical state).

How shall we represent a macroscopic state? The answer is simple: we merely state that there are a_0 systems in state ε_0, a_1 in state $\varepsilon_1, \ldots a_r$ systems in state ε_r, and so on. Hence a macroscopic state is characterized by any sequence of integers

$$a_0, a_1, a_2, \ldots a_r \ldots \qquad (2)$$

Since the number of systems, ν, is a constant, every such sequence is subject to the condition

$$\sum_i a_i = \nu, \qquad (3)$$

and since we are dealing with conservative assemblies,

$$\sum_i a_i \varepsilon_i = E, \qquad (4)$$

where E is the total energy.

Next we consider a microscopic state, defined by a detailed statement as to which systems have energy ε_0, which have energy ε_1, etc. If we number the systems from 1 to ν, a particular microscopic state (often called a complexion) may be represented as follows:

$$\varepsilon_0(4, 9, 32); \quad \varepsilon_1(2, 7, 12, 13); \quad \varepsilon_2(1, 8); \ldots \varepsilon_r(3) \ldots, \qquad (5)$$

---

[1] Physically, an assembly without any interactions between its parts is indeed an absurdity, for it could never change its state of motion. Any state would be an equilibrium state, and there would be no unique distribution laws characterizing the statistical assembly. What we actually mean by absence of appreciable interactions may be illustrated by the case in which the systems are so far apart that the time during which they experience no influences due to others is very large compared to the time in which they exert mutual forces on one another. For instance, if any gas is so rare that the time of flight between impacts is vastly greater than the duration of a collision, the condition is satisfied. Another example would be an assembly consisting of isolated fixed oscillators, like the constituents of a crystal, with a few free atoms, molecules, or electrons flying about between the oscillators, thus producing a unique equilibrium state after a very long time.

5.7 THE METHOD OF DARWIN AND FOWLER

where the numbers in parentheses denote the particular systems having the energy which precedes the numbers. Their order is immaterial. (5) may be said to correspond to one "cell" in the phase space of the whole assembly. Its weight is the product of all the weights of the individual states of the systems, that is

$$g_0^3 g_1^4 g_2^2 \ldots g_r. \qquad (6)$$

The macroscopic state corresponding to (5) would be defined by the sequence

$$a_0 = 3, \quad a_1 = 4, \quad a_2 = 2, \ldots a_r = 1, \ldots.$$

This same macroscopic state is realized by many other microscopic states, in fact, by as many as there are ways of distributing ν systems in groups of $3, 4, 2, \ldots 1, \ldots$ etc. In general, then, the number of microscopic states corresponding to a macroscopic state such as (2) is equal to the number of distributions of ν elements in groups given by (2). Let us find this number, L. One distribution can be changed into another by certain permutations of the elements. The total number of permutations of ν elements is $\nu!$. But not every one of these permutations will produce a new microscopic state since the order of the elements within each group is immaterial. (By interchanging 4 and 7 in (5) we get a new microscopic state, but by interchanging 4 and 9 we do not.) $\nu!$, the total number of permutations, is therefore equal to L, the number of significant permutations, times the number of insignificant permutations, i.e., the number of permutations which do not produce new microscopic states. Every permutation of elements within the same parenthesis is insignificant, and there are altogether $a_0! a_1! a_2! \ldots a_r! \ldots$ such permutations. Hence, since $\nu! = L a_0! a_1! \ldots a_r!$.

$$L = \frac{\nu!}{a_0! a_1! \ldots a_r! \ldots}. \qquad (7)$$

Now every microscopic state belonging to the same macroscopic state has the same weight, (6). Since the weight w of a macroscopic state is equal to the number of cells of phase space (i.e., microscopic states) which it contains multiplied by the weight of each,

$$w = \frac{\nu! \, g_0^{a_0} g_1^{a_1} g_2^{a_2} \ldots}{a_0! a_1! a_2! \ldots}. \qquad (8)$$

Next, how many macroscopic states are there? This is the same as asking how many w's do we have to deal with. The answer is, clearly, as many as there are ways of satisfying eqs. (3) and (4) simultaneously. We must sum over all these states, assigning to each its proper w, when we wish to find averages.

We set out initially to find the distribution law of the assembly, i.e., the average number of systems in any given energy state, say ε_r. For every macroscopic state, this number is a_r, hence its average

$$\bar{a}_r = \frac{\sum_a \frac{a_r \nu! \, g_0{}^{a_0} g_1{}^{a_1} g_2{}^{a_2} \cdots}{a_0! \, a_1! \cdots}}{\sum_a \frac{\nu! \, g_0{}^{a_0} g_1{}^{a_1} g_2{}^{a_2} \cdots}{a_0! \, a_1! \cdots}}. \qquad (9)$$

By $\sum_a$ we mean the sum over all sets $a_0, a_1, a_2 \ldots$ satisfying eqs. (3) and (4), a multiple sum, of course. The remainder of this paragraph deals with the mathematics of evaluating eq. (9), which is both beauful and simple; it is the point of chief attraction in the statistical method of Darwin and Fowler.

We recall a theorem of algebra, known as the polynomial law:

$$(z_0 + z_1 + z_2 + \ldots + z_\rho)^\nu = \sum_a \frac{\nu!}{a_0! \, a_1! \ldots a_\rho!} z_0{}^{a_0} z_1{}^{a_1} \ldots z_\rho{}^{a_\rho},$$

where $\sum_a$ is a sum over all sets $a_0, a_1, a_2 \ldots a_\rho$ satisfying $\sum_i a_i = \nu$. Hence we have

$$\left(\sum_i g_i z^{\varepsilon_i}\right)^\nu = \sum_a \frac{\nu! \, g_0{}^{a_0} g_1{}^{a_1} g_2{}^{a_2} \cdots}{a_0! \, a_1! \cdots} z^{\Sigma a_i \varepsilon_i}. \qquad (10)$$

The coefficient of every power of z on the right of eq. (10) is equal to $\sum_a \frac{\nu! \, g_0{}^{a_0} g_1{}^{a_1} \cdots}{a_0! \, a_1! \cdots}$ where the a's satisfy eq. (3); the coefficient of z^E satisfies both eqs. (3) and (4) and is therefore equal to the denominator of eq. (9). Hence we can calculate the denominator of eq. (9) by finding the coefficient of z^E in $(\sum_i g_i z^{\varepsilon_i})^\nu$. But this can be obtained by a theorem in complex integration which states:

. If γ is a circle of radius less than unity about the origin $z = 0$, then

$$\frac{1}{2\pi i} \int_\gamma z^n dz = \begin{cases} 0 \text{ when } n \neq -1, \\ 1 \text{ when } n = -1. \end{cases}$$

Thus if we integrate $(\sum_i g_i z^{\varepsilon_i})^\nu$ about γ and divide by $2\pi i$ we shall obtain the coefficient of z^{-1}, and the coefficient of z^E will be

$$\frac{1}{2\pi i} \int_\gamma \frac{dz}{z^{E+1}} Z^\nu, \qquad (11)$$

5.7 THE METHOD OF DARWIN AND FOWLER

where Z has been written for the series $\sum_i g_i z^{\varepsilon_i}$. Z is called the "partition function" by Darwin and Fowler (it is essentially the same quantity as Planck's "Zustandssumme"). We shall see that it is an extremely important quantity from which almost all statistical properties of the assembly can be derived. It is a dynamical expression and contains nothing but elementary weights and energies, z being in all cases merely a variable which will disappear on integration.

(11) is the value of the denominator of eq. (9). Let us now evaluate the numerator. Upon canceling the a_r, it takes the form

$$\nu \sum_a \frac{(\nu-1)!\, g_0^{a_0} g_1^{a_1} \ldots}{a_0!\, a_1! \ldots (a_r-1)!\, a_{r+1}! \ldots}, \text{ subject to } \sum_i a_i = \nu, \sum_i a_i \varepsilon_i = E.$$

If we introduce a new set of distribution numbers: $b_0, b_1 \ldots$, defined by: $b_0 = a_0, b_1 = a_1, b_2 = a_2, \ldots b_r = a_r - 1, \ldots$, this may be written

$$\nu g_r \sum_b \frac{(\nu-1)!\, g_0^{b_0} g_1^{b_1} \ldots g_r^{b_r} \ldots}{b_0!\, b_1! \ldots b_r! \ldots}, \text{ subject to } \sum_i b_i = \nu - 1, \sum_i b_i \varepsilon_i = E - \varepsilon_r.$$

But by the previous argument, this is equal to νg_r times the coefficient of $z^{E-\varepsilon_r}$ in the expansion of $(\sum_i g_i z^{\varepsilon_i})^{\nu-1}$, and this is given by

$$\frac{\nu g_r}{2\pi i} \int_\gamma \frac{dz}{z^{E-\varepsilon_r+1}} Z^{\nu-1}.$$

Therefore eq. (9) becomes:

$$\bar{a}_r = \frac{\nu \int_\gamma (g_r z^{\varepsilon_r}/Z) \dfrac{dz}{z^{E+1}} Z^\nu}{\int_\gamma \dfrac{dz}{z^{E+1}} Z^\nu}. \tag{12}$$

The value of this expression can be found by the method of steepest descents which will now be described.

The method of steepest descents (*Sattelpunktmethode*). Consider an integral of the type

$$\frac{1}{2\pi i} \int_\gamma F(z) \cdot [\phi(z)]^\nu \frac{dz}{z}, \tag{13}$$

where

$$\phi(z) = \frac{Z(z)}{z^{E/\nu}}, \tag{14}$$

with

$$Z = \sum_{i=0}^\infty g_i z^{\varepsilon_i}, \tag{15}$$

and $F(z)$ is an analytic function with no other singularities in the unit circle than possibly a pole at the origin. In eq. (15), all g_i are positive constants since they stand for statistical weights; the ε_i will in general be all positive.[1] On the real axis, $\phi(z)$ is therefore certainly infinite both at $z = 0$ and at $z = 1$, and has but one minimum between these points. Let this minimum be at $z = \vartheta$, defined by $\dfrac{d\phi}{dz} = 0$. This equation takes the form

$$\frac{E}{\nu} Z\vartheta^{-1} = \frac{dZ}{dz}\bigg|_\vartheta, \quad \text{or} \quad E = \nu\vartheta \frac{d}{d\vartheta} \log Z(\vartheta). \tag{16}$$

Next we study the behavior of $\phi(z)^\nu$ on the circle $z = \vartheta e^{i\alpha}$. The factor Z^ν takes here the form $\left(\sum_j g_j \vartheta^{\varepsilon_j} e^{i\alpha\varepsilon_j}\right)^\nu$. Every term of the sum in this expression has a complex factor which is unity whenever $\alpha\varepsilon_i = 2\pi n$ (n = zero or an integer), and otherwise its absolute value is smaller. If for any one term in this sum

$$\alpha\varepsilon_i \neq 2\pi n, \tag{17}$$

the absolute value of the sum will be smaller than it is at $z = \vartheta$, and the νth power will then be insignificant compared with the νth power of $Z(\vartheta)$ since ν is an extremely large integer.

Hence the situation is this: on the circle $z = \vartheta e^{i\alpha}$ the expression Z^ν, and therefore ϕ^ν, will have a strong maximum at $\alpha = 0$. This point is known as a "saddle point," for it represents a minimum on the real axis. If, furthermore, (17) is true for every ε_i all along the circle ($0 \leq \alpha < 2\pi$), as will now be assumed, then there will be no other maximum on this circle, in fact, ϕ^ν will vanish everywhere else on our path of integration provided $\nu \to \infty$. Let us now choose for γ the circle $\vartheta e^{i\alpha}$. The only contribution [2] to the integral (13) will then come from the point $z = \vartheta$, and (13) may be written

$$\frac{F(\vartheta)}{2\pi i} \int_\gamma [\phi(z)]^\nu \frac{dz}{z}. \tag{18}$$

[1] In any physical problem, the ε's denote energies of a physical system. These may be reckoned from the lowest possible energy state, in which case they are all positive.

[2] Condition (17) is violated when all energies of the system, ε_i, are connected by a linear relation. As a consequence of this restriction it appears that we are not permitted to treat the case of an assembly consisting of linear simple harmonic oscillators for which $\varepsilon_j = (j + \frac{1}{2})c$, so that Z takes the form $Z = (\vartheta e^{i\alpha})^{\frac{1}{2}c} \sum_j g_j \vartheta^{jc} e^{i\alpha jc}$, c being a constant ($= hf$). Z^ν has then another maximum at $\alpha = 2\pi/c$. Closer inspection shows, however, that this difficulty disappears.

5.7 THE METHOD OF DARWIN AND FOWLER

To evaluate $\bar{a}_r$, the average number of systems in the rth energy state, is now a simple matter. The denominator of (12) divided by $2\pi i$ is identical with (13) if we put $F = 1$, and the numerator takes the same form if $F(z) = \dfrac{\nu g_r z^{\varepsilon_r}}{Z}$. Therefore, because of (18),

$$\bar{a}_r = \frac{\nu}{Z(\vartheta)} g_r \vartheta^{\varepsilon_r}. \tag{19}$$

Before discussing the physical meaning of this result let us finish another detail which we shall use later. The denominator in (9), which was transformed into (11), represents the *sum of the weights for all states of the assembly* which are compatible with (3) and (4), i.e., *with a total energy E*. Let us call this number C. Calculation [1] shows that

$$C = \frac{\vartheta^{-E}[Z(\vartheta)]^\nu}{\left[2\pi\nu\vartheta^2 \dfrac{\phi''(\vartheta)}{\phi(\vartheta)}\right]^{1/2}}, \tag{20}$$

where ϕ is given by (14).

---

[1] To calculate C we expand $[\phi(z)]^\nu$ about the point $z = \vartheta$. Thus, by Taylor's theorem,

$$\log[\phi(z)]^\nu = \nu \log \phi(z) = \nu \left\{ \log \phi(\vartheta) + \frac{\phi'(\vartheta)}{\phi(\vartheta)}(z - \vartheta) \right.$$

$$\left. + \left[\frac{\phi''(\vartheta)}{\phi(\vartheta)} - \frac{\phi'(\vartheta)}{[\phi(\vartheta)]^2}\right] \frac{(z-\vartheta)^2}{2} + \cdots \right\}$$

But $\phi'(\vartheta) = 0$ since ϑ is the minimum value of $\phi(z)$; $z - \vartheta = i\vartheta\alpha$ if $z - \vartheta$ is sufficiently small. Hence

$$\log [\phi(z)]^\nu = \nu \left\{ \log \phi(\vartheta) - \tfrac{1}{2}\vartheta^2\alpha^2 \frac{\phi''(\vartheta)}{\phi(\vartheta)} \right\},$$

and

$$\phi(z)^\nu = \phi(\vartheta)^\nu \exp\left[-\frac{\nu}{2}\vartheta^2\alpha^2 \frac{\phi''(\vartheta)}{\phi(\vartheta)}\right].$$

We see that, if ν is large, the rate of decrease in $[\phi(z)]^\nu$ as we pass from $\alpha = 0$ to some small finite value is enormous. This provides justification for replacing (13) by (18). To calculate (18), put $z = \vartheta e^{i\alpha}$, so that $\dfrac{dz}{z} = id\alpha$, and integrate over α from $-\pi$ to $+\pi$. Since the integrand is practically 0 everywhere except at $\alpha = 0$, any convenient path of integration which passes once through the point $\alpha = 0$ may be substituted for the original one without appreciable effect on the value of the integral. (All this is strictly true only if $\nu \to \infty$, as will be noticed. Several questions of rigor arise at this point which will not be dealt with here. Cf. Fowler, pp. 27, et seq., for a better

Eq. (19) is the equivalent of the Maxwell-Boltzmann distribution law for quantized systems. We observe at once that it is formally correct, for if we sum $\overline{a_r}$ over all values of r we obtain the total number of systems ν as should be the case. This is evident if we recall that, by definition, $Z(\vartheta) = \sum_r g_r \vartheta^{\varepsilon_r}$.

It will be shown later that the parameter ϑ which appears in this work is related to the absolute temperature T by

$$\log \vartheta = -\frac{1}{kT}. \tag{21}$$

Assuming this at present, eq. (19) may be written

$$\overline{a_r} = \frac{\nu}{Z(\vartheta)} g_r e^{-\frac{\varepsilon_r}{kT}}. \tag{22}$$

The Maxwell-Boltzmann law eq. (5.5–18) is to be regarded as a special form of this relation for the case of classical systems, where g_r is proportional to the size of the extension of phase in question, and $\varepsilon_r = \frac{1}{2m}(p_1^2 + p_2^2 + p_3^2) + V(x, y, z)$. With these substitutions, eqs. (22) and (5.5–18) become identical. (The multiplying constants, although they appear to be different, must agree because summation over all possible states gives ν in both cases.)

Other interesting similarities between equations in this section and the previous one begin to appear—cf., for instance, eqs. (5.5–22) and (16)—but we shall not stop to discuss them.

Generalization to an assembly consisting of several kinds of systems. Let there be λ different kinds of systems (oscillators, molecules, or atoms with different energy states) in the whole assembly. Quantities referring to these different kinds will be indicated by *superscripts*. Thus $\varepsilon_0^{(p)}$, $\varepsilon_1^{(p)}$, $\varepsilon_2^{(p)}$ ... represents the series of energies for the pth kind of system, of which there are $\nu^{(p)}$ specimens. Their elementary

attempt at mathematical rigor.) Thus we shall integrate over α from $-\infty$ to $+\infty$. We obtain

$$\frac{1}{2\pi i} \int_\gamma F(z)\phi(z)^\nu \frac{dz}{z} = \frac{F(\vartheta)\phi(\vartheta)^\nu}{2\pi} \int_{-\infty}^\infty \exp\left[-\frac{\nu}{2}\vartheta^2 \frac{\phi''(\vartheta)}{\phi(\vartheta)}\right] \alpha^2 d\alpha$$

$$= F(\vartheta)\phi(\vartheta)^\nu \left[2\pi\nu\vartheta^2 \frac{\phi''(\vartheta)}{\phi(\vartheta)}\right]^{-\frac{1}{2}}$$

Eq. (20) results if in this expression we put $F = 1$ and replace ϕ according to its definition (14).

5.7   THE METHOD OF DARWIN AND FOWLER

weights are $g_0^{(p)}, g_1^{(p)}, g_2^{(p)}, \ldots$, and their number in the rth energy state is $a_r^{(p)}$. The total number of systems in the whole assembly is again ν. An equilibrium state of the assembly is now no longer characterized by eqs. (3) and (4), but is subject to $\lambda + 1$ equations

$$\sum_i a_i^{(1)} = \nu^{(1)}, \sum_i a_i^{(2)} = \nu^{(2)}, \ldots \sum_i a_i^{(\lambda)} = \nu^{(\lambda)};$$

$$\sum_i a_i^{(1)} \varepsilon_i^{(1)} + \sum_i a_i^{(2)} \varepsilon_i^{(2)} + \ldots + \sum_i a_i^{(\lambda)} \varepsilon_i^{(\lambda)} = E. \quad (23)$$

A macroscopic state is now defined by λ sequences

$$a_0^{(1)}, a_1^{(1)}, a_2^{(1)}, \ldots, a_0^{(2)}, a_1^{(2)}, a_2^{(2)}, \ldots, \ldots a_0^{(\lambda)}, a_1^{(\lambda)}, a_2^{(\lambda)}, \ldots \quad (24)$$

a microscopic state by λ assignments like (5). The number of microscopic states (cells of phase space of the total assembly) corresponding to the macroscopic state (24) is now

$$L = \frac{\nu^{(1)}!}{a_0^{(1)}! \, a_1^{(1)}! \, a_2^{(1)}! \ldots} \cdot \frac{\nu^{(2)}!}{a_0^{(2)}! \, a_1^{(2)}! \, a_2^{(2)}! \ldots} \cdots$$

$$\cdot \frac{\nu^{(\lambda)}!}{a_0^{(\lambda)}! \, a_1^{(\lambda)}! \, a_2^{(\lambda)}! \ldots} \cdots,$$

and the weight of each is

$$\Pi g_0^{a_0} g_1^{a_1} g_2^{a_2} \cdots$$

where Π means product over all superscripts (which have been omitted). Consequently the weight of a macroscopic state will be instead of (8)

$$w = \frac{\nu^{(1)}! \, g_0^{(1)a_0^{(1)}} g_1^{(1)a_1^{(1)}} \cdots}{a_0^{(1)}! \, a_1^{(1)}! \ldots} \cdot \frac{\nu^{(2)}! \, g_0^{(2)a_0^{(2)}} g_1^{(2)a_1^{(2)}} \cdots}{a_0^{(2)}! \, a_1^{(2)}! \ldots} \cdots \quad (25)$$

It is hardly necessary to continue to write down these lengthy formulas, for their meaning is quite simple. Going through the previous work we notice the following generalizations:

In eq. (9) we must now specify to what kind of system $\bar{a}_r$ is to refer, i.e., we have to add a superscript, say β. We wish to find $\bar{a}_r^{(\beta)}$, the average number of systems of the kind β in the rth energy state. On the right, we have λ products like the one in eq. (9) appearing after each $\sum_a$, and the summation now extends over all a's that satisfy eq. (23).

Use is made of the polynomial theorem as before. Suppose we write down eq. (10) λ times, each time using a different superscript for the letters involved, and then multiply the corresponding sides of all these expressions. If on the right we now select the term z^E, its coefficient will be precisely the denominator of the expression for a_r. Hence, by the same argument as before, this denominator, which represents the total weight and was called C, can be written

$$C = \frac{1}{2\pi i} \int_\gamma \frac{dz}{z^{E+1}} \cdot [Z^{(1)}]^{\nu^{(1)}} [Z^{(2)}]^{\nu^{(2)}} \ldots [Z^{(\lambda)}]^{\nu^{(\lambda)}}. \qquad (26)$$

The numerator is transformed in the same manner as previously, so that we finally obtain in place of eq. (12)

$$\overline{a_r^{(\beta)}} = \frac{\dfrac{\nu^{(\beta)}}{2\pi i} \int_\gamma (g_r^{(\beta)} z^{\varepsilon_r^{(\beta)}} / Z^{(\beta)}) \dfrac{dz}{z^{E+1}} [Z^{(1)}]^{\nu^{(1)}} [Z^{(2)}]^{\nu^{(2)}} \ldots [Z^{(\lambda)}]^{\nu^{(\lambda)}}}{\dfrac{1}{2\pi i} \int_\gamma \dfrac{dz}{z^{E+1}} [Z^{(1)}]^{\nu^{(1)}} [Z^{(2)}]^{\nu^{(2)}} \ldots [Z^{(\lambda)}]^{\nu^{(\lambda)}}}.$$

The method of steepest descents is applicable as before. This may be seen analytically from the fact that all the properties of the function $\phi(z)$ used in the previous work are also possessed by

$$\Phi(z) = \frac{[Z^{(1)}]^{\frac{\nu^{(1)}}{\nu}} [Z^{(2)}]^{\frac{\nu^{(2)}}{\nu}} \ldots [Z^{(\lambda)}]^{\frac{\nu^{(\lambda)}}{\nu}}}{z^{\frac{E}{\nu}}}.$$

Hence our final result is identical with eq. (19) (except for the superscripts):

$$\overline{a_r^{(\beta)}} = \frac{\nu^{(\beta)}}{Z^{(\beta)}(\vartheta)} g_r^{(\beta)} \vartheta^{\varepsilon_r^{(\beta)}}, \qquad (27)$$

if ϑ is defined as the minimum of $\Phi(z)$ on the real axis, viz.,

$$\left.\frac{d\Phi(z)}{dz}\right|_{z=\vartheta} = 0; \quad E = \nu^{(1)} \vartheta \frac{\partial}{\partial \vartheta} \log Z^{(1)}$$
$$+ \nu^{(2)} \vartheta \frac{\partial}{\partial \vartheta} \log Z^{(2)} + \ldots + \nu^{(\lambda)} \vartheta \frac{\partial}{\partial \vartheta} \log Z^{(\lambda)}.$$

This last expression is very suggestive. It shows that the total energy of the assembly is made up of terms each referring only to one kind of system, which tempts us to interpret each separate term as the average energy of that kind of system to which it relates. We shall verify this in a more adequate way.

In the macroscopic state defined by the sequences

$$a_0^{(1)}, a_1^{(1)} \ldots, \quad a_0^{(2)}, a_1^{(2)}, \ldots, \ldots a_0^{(\lambda)}, a_1^{(\lambda)}, \ldots;$$

5.7 THE METHOD OF DARWIN AND FOWLER

the energy of the βth kind of system has the value $\sum_i a_i^{(\beta)} \varepsilon_i^{(\beta)}$, and the weight of the state is given by eq. (25). Hence

$$\overline{E^{(\beta)}} = \frac{\sum_a (\sum_i a_i^{(\beta)} \varepsilon_i^{(\beta)}) w}{\sum_a w}. \tag{28}$$

Now consider again eq. (10), or rather the λ fold set of equations like (10) with different superscripts. Select the one with superscript β and apply to it the operator $z \dfrac{\partial}{\partial z}$. This will multiply the coefficient of every power of z by $(\sum_i a_i^{(\beta)} \varepsilon_i^{(\beta)})$, the factor needed in eq. (28). Next, multiply this result by the remaining unmodified $\lambda - 1$ equations. If on the right we pick out the coefficient of z^E this will be identical with the numerator of eq. (28). The denominator is of course again C, and given by eq. (26). Hence

$$\overline{E^{(\beta)}} = \frac{\dfrac{1}{2\pi i}\int_\gamma \dfrac{dz}{z^{E+1}} [Z^{(1)}]^{\nu^{(1)}} \cdots z \dfrac{\partial}{\partial z}\left\{[Z^{(\beta)}]^{\nu^{(\beta)}}\right\} \cdots [Z^{(\lambda)}]^{\nu^{(\lambda)}}}{\dfrac{1}{2\pi i}\int_\gamma \dfrac{dz}{z^{E+1}} [Z^{(1)}]^{\nu^{(1)}} \cdots [Z^{(\lambda)}]^{\nu^{(\lambda)}}}$$

$$= \frac{\int_\gamma \left(\nu^{(\beta)} z \dfrac{\partial}{\partial z} \log Z^{(\beta)}\right) \cdot \dfrac{dz}{z^{E+1}} [Z^{(1)}]^{\nu^{(1)}} \cdots [Z^{(\lambda)}]^{\nu^{(\lambda)}}}{\int_\gamma \dfrac{dz}{z^{E+1}} [Z^{(1)}]^{\nu^{(1)}} \cdots [Z^{(\lambda)}]^{\nu^{(\lambda)}}}$$

$$= \nu^{(\beta)} \vartheta \frac{\partial}{\partial \vartheta} \log Z^{(\beta)}(\vartheta). \tag{29}$$

This confirms our expectation.

To find a clue as to the physical meaning of ϑ let us consider again eq. (27). In an assembly containing $\nu^{(1)}$ systems of *one* kind only, the equilibrium distribution is given by $\overline{a_r} = \dfrac{\nu^{(1)}}{Z(\vartheta_1)} g_r^{(1)} \vartheta_1^{\varepsilon_r^{(1)}}$, where ϑ_1 depends on the total energy by eq. (16). Similarly, another assembly containing $\nu^{(2)}$ systems of another kind can have an equilibrium distribution $\overline{a_r} = \dfrac{\nu^{(2)}}{Z(\vartheta_2)} g_r^{(2)} \vartheta_2^{\varepsilon_r^{(2)}}$ with any ϑ_2 not necessarily equal to ϑ_1. If, however, the two were to form one assembly in equilibrium the two values of $\overline{a_r}$ written down would be two special ones of the group (27) with $\beta = 1$ and 2, say. This means that $\vartheta_1 = \vartheta_2$. Consequently two separate assemblies will be in equilibrium on contact when their ϑ's are the same. This, together with the fact that ϑ cannot be negative (cf. the mathematical definition, p. 262) leads us to associate the parameter ϑ with the temperature of the assembly. In the next paragraph we shall investigate more exactly the relation between ϑ and T.

Assembly of simple harmonic oscillators. The formulas derived in the last paragraph will now be applied to a special case of physical interest, that of an assembly consisting of simple harmonic oscillators vibrating in one, two, or three dimensions. These will be dealt with under I, II, and III, respectively. For convenience, we repeat the two most important results so far obtained:

$$\overline{a_r^{(\beta)}} = \frac{\nu^{(\beta)}}{Z^{(\beta)}(\vartheta)} g_r^{(\beta)} \vartheta^{\varepsilon_r^{(\beta)}}, \qquad (27)$$

$$\overline{E^{(\beta)}} = \nu^{(\beta)} \vartheta \frac{\partial}{\partial \vartheta} \log Z^{(\beta)}(\vartheta), \qquad (29)$$

where

$$Z(\vartheta) = \sum_i g_i \vartheta^{\varepsilon_i}. \qquad (14')$$

Let there be two types of oscillators in the assembly, so that β takes the values 1 and 2. Type 1 is supposed to have a natural fundamental frequency f_1; the other type has the fundamental frequency f_2. There are ν_1 oscillators of the first, ν_2 of the second type.

I. By (1), $\varepsilon_n^{(1)} = (n + \frac{1}{2})hf_1$; $\varepsilon_n^{(2)} = (n + \frac{1}{2})hf_2$;

$g_n^{(1)} = 1$; $g_n^{(2)} = 1$ for every n.

Hence

$$Z^{(1)}(\vartheta) = \sum_n \vartheta^{(n+\frac{1}{2})hf_1} = \vartheta^{hf_1/2} \sum_{n=0}^\infty \vartheta^{nhf_1} = \frac{\vartheta^{hf_1/2}}{1 - \vartheta^{hf_1}},$$

and

$$Z^{(2)}(\vartheta) = \sum_n \vartheta^{(n+\frac{1}{2})hf_2} = \frac{\vartheta^{hf_2/2}}{1 - \vartheta^{hf_2}}.$$

It happens that in this case the partition functions can be written in closed form. The distribution law (27) takes the form

$$\overline{a_r^{(1)}} = \nu_1 \vartheta^{rhf_1}(1 - \vartheta^{hf_1}); \quad \overline{a_r^{(2)}} = \nu_2 \vartheta^{rhf_2}(1 - \vartheta^{hf_2}). \qquad (30)$$

The partition of energy between the two kinds of oscillators is given by eq. (29), which now reads after performing the differentiation

$$\overline{E^{(1)}} = \nu_1 \left[\frac{hf_1}{\vartheta^{-hf_1} - 1} + \frac{hf_1}{2} \right]; \quad \overline{E^{(2)}} = \nu_2 \left[\frac{hf_2}{\vartheta^{-hf_2} - 1} + \frac{hf_2}{2} \right]. \qquad (31)$$

These expressions are not quite as simple as the equipartition law of the classical theory; cf. (5.5–24). Nevertheless, when we suppose the force of binding of all the oscillators to become very small, so that $f \to 0$, their behavior should approach that of classical systems, for it is well known that slowly moving things obey classical mechanics. To make the transition, it is to be remembered that, on the average, half the total energy of an oscillator is kinetic, half is potential. Therefore we should have

$$\lim_{f_1 \to 0} \frac{1}{2} \frac{\overline{E^{(1)}}}{\nu_1} = \frac{1}{2} kT. \tag{32}$$

But in the limit,

$$\frac{\overline{E^{(1)}}}{\nu_1} = \frac{hf_1}{(e^{-\log \vartheta})^{h/1} - 1} = \frac{1}{-\log \vartheta}.$$

To make this true, we must have

$$\vartheta = e^{-\frac{1}{kT}}. \tag{33}$$

A more convincing proof of this relation will be given later. For the present we shall use it in eqs. (30) and (31) (where we shall now omit the subscripts referring to the two different kinds of system):

$$\overline{a_r} = \nu e^{-\frac{rhf}{kT}} \left(1 - e^{-\frac{hf}{kT}}\right) \tag{30'}$$

$$\overline{E} = \nu \left[\frac{hf}{e^{hf/kT} - 1} + \frac{hf}{2}\right]. \tag{31'}$$

These are the famous distribution laws found by Planck (in a slightly different form). They were discovered in connection with the problems of black-body radiation, and this discovery marked the inception of quantum mechanics (cf. 9.1).

The second term on the right of eq. (31'), $+hf/2$, is of interest. When the temperature becomes very small, the first term vanishes. Hence, except for the added term in question, the average energy of a system of oscillators would be zero at the absolute zero of temperature. The second term prevents this and provides what is called the "zero point energy." This term did not appear in the first formulations of quantum mechanics. It is due to the fact that the energy levels of a simple oscillator are given by $(n + \frac{1}{2})hf$, not by nhf as was first supposed (Bohr's theory). But the correctness of eq. (31') including this term seems now beyond doubt.

II. Almost identical results will be obtained for the oscillator in two dimensions. We shall omit superscripts again since the formulas

are the same for every kind of oscillator constituting the assembly. By (1)
$$\varepsilon_n = (n+1)hf, \quad g_n = n+1.$$
Hence
$$Z(\vartheta) = \sum_{n=0}^{\infty}(n+1)\vartheta^{(n+1)hf} = \vartheta^{hf}\sum_{n=0}^{\infty}(n+1)\vartheta^{nhf} = \frac{\vartheta^{hf}}{(1-\vartheta^{hf})^2}.$$

Consequently
$$\overline{E} = \nu\left[\frac{2hf}{\vartheta^{-hf}-1} + hf\right] = \nu\left[\frac{2hf}{e^{hf/kT}-1} + hf\right].$$

III. For the three-dimensional oscillator
$$\varepsilon_n = (n+\tfrac{3}{2})hf, \quad g_n = \tfrac{1}{2}(n+1)(n+2).$$
Therefore
$$Z(\vartheta) = \sum_{n=0}^{\infty}\tfrac{1}{2}(n+1)(n+2)\vartheta^{(n+3/2)hf} = \vartheta^{3/2 hf}\sum_{n=0}^{\infty}\tfrac{1}{2}(n+1)(n+2)\vartheta^{nhf}$$
$$= \frac{\vartheta^{\frac{3hf}{2}}}{(1-\vartheta^{hf})^3},$$
so that
$$\overline{E} = \nu\left[\frac{3hf}{\vartheta^{-hf}-1} + \frac{3}{2}hf\right] = \nu\left[\frac{3hf}{e^{hf/kT}-1} + \frac{3}{2}hf\right].$$

We observe that an increase in the number of degrees of freedom of each oscillator affects the average energy in the same way as the same relative increase in the number, ν, of oscillators. That is to say, an assembly of ν two-dimensional oscillators has twice as much energy as an assembly of ν one-dimensional oscillators at the same temperature.

The great analytic power of the method is already apparent; we shall not stop to illustrate it further by an application to other quantized systems. More examples will be found in Fowler's "Statistical Mechanics." A new point appears when we consider an assembly of systems which, unlike the oscillators in the preceding examples, are free to move in a given portion of space. For then there is introduced a new variable, the volume τ, which has played no rôle in connection with our fixed oscillators. The considerations up to this point, therefore, are inapplicable to an assembly of gas molecules. Let us see how they can be extended to include the treatment of

5.7 THE METHOD OF DARWIN AND FOWLER

gaseous assemblies. The principles involved are best discussed in connection with an ideal gas.

The ideal gas. The assembly will be supposed to consist of ν mass points, each of mass m, without mutual interactions, confined to a volume τ. Mathematically, this confinement can be represented as follows: introduce a potential energy $V(x,y,z)$ for every mass point, and assume that $V = 0$ everywhere within the volume of the gas, while it rises suddenly to $+\infty$ at the walls of the container and maintains this value everywhere outside. Then, since no mass points have an infinite kinetic energy, none can leave the volume; the function V is a symbol of their imprisonment. It introduces no undesirable forces because it is constant everywhere inside the gas.

To find the partition function we proceed exactly as we did for the quantized systems:

$$Z = \sum_i g_i z^{\varepsilon_i}.$$

The question is: how are we to find our sequence of ε_i? As will be remembered, the ε_i are the energy states of a single system, and these form a continuous sequence in this case. The energy of one mass point is simply

$$\varepsilon = \frac{1}{2m}(p_x^2 + p_y^2 + p_z^2) + V(x,y,z);$$

it depends on the p's and the q's (for which we have here written for convenience x, y, z). In order to be able to apply this method we divide the six-dimensional continuum of the p's and q's of a single mass point (the phase space of a single mass point, to use Gibbs's terminology) into small but discrete cells $\Delta p_x \Delta p_y \Delta p_z \Delta x \Delta y \Delta z$, not necessarily of equal size, and number them $\Delta_1, \Delta_2, \Delta_3, \ldots$. We select one point from the inside of every cell and label the values of $p_x \ldots z$ corresponding to it $p_{xi}, p_{yi}, p_{zi}, x_i, y_i, z_i$, i being the index of the particular cell. We have now produced artificially a discrete set of energies

$$\varepsilon_i = \frac{1}{2m}(p_{xi}^2 + p_{yi}^2 + p_{zi}^2) + V(x_i, y_i, z_i).$$

The elementary weight corresponding to each ε_i is clearly that corresponding to the size of the element of phase space Δ_i, viz.,

$$g_i = \frac{\Delta_i}{h^3}.$$

Hence

$$Z' = \sum_i \frac{\Delta_i}{h^3} z^{\frac{1}{2m}(p_{xi}^2+p_{yi}^2+p_{zi}^2)+V(x_i,y_i,z_i)}$$

If now the sizes of the cells Δ_i shrink to 0, the sum over all i must be replaced by an integration over the 3 p's and the 3 q's (x,y,z), while Δ_i becomes $dp_x dp_y dp_z\, dxdydz$. Consequently, in the limit,

$$Z' \rightarrow Z = \frac{1}{h^3} \int \overset{(6)}{\cdots} \int z^{\frac{1}{2m}(p_x^2+p_y^2+p_z^2)+V(x,y,z)} dp_x dp_y dp_z\, dxdydz. \quad (34)$$

As we have seen before, partition functions diverge rapidly when $z \rightarrow 1$, for they are power series in z. Our interest in them is always confined to values of $z < 1$. With this in mind the integration over the coordinates in eq. (34) can be performed very easily. The term $z^V = 1$ inside the volume τ; it is zero everywhere outside because a fraction, raised to the power ∞, vanishes. Hence

$$Z = \frac{\tau}{h^3} \int\int\int z^{\frac{1}{2m}(p_x^2+p_y^2+p_z^2)} dp_x dp_y dp_z = \frac{(2\pi m)^{3/2} \tau}{h^3 (\log \frac{1}{z})^{3/2}}. \quad (35)$$

The average number of mass points within the element of phase $dp_x dp_y dp_z\, dxdydz$ is given by eq. (27). Dividing this expression by the total number of mass points, ν, we get the probability, P_1, of finding a given mass point within this element of phase. Thus

$$P_1 = \frac{1}{Z(\vartheta)} \cdot g \cdot \vartheta^e = \frac{h^3 \left(\log \frac{1}{\vartheta}\right)^{3/2}}{(2\pi m)^{3/2} \tau} \cdot \frac{dp_x dp_y dp_z\, dxdydz}{h^3} \cdot \vartheta^{\frac{1}{2m}(p_x^2+p_y^2+p_z^2)}.$$

This probability is seen to be independent of position. We find the probability P_1' that the mass point shall have a given triple of momentum components, $p_x p_y p_z$, regardless of its position, by integrating over $dxdydz$. Thus

$$P'_1 = \int\int\int P_1\, dxdydz = \frac{\left(\log \frac{1}{\vartheta}\right)^{3/2}}{(2\pi m)^{3/2}} \vartheta^{\frac{1}{2m}(p_x^2+p_y^2+p_z^2)} dp_x dp_y dp_z,$$

and if we substitute eq. (33) as well as $p = mv$,

$$P_1' = \left(\frac{m}{2\pi kT}\right)^{3/2} e^{-\frac{\frac{1}{2}m(v_x^2+v_y^2+v_z^2)}{kT}} dv_x dv_y dv_z.$$

This is again Maxwell's law of distribution of velocities which was derived previously by Gibbs's method (5.5–18a).

5.7 THE METHOD OF DARWIN AND FOWLER

By applying eq. (29) to our assembly we expect to arrive at the equipartition theorem, eq. (5.5-24), for an ideal gas. This is actually the case, for if we take $\frac{\partial}{\partial \vartheta} \log Z(\vartheta)$ according to eq. (35) the result is $\frac{\frac{3}{2}\vartheta^{-1}}{\log \frac{1}{\vartheta}}$, so that

$$\overline{E} = \nu \vartheta \frac{\partial}{\partial \vartheta} \log Z(\vartheta) = \frac{\frac{3}{2}\nu}{\log \frac{1}{\vartheta}} = \frac{3}{2}\nu kT = \frac{n}{2} kT,$$

if n is the number of degrees of freedom.

If it were desirable to derive again the relation $\vartheta = e^{-\frac{1}{kT}}$ we could have proceeded conversely: if we know, on the one hand, that $\overline{E} = \frac{3}{2}\nu \left(\log \frac{1}{\vartheta}\right)^{-1}$, and on the other hand, $\overline{E} = \frac{3}{2}\nu kT$, this relation follows.

Gibbs's phase integral and Darwin-Fowler's partition function. There is a remarkable mathematical similarity between the phase integral of Gibbs

$$I = \int e^{-\frac{H}{\vartheta}} d\phi = \int e^{-\frac{H}{kT}} d\phi \qquad (36)$$

and the partition function of Darwin and Fowler

$$Z(\vartheta) = \sum_i g_i \vartheta^{\varepsilon_i} = \sum_i g_i e^{-\frac{\varepsilon_i}{kT}}, \qquad (37)$$

although the logical implications are entirely different. While we are here emphasizing formal similarities it must not be forgotten that eq. (36) presupposes the existence of a whole "ensemble of assemblies," while eq. (37) implies the existence of only one. H in eq. (36) is the total energy of the whole assembly which is in all cases a significant quantity; ε_i in eq. (37) is the energy of one component system of the assembly, i.e., a quantity which has meaning as long as the energy of an individual system can be specified, and loses it when interactions become so strong that this is no longer possible. Hence we expect eq. (37), also, to become meaningless when interactions between individual systems become strong. We shall return to this shortly and consider at present the condition in which the assembly may be said to consist of practically isolated systems in the sense previously outlined.

In that case one may regard eq. (37) as a formal generalization of Gibbs's phase integral to include quantized systems. More precisely, $[Z(\vartheta)]^\nu$ must be looked upon as a generalized phase integral. For, suppose that the system in question is a classical one. Then, as was shown, eq. (37) takes the form

$$Z(\vartheta) = \int \overset{(6)}{\cdots} \int e^{-\frac{1}{2mkT}(p_1^2+p_2^2+p_3^2) - \frac{V(q_1, q_2, q_3)}{kT}} dp_1 dp_2 dp_3 dq_1 dq_2 dq_3,$$

where we are now writing q's for x's. We can raise this to the νth power by making the integration 6ν fold and multiplying the integrand by $6\nu - 1$ similar expressions with different indices on the p's and q's. But then we get exactly I of eq. (36).

It is for this reason, of course, that the Darwin-Fowler

$$\bar{E} = \nu\vartheta \frac{\partial}{\partial \vartheta} \log Z(\vartheta) \tag{29}$$

is the same as Gibbs's

$$\bar{H} = -\frac{\partial}{\partial\left(\frac{1}{\theta}\right)} \log I,$$

which is seen immediately on substituting $[Z(\vartheta)]^\nu = I$, and $\log \vartheta = -\frac{1}{\theta}$.

Now let us consider a real gas at high pressure, that is, a case in which the interactions between the systems are so strong that no individual ε_i can be specified. There is no reason why we should not now define a $Z(\vartheta)$ which refers to the whole body of the gas, treating the whole assembly as a single system. This would read

$$Z(\vartheta) = \sum_i g_i e^{-\frac{H_i}{kT}}, \tag{38}$$

where the H_i's are the energy values of the whole assembly and the g_i's the corresponding elementary weights, both determined by the rules of quantization. This Z would be an exact transcription of Gibbs's phase integral. Unfortunately, however, it has no meaning in connection with the reasoning of Darwin and Fowler, for there is in this case no great number of similar systems to be used in averaging, nor are there composite states of a " super " assembly whose weights are determined by means of this function. Darwin and Fowler do employ " partition " functions relating to entire assemblies as individual systems, i.e., functions like eq. (38). They are undoubtedly convenient mathematical constructs, but we feel that the justification of their use requires an appeal to the Gibbsian notion of ensembles and cannot be made by reference to a really conservative dynamical assembly which otherwise forms the basis of these authors' investigations.

Entropy. No theory of statistical mechanics is complete unless it is fully competent to deal with the question of entropy. We shall satisfy ourselves in this regard by proving that it is possible, with the use of concepts discussed in this section, to define a quantity which has all the significant properties that the empirical entropy is known to possess, viz.: to increase when the assembly undergoes certain changes, to be a variable of state, and to satisfy the equations of thermodynamics.

In Gibbs's theory, we agreed to regard $-\bar{\eta}$ as the equivalent of entropy. This quantity could be calculated whenever the distribution in the ensemble was known, even in the case of non-canonical distributions. In one sense this fact is disturbing, for empirical entropy, as pointed out in the beginning of this chapter, has no meaning in connection with non-equilibrium states. Hence Gibbs's theory may be said to provide more information on this point than experiment can verify. But this state of affairs is not a very unusual one; it is always tolerated in physics unless it leads to definite contradictions. Few theories do not "overpredict" experience.

The reasoning of Darwin and Fowler does not permit an entropy definition which is applicable to non-equilibrium states and is therefore perhaps more satisfactory. Their results do not refer to any one macroscopic state that may be specified at will; they refer to the average over all macroscopic states and are derived by passing them all in review, as it were. This "average" state is then interpreted as the normal state of the assembly, i.e., the one which it assumes in equilibrium. Entropy, too, can be defined only with reference to this "average" state. As a consequence of this significant difference between the entropy analogy of Gibbs and the entropy of Darwin and Fowler—however it may be defined in detail—we cannot now prove the increasing property by showing that the entropy of one given system increases as time goes on, as we did for $-\bar{\eta}$. We can now deal only with equilibrium states and must limit our proof to the increase in entropy when one assembly is combined with another in space, each being in equilibrium initially. Let us turn to the mathematical formulation of this matter.

We *define* the entropy S as follows:

$$S = k \log C; \qquad (39)$$

k is Boltzmann's constant; and C, it will be remembered, is the sum of the weights for all states of the assembly compatible with a total energy E. C was calculated in eq. (20) for the case of an assembly containing but one kind of system. From this relation we see at once that

$$\log C = -E \log \vartheta + \nu \log Z(\vartheta) - \frac{1}{2} \log \left[2\pi \nu \vartheta^2 \frac{\phi''(\vartheta)}{\phi(\vartheta)} \right].$$

Substitution of eq. (16) for E shows that the first and second terms on the right have ν, a very large number, multiplying the logarithm, and this leads us to suspect that the third term is small as compared with the other two. This can actually be verified by closer study. We may therefore write

$$\log C = -E \log \vartheta + \nu \log Z(\vartheta). \tag{40}$$

In the general case of several kinds of systems, treated previously, it is only necessary to replace $Z(\vartheta)$ by the product

$$Z^{(1)\nu_1/\nu} Z^{(2)\nu_2/\nu} \ldots Z^{(\lambda)\nu_\lambda/\nu}, \quad \text{and} \quad E \text{ by } \overline{E^{(1)}} + \overline{E^{(2)}} + \ldots + \overline{E^{(\lambda)}}.$$

(Cf. eq. (26), et seq.) Hence, if the assembly contains two kinds of systems,

$$\log C = -E \log \vartheta + \nu_1 \log Z^{(1)}(\vartheta) + \nu_2 \log Z^{(2)}(\vartheta) \tag{41}$$

A. We shall show first that the entropy as defined by eq. (39) increases when two assemblies at different temperatures are joined into one. The entropies of the two separate assemblies are

$$S_1 = k \{ \nu_1 \log Z^{(1)}(\vartheta_1) - E_1 \log \vartheta_1 \}$$
$$S_2 = k \{ \nu_2 \log Z^{(2)}(\vartheta_2) - E_2 \log \vartheta_2 \}.$$

Putting $\dfrac{\partial S_1}{\partial \vartheta_1} = 0$, the value of ϑ_1 which makes S_1 a minimum is found to satisfy the relation $E_1 = \nu_1 \vartheta_1 \dfrac{\partial}{\partial \vartheta_1} \log Z^{(1)}(\vartheta_1)$. But this is exactly the equation defining the equilibrium value of E_1; cf. eq. (16). Hence we see that S_1 is a minimum for that value of ϑ_1 which the assembly assumes in equilibrium for a given total energy E_1. Similar remarks apply to S_2 and ϑ_2.

Now let ϑ be the common temperature (on the $e^{-\frac{1}{kT}}$ scale!) of the two assemblies after they have been joined. The entropy S of this composite assembly is given by eq. (41) multiplied by k, with

$$E = E_1 + E_2.$$

This may be written

$$S = S_1(\vartheta) + S_2(\vartheta).$$

However, unless $\vartheta = \vartheta_1 = \vartheta_2$,

$$S_1(\vartheta_1) < S_1(\vartheta), \quad \text{and} \quad S_2(\vartheta_2) < S_2(\vartheta),$$

therefore

$$S(\vartheta) > S_1(\vartheta_1) + S_2(\vartheta_2).$$

B. Having proved the increasing property of S, we shall now convince ourselves that the definition of S by eq. (39) satisfies the fundamental thermodynamic equation (5.4-4) which will be written in the form

$$dE + dW = TdS. \tag{42}$$

Let us first calculate the left-hand side of this equation. For simplicity, the assembly will be supposed to contain systems of only one kind. The infinitesimal change represented by eq. (42) will be regarded as due to: (1) a change in temperature, i.e., in the parameter ϑ; (2) changes in the external configuration of bodies during which the assembly does work against outside forces. In general, every elementary energy state of a system composing the assembly is a function of external coordinates $\xi_1, \xi_2, \ldots$ and should therefore be written $\varepsilon_i(\xi_1, \xi_2 \ldots)$. It is only when the ξ's remain constant (as in our previous considerations) that this dependence may be neglected. For the same reason the partition function Z will also contain the coordinates, so that $Z = Z(\vartheta, \xi_1, \xi_2 \ldots)$.

Bearing this in mind we differentiate

$$E = \nu\vartheta \frac{\partial}{\partial \vartheta} \log Z(\vartheta, \xi_1, \xi_2, \ldots), \tag{43}$$

obtaining

$$dE = \nu \left\{ \left[\frac{\partial}{\partial \vartheta} \log Z + \vartheta \frac{\partial^2}{\partial \vartheta^2} \log Z \right] d\vartheta + \sum_\lambda \vartheta \frac{\partial^2}{\partial \vartheta \partial \xi_\lambda} (\log Z) d\xi_\lambda \right\}. \tag{44}$$

In calculating dW it is to be observed that this quantity is the work done by the forces exerted by all systems in the assembly. Consider the systems in state i, each of which has an energy $\varepsilon_i(\xi_1, \xi_2, \ldots)$.

The force called into play by one such system when the λth external coordinate is slightly changed will be $-\dfrac{\partial}{\partial \xi_\lambda}\varepsilon_i$. Hence the total work done by one such system is $-\sum_\lambda \dfrac{\partial}{\partial \xi_\lambda}\varepsilon_i d\xi_\lambda$. Now there are, on the average, $\bar{a}_i = \dfrac{\nu g_i \vartheta^{\varepsilon_i}}{Z(\vartheta)}$ systems in the state i. Their contribution to the work is

$$-\frac{\nu g_i \vartheta^{\varepsilon_i}}{Z}\sum_\lambda \frac{\partial}{\partial \xi_\lambda}\varepsilon_i d\xi_\lambda,$$

and the total contribution of the systems in all states

$$dW = -\frac{\nu}{Z}\sum_{i,\lambda} g_i \vartheta^{\varepsilon_i} \frac{\partial}{\partial \xi_\lambda}\varepsilon_i d\xi_\lambda.$$

This may be put in a more convenient form. Since $Z(\vartheta) = \sum_i g_i \vartheta^{\varepsilon_i}$,

$$\sum_\lambda \frac{\partial}{\partial \xi_\lambda}\log Z(\vartheta, \xi_1, \xi_2, \ldots) = \frac{1}{Z(\vartheta, \xi_1, \xi_2, \ldots)}\sum_{i,\lambda} g_i \vartheta^{\varepsilon_i} \log \vartheta \frac{\partial}{\partial \xi_\lambda}\varepsilon_i.$$

Hence

$$dW = -\frac{\nu}{\log \vartheta}\sum_\lambda \frac{\partial}{\partial \xi_\lambda}\log Z(\vartheta, \xi_1, \xi_2, \ldots)d\xi_\lambda.$$

Adding this to eq. (44),

$$dE + dW = \nu\left\{\left[\frac{\partial}{\partial \vartheta}\log Z + \vartheta \frac{\partial^2}{\partial \vartheta^2}\log Z\right]d\vartheta \right.$$

$$\left. + \sum_\lambda \left[\vartheta \frac{\partial^2}{\partial \vartheta \partial \xi_\lambda}\log Z - \frac{1}{\log \vartheta}\frac{\partial}{\partial \xi_\lambda}\log Z\right]d\xi_\lambda\right\} \qquad (45)$$

On the other hand, consider eq. (39). In view of eqs. (40) and (43) this may be stated thus:

$$S = k\nu\left[\log Z - \vartheta \log \vartheta \cdot \frac{\partial}{\partial \vartheta}\log Z\right].$$

If we take the total differential of this expression, and then divide it

5.7 THE METHOD OF DARWIN AND FOWLER

by $(-\log \vartheta)$, the result is k times the right-hand side of eq. (45). But $-\log \vartheta = 1/kT$, as was shown previously. Therefore

$$kTdS = k(dE + dW),$$

which is identical with eq. (42). This completes the proof.

C. No further work is involved in showing that $k \log C$ is a variable of state, for one observes that this expression, by eq. (40), is a function only of ϑ and $Z(\vartheta_1, \xi_1, \xi_2, \ldots)$. Now ϑ is simply related to the temperature, and Z is fixed when all ξ's are given. The knowledge of all ξ's, however, is tantamount to a knowledge of the confinement of the systems, i.e., to a knowledge of the volume. Consequently $k \log C$ is a function of two variables of state, viz., temperature and volume.

This concludes our discussion of the method of Darwin and Fowler in developing the theory of statistical mechanics. Our principal aim has been to expose its logical foundation. We have naturally been unable, in this brief space, to do full justice to its analytic power, which is apparent in its many applications to problems of recent physical interest. As a physical theory it is particularly interesting from the point of view of the present book, for it is sufficiently elaborate and useful to make its close inspection very much worth while; yet it is still unfamiliar enough so that its logical elements appear in sharp outline, undimmed by the hazy self-evidence which theories assume when they are widely used.

In Chapter IX we shall again make use of the formalism of Darwin and Fowler in deriving distribution laws for new types of statistics in which different assignments of weights are made from the one considered in this section.

CHAPTER VI

THE PHYSICS OF CONTINUA

6.1. Concept of the Continuum in Physics. In our treatment of the foundations of mechanics we introduced the concept of the material particle as the fundamental entity in the study of motion. The motion of a single isolated particle, however, is a barren subject for discussion. In fact, we found that our definitions of mass and force are dependent on the existence of a plurality of particles. If the interaction of an aggregate of particles is describable by central forces, several interesting and valuable theorems may be derived concerning the momentum, energy, and virial of such an aggregate. A particular case of much importance is that of a rigid body, which may be considered to be composed of a large number of particles constrained to move so that the distance between every pair of particles remains unaltered. For the solution of problems arising in the motion of such bodies all the methods described in Chapter III are at once available, although the solution is much facilitated by the introduction of several new concepts, viz.: torque or force moment, center of mass, center of gravity, moment of inertia, angular momentum, etc. However, no essentially new methods of attack are required, and hence from the point of view of foundations it is unnecessary to discuss further the mechanics of rigid bodies. The motion of aggregates of particles not subject to the constraints of rigidity has been the subject-matter of Chapter V, where we have seen the importance of the statistical method.

Particle versus continuum. It might seem, then, as if the particle concept and particle mechanics were entirely fitted to handle all cases of the motions of bodies. This, however, is not the situation actually encountered in physics, and we have now to consider the reason. It will be recalled that the particle or atomic concept was not without a rival in antiquity. The notion of matter as being continuous in nature long held sway and was not completely dislodged by the efforts of the atomists. It is true that the two contrasting views of discrete and continuous matter were of most interest to the early philosophers from the standpoint of the ultimate constitution of things. Nevertheless, the idea of the continuum, once having been firmly rooted in men's minds, continued to impress physicists with its importance.

6.1 CONCEPT OF THE CONTINUUM IN PHYSICS 281

During the eighteenth and nineteenth centuries much attention was paid to the motion of elastic media and in particular to fluids considered as purely continuous media. In Chapter I we illustrated our analysis of the method of elementary abstraction by applying it to the case of a fluid and there developed the notion of a physical continuum. We now wish to elaborate on that discussion. The basic idea of a continuum is that of a region or medium in which any volume element no matter how small still contains points of the region or material of the medium.[1] The continuum is supposed to occupy a finite space which in classical physics is assumed to be Euclidean and three-dimensional, although there is no *a priori* reason for excluding other types of space if occasion should demand them.

We have first to note the character of the physical quantities which specify the behavior of such a medium; as we have already seen in Chapter I, these show in certain respects considerable contrast to the quantities characteristic of the mechanics of particles. In particle mechanics the essential problem is the determination of the position and velocity of each particle as a function of the time, which thus appears as the sole independent variable; the differential equations of particle mechanics are fundamentally ordinary in nature. This statement may at first seem to be contradicted by the existence of the equations of Lagrange and Hamilton with their partial derivatives of L and H with respect to the generalized coordinates, velocities, or conjugate momenta (Chapter III). However, when the indicated partial differentiations have been performed the resulting equation is invariably an ordinary one with the time as the fundamental independent variable, and the problem reduces to the evaluation of the generalized coordinates, etc., as functions of the time. Hence though the equations of Lagrange and Hamilton contain partial derivatives they are not treated mathematically as partial differential equations: we do not try to solve them for H and L. On the other hand, we have seen that the transformation theory of mechanics does involve a genuine partial differential equation, namely, the Hamilton-Jacobi equation for the transformation or substitution function S (Sec. 3.15). In this case the ordinary equations are, so to speak, elided, and the problem solved in terms of the complete integral of a partial differential equation. From this integral the equations which implicitly contain the dependence of the coordinates on the time are obtained by differ-

---

[1] This should be extended to include local discontinuities or singularities. It will be assumed, however, that if such exist they form a discrete set. We wish, for example, to be able to contemplate the existence of material particles or rigid bodies in the continuum without thereby causing it to lose its significance.

entiation with respect to the constants of integration. This provides an explicit exception to our dogmatic assertion immediately above. It is not a serious one, however, for it does not disturb the really important thesis that the one fundamental independent physical variable in particle mechanics is the time.[1] This statement must of course be understood in the light of the remarks on the nature of time as used in physics in Sec. 2.5. Time is a parameter which relates all physical systems to a standard system, and a transformation of this parameter involving a transformation of the standard system is perfectly feasible. In this way it is possible, for example, to take away from the symbol t (representing physical time in the usual sense—viz., as measured by an astronomical clock) its peculiar rôle in the equations of motion and indeed to give it the form of a canonical variable in the Hamiltonian equations of motion, with the Hamiltonian itself as its conjugate momentum. This transformation[2] has appealed particularly to some since it gives the canonical equations a highly symmetrical form. Nevertheless in all the common applications of mechanics the way in which the equations are actually used justifies the statement made above.

Now the situation with respect to continua is quite otherwise, for our knowledge of the behavior of a physical continuum depends on our ability to express what is happening in all its parts as a function of the time. Consequently all physical quantities appropriate for the description of continua must be functions of the space coordinates as well as the time. Such quantities, for example, are the *density* of the medium, the *displacement* from equilibrium position, the *pressure*, etc. It is clear that the differential equations fundamental for the description of continua must contain both space coordinates and time as independent variables: hence they must be partial differential equations. Recall the equation of continuity in fluid motion, as well as the wave equation (Secs. 1.7 and 1.8) as illustrations of this fact. We have already noted the multiplicity of solutions of such equations and the fact that they involve arbitrary functions; further, we have emphasized the significance of boundary conditions in supplying the concreteness (and sometimes discreteness) necessary for the meaningful application of the solutions to actual physical situations. These points do not need to be stressed further here.

---

[1] The equations of the potential theory, e.g., that of Laplace ($\nabla^2 V = 0$) are certainly important in particle mechanics. They will not be forgotten and are treated later in this section.

[2] See, for example, "Handbuch der Physik" (Springer, Berlin, 1927), Vol. 5, pp. 100 ff.

The field concept in particle dynamics. For the moment, however, we wish to point out that the concept of the continuum is capable of more extensive application than our previous remarks might indicate. Curiously enough, it also proves possible to use this concept to advantage in the study of particle motion, in which one might at first suppose it to be of little value. Here the continuum enters in the guise of *field*. Suppose we have a particle. It is acted on by a force, presumably due ultimately to other particles: if we insert the proper function of space and time for **F** in the equation $\mathbf{F} = m\ddot{\mathbf{r}}$, integration and use of boundary conditions yield the path of the particle in space and time. Now there is often a gain in physical perspicuousness in forgetting the "origin" of the force acting on the particle and considering simply that the latter moves in a *field of force*, i.e., a continuous region of space at every point of which there is defined the force which would act on a "standard" particle placed there (e.g., particle of unit mass, unit electric charge, etc.). Instances of such fields of force are familiar to all, viz., gravitational, electrostatic, magnetic, and electromagnetic fields. Thus perhaps the simplest type of gravitational field is a continuum of unlimited extent in which a particle of unit mass is acted on by a force varying in magnitude inversely as the square of the distance from some fixed point, and directed along the line joining the particle to that point. Similarly, an illustration of a magnetic field is a limitless continuum in which a unit magnetic pole is acted on by a force varying in magnitude inversely as the square of the distance from some fixed point and directed along the line joining the pole to the point. This is of course an idealized case: magnetic fields encountered in practice are not so simply described analytically, particularly with respect to direction; and the same is true of electrostatic and electromagnetic fields. For this reason, and to provide a convenient pictorial representation of fields, graphical methods are often resorted to. The concept of *line of force* is a useful tool in the visualization of fields. Such a line is a curve whose direction at any point coincides with that of the force acting at that point. Thus if the components of the force along the x, y, z axes are given by F_x, F_y, F_z, the differential equations of the lines of force are

$$\frac{dx}{F_x} = \frac{dy}{F_y} = \frac{dz}{F_z}. \tag{1}$$

The mapping of a force field by means of lines of force not only provides a picture of force direction but also can be made to describe the *magnitude* of the force at any point from the number of lines of force which pass normally through unit area at the point.

Force is not the only quantity which may characterize a field. The idea of potential immediately comes to mind as being often more useful. Whereas force is a vector quantity, the potential is in general a scalar. It may be worth while to illustrate its significance by reviewing the special case of gravitational potential, defined as the *potential energy* per unit mass for a particle in a gravitational field. Thus the potential is a function of the space coordinates being used and may also be an explicit function of the time. In the latter case the field may be called variable, non-static, or non-conservative. For the moment let us neglect such cases and confine our attention to the static potential. Its most important characteristic is that its gradient is the negative of the force per unit mass, often called the *intensity* of the field. Thus if V denotes the potential and $\mathbf{F}$ the intensity,

$$\mathbf{F} = -\left[\mathbf{i}\frac{\partial V}{\partial x} + \mathbf{j}\frac{\partial V}{\partial y} + \mathbf{k}\frac{\partial V}{\partial z}\right] = -\nabla V. \tag{2}$$

The use of the negative sign has its source in the definition of potential energy. From the definition of work, it is seen that the difference in potential between any two points in the field is the work done when unit mass moves from one point to another. Since V is a function of the coordinates only, this work is independent of the path taken between the two points. The definition of electrostatic and magnetic potentials follows in similar fashion with unit mass replaced by unit charge and unit pole, respectively.

The gravitational potential gains added significance from the law of force flux due to Gauss. The flux of gravitational force through any closed surface is defined as the surface integral of the normal component of the intensity of the gravitational field. Thus if Φ denotes this flux, we have

$$\Phi = \int_S \mathbf{F} \cdot d\mathbf{S}. \tag{3}$$

Here $d\mathbf{S}$ denotes a vector normal to the surface dS and of magnitude equal to the area of dS. The integral is taken over the whole surface. Gauss's law now says

$$\Phi = -4\pi GM, \tag{4}$$

where M is the total mass enclosed by the surface and G is the constant of gravitation (6.67×10^{-8} cm^3/gram sec^2). The surface integral alone can be transformed into a volume integral by means of the divergence theorem. Thus

$$\Phi = \int_S \mathbf{F} \cdot d\mathbf{S} = \int_\tau \nabla \cdot \mathbf{F} d\tau, \tag{5}$$

where $\nabla \cdot \mathbf{F} = \dfrac{\partial F_x}{\partial x} + \dfrac{\partial F_y}{\partial y} + \dfrac{\partial F_z}{\partial z}$, and $d\tau$ is the element of volume.[1]

Now using eqs. (2) and (4) we have

$$\int_\tau \nabla \cdot \nabla V \, d\tau = 4\pi G M = 4\pi G \int_\tau \rho \, d\tau, \tag{6}$$

where ρ is the density of mass distribution enclosed by the surface and a function of the space coordinates. Since eq. (6) holds for any closed surface in the field, it is clear that we must have

$$\nabla^2 V = 4\pi G \rho, \tag{7}$$

replacing $\nabla \cdot \nabla V$ by $\nabla^2 V$, where in Cartesian notation

$$\nabla^2 V = \frac{\partial^2 V}{\partial x^2} + \frac{\partial^2 V}{\partial y^2} + \frac{\partial^2 V}{\partial z^2}, \tag{8}$$

and is called the *Laplacian* of V. Eq. (7) then says that the Laplacian of the potential of a gravitational field is equal to a constant times the density of the distribution of gravitating matter in the field. The nature of the field may then be thought of as completely determined by the solution of this second order partial differential equation, which is called Poisson's equation. In a region of the field where ρ is everywhere zero, the equation reduces to the form,

$$\nabla^2 V = 0, \tag{9}$$

which is named for Laplace, and describes completely the character of the field outside gravitating matter. From its nature as a partial differential equation, it has very general solutions which must be made specific by the application of boundary conditions. It might be thought that even with the latter it would be difficult to get solutions for special cases. However, there is an important uniqueness theorem governing the solutions of Laplace's equation. It can be shown that a solution for which the potential has an assigned value at every point of the boundary of a given region is unique, i.e., in this case V is unambiguously defined everywhere in space.[2] Laplace's and Poisson's equations are without doubt two of the most important field equations in physics, describing as they do the nature of not only

---

[1] For the derivation of Gauss's law and the divergence theorem, any text on analytical mechanics may be consulted, e.g., Lindsay, "Physical Mechanics" (D. Van Nostrand, New York, 2nd edition, 1949.)

[2] See, for example, L. Page, "Introduction to Theoretical Physics" (D. Van Nostrand, New York, 2nd edition), p. 266.

gravitational but also electrostatic and magnetic fields. We have further noticed that Laplace's equation enters into hydrodynamics (Sec. 1.7) as the equation of continuity of a perfect, homogeneous, incompressible fluid. Here V becomes the so-called velocity potential, i.e., the flow velocity of the fluid is the gradient of V (or the negative of this in the choice of many writers).

We have already noted the graphical representation of a field by means of lines of force. A similar and often useful plot can be constructed of the surfaces of constant potential, the so-called equipotential surfaces, to which the lines of force are normal trajectories. These are particularly important in electrostatics, since the surfaces of conductors are equipotential surfaces. In the same connection there come to mind numerous other examples of the mapping of fields by means of equal-value curves and surfaces, e.g., the isoclinic, isogonic, and isodynamic lines of the terrestrial magnetism map, and the isobaric lines of the weather map. For many purposes this type of graphical representation of a field provides a picturesque completeness which is lacking in a purely analytical description, and its utility is attested by its widespread use in practical physics.

A question which now arises is: when one stresses the field concept in particle mechanics, what part does the particle play? In the first place, we note that the presence of a particle must always be considered to disturb the field into which it is brought. This introduces a certain difficulty into the definition of the physical quantities characteristic of the field, which are always defined in terms of some effect on a unit particle. Strictly speaking, it is therefore impossible to measure precisely any field quantity. Theoretically the definition must be carried out by the use of the limit concept: the intensity of the field becomes the limit of the force per unit quantity (charge, pole, mass, etc.) as the quantity goes to zero. In practice this difficulty is not troublesome, for most such quantities are measured indirectly. However, the whole problem of the disturbance of physical systems (whether considered as continua or discrete systems of particles) by the instruments of measurement is an important one; it will be discussed in the chapter on quantum mechanics, for it is in connection with atomic measurement that it has become most acute.

The second part of the answer to the question mentioned at the beginning of the previous paragraph is that it is perfectly possible to find a place for a particle (without extension) in a continuum: it becomes a *singularity*, i.e., a place where at least some of the field quantities become discontinuous. In the neighborhood of an isolated mass particle the potential varies as $1/r$, where r is the distance to the particle. Hence $V \doteq \infty$ as $r \doteq 0$. It is therefore clear that particle mechanics can be considered a branch of field or continuum physics if we are willing to consider continua with singularities. The equations of motion for particles become the equations of motion of the singularities of a field. With regard to mechanics we can then adopt formally two extreme points of view: (1) there is no such thing as a continuum—all physical systems are ultimately aggregates of particles; (2) the only physical entities are continua containing singularities whose motion may be interpreted as that of particles. The first is, of course, fundamentally the atomic hypothesis; the second emphasizes the purely ideal nature of the particle concept and prefers to look upon the continuum as more fundamental. Actually, physics has progressed without fanatical adherence to either doctrine. A middle course has been pursued as various problems seemed more approachable by one method than by the other. If the contemporary era seems wholly an age of atomism we must not forget that the theory of relativity is essentially a field theory.

6.2. Scalar and Vector Fields. We have seen that a field is characterized by the values associated with certain physical quantities at every point and at every instant of time. There are several types of fields, depending on the nature of these quantities. A physical quantity which has a numerical value at every point in a region independent of direction is a *scalar* quantity, and a field characterized by such is called a *scalar* field. The gravitational potential is a scalar, for its value at any point has no direction associated with it. Moreover, if the gravitational field is due to an aggregate of mass particles, the total value of the potential at a given point is the algebraic sum of the values due to each particle separately. In so far as the gravitational field is characterized by the potential, it is a scalar field. In similar fashion the velocity potential which characterizes a fluid medium under certain conditions (irrotational motion) is a scalar, and the fluid under these conditions and to this extent may be considered a scalar field. Other illustrations of scalars and scalar fields are provided by density and temperature.

Transformations of coordinates. Meaning of invariance. Many physical quantities are characterized not only by their magnitude but also by direction in space. Such are the mechanical quantities position,[1] displacement, velocity, acceleration, force. These are called *vectors*, and a field characterized by such quantities is called a *vector* field. So far as a gravitational field is described by the gravitational

---

[1] See Chapter III, Secs. 3.1, 3.2, 3.3 for details.

intensity it is a vector field. The magnetic, electrostatic, and electromagnetic fields provide similar illustrations. A more important criterion for the meaning of a vector is to be found in its transformation properties. Let us consider this question.

Suppose first we take a scalar field quantity, say the density ρ. This will be a function of the rectangular coordinates, x, y, z, in a given inertial system, and perhaps of t, although for simplicity we omit the t at present. Suppose now that there is a transformation of coordinates represented by

$$x' = x'(x, y, z), \quad y' = y'(x, y, z), \quad z' = z'(x, y, z)$$

$$x = x(x', y', z'), \quad y = y(x', y', z'), \quad z = z(x', y', z').$$

On substitution of x, y, z in terms of x', y', z' into ρ, we get

$$\rho(x, y, z) = \rho'(x', y', z').$$

The *form* of the functional dependence of ρ on x, y, z is in general changed by the transformation, but the numerical value of ρ is not changed, no matter how general the transformation, as long as the same units are used in both cases. The scalar quantity has the property of invariance in value under any transformation of coordinates, though in general it does not display invariance in functional form.

Next consider the straight-line segment from the origin of a system of rectangular coordinates to the point $P(x, y, z)$. The length of this segment is $\sqrt{x^2 + y^2 + z^2}$, and its direction with respect to the system of reference is given by its direction cosines or more simply by its projections on the three axes respectively, viz., x, y, z. In a certain sense these three quantities are entirely sufficient to represent the line segment, since if we know them, we know the length of the segment and also its direction. Now let us introduce the linear transformation of coordinates defined in the following way

or

$$\left.\begin{aligned}
x' &= xl_1 + yl_2 + zl_3, \\
y' &= xm_1 + ym_2 + zm_3, \\
z' &= xn_1 + yn_2 + zn_3, \\
x &= x'l_1 + y'm_1 + z'n_1, \\
y &= x'l_2 + y'm_2 + z'n_2, \\
z &= x'l_3 + y'm_3 + z'n_3,
\end{aligned}\right\} \quad (1)$$

in which the coefficients of the x, y, z are not just any numbers but satisfy the conditions expressed by the following equations:

$$\left.\begin{array}{c} \sum_{j=1}^{3} l_j^2 = 1 = l_1^2 + m_1^2 + n_1^2 \\ \sum_{j=1}^{3} m_j^2 = 1 = l_2^2 + m_2^2 + n_2^2 \\ \sum_{j=1}^{3} n_j^2 = 1 = l_3^2 + m_3^2 + n_3^2 \\ \sum_{j=1}^{3} l_j m_j = \sum_{j=1}^{3} m_j n_j = \sum_{j=1}^{3} n_j l_j = 0 \\ = l_1 l_2 + m_1 m_2 + n_1 n_2 = l_2 l_3 + m_2 m_3 + n_2 n_3 \\ = l_3 l_1 + m_3 m_1 + n_3 n_1. \end{array}\right\} \quad (2)$$

Definition of a vector. The transformation represented by (1) is a rotation without translation of the coordinate axes so that the x' axis (the x axis after rotation) has the direction cosines l_1, l_2, l_3 with respect to the old reference system; the y' axis, direction cosines m_1, m_2, m_3; and the z' axis, direction cosines n_1, n_2, n_3. Similarly the direction cosines of the $x, y,$ and z axes with respect to the new axes are l_1, m_1, n_1; l_2, m_2, n_2; l_3, m_3, n_3, respectively. Any set of three quantities which transform in accordance with (1) and (2) constitute a *vector*, or more specifically a three-dimensional vector or 3-vector. Suppose we have two 3-vectors (A_x, A_y, A_z) and (B_x, B_y, B_z). Each of the quantities $A_x, A_y, A_z, B_x, B_y, B_z$ may be a function of the coordinates x, y, z. If the axes are rotated the first vector becomes (A_x', A_y', A_z') and the second (B_x', B_y', B_z'). In order that the two triplets may be 3-vectors, we must then have

$$\left.\begin{array}{l} A_x' = A_x l_1 + A_y l_2 + A_z l_3, \text{ etc.} \\ B_x' = B_x l_1 + B_y l_2 + B_z l_3, \text{ etc.} \end{array}\right\} \quad (3)$$

What does this mean? Imagine that there exists some relation between the two vectors. For example, let us suppose that

$$(A_x, A_y, A_z) = (B_x, B_y, B_z),$$

which is a formal way of writing $A_x = B_x$, $A_y = B_y$ and $A_z = B_z$. If we use the commonly accepted notation and write $(A_x, A_y, A_z) =$ **A** and $(B_x, B_y, B_z) =$ **B** this becomes

$$\mathbf{A} = \mathbf{B}.$$

Then after transformation we must have because of (2), $A_x' = B_x'$, etc., and hence $(A_x', A_y', A_z') = (B_x', B_y', B_z')$, or more concisely

$$\mathbf{A'} = \mathbf{B'}.$$

In other words the original relationship between the two vectors remains true after the transformation. The same result will clearly follow for any other relationship of **A** and **B**. It is this aspect of the vector concept, namely, the invariance of vector relations under transformation, which is perhaps its most important feature. If our physical equations were not to remain invariant under transformation of coordinates it would mean that physical laws depend on the choice of axes. Now we have already noted in our discussion of the equations of motion in mechanics that we cannot expect these equations to keep the same form in all reference systems, particularly if the reference systems in question represent transformations in time as well as in space. Thus the Newtonian equations of motion retain the form $\mathbf{F} = m\mathbf{a}$ only in *inertial* systems, i.e., those which move with constant velocity with respect to each other and the primary inertial system (cf. Sec. 3.7). But transformations into which the time does not enter would seem to be in a somewhat different category. If physical laws change their form under a transformation involving space coordinates alone the indication is that physical space is not isotropic or homogeneous. There is nothing *a priori* impossible about this, for we have already seen that the private space of the individual person is actually non-isotropic and non-homogeneous (Chapter II). However, in classical physics it has been felt desirable to use what might be called amorphous space, and it is this feeling which is responsible for our use of vector equations.

Now it must be admitted at once that the precise type of transformation is not important for the invariance which has just been discussed. Invariance will clearly be secured as long as both **A** and **B** transform according to the same rule, whether this is eqs. (3) or not. In fact, the l's, m's, and n's may be very arbitrary functions of the coordinates and may not satisfy the conditions (2). Still as long as eqs. (3) hold, the invariance of the relation between **A** and **B** will subsist also. It is clear that a 3-vector, as just defined, is only one of a large class of quantities such that equations between quantities which transform in the *same* way remain invariant in form under transforma-

6.2 SCALAR AND VECTOR FIELDS

tion of coordinates. The class of 3-vectors is the class of all quantities which transform the way a line segment transforms under a rotation of axes. The reason these have proved so useful in classical physics is of course that all mechanical quantities which are not scalars are essentially derivatives of the fundamental quantity, displacement, which is most readily represented by a line segment, coupled of course with the *time*.

The definition of a 3-vector which has been discussed above may seem unduly abstract to one who thinks of a vector in physics as being a quantity having direction as well as magnitude. The advantage of the definition given is that it emphasizes the most important aspect of a vector, namely that of the invariance of vector relations under transformation, while the notion of direction is of course only relative to a given reference system. Moreover, the definition we have given is open to much more immediate generalization: we need not confine our attention to sets of three numbers, but may introduce 4-vectors, etc., as we require them, without hampering ourselves with the three-dimensional geometrical restriction implied in "direction." Even in the three-dimensional case the present definition is often more illuminating. As an illustration suppose that in the course of investigation of the motion of a particle with coordinates x, y, z we encounter a set of expressions of the form

$$y\dot{z} - z\dot{y}, \quad z\dot{x} - x\dot{z}, \quad x\dot{y} - y\dot{x},$$

which at first solely for convenience we may denote by $\Omega_x, \Omega_y, \Omega_z$. Let us see how these transform under rotation of axes. Using (1) and (2) we get

$$\Omega_x' = y'\dot{z}' - z'\dot{y}'$$
$$= (y\dot{z} - z\dot{y})(m_2 n_3 - n_2 m_3) + (z\dot{x} - x\dot{z})(m_3 n_1 - m_1 n_3)$$
$$\quad + (x\dot{y} - y\dot{x})(m_1 n_2 - m_2 n_1)$$
$$= \Omega_x l_1 + \Omega_y l_2 + \Omega_z l_3.$$

Similarly we find that

$$\Omega_y' = \Omega_x m_1 + \Omega_y m_2 + \Omega_z m_3,$$
$$\Omega_z' = \Omega_x n_1 + \Omega_y n_2 + \Omega_z n_3.$$

In other words, the triplet $(\Omega_x, \Omega_y, \Omega_z)$ transforms under a rotation of axes precisely as a line segment. According to our definition it is therefore a 3-vector. We may call it Ω and consider Ω_x, Ω_y, and Ω_z as its components along the x, y, z axes, respectively. Ω_x when multiplied by the mass m is called the moment of momentum or angular momentum of the particle about the x axis, and similarly for Ω_y and Ω_z. The vector $m\Omega$ then appears as the total moment of momentum of the particle. The fact that a quantity constructed in this way turns out to be a vector enables us to visualize it geometrically in terms of a line segment with definite length, viz., $m\sqrt{\Omega_x^2 + \Omega_y^2 + \Omega_z^2}$, and with direction cosines $\Omega_x/\sqrt{\Omega_x^2 + \Omega_y^2 + \Omega_z^2}$, etc. As a matter of fact, anyone familiar with the elements of vector analysis will recognize at once that $m\Omega$ is the so-called *vector* product of $\mathbf{r}$ and $m\dot{\mathbf{r}}$, where $\mathbf{r}$ represents the position vector of the particle, i.e., the vector (x, y, z), and $\dot{\mathbf{r}}$ the velocity vector (see Sec. 3.3). Thus

$$m\Omega = m\mathbf{r} \times \dot{\mathbf{r}}$$

to use the Gibbs notation for the vector product.

Our discussion so far has been restricted to the case of 3-vectors. This suffices for a good deal of classical physics. That it does not go far enough, however, is clear from the illustration which introduces the next section.

6.3. Deformable Media and Tensor Fields. We have already had occasion to point out that the rigid body of mechanics is a highly idealized concept and that all actual bodies can be deformed. In the discussion of the deformation of a continuous medium an interesting generalization of the vector concept presents itself. Consider the elementary volume element $dxdydz$, indicated in the following diagram (Fig. 6.1), where for simplicity one vertex is located at the origin of the system of coordinates. We suppose that the element is acted on by a certain stress (force per unit area) which we resolve into the following components. The tensile stress normal to the face $OCGD$ we shall call X_{11}; the tangential stresses (shears) on this face along the y and z directions respectively will be denoted by X_{12} and X_{13}. The corresponding quantities for the face $OAED$ are X_{22}, X_{21}, X_{23}; those for the face $OABC$ are X_{33}, X_{31}, X_{32}. It is thus seen that the complete stress description involves an aggregate of *nine* quantities, which in general will be functions of x, y, z, and t. This aggregate is called the

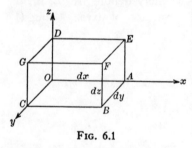

FIG. 6.1

stress tensor [1] and may conveniently be written in the form of a matrix or array, thus

$$S = \begin{Vmatrix} X_{11} & X_{12} & X_{13} \\ X_{21} & X_{22} & X_{23} \\ X_{31} & X_{32} & X_{33} \end{Vmatrix} \quad (1)$$

The above expression must not be confused with a determinant. It is merely a convenient method of arrangement. The separate quantities X_{ik} are called the components of the tensor.

Strain components and strain tensor. Before we investigate the properties of such a tensor let us look further into the physical problem presented by the continuous medium. We shall consider the nature of the deformation associated with the stress. Denote the rectangular components of the displacement of the point O in Fig. 6.1 by ξ, η, ζ, respectively. The displacement in the x direction at A then becomes [2]

$$\xi + \frac{\partial \xi}{\partial x} dx.$$

Hence the net elongation of the volume element in question in the x direction is $\frac{\partial \xi}{\partial x} dx$. The elongation per unit length is $\frac{\partial \xi}{\partial x}$. This is called the *linear* strain in the x direction. It will be denoted by δ_{11}. We therefore have for the linear strains along the x, y, z axes, respectively

$$\delta_{11} = \frac{\partial \xi}{\partial x}, \quad \delta_{22} = \frac{\partial \eta}{\partial y}, \quad \delta_{33} = \frac{\partial \zeta}{\partial z}. \quad (2)$$

These are not the only strains. Consider the face $OAED$ of the volume element in Fig. 6.1. It is represented in Fig. 6.2. A deformation into the parallelogram $OAE'D'$ is a shear, and its measure is the shearing strain DD'/DO, which for an infinitesimal deformation is $\partial \xi/\partial z$, since $DD' = d\xi$ and $OD = dz$. Another possible deformation is that into the parallelogram $OA'E''D$ for which the shearing strain is $\partial \zeta/\partial x$. Both of these are strains for which nothing

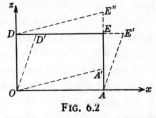

FIG. 6.2

---

[1] This must not be taken to mean that every set of nine quantities constitutes a tensor. Only those aggregates which have certain transformation properties are known by this name. These properties are discussed in detail in the present section.

[2] Refer here to the general discussion in Sec. 1.7.

happens along the y axis. We may therefore think of the total shearing strain about the y axis as the sum of these, viz.,

$$\frac{\partial \xi}{\partial z} + \frac{\partial \zeta}{\partial x}.$$

Following the notation adopted above we might naturally call this expression $\delta_{13} = \delta_{31}$. However, it is more convenient to write

$$\delta_{13} = \frac{1}{2}\left(\frac{\partial \xi}{\partial z} + \frac{\partial \zeta}{\partial x}\right) = \delta_{31}, \quad (3)$$

and we shall follow the latter choice although it is not the conventional one. Similarly the shearing strains about the z and x axes respectively will be written in our notation

$$\left.\begin{array}{l} \delta_{12} = \dfrac{1}{2}\left(\dfrac{\partial \eta}{\partial x} + \dfrac{\partial \xi}{\partial y}\right) = \delta_{21}, \\[2mm] \delta_{23} = \dfrac{1}{2}\left(\dfrac{\partial \zeta}{\partial y} + \dfrac{\partial \eta}{\partial z}\right) = \delta_{32}. \end{array}\right\} \quad (4)$$

We are again confronted with an aggregate of nine quantities (of which, to be sure, only six are independent) which describe completely the deformation of the medium and which may be written in the form of the array

$$\mathbf{D} = \left\| \begin{array}{ccc} \delta_{11} & \delta_{12} & \delta_{13} \\ \delta_{21} & \delta_{22} & \delta_{23} \\ \delta_{31} & \delta_{32} & \delta_{33} \end{array} \right\|. \quad (5)$$

This is called the *deformation* or *strain tensor*. Owing to the identity of δ_{12} and δ_{21}, etc., it is a so-called *symmetric* tensor. An examination of the stress tensor (1) shows that it is also symmetric. For, to consider X_{12} and X_{21}, $(X_{12} - X_{21})\,dxdydz$ represents twice the moment of stress force on the volume element $dxdydz$ (see again Fig. 6.1) about an axis through its center of mass parallel to the z axis, and is therefore equal to the moment of inertia about this axis multiplied by the component angular acceleration about the same axis. The moment of inertia contains terms in dx^2 and dy^2 which go to zero as the element is made smaller; hence in the limit $X_{12} = X_{21}$, since the acceleration must remain finite. The same argument applies to the other X_{ij} terms.

By the use of the law of Hooke, relations can be set up between the components of the stress tensor and those of the deformation tensor.

6.3 DEFORMABLE MEDIA AND TENSOR FIELDS

These are the equations which lead to a complete understanding of the behavior of an elastic medium. Before proceeding to a review of these, however, let us examine the expression **D** more carefully.

Consider any two points of the medium $P_1(x, y, z)$ and $P_2(x + dx, y + dy, z + dz)$ separated originally by the infinitesimal distance ds, where

$$ds^2 = dx^2 + dy^2 + dz^2.$$

After the medium is deformed, let us imagine that P_1 has moved to $P'_1(x + \delta x, y + \delta y, z + \delta z)$ and P_2 has moved to

$$P'_2(x + dx + \delta(x + dx), y + dy + \delta(y + dy), z + dz + \delta(z + dz)),$$

where the δ is now used to denote the variation in position due to deformation. As in Chapter III we have

$$\delta dx = d\delta x, \text{ etc.},$$

so that the alteration in the separation of P_1 and P_2 may be described by

$$\begin{aligned}
\tfrac{1}{2}\delta(ds^2) &= \left(\frac{\partial \delta x}{\partial x} dx^2 + \frac{\partial \delta x}{\partial y} dy dx + \frac{\partial \delta x}{\partial z} dz dx\right) \\
&+ \left(\frac{\partial \delta y}{\partial x} dx dy + \frac{\partial \delta y}{\partial y} dy^2 + \frac{\partial \delta y}{\partial z} dz dy\right) \\
&+ \left(\frac{\partial \delta z}{\partial x} dx dz + \frac{\partial \delta z}{\partial y} dy dz + \frac{\partial \delta z}{\partial z} dz^2\right) \\
&= \sum_{i,k=1}^{3} \frac{\partial \delta x_i}{\partial x_k} dx_i dx_k,
\end{aligned} \qquad (6)$$

to use a more compact notation in which $x_1 = x$, $x_2 = y$ and $x_3 = z$. It is clear that eq. (6) may equally well be written in the form

$$\tfrac{1}{2}\delta(ds^2) = \frac{1}{2}\sum_{i,k=1}^{3}\left(\frac{\partial \delta x_i}{\partial x_k} + \frac{\partial \delta x_k}{\partial x_i}\right) dx_i dx_k. \qquad (7)$$

Now when we recall that $\delta x = \xi$, $\delta y = \eta$, $\delta z = \zeta$ in the notation used at the beginning of this section, we see that the aggregate of quantities

$$\frac{1}{2}\left(\frac{\partial \delta x_i}{\partial x_k} + \frac{\partial \delta x_k}{\partial x_i}\right) \qquad (8)$$

as i and k take integral values from 1 to 3 forms an array very similar to **D**. In fact, writing them in the previous notation we have

$$\mathbf{D'} = \begin{Vmatrix} \delta_{11} & \frac{1}{2}(\delta_{12} + \delta_{21}) & \frac{1}{2}(\delta_{13} + \delta_{31}) \\ \frac{1}{2}(\delta_{21} + \delta_{12}) & \delta_{22} & \frac{1}{2}(\delta_{23} + \delta_{32}) \\ \frac{1}{2}(\delta_{31} + \delta_{13}) & \frac{1}{2}(\delta_{32} + \delta_{23}) & \delta_{33} \end{Vmatrix}. \tag{9}$$

This may be called the modified deformation or strain tensor. For practical purposes it is very convenient to represent the whole tensor by the type term $\frac{1}{2}\left(\frac{\partial \delta x_i}{\partial x_k} + \frac{\partial \delta x_k}{\partial x_i}\right)$, and we shall follow this practice in the future.

Transformation property of a covariant tensor. In order the better to understand the nature of this tensor, let us introduce a transformation of coordinates, viz.,

$$x_i = x_i(x_1', x_2', x_3'). \tag{10}$$

Let us see how the quadratic form (7) behaves under this transformation. We have

$$dx_i = \sum_{j=1}^{3} \frac{\partial x_i}{\partial x_j'} dx_j'; \; dx_k = \sum_{l=1}^{3} \frac{\partial x_k}{\partial x_l'} dx_l', \tag{11}$$

whence, letting $\frac{1}{2}\left(\frac{\partial \delta x_i}{\partial x_k} + \frac{\partial \delta x_k}{\partial x_i}\right) = a_{ik}$ for brevity

$$\sum_{i,k=1}^{3} a_{ik} dx_i dx_k = \sum_{i,k=1}^{3} \sum_{j,l=1}^{3} a_{ik} \frac{\partial x_i}{\partial x_j'} \frac{\partial x_k}{\partial x_l'} dx_j' dx_l'. \tag{12}$$

By interchanging the indices of summation this becomes

$$\sum_{i,k=1}^{3} a_{ik} dx_i dx_k = \sum_{i,k=1}^{3} \sum_{j,l=1}^{3} a_{jl} \frac{\partial x_j}{\partial x_i'} \frac{\partial x_l}{\partial x_k'} dx_i' dx_k'. \tag{13}$$

On comparing the two sides it is seen that the expression which in the new quadratic form replaces a_{ik} of the original quadratic form is

$$\sum_{j,l=1}^{3} a_{jl} \frac{\partial x_j}{\partial x_i'} \frac{\partial x_l}{\partial x_k'}. \tag{14}$$

We therefore look upon (14) as the transformed a_{ik} and denote it by

6.3 DEFORMABLE MEDIA AND TENSOR FIELDS

(a_{ik}). The transformation may be considered as having changed the aggregate of quantities $\frac{1}{2}\left(\frac{\partial \delta x_i}{\partial x_k} + \frac{\partial \delta x_k}{\partial x_i}\right) = a_{ik}$ into the aggregate

$$\sum_{j,l=1}^{3} \frac{1}{2}\left(\frac{\partial \delta x_j}{\partial x_l} + \frac{\partial \delta x_l}{\partial x_j}\right) \frac{\partial x_j}{\partial x_i{'}} \frac{\partial x_l}{\partial x_k{'}} = (a_{ik}).$$

A set of quantities which transform according to this rule is called a *covariant tensor* and indeed one of the *second order*. The strain tensor is therefore a covariant tensor of the second order, and the deformable medium may be said to constitute a tensor field of this character.

We can readily find a more simple illustration of a covariant tensor. In hydrodynamics we have to make use (Sec. 1.7) of the velocity potential ϕ whose gradient is the fluid velocity. Thus

$$\mathbf{v} = \nabla \phi = \mathbf{i}\frac{\partial \phi}{\partial x_1} + \mathbf{j}\frac{\partial \phi}{\partial x_2} + \mathbf{k}\frac{\partial \phi}{\partial x_3},$$

to use the notation of the present section. Consider the quantity which is the scalar product of $\nabla \phi$ and $d\mathbf{r}$, viz.,

$$d\mathbf{r} \cdot \nabla \phi = \sum_{k=1}^{3} \frac{\partial \phi}{\partial x_k} dx_k. \tag{15}$$

When this is subjected to a transformation of the form (10) there results

$$\sum_{k=1}^{3} \frac{\partial \phi}{\partial x_k} dx_k = \sum_{k,l=1}^{3} \frac{\partial \phi}{\partial x_k} \frac{\partial x_k}{\partial x_l{'}} dx_l{'} = \sum_{l,k=1}^{3} \frac{\partial \phi}{\partial x_l} \frac{\partial x_l}{\partial x_k{'}} dx_k{'},$$

whence we represent the transformed derivatives as

$$\left(\frac{\partial \phi}{\partial x_k}\right) = \sum_{l=1}^{3} \frac{\partial \phi}{\partial x_l} \frac{\partial x_l}{\partial x_k{'}}. \tag{16}$$

The aggregate of quantities $\partial \phi / \partial x_k$ (three in number) which transform according to eq. (16) form what is called a covariant tensor of the first order. But we already recognize this aggregate as a 3-vector. Hence we see that a 3-vector is a special case of a covariant tensor.

We have just considered covariant tensors which are aggregates of three and nine quantities, each a function of three independent variables respectively, (x_1, x_2, x_3). It is possible to construct more elaborate sets which transform according to the above rule. The covariant tensor of first order may be thought of as a set of quantities distinguished by *one* index (as $\frac{\partial \phi}{\partial x_1}, \frac{\partial \phi}{\partial x_2}, \frac{\partial \phi}{\partial x_3}$, above); that of second order may be thought of as distinguished by *two* indices (as a_{ik}, etc., above). The number of quantities in a covariant tensor of the first order is therefore the number of permutations with repetition of three quantities taken one at a time, which is *three*. Similarly in the case of a covariant tensor of the second order the number is that of the permutations with repetition of three quantities taken two at a time, which is *nine*. In like fashion one can imagine the construction of a covariant tensor which is a set of quantities each distinguished by *three* indices to the number of twenty-seven in all, i.e., the number of permutations with repetition of three things taken three at a time. This, however, does not mark the limit of generality. Suppose we consider a set of physical quantities which are functions of the n independent variables $x_1, x_2, \ldots x_n$. These quantities may be distinguished by m indices and may thus be represented by

$$X_{k_1 k_2 \ldots k_m}, \tag{17}$$

where the $k_1 \ldots k_m$ are all integers, each one running from 1 to n, with all permutations possible and all repetitions allowed. The total number of quantities in such a set is clearly n^m. If the set transforms according to the rule

$$(X_{k_1 \ldots k_m}) = \sum_{j_1 \ldots j_m = 1}^{n} X_{j_1 \ldots j_m} \frac{\partial x_{j_1}}{\partial x_{k_1}'} \frac{\partial x_{j_2}}{\partial x_{k_2}'} \ldots \frac{\partial x_{j_m}}{\partial x_{k_m}'}, \tag{18}$$

it will be called a covariant tensor of order m. Such a tensor is an aggregate of n^m quantities. Thus if $m = 2$ and $n = 3$ we have a tensor with nine components like our **D**$'$. $m = 1$ and $n = 4$ yield a 4-vector, which we shall meet later in the discussion of. relativity. If $m = 2$ with $n = 4$, there results a tensor of the second order with sixteen components and for $m = 3$ with $n = 4$ a tensor of the third order with sixty-four components. Such tensors prove to be of value in the general theory of relativity.

Contravariant tensors. Up to this point we have dealt exclusively with covariant tensors, which is the type of most importance in the physics of deformable media. There is, however, another kind of some interest which should be mentioned briefly. Consider the displacement vector

$$d\mathbf{r} = \mathbf{i}dx_1 + \mathbf{j}dx_2 + \mathbf{k}dx_3,$$

with the three components dx_1, dx_2, dx_3, and introduce the transformation $x_k = x_k(x_1', x_2', x_3')$. Then we have at once

$$dx_k' = \sum_{j=1}^{3} \frac{\partial x_k'}{\partial x_j} dx_j. \tag{19}$$

The set dx_1, dx_2, dx_3 has been transformed into the set dx_1', dx_2', dx_3'. A set which transforms according to eq. (19) is called a *contravariant tensor*. The present is an illustration of a contravariant tensor of the first order, i.e., a contravariant 3-vector. This result can be generalized to any number of coordinates. Moreover, it is easy to construct contravariant tensors of higher order and finally to build tensors which are mixed, i.e., transform partly in one way and partly in the other. However, we shall not investigate these further at this place.

It has been seen that according to definition the components of strain form a tensor. It remains to show that the same thing is true of the stress. We have indeed assumed this at the beginning of the present section. Let us go back to Fig. 6.1 and assume a transformation of coordinates $x_i' = x_i'(x_1, x_2, x_3)$ which is equivalent to the introduction of new axes $x_1', x_2', x_3'(x', y', z'$ in the usual notation). If this transformation is linear, its most general form involves only rotation and translation of axes. We shall tacitly assume this type in our discussion. However, the treatment for general point transformations has been carried out by Thirring,[1] to whom reference may be made for details. Consider the face $OCGD$ normal to the x_1 axis. The component along the x_1' axis of the stress X_{11} is clearly $X_{11} \frac{\partial x_1}{\partial x_1'}$. Likewise the components along the x_1' axis of the stresses X_{12} and X_{13} are $X_{12} \frac{\partial x_2}{\partial x_1'}$ and $X_{13} \frac{\partial x_3}{\partial x_1'}$, respectively. Hence the total force on a *unit area* at the origin *normal* to x_1' due to these three components is

$$\left(X_{11} \frac{\partial x_1}{\partial x_1'} + X_{12} \frac{\partial x_2}{\partial x_1'} + X_{13} \frac{\partial x_3}{\partial x_1'} \right) \frac{\partial x_1}{\partial x_1'}.$$

For the sets of components X_{21}, X_{22}, X_{23} and X_{31}, X_{32}, X_{33} we obtain similar expressions, and hence the total stress on a surface normal to x_1', namely (X_{11}), is

$$(X_{11}) = \sum_{j, l=1}^{3} X_{jl} \frac{\partial x_j}{\partial x_1'} \frac{\partial x_l}{\partial x_1'}.$$

[1] H. Thirring, *Phys. Zs.*, 26, 518, 1925.

In general we have

$$(X_{ik}) = \sum_{j,l=1}^{3} X_{jl} \frac{\partial x_j}{\partial x_i'} \frac{\partial x_l}{\partial x_k'}. \tag{20}$$

This, however, is precisely the law of transformation for a covariant tensor of the second order, and hence the stress tensor really has the character previously claimed for it.

Physical significance of tensor equations. We ought now to inquire what physical significance may be attached to tensors. For the purpose of this inquiry let us go back to the behavior of an elastic medium described by means of the stress and strain tensors, **S** and **D**′. We first review the relations between the stress and strain components, which are so well known from the classical theory of elasticity that we need not derive them here.[1] We shall write for the "total" dilatation

$$\Theta = \delta_{11} + \delta_{22} + \delta_{33}. \tag{21}$$

Denoting the shear modulus, the bulk modulus, and Poisson's ratio (the ratio of lateral to longitudinal strain) by μ, E, and σ, respectively, and using $\lambda = E\sigma/(1+\sigma)(1-2\sigma)$, we write the fundamental elastic relations in the form

$$\left.\begin{array}{l} X_{11} = \lambda\Theta + 2\mu\delta_{11} \\ X_{22} = \lambda\Theta + 2\mu\delta_{22} \\ X_{33} = \lambda\Theta + 2\mu\delta_{33} \end{array}\right\} \tag{22}$$

and

$$\left.\begin{array}{l} X_{12} = X_{21} = \mu(\delta_{12} + \delta_{21}) \\ X_{23} = X_{32} = \mu(\delta_{23} + \delta_{32}) \\ X_{31} = X_{13} = \mu(\delta_{31} + \delta_{13}). \end{array}\right\} \tag{23}$$

These relations may be at once replaced by a single relation between the stress and strain tensors, for on substitution we get

$$\mathbf{S} = \left\| \begin{array}{ccc} \lambda\Theta + 2\mu\delta_{11}, & \mu(\delta_{12}+\delta_{21}), & \mu(\delta_{13}+\delta_{31}) \\ \mu(\delta_{21}+\delta_{12}), & \lambda\Theta + 2\mu\delta_{22}, & \mu(\delta_{23}+\delta_{32}) \\ \mu(\delta_{31}+\delta_{13}), & \mu(\delta_{23}+\delta_{32}), & \lambda\Theta + 2\mu\delta_{33} \end{array} \right\|, \tag{24}$$

which may be represented symbolically in tensor notation

$$\mathbf{S} = 2\mu\mathbf{D}' + \lambda\mathbf{D}''. \tag{25}$$

---

[1] See, e.g., Love, "Theory of Elasticity" (Cambridge University Press, third edition, 1920), pp. 100, 123, etc.

6.3 DEFORMABLE MEDIA AND TENSOR FIELDS

Here $\mathbf{D}''$ is the *diagonal* tensor

$$\mathbf{D}'' = \begin{Vmatrix} \Theta & 0 & 0 \\ 0 & \Theta & 0 \\ 0 & 0 & \Theta \end{Vmatrix}. \tag{26}$$

Eq. (25) is a *tensor equation*. It will be recognized at once as the generalized expression of Hooke's law.

What is the physical significance of such a tensor equation? This is found in the fact that if we introduce any transformation of coordinates, since both sides of the equation transform in the same way, the equation will continue to be true in the transformed coordinates. To assure ourselves of this in the special case under consideration, we have only to note again the transforms of the tensor components. Thus, from eq. (20), the fact that

$$(\tfrac{1}{2}(\delta_{ik} + \delta_{ki})) = \sum_{j,l} \tfrac{1}{2}(\delta_{jl} + \delta_{lj}) \frac{\partial x_j}{\partial x_i'} \frac{\partial x_l}{\partial x_k'},$$

and

$$X_{jl} = 2\mu \tfrac{1}{2}(\delta_{jl} + \delta_{lj}) + \lambda D_{jl}'', \tag{27}$$

there follows

$$(X_{ik}) = 2\mu(\tfrac{1}{2}(\delta_{ik} + \delta_{ki})) + \lambda(D_{ik}''). \tag{28}$$

It should be noted that

$$(D_{ik}'') = \sum_{j,l} D_{jl}'' \frac{\partial x_j}{\partial x_i'} \frac{\partial x_l}{\partial x_k'} = D_{ik}'', \tag{29}$$

as is seen at once from eq. (26) and analytic geometry. Thus the tensor $\mathbf{D}''$ really reduces to a scalar quantity having only numerical value and might have been written as such (without indices) in the preceding equations.

The general result of which we have noted a special case is of great importance in physics, for, in order that the equations representing physical laws shall continue to be true no matter in what coordinates they are expressed, it is essential that the quantities on the two sides of the equation shall transform in the same way. This means effectively that the equations shall be tensor equations, since the transformation rules involved in the definition of covariant and contravariant tensors are those most commonly encountered in dealing with physical quantities. This fundamental significance of tensor equations has been brought very much to the fore by the general theory of relativity, which is discussed in Chapter VIII.

We shall now dismiss the theory of elasticity, the further development of which is readily found in numerous treatises.[1] It provides a concrete illustration of the tensor field. Further illustrations may be found in the development of the equations of hydrodynamics and the elastic solid theory of optics.

6.4. The Electromagnetic Field. If one were to inquire as to the most beautiful and significant illustration of field theory in classical physics the answer without question would be the theory of the electromagnetic field. As we survey the history of science we cannot help feeling that this was the crowning achievement of nineteenth-century physics. It marked the culmination of the classical method—that of elementary abstraction, of differential laws, of the continuum point of view. It failed, of course, to meet the demands of the flood of electrical phenomena discovered during the last few decades of the century, but in its very failure it has profoundly influenced the course of present-day physics.

The equations of the electromagnetic field as presented by Maxwell form a theoretical structure having as its background the experimental observations in electrostatics and magnetostatics, essentially as old as the Greeks but attaining more quantitative meaning during the eighteenth century, together with the nineteenth-century discoveries of Oersted and Faraday of the relations between electric currents and magnetism. It is not our purpose here to recount the story of these discoveries, which form a fascinating chapter in the history of the correlation of at first apparently dissimilar physical phenomena. We shall merely recall that Oersted was the first (1820) to observe that associated with an electric current there is always a magnetic field, and that in 1831 Faraday (and, independently at about the same time, Henry) observed the production of an electric current by the relative motion of a conductor and a magnetic field. It is of interest to note that Faraday preferred to look upon these phenomena as manifestations of stresses and strains in a continuous medium which later came to be called the *electromagnetic ether*. This was a natural result of the previous success of " medium " or field theories in the description of fluids, elasticity, and even light (elastic solid theory) as contrasted to " action at a distance " theories which classical particle mechanics involved. Faraday's geometrical intuition was later refined by Maxwell's brilliant analysis, and the electromagnetic ether became the medium for the propagation of light as well as of large-scale electromagnetic phenomena. The advent of relativity showed that the ether involved too many inconsistencies to serve longer as a useful physical

---

[1] See, for example, Love, *op. cit.* An interesting treatment of Cartesian tensors will be found in " Cartesian Tensors," H. Jeffreys (Cambridge University Press, 1931).

concept. However, this has not destroyed the utility of the field concept in electromagnetism.

The field equations in free space. Let us now investigate more closely the nature of this field and likewise the nature of the quantities which are used to describe it. We shall at first confine our attention to the classical point of view and to a given region in free space, i.e., a vacuum. Every point of this region at every instant of time is conceived to have associated with it two 3-vector quantities denoted by **E** and **H** which satisfy the following equations

$$\left.\begin{array}{c} \nabla \times \mathbf{E} = -\frac{1}{c}\dot{\mathbf{H}} \\ \nabla \times \mathbf{H} = +\frac{1}{c}\dot{\mathbf{E}} \\ \nabla \cdot \mathbf{H} = 0 \\ \nabla \cdot \mathbf{E} = 0. \end{array}\right\} \quad (1)$$

Here c appears as the ratio between the electromagnetic and the electrostatic units of charge, viz., approximately 3×10^{10}. **E** and **H** are vector functions of space and time which, together with their derivatives, are assumed to be continuous and differentiable. At first we do not attempt to give physical meaning to these symbols. We merely say: let us assume that there exist physical quantities represented by symbols having the indicated properties and see what these equations say about the quantities. From the last two equations it is clear that **E** and **H** are *solenoidal* vectors, and hence, for any volume of space over which they are defined, we have by the divergence theorem

$$\int \nabla \cdot \mathbf{H} d\tau = \int \mathbf{H} \cdot d\boldsymbol{\sigma} = 0,$$
$$\int \nabla \cdot \mathbf{E} d\tau = \int \mathbf{E} \cdot d\boldsymbol{\sigma} = 0. \quad (2)$$

Volume and surface elements are represented by $d\tau$ and $d\boldsymbol{\sigma}$, respectively. Thus the normal flux of **E** and **H** through any closed surface in space is zero. The vector lines for **E** and **H** are closed lines. The first two equations tell us that, if **E** and **H** happen to be independent of the time, they are also *irrotational* vectors. In this case Stokes's theorem yields

$$\int \nabla \times \mathbf{E} \cdot d\boldsymbol{\sigma} = \oint \mathbf{E} \cdot d\mathbf{s} = 0,$$
$$\int \nabla \times \mathbf{H} \cdot d\boldsymbol{\sigma} = \oint \mathbf{H} \cdot d\mathbf{s} = 0. \quad (3)$$

Here the $\oint \mathbf{E} \cdot d\mathbf{s}$ and $\oint \mathbf{H} \cdot d\mathbf{s}$ are the line integrals of $\mathbf{E}$ and $\mathbf{H}$, respectively, about any closed curve in space, and the surface integrals are taken over any surface bounded by the curve. If $\mathbf{E}$ and $\mathbf{H}$ were to be interpreted physically as *forces* acting on unit particles of some kind, eqs. (3) would mean that the work done by the forces in the motion of these unit particles about any closed curve in space is zero.[1] This is the situation in the static case, and it will be observed that there is complete symmetry throughout with respect to $\mathbf{E}$ and $\mathbf{H}$. If the equations actually describe physical phenomena, they say nothing about the phenomena associated with $\mathbf{E}$ which they do not likewise say about $\mathbf{H}$. The two vectors are on absolutely the same footing. The situation is quite different when $\mathbf{E}$ and $\mathbf{H}$ are supposed to be variable in time. They are no longer irrotational vectors; the first two equations in (1) imply definite relations between them. We cannot expect the same physical significance to be attached to the first as to the second; the interchange of $\mathbf{E}$ with $\mathbf{H}$ in the first does not yield the second, owing to the very interesting difference in sign. The equations cease to represent a static state of affairs and become dynamical. Can we learn any more about $\mathbf{E}$ and $\mathbf{H}$ directly from them? We can do this by the very simple mathematical process of taking the curl of either equation. Thus from the first equation and the vector expansion of curl curl we have

$$\nabla \times \nabla \times \mathbf{E} = \nabla(\nabla \cdot \mathbf{E}) - \nabla^2 \mathbf{E} = -\frac{1}{c} \nabla \times \dot{\mathbf{H}}.$$

Substitution of the second equation and the use of the last equation yield

$$\nabla^2 \mathbf{E} = \frac{1}{c^2} \ddot{\mathbf{E}}. \tag{4}$$

A precisely similar process gives likewise

$$\nabla^2 \mathbf{H} = \frac{1}{c^2} \ddot{\mathbf{H}}. \tag{5}$$

These two equations tell us that $\mathbf{E}$ and $\mathbf{H}$ are propagated in free space as waves (cf. Sec. 1.8) with the velocity c; in other words, $\mathbf{E}$

---

[1] In the present case no particles are supposed to be present. However, it may be still possible to interpret $\mathbf{E}$ and $\mathbf{H}$ as forces which *would* act on unit particles *were* they present.

and **H** are wave functions of space and time. The equations can tell us more than this. For we have

$$\mathbf{H}\cdot(\nabla\times\mathbf{E}) = -\frac{1}{c}\mathbf{H}\cdot\dot{\mathbf{H}},$$

$$\mathbf{E}\cdot(\nabla\times\mathbf{H}) = \frac{1}{c}\mathbf{E}\cdot\dot{\mathbf{E}},$$

whence by subtraction

$$\mathbf{H}\cdot(\nabla\times\mathbf{E}) - \mathbf{E}\cdot(\nabla\times\mathbf{H}) = -\frac{1}{c}(\mathbf{H}\cdot\dot{\mathbf{H}} + \mathbf{E}\cdot\dot{\mathbf{E}}),$$

which by the use of vector identities becomes

$$c\nabla\cdot(\mathbf{E}\times\mathbf{H}) = -\frac{\partial}{\partial t}\left[\frac{1}{2}(H^2 + E^2)\right].$$

Let us integrate over a finite region of space and employ the divergence theorem. We are finally led to

$$c\int(\mathbf{E}\times\mathbf{H})\cdot d\boldsymbol{\sigma} + \frac{\partial}{\partial t}\int\frac{1}{2}(E^2 + H^2)d\tau = 0, \qquad (6)$$

where the surface integral is taken over the surface bounding the given volume. What physical meaning can be attributed to this result? Suppose for a moment that **E** and **H** vanish on the bounding surface. We are therefore left in this special case with

$$\frac{\partial}{\partial t}\int\frac{1}{2}(E^2 + H^2)d\tau = 0. \qquad (7)$$

When a physicist encounters the vanishing of the time rate of change of an extensive quantity he is apt to look to see whether or not the quantity in question may plausibly be interpreted as *energy*. The quantity $\frac{1}{2}(E^2 + H^2)$ would then appear as the energy per unit volume associated with the presence of the electric and magnetic fields, and this energy would be conserved inside the given volume as long as **E** and **H** are zero on the boundary. If **E** and **H** are actually the forces per unit particle in the two fields respectively and if these forces obey an inverse square law, it indeed follows by direct calculation that the quantity $\frac{1}{2}(E^2 + H^2)$ is the energy per unit volume in the resultant field. Let us go back to eq. (6) and no longer suppose that **E** and **H** vanish on the boundary of the space considered. The integral $c\int(\mathbf{E}\times\mathbf{H})\cdot d\boldsymbol{\sigma}$ must then be the rate of change of the energy in the field inside the region. If the field is still conservative, this can only mean that $c\int (\mathbf{E} \times \mathbf{H}) \cdot d\sigma$ is the rate at which energy flows normally across the boundary. Hence $c(\mathbf{E} \times \mathbf{H})$ must be the flow per unit area per unit time. It is the so-called *Poynting flux*, and is a vector quantity which is normal to both $\mathbf{E}$ and $\mathbf{H}$. Therefore the wave propagation of $\mathbf{E}$ and $\mathbf{H}$ is accompanied by an energy propagation. From what has just been stated this must be a transverse wave. It does not exist unless both $\mathbf{E}$ and $\mathbf{H}$ exist: it is strictly an electromagnetic wave. In the hands of Maxwell it formed the beginning of the electromagnetic theory of light.

The above discussion emphasizes that the quantities $\mathbf{E}$ and $\mathbf{H}$ satisfying eqs. (1) have some very interesting properties which serve as an adequate description for a wide variety of natural phenomena. In a very real sense, therefore, these equations may be said to constitute a *definition* of $\mathbf{E}$ and $\mathbf{H}$.

The equations as we have written them are of course too restricted to describe all electrical and magnetic phenomena. However, it is rather remarkable how little they need to be changed in order to get something which will serve as a general basis for electrodynamics. All that is necessary is the introduction of the concept of electric charge. In logical strictness this involves a quantity which is defined as the charge density, ρ, i.e., charge per unit volume, and a further quantity $\rho\mathbf{v}$, where $\mathbf{v}$ is a velocity. If the charge distribution is considered to be continuous, $\mathbf{v}$ is the hydrodynamic velocity of flow (see Chapter I). If the charges are discrete particles (as in the contemporary electron theory) $\mathbf{v}$ is the particle velocity. Let us introduce these quantities tentatively in the following way:

$$\left.\begin{aligned} \nabla \times \mathbf{E} &= -\frac{1}{c}\dot{\mathbf{H}} \\ \nabla \times \mathbf{H} &= \frac{1}{c}(\dot{\mathbf{E}} + \rho\mathbf{v}) \\ \nabla \cdot \mathbf{H} &= 0 \\ \nabla \cdot \mathbf{E} &= \rho. \end{aligned}\right\} \quad (8)$$

We shall see that these equations hold for the case of non-material media. According to them, $\mathbf{H}$ remains a solenoidal vector, while $\mathbf{E}$ ceases to be one. The total normal flux of $\mathbf{E}$ through any closed surface is now equal to the total charge enclosed by the surface. Even now the equations are not quite complete. However, let us postpone discussion of this question for a moment, and note an interesting result of eqs. (8). From the second equation

$$\nabla \cdot \nabla \times \mathbf{H} = \frac{1}{c}(\nabla \cdot \dot{\mathbf{E}} + \nabla \cdot (\rho \mathbf{v})) = 0,$$

since the divergence of the curl of any vector is identically zero. With the use of the fourth equation, however, this becomes

$$\nabla \cdot (\rho \mathbf{v}) + \dot{\rho} = 0, \tag{9}$$

which is in precisely the form of the equation of continuity of hydrodynamics. In other words, the equations say that electrically charged media move like a perfect fluid. The variation of $\mathbf{E}$ and $\mathbf{H}$ in space and time now becomes decidedly more complicated and we are no longer led to the simple wave equations (4) and (5).

In order to complete the list of eqs. (8) it is necessary to express the force acting on a unit charge. We have already hinted at this as the possible meaning to be attached to $\mathbf{E}$. However, if the charged media move, such an assignment would lead to disagreement with observation. To get a proper description of experience it is necessary to let the force on unit charge, which we shall denote by $\mathbf{F}$, be given by

$$\mathbf{F} = \mathbf{E} + \frac{1}{c} \mathbf{v} \times \mathbf{H}. \tag{10}$$

This expression—sometimes called the force equation of electrodynamics—must be added to the four equations (8) to obtain a complete description.

The five equations given in (8) and (10) suffice to describe electromagnetic phenomena in a space occupied only by non-magnetic conductors and otherwise free. The general case of material media is handled by replacing the $\mathbf{E}$ and $\mathbf{H}$ in the last two equations (8) and on the right-hand sides of the first two equations by the quantities $\mathbf{D}$ and $\mathbf{B}$, respectively, where

$$\mathbf{D} = \kappa \mathbf{E}, \quad \mathbf{B} = \mu \mathbf{H}. \tag{11}$$

Here κ is the dielectric constant of the medium and μ is the magnetic permeability. Both are equal to unity in free space. We shall not try to go into a detailed study of material media, for the complications are many and would lead too far afield. Fortunately the logical study of the nature of the field equations does not necessitate a careful investigation of this special problem.

Historical retrospect. We have now written the electrodynamical equations in a variety of forms and have noted the more important properties which they confer on the quantities entering into them. The questions now arise: where do the equations come from, and just how do they describe physical phenomena? The answer to the first question demands some historical retrospect. The equations might have come out of the air; actually they did not. They arose in the endeavor to describe electromagnetic observations, and in their development they have undergone some very interesting vicissitudes. Let us look for a moment at eqs. (8), keeping the historical attitude uppermost in mind. The third equation, which says that **H** is a solenoidal vector, describes symbolically the well-known experimental fact that while magnetic poles act on each other according to the inverse square law, no magnet exists having but one pole; no matter what closed surface is considered in space, whether filled with magnets or not, the magnetic flux through the surface is zero: magnetic lines are *closed* lines. On the other hand, the fourth equation expresses the possibility of the existence of isolated electric charges which act on each other according to Coulomb's law. The units are, of course, Heavisidian, the unit of charge being $1/\sqrt{4\pi}$ times the electrostatic unit. These two equations may be considered the fundamental equations of magnetostatics and electrostatics. The dynamics of electromagnetism enters with the first two equations. The first, indeed, is a variety of Faraday's law of electromagnetic induction, expressed originally in the symbolic form

$$\oint \mathbf{E} \cdot d\mathbf{s} = -\frac{d}{dt} \int \mathbf{H} \cdot d\boldsymbol{\sigma}, \qquad (12)$$

viz., the line integral of **E** (the electric field intensity) about any closed curve in the field is equal to the negative of the time rate of change of the total normal flux of **H** (magnetic flux) through any surface bounded by the curve. On the customary elementary interpretation of **E**, the line integral represents the work involved in the motion of a unit (electromagnetic) of charge once about the curve. This is defined as the *electromotive force* in the circuit, and the statement of eq. (12) is equivalent to the usual one that the electromotive force induced in a circuit is equal numerically to the time rate of change of the total normal magnetic flux through the circuit. The negative sign is the expression for Lenz's law, which states that the direction of the induced electromotive force is such as to oppose the change of flux which induces it. Eq. (12), which is, so to speak, an integral representation of Faraday's law, can be replaced by the corresponding differ-

6.4 THE ELECTROMAGNETIC FIELD

ential expression in (8) by the use of Stokes's theorem, i.e., eqs. (3). We then get

$$\nabla \times \mathbf{E} = -\dot{\mathbf{H}},$$

in which, of course, $\mathbf{H}$ is measured in *magnetic* units (i.e., gauss) and $\mathbf{E}$ in the corresponding *electromagnetic* units based on the gauss. If $\mathbf{E}$ is to be measured in *electrostatic* units we must multiply $\nabla \times \mathbf{E}$ by c, the number of electromagnetic units of field intensity in one electrostatic unit. This gives the first equation in (8). The quantity c is numerically approximately 3×10^{10}, which in turn is very nearly the velocity of light in free space in centimeters per second.

The second equation in (8) goes back to the fundamental experiments of Oersted showing the existence of a magnetic field in the neighborhood of every conductor carrying an electric current. Ampère showed that this field is that which would be produced by a magnetic shell with a boundary coinciding with the circuit and with a strength proportional to the current. From this he was then immediately led to the result that the work to take a unit pole once around a current-carrying wire is equal to 4π times the current in electromagnetic units. Expressed symbolically,

$$\oint \mathbf{H} \cdot d\mathbf{s} = 4\pi I. \tag{13}$$

Transforming to Heavisidian units and using electrostatic units for charge density this becomes

$$\oint \mathbf{H} \cdot d\mathbf{s} = \frac{1}{c} \int (\rho \mathbf{v}) \cdot d\boldsymbol{\sigma},$$

the surface integral being taken over any closed surface bounded by the path of the pole; employing again Stokes's theorem, we are led to

$$\nabla \times \mathbf{H} = \rho \mathbf{v}/c. \tag{14}$$

This is not in the exact form of (8) since the $\dot{\mathbf{E}}/c$ is lacking. Maxwell observed that, although eq. (14) tells part of the truth about the magnetic fields associated with currents, it does not tell the whole truth. Since $\nabla \cdot \nabla \times \mathbf{H} = 0$, it implies indeed that

$$\nabla \cdot (\rho \mathbf{v}) = 0, \tag{15}$$

which is in disagreement with the equation of continuity (9). It would be out of the question in any case in which the charge density varies with the time, as for example in the discharge of a condenser.

With this realization Maxwell sought for a way to modify eq. (14) to make it consistent with the equation of continuity in the general case. It is clear that this end can be attained if one adds $\dot{\mathbf{E}}/c$ to $\rho\mathbf{v}/c$, leading thus to the second equation in (8). Of course, one might also add any vector whose divergence vanishes, since then eq. (9) would still be satisfied. However, presumably Maxwell saw no point in encumbering the equation more than was absolutely necessary—a good application of William of Occam's famous razor. He was doubtless more interested in the physical interpretation of the added $\dot{\mathbf{E}}/c$. It is clear from the discussion in the earlier part of this section that without this term the existence of electromagnetic waves would not appear in the theory at all. It is also evident that $\dot{\mathbf{E}}$ must have the dimensions of $\rho\mathbf{v}$, i.e., current density. Maxwell interpreted $\dot{\mathbf{E}}$ as a *displacement* current density. In free space this is not a very obvious physical concept, for it involves the assumption that when charge flows through a medium there is a current supplementary to that represented by the motion of the charges and somehow connected with a *displacement* of the medium. Maxwell was able to lend some plausibility to the assumption by treating the electromagnetic medium—the so-called ether—as an elastic medium subject to stresses and strains. However, in view of the general attitude which we have taken throughout our discussion it seems more satisfactory to look upon the introduction of the displacement current as a definite *a priori* hypothesis to be justified by its success. It is a vital part of Maxwell's electromagnetic theory.

Physical meaning of **E** *and* **H**. In giving the above brief historical retrospect of the field equations in order to make clear their origin we have also to a considerable extent really answered our other question as to how these equations describe phenomena. We have, however, neglected one vitally important matter: Just what precise physical meaning is to be attached to the quantities **E** and **H** which enter into the equations? To be sure, we mentioned the idea of them as forces on unit particles: **E** as the force exerted by the field on unit electric charge and **H** the force on a unit pole. These are the conventional definitions of **E** and **H** associated with the mode of " derivation " of the equations which, as in our historical review, begins with the fundamental experimental laws. There are certain difficulties with all such definitions of field-characteristic quantities. Suppose we follow the definition literally in order to measure the intensity of an electric field at some point, i.e., we place a unit charge at the point and observe the force on it. This is not the intensity of the original field, for the latter has been modified by the introduction of the test charge. Of course, if the test charge is small enough so that its influence on the field is small, this logical difficulty is of little practical import. Moreover, we can imagine the possibility of compensating for the effect by a calculation of the field produced by the test charge, etc. As long as one is dealing with macroscopic or relatively large-scale fields, such considerations are satisfactory. In fact, one can apparently avoid the logical difficulty altogether by defining the field as the limit of the ratio of force to charge as the charge approaches zero. This is reminiscent of the definitions of density and pressure in a continuous medium. In so far as electrodynamics rests on a strictly continuum basis, it is perfectly satisfactory. The real trouble occurs when we come to a discrete picture such as that afforded by the electron theory, for here the test charge cannot be less than one electron and the limit process cannot logically be carried out unless we wish to endow the electron with a structure and imagine the possibility of its infinite subdivision. This, presumably, would involve a variation of the field intensity over the region occupied by a single electron, and the question arises as to what meaning may be attached to this. Considerations of this kind play an important rôle in Bohr's interpretation of the Heisenberg indeterminacy principle which is discussed in Chapter IX. We need not elaborate on the matter here. All we need to do is to indicate a simple way out of the difficulty: we may treat the equations themselves as defining **E** and **H**. This means that the only real importance of the quantities is involved in the fact that they satisfy the field equations. As a matter of fact, it has been pointed out (notably by Ritz [1] and Swann) that the only use which we make of **E** and **H** in atomic problems is their calculation from the equations, assuming assigned values of ρ and ρv. In atomic problems the definitions in terms of unit charge and pole are wholly unnecessary. It seems most logical to go the whole way and treat **E** and **H** as defined by the field equations in all cases. The commoner definitions can then be looked upon as mere picturizations. When we use **E** and **H**, all we are really interested in is a description in simplest terms of certain phenomena. The equations assure us that there exist such quantities with certain well-defined properties. Under various conditions values can be assigned to these quantities, and from these values other things such as the motion of circuits or charged particles can be computed. If we stop to think of how we use the field vectors even

[1] See the remarkable critical paper "Recherches critiques sur les théories électrodynamique de Cl. Maxwell et H. A. Lorentz," *Archives de Genève*, 4th Period, Vol. 26, 1908. Also W. F. G. Swann, "Relativity and Electrodynamics," *Reviews of Modern Physics*, 2, 243, 1930.

in practical problems it is clear that the idea of them in connection with unit charge and unit pole never enters into our measurements. When we talk about the electric field existing between the plates of a condenser we think of it in terms of the gradient of the potential rather than in terms of force per unit charge. In similar fashion the magnetic field which we calculate at the center of a circular current-carrying coil, for example, is used in the calculation of the force action between this coil and other current-carrying conductors or magnetic material, or perhaps in the calculation of the deviation of a stream of charged particles in the vicinity of the coil. The unit pole rarely if at all enters into our calculations.

Solution of the field equations. To revert to the field equations themselves, the point of view that they constitute a definition of $\mathbf{E}$ and $\mathbf{H}$ will assume increased plausibility from a general explicit solution of the equations in which these quantities are expressed in terms of the distribution of ρ and $\rho \mathbf{v}$. This can be done by means of the vector and scalar potentials $\mathbf{A}$ and V. Since $\mathbf{H}$ is solenoidal it is expressible as the curl of another vector. We call the latter $\mathbf{A}$ and write

$$\mathbf{H} = \nabla \times \mathbf{A}. \tag{16}$$

In a static conservative field $\mathbf{E}$ would be expected to appear as the negative gradient of a scalar—the electrostatic potential, V. In the general case, however, since $\mathbf{E}$ depends on the time variation in $\mathbf{H}$, we write

$$\mathbf{E} = -\nabla V - \frac{1}{c}\dot{\mathbf{A}}. \tag{17}$$

In order that the field equations may be satisfied by (16) and (17) we must have, as is clear by substitution,

$$\nabla^2 \mathbf{A} - \frac{1}{c^2}\ddot{\mathbf{A}} = -\frac{1}{c}\rho\mathbf{v}, \tag{18}$$

$$\nabla^2 V - \frac{1}{c^2}\ddot{V} = -\rho, \tag{19}$$

in addition to the relation between $\mathbf{A}$ and V

$$\nabla \cdot \mathbf{A} = -\frac{\dot{V}}{c}. \tag{20}$$

6.4 THE ELECTROMAGNETIC FIELD

Eqs. (18) and (19) can be solved, yielding the results

$$\mathbf{A} = \frac{1}{4\pi c}\int \frac{[\rho \mathbf{v}]}{r}d\tau, \tag{21}$$

$$V = \frac{1}{4\pi}\int \frac{[\rho]}{r}d\tau. \tag{22}$$

The integrals are volume integrals taken over all space. The meaning of the square brackets in the integrand is as follows: Suppose we wish to evaluate V at some point P at time t. We take ρ at the volume element $d\tau$ distant r from P at the time $t - r/c$ and do the same for all volume elements. The result of the integration is known as a *retarded* potential. In similar fashion $[\rho \mathbf{v}]$ refers to the value of $\rho \mathbf{v}$ at the volume elements at the time $t - r/c$. This solution of eqs. (18) and (19) can be obtained most simply by the use of the Fourier integral. For convenience let us consider merely the scalar equation. If we have a function $f(t)$ subject to certain conditions the Fourier integral theorem assures us of the possibility of representing it by the expression

$$f(t) = \frac{1}{2\pi}\int_{-\infty}^{+\infty}d\alpha \int_{-\infty}^{+\infty}f(\lambda)e^{i\alpha(t-\lambda)}d\lambda, \tag{23}$$

where α and λ are arbitrary parameters. Applying this to $V(x, y, z, t)$ of our eq. (19) (recalling that the Laplacian is a space operator only),

$$\nabla^2 V(x, y, z, t) = \frac{1}{2\pi}\int_{-\infty}^{+\infty}d\alpha \int_{-\infty}^{+\infty}\nabla^2 V(x, y, z, \lambda)e^{i\alpha(t-\lambda)}d\lambda.$$

Moreover,

$$\ddot{V}(x, y, z, t) = -\frac{1}{2\pi}\int_{-\infty}^{+\infty}d\alpha \int_{-\infty}^{+\infty}\alpha^2 V(x, y, z, \lambda)e^{i\alpha(t-\lambda)}d\lambda$$

and

$$\rho(x, y, z, t) = \frac{1}{2\pi}\int_{-\infty}^{+\infty}d\alpha \int_{-\infty}^{+\infty}\rho(x, y, z, \lambda)e^{i\alpha(t-\lambda)}d\lambda.$$

The use of these results transforms our original equation into

$$\frac{1}{2\pi}\int_{-\infty}^{+\infty}d\alpha \int_{-\infty}^{+\infty}e^{i\alpha(t-\lambda)}\left\{\nabla^2 V(\lambda) + \frac{\alpha^2}{c^2}V(\lambda) + \rho(\lambda)\right\}d\lambda = 0, \tag{24}$$

where the space dependence of V is understood but omitted for simplicity. It is clear that eq. (24) will be satisfied if we can find a function $V(\alpha, \lambda)$ such that for all values of x, y, z, α, λ,

$$\nabla^2 V(\alpha, \lambda) + \frac{\alpha^2}{c^2} V(\alpha, \lambda) + \rho(\lambda) = 0. \tag{25}$$

Such a function is

$$V(\alpha, \lambda) = \frac{1}{4\pi} \int \rho(\lambda) \frac{e^{\pm \frac{i\alpha r}{c}}}{r} d\tau, \tag{26}$$

where the integration is extended over an infinite volume about the point whose coordinates appear in V.[1]

[1] To obtain this function we consider

$$\nabla^2 u + k^2 u = 0$$

with a particular integral, $u = \dfrac{e^{\pm ikr}}{r}$, where r is the distance from some fixed point P and $k = \alpha/c$. If u and v are two arbitrary scalar space functions of suitable differentiability, one form of Green's theorem assures us that

$$\int (u\nabla^2 v - v\nabla^2 u) d\tau = \int (u\nabla v - v\nabla u) \cdot d\sigma,$$

the surface integration on the right side being taken over the surface bounding the volume over which the integral on the left extends. (Apply the divergence theorem to $u\nabla v$ and $v\nabla u$ in turn and subtract.) Draw about P a sphere of radius a and let the volume integration be taken throughout the space *outside* this sphere included by a sphere of arbitrarily large radius a'. Letting $u = \dfrac{e^{\pm ikr}}{r}$, we get

$$\int \frac{e^{\pm ikr}}{r}(\nabla^2 v + k^2 v)d\tau = \int \left\{ \frac{e^{\pm ikr}}{r} \nabla v - v\nabla\left(\frac{e^{\pm ikr}}{r}\right) \right\} \cdot d\sigma$$
$$+ \int \left\{ v \frac{\partial}{\partial a}\left(\frac{e^{\pm ika}}{a}\right) - \frac{e^{\pm ika}}{a}\frac{\partial v}{\partial a} \right\} a^2 d\omega. \tag{A}$$

Here the first integral on the right comes from the outer integration whereas the second is taken over the sphere of radius a (with $d\sigma = a^2 d\omega$, etc.), and we have written $\mathbf{n}\cdot\nabla v = \dfrac{\partial v}{\partial n}$. Letting the sphere about P shrink to zero and the outer sphere go to infinity, if v vanishes for large r at least as strongly as $1/r$, the first integral on the right will vanish. The second integral can be expressed in the form

$$-\int v e^{\pm ika} d\omega \pm ik \int ave^{\pm ika} d\omega - \int a \frac{\partial v}{\partial a} e^{\pm ika} d\omega,$$

where a goes to zero. The only non-vanishing term is the first, which becomes $-4\pi v_P$, where v_P is the value of v at the point P. We therefore obtain from (A)

$$\int \frac{e^{\pm ikr}}{r}(\nabla^2 v + k^2 v)d\tau = -4\pi v_P.$$

Originally v was arbitrary. Let us choose $v = V$ in eq. (25). We then get $V(\alpha, \lambda)$ as in eq. (26).

This may be substituted into the equation

$$V_P(t) = \frac{1}{2\pi}\int_{-\infty}^{+\infty} d\alpha \int_{-\infty}^{+\infty} V_P(\lambda)e^{i\alpha(t-\lambda)}d\lambda,$$

with the result that

$$V_P(t) = \frac{1}{2\pi}\int \frac{d\tau}{4\pi r}\int_{-\infty}^{+\infty} d\alpha \int_{-\infty}^{+\infty} \rho(\lambda)e^{i\alpha(t-\lambda \pm r/c)}d\lambda. \quad (27)$$

But if we compare eq. (27) with

$$\rho(t) = \frac{1}{2\pi}\int_{-\infty}^{+\infty} d\alpha \int_{-\infty}^{+\infty} \rho(\lambda)e^{i\alpha(t-\lambda)}d\lambda$$

we see that we can write

$$V_P(t) = \frac{1}{4\pi}\int \frac{\rho(t \pm r/c)}{r} d\tau. \quad (28)$$

The choice of the plus sign in the parenthesis would correspond to an *advanced* potential, of some mathematical interest but of doubtful physical value. The choice of the minus sign corresponds to a *retarded* potential, and with this choice eq. (22) becomes perfectly clear. The derivation of eq. (21) follows precisely the same argument.

Interpretations of the field equations. We shall conclude our brief survey of field theory electrodynamics by asking ourselves the question, typical of the method we have tried to pursue throughout this work: what should be our basic attitude toward the field equations? In the first place, although they are naturally closely connected with the fundamental experimental observations, they can hardly be termed physical laws in the sense in which we have used this designation. It appears more logical to consider them as principles, that is, the expression of the basic postulates of the theory of electrodynamics. This conception is brought out the more strongly when we recall the introduction of the displacement current by Maxwell. We have already noted the view that the equations serve to define **E** and **H**; this is not inconsistent with their postulational nature, for we have previously seen that the fundamental hypotheses of mechanics involve the definitions of force and mass. It is not surprising, however, that many different points of view have been maintained with respect to the field equations. An extremely interesting one is that of Planck,[1]

---

[1] M. Planck, "Theory of Electricity and Magnetism" (trans. from the German) (London, Macmillan, 1932), pp. 20 ff.

who approaches the formulation of the equations from the standpoint of energy. His method may be thus described. The presence of an electromagnetic field in a certain region implies the localization of electromagnetic energy in this region. If conservation is postulated, how can this energy change? Only through its transformation into other forms inside the region or the passage of the electromagnetic energy through the surface enclosing the region. Hence if we write the expression for the total electromagnetic energy in the given region, its time rate of change should be equal to the rate at which energy flows through the surrounding surface (i.e., the surface integral of the normal component of the Poynting flux) plus the rate at which the energy is being dissipated in the region into the form of heat. By the use of the divergence theorem this equation may be transformed into differential form. Placing the coefficients of the components of **E** and **H** equal to zero separately then yields the field equations, or at any rate, the first two in (8). In most discussions, as in the first part of this section, it is customary to show that these equations are consistent with the principle of the conservation of energy. Planck reverses the procedure by treating that principle as the fundamental postulate and deducing the equations. The details will not be presented, as they are readily accessible in the book above referred to.

The number of ways in which one may interpret the electromagnetic field equations seems almost endless. A rather interesting connection between them (in the case of free space) and the equations for small disturbances from equilibrium in a perfect fluid may be seen as follows. For a fluid in which the motion can be described by a velocity potential ϕ, the behavior for small disturbances conforms to the equations [1]

$$\nabla^2 \phi = - \dot{s} \tag{29}$$

$$\dot{\phi} = - c^2 s. \tag{30}$$

Here s is the so-called *condensation*, defined by the relation $s = (\rho - \rho_0)/\rho_0$, where ρ_0 is the equilibrium density and ρ the actual density at any place at any moment. The quantity c is equal to $\sqrt{dp/d\rho}$, and is thus connected with the pressure density relation in the fluid. The equation (29) will be recognized as the equation of continuity in a somewhat disguised form, while eq. (30) is really the form assumed by the hydrodynamic equation of motion in the case of small

---

[1] See, for example, "Acoustics" by Stewart and Lindsay (D. Van Nostrand Co., New York, 1930), pp. 20 ff.

disturbances. The elimination of s between the two equations yields at once the wave equation

$$\nabla^2 \phi = \ddot{\phi}/c^2 \qquad (31)$$

describing propagation with velocity c. Let us now revert to the field equations (1) for free space and use the vector and scalar potentials **A** and V, respectively. From eq. (17) we have at once

$$\dot{\mathbf{A}} = -c^2 \left[\frac{1}{c}(\mathbf{E} + \nabla V) \right], \qquad (32)$$

and from the second equation of (1) with eq. (16)

$$\nabla \times \nabla \times \mathbf{A} = -\nabla^2 \mathbf{A} + \nabla(\nabla \cdot \mathbf{A}) = \frac{1}{c}\dot{\mathbf{E}}.$$

But since from eq. (20) $\nabla \cdot \mathbf{A} = -\dfrac{\dot{V}}{c}$, this last equation can be written in the form

$$\nabla^2 \mathbf{A} = -\frac{\partial}{\partial t}\left[\frac{1}{c}(\mathbf{E} + \nabla V)\right]. \qquad (33)$$

Comparings eqs. (32) and (33) with (29) and (30), we note at once the possibility of considering the quantity $\dfrac{1}{c}(\mathbf{E} + \nabla V)$ as a kind of condensation in the field space while the vector potential **A** plays the rôle of "velocity" potential. Of course eqs. (32) and (33) are vector equations while (29) and (30) are scalar. Nevertheless, elimination of $\dfrac{1}{c}(\mathbf{E} + \nabla V)$ between eqs. (32) and (33) yields the wave equation

$$\nabla^2 \mathbf{A} = \frac{1}{c^2}\ddot{\mathbf{A}}$$

for wave propagation with velocity c. In any case we may say that in free space the field equations can be taken as representing the phenomena associated with the small irrotational disturbances from equilibrium of a perfect fluid in which the condensation is the vector quantity $(\mathbf{E} + \nabla V)/c$ and the velocity potential the vector potential **A**. Such a fluid might be hard to realize, and hence the analogy is presumably of doubtful value. However, it is well to emphasize that in this respect it is probably no more useless than the other attempts to give the field equations objective significance in terms of stress-strain phenomena in elastic media, i.e., as in the so-called ether theories. In every case the fundamental basis of the analogy rests on the mathematical form of the equations.

It is well known that Maxwell was able to cast the general field equations into dynamical form through the replacement of the seat of the electrodynamical phenomena by a dynamical system; further, by the appropriate choice of generalized coordinates and the expression of the kinetic energy in terms of the time derivatives of these coordinates (the electric currents), he was able to show that Lagrange's equations in these generalized coordinates contain in themselves equations equivalent to $\nabla \times \mathbf{E} = -\frac{1}{c}\dot{\mathbf{H}}$ for the various conductors present, as well as the equivalent of the force equations (10). Maxwell regarded the equation $\nabla \times \mathbf{H} = \frac{1}{c}(\dot{\mathbf{E}} + \rho \mathbf{v})$ as a definition of $\mathbf{H}$ in terms of the currents in the system. We shall not go into the details here.[1] They are probably sufficiently familiar to many readers.

We have pursued far enough our survey of the classical theory of the electromagnetic field. It will next be in order to examine the situaation brought about by the introduction of the theory of electrons, i.e., the assumption of the discrete nature of electricity. This forms a natural bridge between classical electrodynamics and the theory of relativity, which marks the most advanced stage yet attained by field theory in physics. It will therefore be suitable material for the next chapter.

[1] See Maxwell, "Treatise on Electricity and Magnetism" (third edition, Oxford, 1904), Vol. II, pp. 199 ff.

CHAPTER VII

THE ELECTRON THEORY AND SPECIAL RELATIVITY

7.1. Electron Theory of Electrodynamics. The electron theory of matter assumes that all matter is composed of particles of positive and negative charge, known as protons and electrons, respectively. Since we are here still concerned with classical theory we shall not consider the more recently discovered elementary particles, the neutron and the positron. The electron has a charge of 4.80×10^{-10} esu or $4.80 \times 10^{-10}\sqrt{4\pi}$ Heaviside units which are the kind used in the discussion of the previous chapter. This is assumed to be the smallest charge of electricity which can exist separated from other charges. The proton has the same charge but of opposite sign. Its mass is approximately 1840 times that of the electron. The electron theory assumes the existence of no intrinsic magnetic material: all magnetic fields are supposed to originate in the motions of charged particles which alone are considered to exist in matter. The question now arises as to what form the equations of electrodynamics will take in a theory of this sort. Clearly we have no longer a field theory in the sense of Chapter VI, for there is no longer a continuum. Nevertheless, the fact that the classical field equations of Maxwell work so well in describing macroscopic electromagnetic phenomena would lead one to try their use in the electron theory with the smallest possible amount of modification. Certainly the electron theory equations must reduce to the classical theory equations in free space at points not occupied by matter. What is usually assumed is that the same equations, namely (6.4–8), are of universal validity. In them ρ still means charge per unit volume and is assumed to be a continuous function of the coordinates, i.e., there are no abrupt changes of charge density. The same is assumed to be true of $\mathbf{v}$, the charge velocity. What then shall we say about the charge distribution on an electron? It must be continuous, but beyond that a variety of assumptions are possible. In the theory of Lorentz,[1] the electron is assumed to be a spherical

---

[1] It is of historical interest to note that Lorentz's theory was not the first attempt to describe electromagnetic phenomena in terms of charged particles. One recalls the mid-nineteenth-century theories of Weber, Riemann, and Clausius in which

shell (with a continuous distribution of charge) when observed in a reference system in which it is at rest. The electromagnetic force on an element of charge $\rho d\tau$ is defined to be $\mathbf{E}\rho d\tau$, as measured in the reference system in which the charge element at the moment is at rest; the electromagnetic force on the element $\rho d\tau$ as measured in the reference system with respect to which the element has a velocity $\mathbf{v}$ at the instant considered is defined to be $\left[\mathbf{E} + \dfrac{1}{c}(\mathbf{v} \times \mathbf{H})\right]\rho d\tau$. The significance of the latter which is the more general expression has already been indicated in Sec. 6.4, eq. (6.4–10).

Dynamical equation of the electron. Now to discuss the motion of electrons it is necessary to have a dynamical equation of motion which will serve the same purpose for electrons as the principles of mechanics for ordinary macroscopic material particles. One might at first suppose that the latter would also suffice for the present purpose. We have, to be sure, defined the force on an electron, which from the preceding paragraph is

$$\int \left[\mathbf{E} + \frac{1}{c}(\mathbf{v} \times \mathbf{H})\right]\rho d\tau, \tag{1}$$

the integration being extended over the whole distribution of charge which constitutes the electron. The difficulty arises with respect to *mass*. In the mechanical case the definition of mass was based on certain assumptions about macroscopic particles which we have no assurance whatever will apply to an electron. Hence it appears that the dynamics of the electron must be approached from a different point of view. This is the function of the so-called *dynamical assumption*, which may be stated as follows: an electron moves in such a way that the total electromagnetic force on it as measured in any reference system is equal to zero. This means that the integral in (1) shall be placed equal to zero, and that this constitutes the equation of motion of the electron. One or two comments are necessary here. The phrase "any reference system" is too sweeping. It turns out that we must restrict ourselves to systems which have the same geometry and the same devices for the measurement of time and distance. These have been called *reciprocal* systems. Thus we must avoid using any reference system in which the geometry, etc., is different from that of the

charged particles were fundamental. In all these theories, however, the particles were supposed to act on each other at a distance, whereas in that of Lorentz the electrons interact with the medium in which they are embedded. To this extent Lorentz's theory is still a field theory.

7.1 ELECTRON THEORY OF ELECTRODYNAMICS

reference system in which the electron is momentarily at rest—its *proper* system. If the total electromagnetic force on the electron is zero in such a system, it will not necessarily be zero in another system of quite different geometry. The term "total" force also needs some explanation. This means the force due to the impressed field as well as that due to the electron's own field. Only electrical forces are here considered: we expressly exclude forces (such as gravitational) which are outside this category. One further matter needs attention: the **E** and **H** occurring in the integral (1) are supposed to refer to that system in which the electron is momentarily at rest.

We have now to see how the dynamical assumption leads to a complete understanding of the motion of an electron. It is simplest to obtain first the equation of motion for that system with respect to which the electron is momentarily at rest. If $\mathbf{F}_0$ is the total force due to the impressed field and $\mathbf{F}_i$ that due to the electron's own field, the dynamical assumption says that

$$\mathbf{F}_0 = - \mathbf{F}_i, \qquad (2)$$

where

$$\mathbf{F}_0 = \int \rho \mathbf{E}_0 d\tau = e\mathbf{E}_0, \qquad (3)$$

if it is assumed that the impressed field intensity $\mathbf{E}_0$ is constant over the electron, and we write $\int \rho d\tau = e$, the charge of the electron. The evaluation of $\mathbf{F}_i$ is based on the expression for the simultaneous field of a point charge which is too complicated for reproduction here.[1] Essentially what one does is to write the simultaneous field for an element of charge de of the electron and so find the force which the latter exerts on another element of charge of the electron, say de_1. Integration with respect to de_1 gives the force on the electron due to the field produced by de. Finally integration of the result with respect to de gives $\mathbf{F}_i$. We shall merely quote the result, in the form of a series of which only the first three terms are here retained.

$$\mathbf{F}_i = - \frac{e^2}{6\pi a c^2} \mathbf{f} + \frac{e^2}{6\pi c^3} \dot{\mathbf{f}} - \frac{e^2 a}{9\pi c^4} \ddot{\mathbf{f}} \ldots \qquad (4)$$

In this expression **f** is the acceleration of the electron (its variation from point to point over the electron proves to be negligible to the

---

[1] See, for instance, Leigh Page, "Introduction to Electrodynamics" (Boston, 1922), Chapters III and IV.

THE ELECTRON THEORY AND SPECIAL RELATIVITY

indicated approximation) and a is its radius. The dynamical equation then takes the form:

$$e\mathbf{E}_0 = \frac{e^2}{6\pi a c^2}\mathbf{f} - \frac{e^2}{6\pi c^3}\dot{\mathbf{f}} + \frac{e^2 a}{9\pi c^4}\ddot{\mathbf{f}} \ldots \quad (5)$$

We may now transform to the system with respect to which the electron has the velocity $\mathbf{v}$. The transformation equations are those of the theory of relativity, but we shall for the moment ignore the significance of this fact and indicate the result, which is to two terms

$$e\left[\mathbf{E}_0 + \frac{1}{c}(\mathbf{v} \times \mathbf{H}_0)\right] = \frac{e^2}{6\pi a c^2}\mathbf{f} - \frac{e^2}{6\pi c^3}\dot{\mathbf{f}} \ldots \quad (6)$$

In order for this result to be valid in general, one may show that it is necessary to define the acceleration $\mathbf{f}$ as

$$\mathbf{f} = \frac{d}{dt}\left(\frac{\mathbf{v}}{\sqrt{1 - v^2/c^2}}\right)$$

$$= \frac{\dot{\mathbf{v}}}{\sqrt{1 - v^2/c^2}} + \frac{\mathbf{v}\cdot\dot{\mathbf{v}}/c^2}{(1 - v^2/c^2)^{3/2}}\mathbf{v}. \quad (7)$$

Transverse and longitudinal mass. Now the velocity $\mathbf{v}$ may be written in the form

$$\mathbf{v} = v\mathbf{v}_1, \quad (8)$$

where $\mathbf{v}_1$ is a unit vector in the direction of $\mathbf{v}$ and v is the magnitude of $\mathbf{v}$. Then

$$\dot{\mathbf{v}} = \dot{v}\mathbf{v}_1 + v\dot{\mathbf{v}}_1, \quad (9)$$

where $\dot{\mathbf{v}}_1$ is the vector representing the rate of change in the direction of $\mathbf{v}$. It is normal to $\mathbf{v}_1$. We may look upon $\dot{v}\mathbf{v}_1$ as the component of $\dot{\mathbf{v}}$ along the velocity while $v\dot{\mathbf{v}}_1$ is its component perpendicular to $\mathbf{v}$. Substituting from eq. (9) into (7) we get after some reduction

$$\mathbf{f} = \frac{\dot{v}\mathbf{v}_1}{(1 - v^2/c^2)^{3/2}} + \frac{v\dot{\mathbf{v}}_1}{\sqrt{1 - v^2/c^2}}. \quad (10)$$

Calling $\frac{e^2}{6\pi a c^2} = m_0$, the force equation (6) becomes, if we confine our attention to the first term,

$$\mathbf{F}_0 = \frac{m_0}{(1 - v^2/c^2)^{3/2}}\dot{v}\mathbf{v}_1 + \frac{m_0}{\sqrt{1 - v^2/c^2}}v\dot{\mathbf{v}}_1. \quad (11)$$

7.2 INVARIANCE OF FIELD EQUATIONS OF ELECTRON THEORY

To this approximation the result is in the form of the mechanical equation of motion $\mathbf{F} = m\mathbf{a}$, save that the mass associated with the ordinary mechanical acceleration in the direction of the velocity is different from the mass associated with the acceleration normal to the velocity. The electron thus appears to have two distinct masses. These are referred to as longitudinal and transverse, respectively. Writing them m_l and m_t we have

$$m_l = \frac{m_0}{(1 - v^2/c^2)^{3/2}}, \quad m_t = \frac{m_0}{\sqrt{1 - v^2/c^2}}. \tag{12}$$

Another striking result of the analysis is the dependence of the electron mass on its velocity. As $v \to c$, m_l and m_t both grow infinitely great, while for velocities small compared with c, m_l and m_t reduce approximately to m_0, which is therefore called the *rest mass*.

In the evaluation of m through the well-known measurement of e/m by the method of the effect on an electron stream (of *small* velocity) of crossed electric and magnetic fields and the independent measurement of e by Millikan's method it is really m_0 which results. The best value of m_0 obtained in this way is $(8.994 \pm 0.014) \times 10^{-28}$ gram. This differs slightly from that obtained by spectroscopic methods, which is $(9.035 \pm 0.010) \times 10^{-28}$ gram. Millikan's value of e is $(4.80 \pm 0.005) \times 10^{-10}$ esu. Using $c = (2.99796 \pm 0.00004) \times 10^{10}$ cm/sec we have from

$$m_0 = \frac{e^2}{6\pi a c^2} \tag{13}$$

the approximate value

$$a = 1.88 \times 10^{-13} \text{ cm}$$

for the radius of the Lorentz electron.

The formula for the transverse mass has received experimental verification by the experiments of Bucherer and others in which e/m was measured for electrons of varying velocities of the order of magnitude of c. It is interesting that the theory of relativity leads independently of electron theory to similar results.

7.2. Invariance of the Field Equations of Electron Theory. We must now retrace our steps somewhat and recall that the nineteenth century physicists of the Maxwell school were firm believers in the existence of a medium in which all electromagnetic effects take place and through which they propagate themselves. This medium was supposed to pervade all space, even material bodies, and received the name *ether*. The theoretical prediction by Maxwell of the existence of electromagnetic waves and the subsequent experimental verification by Hertz strengthened the belief in the existence of this ether, for to the physicists of that epoch a wave without a medium for its propagation was unthinkable. At the same time physicists could not remain contented with the existence of a medium without discussing its properties. Possibly they would have been happier in the long run if they had refrained from such inquiries, for much confusion resulted. Yet these investigations later proved their worth in the establishment of the theory of relativity.

One of the questions which puzzled the ether advocates was this: when material bodies move, does the ether remain stationary or is it carried along with the bodies? The phenomenon of aberration, i.e., the apparent displacement of the position of stars which cannot be explained as due to parallax, discovered by Bradley in 1726, seemed to demand for its explanation an ether which is not carried along by the earth. On the other hand the experiments of Fizeau (1851) on the velocity of light in flowing water indicated that the ether is to a certain extent dragged along with the water though it does not have the same velocity. During the late nineteenth century a great many experiments were performed for the purpose of trying to detect the motion of the earth through the ether. We shall discuss only one of these, the famous experiment of Michelson (1881), usually known by the names of Michelson and Morley since they repeated the original test more carefully in 1887. There have been numerous repetitions since that time, which we shall consider later.

Michelson-Morley-Miller experiment. The theory of the experiment rests on a suggestion of Maxwell to the effect that, if the earth moves with respect to a stationary ether, a light signal sent a certain distance in the direction of the earth's motion and reflected back will require for its total trip a time slightly greater than will be needed by a similar signal sent out an equal distance in a direction at right angles to the direction of motion. The following diagram of the experiment as arranged to use the interferometer of Michelson will make clear the fundamental idea.

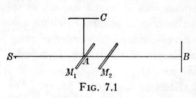

Fig. 7.1

S is a source of light, M_1 and M_2 are two plane parallel slabs of glass of which M_1 is half-silvered on the side toward M_2. Light from the source is incident on M_1 at A and part is reflected from the back surface to the mirror C ($AC = l_1$) while the rest passes through M_2 and proceeds to $B(AB = l_2)$ where it is reflected. AC is

7.2 INVARIANCE OF FIELD EQUATIONS OF ELECTRON THEORY

assumed to be perpendicular to AB. The light from the two mirrors is joined again at A and observed from O. If the apparatus is assumed to be stationary in the ether and if $l_1 = l_2$, the time taken by the two beams is the same. (Note that the presence of the parallel mirror M_2 compensates for the passage of the light through M_1.) Consider now the situation when the apparatus moves in the direction AB with the constant velocity v. The total time for a light signal to travel from A to B and back is

$$\frac{l_2}{c-v} + \frac{l_2}{c+v} = \frac{2cl_2}{c^2-v^2}, \tag{1}$$

c being, as usual, the velocity of light. On the other hand, the time taken by the signal sent to C to return turns out to be

$$\frac{2c\sqrt{1-v^2/c^2}\,l_1}{c^2-v^2}. \tag{2}$$

The difference is

$$(\Delta t)_1 = \frac{2}{c(1-v^2/c^2)}[l_2 - \sqrt{1-v^2/c^2}\,l_1]. \tag{3}$$

Now suppose that the apparatus is rotated through 90 deg. so that AC is in the direction of the motion. The time difference between the two paths becomes

$$(\Delta t)_2 = \frac{2}{c(1-v^2/c^2)}[l_2\sqrt{1-v^2/c^2} - l_1]. \tag{4}$$

There is thus in general a difference between $(\Delta t)_1$ and $(\Delta t)_2$ which we may denote as $\Delta^2 t$ and write

$$\Delta^2 t = \frac{2(l_1+l_2)}{c(1-v^2/c^2)}[1 - \sqrt{1-v^2/c^2}]. \tag{5}$$

For l_1 and l_2 of the order of magnitude of 1 meter and for v of the order of 30 km/sec, the approximate velocity of the earth in its orbit about the sun (relative to axes fixed in the sun), we expect therefore $\Delta^2 t$ to be of the order of 10^{-16} sec. Michelson's interferometer method was sensitive enough to detect such a change by the shift in the interference fringes brought about by the rotation of the apparatus. When the experiment was first performed in 1881 the values of $\Delta^2 t$ obtained were of the order only of one-tenth to one-fourth of the expected value on the basis of the above choice of v. The results were attributed to experimental error, and it was concluded that the experiment was inconsistent with the assumption of a stationary ether with respect to which the earth moves. Alternatively one could say that even if the earth does move through the ether the experiment might, for some reason, be incompetent to detect this motion. When the experiment was repeated more carefully and with more elaborate apparatus by Michelson and Morley in 1887 the result was that the indicated relative velocity of the earth and ether did not exceed one-fourth of the earth's orbital velocity. As D. C. Miller has pointed out,[1] this was not strictly a null result. Nevertheless, it has been so interpreted for many years. The experiment has been repeated with various modifications and refinements a great many times. The work of Miller has been most extensive and appears to indicate a genuine positive result with a time difference corresponding to a maximum velocity of about 10 km/sec, which, however, has a slight seasonal variation. Other observers (mentioned in Miller's paper, to which reference should be made for details) have reported results corresponding to velocities as low as 1 km/sec and have interpreted them as indicative of a null effect. It seems clear that the whole question is still an open one and further work should be carried out. The situation is interesting, for a great deal of theorizing during the last years of the nineteenth century and the first part of the twentieth was based definitely on the existence of a null result of the experiment. Perhaps the neglect of the small positive result and its attribution to experimental error may have been due to the outcome of other experiments such as that of Trouton and Noble, designed to show the existence of motion through the ether by the torque on a suspended charged condenser. These experiments failed to show the indicated effect, at least to the expected magnitude, and this undoubtedly strengthened the feeling that experiments of an electromagnetic nature are incompetent to detect motion of the earth through the ether.

Invariance of the electromagnetic equations. The theoretical implication of the result just stated is that the equations which describe electromagnetic phenomena must by their very form reflect the indifference of these phenomena when referred to two reference systems moving with respect to each other with constant velocity. But this must mean that the form of the equations must remain invariant under the transformation from one such reference system to another. Let us see precisely what this means. We have already examined the meaning of invariance in connection with vectors and tensors in the preceding chapter. We have seen that vector (explicitly 3-vector) equations remain invariant in form under a space transformation that

---

[1] See his very complete account, "The Ether Drift Experiment and the Determination of the Absolute Motion of the Earth," *Reviews of Modern Physics*, 5, 203, 1933.

7.2 INVARIANCE OF FIELD EQUATIONS OF ELECTRON THEORY

is equivalent to a rotation of axes and that tensor equations show similar invariance under more general types of transformation. The question now arises: what kind of transformation connects two systems of reference moving with respect to each other with constant velocity u along their common x axis? This is the question which Lorentz [1] asked himself. More specifically, he wished to know just how the quantities $x, y, z, t, \mathbf{E}, \mathbf{H}, \rho$, and $\mathbf{v}$ of the one system must be related to the corresponding quantities $x', y', z', t', \mathbf{E}', \mathbf{H}', \rho'$, and $\mathbf{v}'$ of the other system, so that the fundamental field equations

$$\left. \begin{array}{l} \nabla \times \mathbf{E} = -\dfrac{1}{c}\dfrac{\partial \mathbf{H}}{\partial t} \\[4pt] \nabla \times \mathbf{H} = \dfrac{1}{c}\left(\dfrac{\partial \mathbf{E}}{\partial t} + \rho \mathbf{v}\right) \\[4pt] \nabla \cdot \mathbf{E} = \rho \\[4pt] \nabla \cdot \mathbf{H} = 0 \end{array} \right\} \quad (6)$$

shall after the transformation have the form

$$\left. \begin{array}{l} \nabla' \times \mathbf{E}' = -\dfrac{1}{c}\dfrac{\partial \mathbf{H}'}{\partial t'} \\[4pt] \nabla' \times \mathbf{H}' = \dfrac{1}{c}\left(\dfrac{\partial \mathbf{E}'}{\partial t'} + \rho' \mathbf{v}'\right) \\[4pt] \nabla' \cdot \mathbf{E}' = \rho' \\[4pt] \nabla' \cdot \mathbf{H}' = 0, \end{array} \right\} \quad (7)$$

where $\nabla' \times E'$ is the vector having the components $\left(\dfrac{\partial E_z'}{\partial y'} - \dfrac{\partial E_y'}{\partial z'}\right)$, etc.

One further important assumption was added to the foregoing, namely, that in the transformation the charge carried by any particular volume element shall be unaltered. Lorentz found the transformation equations satisfying these conditions to be of the form (letting $\gamma = 1/\sqrt{1 - u^2/c^2}$)

$$\left. \begin{array}{l} x' = \gamma(x - ut) \\ y' = y \\ z' = z \\ t' = \gamma\left(t - \dfrac{ux}{c^2}\right) \end{array} \right\} \quad (8)$$

[1] H. A. Lorentz, *Proc. Amsterdam Acad.*, 6, 809, 1904.

$$v_x' = \frac{v_x - u}{1 - v_x u/c^2}; \quad v_y' = \frac{v_y}{\gamma(1 - v_x u/c^2)}; \quad v_z' = \frac{v_z}{\gamma(1 - v_x u/c^2)} \quad (9)$$

$$\left.\begin{aligned} E_x' &= E_x, \quad E_y' = \gamma\left(E_y - \frac{uH_z}{c}\right), \quad E_z' = \gamma\left(E_z + \frac{uH_y}{c}\right) \\ H_x' &= H_x, \quad H_y' = \gamma\left(H_y + \frac{uE_z}{c}\right), \quad H_z' = \gamma\left(H_z - \frac{uE_y}{c}\right) \\ \rho' &= \gamma\rho\left(1 - \frac{v_x u}{c^2}\right). \end{aligned}\right\} \quad (10)$$

To obtain these transformations is a perfectly definite mathematical problem, the solution of which we shall not present in detail. We have dealt with similar problems in Chapter VI when, for instance, we determined the conditions under which equations will remain invariant under a rotation of axes and found that the quantities involved must transform like 3-vectors.

But the physical meaning of eqs. (8) to (10) requires comment. Let us first consider eqs. (8), which express the coordinates in the moving reference system in terms of those in the stationary one. If one who was well acquainted with kinematics were asked to write what he thought were the natural transformation formulas connecting two systems of reference moving with respect to each other along their common x axes in the positive direction with constant velocity u, it is fairly certain that he would reply

$$x' = x - ut, \quad y' = y, \quad z' = z, \quad t' = t. \quad (11)$$

It is true that he might have a few doubts about the invariance of t, but it is probable that he would consider it the most reasonable assumption that could be made since there are no kinematic reasons for supposing a dependence of time on motion. This, of course, was Newton's view, as has already been mentioned in Chapter II. As far as the equations of classical mechanics are concerned we have seen that they remain invariant under the transformation (11), and this would probably strengthen one's confidence in them. Hence the fact that the equations of electrodynamics do not remain invariant under such a transformation is at first rather surprising and somewhat disconcerting. The practically minded person, inspecting again eqs. (8), etc., observes, however, that if $u << c$, as is the case with all macroscopic terrestrial motions, the radical $\gamma \doteq 1$ and (8) reduce approximately to (11), while (9) become to the same approximation $v_x' = v_x - u$, $v_y' = v_y$, $v_z' = v_z$, etc. We shall encounter eqs. (8) again

7.2 INVARIANCE OF FIELD EQUATIONS OF ELECTRON THEORY

in the next sections, where they will be discussed from a somewhat different point of view in connection with the theory of special relativity. It will there be shown that they imply a contraction of moving objects in the direction of their motion. This is the famous hypothesis put forward in 1892 by Fitzgerald to explain the assumed null result of the Michelson-Morley experiment: the failure to observe the time difference for the light signal in and perpendicular to the direction of motion is here attributed to the actual shortening of the apparatus in the direction of the motion by an amount exactly sufficient to compensate for the greater time calculated on the assumption of the independence of length on motion. From this point of view the Lorentz electron in the system with respect to which it is moving is no longer a sphere but an ellipsoid with constantly shrinking minor axis as its velocity approaches that of light.

Eqs. (9) can be derived from (8), as will also be shown in the next section.

One interesting feature of eqs. (10) is the manner in which the E's and H's are intermingled. This clearly indicates the interdependence of electric and magnetic field intensities, since what appears as a purely electric field intensity in one reference system appears as a combination of both electric and magnetic intensities in another system, and vice versa for the magnetic intensity. It emphasizes the unitary nature of the electromagnetic field: the separation of it into electric and magnetic components is purely arbitrary, dependent as it is on the system of reference. The precise form of E_y', H_y', etc., becomes clearer when we recall the so-called force equation (6.4-10): motion of a charged medium or collection of charged particles with velocity $\mathbf{u}$ through a magnetic field involves the addition to the electric intensity of the term $\frac{1}{c}\mathbf{u} \times \mathbf{H}$, and the reference to the moving system is equivalent to such motion. Similarly the magnetic field produced by a current (i.e., here a moving charge) as expressed by the field equation

$$\nabla \times \mathbf{H} = \frac{1}{c}(\mathbf{E} + \rho\mathbf{v})$$

accounts for the precise form of H_y' in terms of the unprimed quantities, etc., as a straightforward calculation shows. Thus the reference to a moving system serves of itself to introduce an additional magnetic field. These considerations recall the celebrated experiment of Rowland in 1876 which showed conclusively the magnetic field associated with moving "static" charges.

The interpretation of eqs. (10) may also be made in another way. By means of the solution of the field equations (6.4–16) to (6.4–22) it is possible to express the field intensities in terms of ρ and $\mathbf{v}$. This means that E_y' for example is expressible in terms of density and velocity in the primed system and hence by means of the purely kinematic transformation equations (8) in terms of ρ and $\mathbf{v}$; and so finally back to the field quantities for the unprimed system. For consistency the result must agree with (10). As a matter of fact, the calculation shows that it does. This may be said to reduce the whole problem to a kinematic one.

It is worth noting that the invariance of the electromagnetic field equations under such a Lorentz transformation as we have been considering at once casts considerable discredit on the concept of the luminiferous ether. A medium of such a nature that you cannot tell whether you are moving with respect to it, or at least not by means of any electromagnetic phenomenon (including of course light), is not a very useful medium in the material sense in which classical physicists viewed media. The feeling therefore arises that one might as well dispense with the ether concept altogether. But the inquiry may then be made: how is one going to talk about the propagation of electromagnetic waves if there is no medium? The answer is that after all in such propagation the medium is the least important thing. From hydrodynamics, elasticity, etc., we have grown too much accustomed to thinking that a wave is a moving disturbance in a medium. Actually we may as well admit that the only relevant feature about wave motion is that there exist physical quantities which are functions of space and time in the form $f(x \pm vt)$. Working with this idea and its developments one can cast aside the medium notion as irrelevant embroidery which is of value only in so far as pictures are valuable and which ceases to be valuable when the pictures break down.

7.3. The Special Theory of Relativity. *Transformation equations.* In our discussion of the electrodynamics of electron theory in the preceding sections we have set the stage, so to speak, for the theory of relativity introduced by Einstein in 1905. The historical development immediately antecedent to this theory was concerned almost entirely with optics and electrodynamics—the former being since Maxwell's time really a branch of the latter in spite of the endeavors of Kelvin and others to make something out of the elastic solid theory. It is true that ideas of relative motion go back to very early times—Einstein's theory was not the first theory of relativity. As was seen in an earlier chapter Newton had a well-defined theory of relativity which consisted in the assumption that the laws of mechanics have the same form in all reference systems moving with respect to each other with constant velocity. If the principles of mechanics are expressible in terms of differential equations of the second order with respect to the time, then this theory of Newton is equivalent to the assumption that the transformation equations between such reference systems as are here considered (namely, *inertial* systems) have the form (7.2–11). Essentially therefore we may say that the Newtonian theory of relativity expressed the impossibility of detecting the motion of an inertial system by any mechanical means. No mechanical experiment that can be performed, for example, on a train moving at constant speed can divulge to us that the train is really moving, for the results of all such experiments *relative* to the train are precisely the same as those which would have been obtained by the performance of similar experiments at the station or on the roadside. One unfamiliar with our brand of mechanics would doubtless be tempted to ask why we must restrict this observation to a train moving at constant velocity? We know from experience that when the train is suddenly brought to rest, i.e., decelerated, the statement is no longer true: passengers tend to lurch forward in their seats, and in general we say that objects inside the train experience forces which we ascribe to the deceleration of the train. Alternatively we can look upon the existence of such forces (or rather more appropriately the otherwise unaccountable accelerated motions of objects inside the train) as an indication of the accelerated motion of the train; in other words we say that although we cannot detect uniform motion by mechanical means we can detect non-uniform motion by this method.

Another at first sight puzzling case is that of uniform rotation. Can an observer detect by mechanical means that he is rotating uniformly about some axis? Suppose, for example, that no other objects in the universe are visible to the observer. How could he observe his motion? The answer according to Newton, is very simple: he would detect his absolute rotation by the observation of the ordinary centripetal acceleration and the Coriolis acceleration which are familiar in the study of rotating axes. These accelerations would be detected as forces by the observer, and on their detection he would conclude that he was rotating. And if we ask, rotating with respect to what, Newton would answer, rotating in absolute space and not relative to any body or bodies. Newton felt that he had made clear the difference between relative and absolute rotation by the famous experiment in which a pail of water is suspended by a twisted rope and is then allowed to rotate while the rope untwists. It is observed that when the pail begins to move the water stands still and preserves its plane surface. After a time, however, the water also begins to rotate with the pail, whereupon its surface assumes the characteristic paraboloidal shape. The interesting and important thing, according to Newton, is that when the *relative* rotational velocity of the water and pail is greatest, one can detect no change in the water surface; it is when the relative velocity is practically *zero*, that the change in surface is most marked. The latter may then be taken to illustrate absolute rotation. Whenever one observes the existence of centripetal acceleration and force, one is justified in assuming absolute rotation about an axis through the center in question. This is Newton's view.

What attitude should we take with respect to it? There appear to be essentially two possibilities. In the first place, the observer recognizing the existence of centripetal and Coriolis accelerations might prefer to attribute them to the existence of forces having nothing essentially to do with rotation. There can be no logical objection to this stand. He might well account for the centripetal force, for example, by a special kind of attraction between material bodies. It is true that the attraction would be a peculiar one, for from the nature of centripetal acceleration ($a_c = \omega^2 r$) it would increase with the distance of separation; that is, bodies far away from the observer would exert greater force than bodies near him, which might conflict with his natural feeling that the influence of bodies should decrease with the distance. The Coriolis acceleration would also probably cause some embarrassment, though the observer might satisfy himself by attributing it to a kind of viscous drag of a medium through which he was moving. As a matter of fact, if he once made the calculation of the kinematical transformation from a fixed to a rotating reference system he would immediately encounter the accelerations in question cropping up in his equations and would probably decide that after all it was simpler to conclude that he was rotating. However, it is important to emphasize that this decision would be solely one of convenience. It is not something which he would be forced to do, nor could he logically turn about and say that by the observation of the centripetal and Coriolis acceleration he had proved his rotation. Even the embarrassment which he might feel in otherwise accounting for these accelerations could be easily mitigated by a very slight change in the principles of his mechanics—a change which would be too slight to be considered in most practical cases and which would therefore put mechanics on a par with all his other physical theories in which after all the mathematically simple laws are only approximate. Thus he could secure consolation for his troubles at a fairly cheap price.

But suppose he felt one day a suspicion that he was fooling himself. He has still another way out. He could logically hold that he was not rotating but that the rest of the universe was rotating about him and by this rotation was producing the observed accelerations. And if one were to object that Newton's pail-of-water experiment definitely ruled out this conclusion, he could legitimately answer that this experiment does not really prove Newton's conclusion: in order to do this, one must use a bigger pail, and if a pail with walls as thick as the distance from the earth to the sun, for example, were used, the rotating pail would then produce the change in the shape of the water surface which Newton used as an indication of absolute rotation. This is the standpoint of Ernst Mach and is a direct denial of the existence of absolute rotation. On this view there is no distinction between the statement that the earth rotates and the rest of the universe is at rest and the statement that the earth is at rest and the remainder of the universe rotates about it.

Einstein decided to discuss the problem of relativity from a more general standpoint than had been previously taken. He lays down the following general postulates:[1]

1. Physical laws and principles are of the same form in all inertial systems, that is, in all reference systems which differ only in the fact that they are moving with constant velocity with respect to each other.

2. The velocity of light has the same value in all inertial systems (i.e., assuming of course the use of the same units of space and time).

The first statement clearly extends Newton's mechanical relativity and Lorentz's electromagnetic relativity to include all physical laws. The second statement is really a special case of the first if it is assumed as a physical law that in some inertial system the velocity of light is constant. All the consequences of the special or restricted theory of relativity flow from these two postulates. It is therefore desirable to be sure of their meaning. Let us consider the second one first. The usual interpretation asks us to imagine two observers, one of whom refers all events to a reference system S in which coordinates are designated by x, y, z and to a clock the measure of time on which is denoted by t, while the other uses S' as his reference system with coordinates and time denoted by x', y', z', t'. Actually the introduction of observers in different reference systems is apt to introduce

---

[1] A. Einstein, *Annalen der Physik*, 17, 891, 1905.

confusion into the discussion, and although this has been the historical method of developing the theory it seems wiser to forego this scheme and talk simply about *one* observer (i.e., the experimenter) who may use various reference systems as he finds them more or less convenient. In S he writes his equations in terms of unprimed coordinates and time, in S' with primed coordinates. When the experimenter wishes to measure velocity he must have a measure of time and must also know how to make simultaneous time measurements, that is, he must be able to note simultaneously the event consisting of the coincidence of a pointer on a scale and that consisting of the coincidence of the hands of a clock on the dial. We have already discussed (Chapter II) the difficulties inherent in this notion of simultaneity, particularly when the event in question takes place at a point in space far distant from the position of the clock. We have seen that in general any definition of simultaneity must be arbitrary, but we have noted Einstein's plausible definition which for convenience will be restated here.

The time of an event at any point in space is to be recorded on a clock at that place. We must then think of a clock attached to every point in space. All these clocks are assumed to be alike when the observer examines them at one place. Before we can use them, however, they must be synchronized. This may be done as follows. Suppose we consider two points P_1 and P_2 with their respective clocks at rest there. Suppose a light signal leaves P_1 at t_1 (i.e., time as indicated on the clock at P_1) and arrives at P_2 at time t_2 (time as indicated on the clock at P_2. It is then reflected back to P_1, arriving at the time t'_1 (again as indicated, of course, on P_1's clock). The two clocks will be said to run in synchronism if

$$t_2 - t_1 = t_1' - t_2$$

or

$$t_2 = \tfrac{1}{2}(t_1' + t_1).$$

The reading of the clock at P_2 when the signal reaches there is thus taken as the arithmetical mean of the readings of the clock at P_1 when the signal left and returned there respectively. To insure the consistency of the definition, it is necessary to assume that synchronism is a reciprocal property, that is, if the clock at P_1 is synchronous with that at P_2, then the latter is synchronous with that at P_1. Moreover, it will be further assumed that if the clock at P_1 synchronizes with one at P_2 and one at P_3, then those at P_2 and P_3 run in synchronism with each other. We then define the time of any event as the reading of a clock which is at rest at the place where the event takes place, it being assumed that the clock in question is synchronized with respect to all the other clocks resting at all other points in space. It is essential that we observe that the clocks are here all assumed to be at rest in space, that is, of course, with respect to some reference system, viz., the one in which the events are being described.

Einstein found that, to satisfy the assumptions (1) and (2), x', y', z', and t' for one inertial system must be related to x, y, z, and t of the other inertial system moving with respect to the first with the constant velocity u, by the Lorentz transformation equations (7.2-8). These are now usually known as the Lorentz-Einstein transformation equations and form the analytical kernel for all the results of special relativity. Many deductions of the formulas have been given, and of course one should be able to deduce them by starting with the assumed invariance in form of the mathematical expression of any physical law whatever. Lorentz's method was to start with the equations and prove the invariance of the electromagnetic field equations under such a transformation (with the added assumptions (7.2-9, 10)). We shall proceed in the reverse direction and from the principle of relativity derive the equations.

A comparatively simple manner of procedure is as follows. We consider two inertial systems moving along their common x axis with constant velocity u. Since the motion is in the x direction, two of the transformation equations will be $y = y'$, $z = z'$, at once. We wish to find the relation between x, x', t, and t'. It will be assumed (and this is matter of some importance) that the transformation equations are *linear*.[1] Moreover, for the sake of convenience the two x axes will be supposed to slide along each other, and for $t = t' = 0$ we take $x = x' = 0$, $y = y' = 0$, $z = z' = 0$. This fixes the time origin: the clock attached to the origin of S is assumed to record zero when it coincides with the origin of S' and the clock fixed at the origin of S' has the same reading then. Consider the origin of S'. In the S system its x coordinate at time t (as indicated on a clock fixed in S, located at x and synchronized with the clock at the origin of S and indeed all the other clocks in S) is given by $x = ut$; hence the kinematic equation describing its motion is $x - ut = 0$. But in the S' system the motion of the origin of S' is given by $x' = 0$. These two equations are equivalent, since both describe the same thing. Hence when one holds the other must, and therefore we write, since we are assuming a linear relationship,

$$x' = \gamma(x - ut) \tag{1}$$

[1] If this were not so, one event in S would correspond to two or more in S', and vice versa.

where γ is a constant presumably dependent on u. If now we apply a similar argument to the motion of the origin of S, which in S is given by $x = 0$ and in S' is given by $x' + ut' = 0$, we get

$$x = \gamma(x' + ut'). \tag{2}$$

The principle of relativity, in particular postulate 1, requires that the parameter shall be the same, i.e., γ, in both cases. On the basis of Newtonian relativity $\gamma = 1$. We can no longer assume this, but must make eqs. (1) and (2) satisfy the condition that the velocity of light is the same in both systems. First let us get the relation between t and t'. On substituting for x' in eq. (2) from eq. (1) we get

$$t' = \gamma t + \left(\frac{1 - \gamma^2}{u\gamma}\right)x. \tag{3}$$

Before discussing further the physical meaning of these equations, let us find γ. We write for the differentials of x', t', etc.

$$\left. \begin{array}{l} dx' = \gamma(dx - u\,dt) \\[6pt] dt' = \gamma dt + \left(\dfrac{1 - \gamma^2}{u\gamma}\right) dx. \end{array} \right\} \tag{4}$$

By definition dx'/dt' is the expression for instantaneous velocity with reference to the system S', and dx/dt has the same meaning with respect to S. From eqs. (4) we therefore have

$$\frac{dx'}{dt'} = \frac{dx - u\,dt}{dt + \left(\dfrac{1 - \gamma^2}{u\gamma^2}\right)dx} = \frac{\dfrac{dx}{dt} - u}{1 + \left(\dfrac{1 - \gamma^2}{u\gamma^2}\right)\dfrac{dx}{dt}} \tag{5}$$

as the relation between dx/dt and dx'/dt'. This should apply in general. Suppose we take the special case of a light particle or photon moving along the x axis. In system S it has the velocity c. But in system S' it must also have the same velocity c. Hence for this particular case $dx'/dt' = dx/dt = c$. Substituting into eq. (5) we get

$$c = \frac{c - u}{1 + \left(\dfrac{1 - \gamma^2}{u\gamma^2}\right)c},$$

whence one immediately obtains

$$\gamma = \frac{1}{\sqrt{1 - \beta^2}}, \tag{6}$$

where we have for simplicity let $\beta = u/c$. The Lorentz-Einstein transformation equations are then at once obtainable by substitution back into eqs. (1) and (3). We shall write them again for reference

$$x' = \frac{1}{\sqrt{1-\beta^2}}(x - ut),\ y' = y,\ z' = z,\ t' = \frac{1}{\sqrt{1-\beta^2}}\left(t - \frac{\beta}{c}x\right). \quad (7)$$

The reciprocal relations giving the unprimed coordinates in terms of the primed may be got by solving the above or more simply by noting that the new relations must have the same form as the above with u changed to $-u$. Thus

$$x = \frac{1}{\sqrt{1-\beta^2}}(x' + ut'),\ y = y',\ z = z',\ t = \frac{1}{\sqrt{1-\beta^2}}\left(t' + \frac{\beta}{c}x'\right). \quad (8)$$

We are now ready to note some of the deductions from the transformation equations. Perhaps it will be better first, however, to look at the equations themselves for a moment. From the standpoint of a single observer, which we are here maintaining, the first equation in (7) means that, if he is observing an event at a place with x coordinate x and time t as indicated by his clock (at the place with coordinate x) which is at rest with respect to him (i.e., he thinks of himself at rest in system S), the corresponding coordinate x' which he must assign to this event (which may be the collision of two particles at this point or merely the reaching of this position by a certain particle) is that given by $x' = \frac{1}{\sqrt{1-\beta^2}}(x - ut)$. Similarly the time which he must assign to this same event as read from a clock fixed at the origin of the second system (S') is given by $t' = \frac{1}{\sqrt{1-\beta^2}}\left(t - \frac{u}{c^2}x\right)$. The reason for assigning these values is simply to insure that the principle of relativity shall be satisfied. If it is asked: does not this mean that the clock fixed in S' will *actually* read t' under the given circumstances, the answer is yes, but this is obtained inferentially, for the experiment itself is clearly out of the question.

Relativity kinematics. Let us now consider the measurement of length in the two systems. We suppose that the physicist wishes to measure the length of a rod AB which is at rest in S and lies along the x axis. Then $x_B - x_A = l =$ length of the rod, and it makes no difference at what times the two ends of the rod are noted. From eqs. (8)

$$x_B - x_A = \frac{1}{\sqrt{1-\beta^2}}[(x_B' - x_A') + u(t_B' - t_A')],$$

or if we denote $x_B' - x_A'$ by l', the length in the system S', we get

$$l' = l\sqrt{1 - \beta^2} - u(t_B' - t_A').$$

It is thus seen that the length in S' is indefinite in that it depends on $t_B' - t_A'$, where t_A' is the time in S' corresponding to the measurement of x_A', and similarly for t_B'. The most natural assumption to make with regard to the measurement of the length of the rod in S' with respect to which it is moving with constant velocity $-u$, is to take $t_B' = t_A'$. If this is done, we have

$$l' = l\sqrt{1 - \beta^2}, \tag{9}$$

that is, the length in the system S' is shorter than that in the system S in the ratio $\sqrt{1 - \beta^2} : 1$. This is the Einstein deduction of the Fitzgerald contraction which has already been mentioned as a means of accounting for an assumed null result of the Michelson-Morley experiment.

Another way of looking at the question of the relation between lengths in two inertial systems is to consider a rigid sphere of radius a with its center at the origin of S' with reference to which it is at rest. The equation of this surface is then

$$x'^2 + y'^2 + z'^2 = a^2. \tag{10}$$

From the standpoint of S' it remains always a sphere. To get its form in S, we must transform by means of (7), obtaining

$$\frac{(x - ut)^2}{1 - \beta^2} + y^2 + z^2 = a^2.$$

This represents a surface moving along x with velocity u. At $t = 0$, when the origins of the two systems pass each other, the equation becomes that of the ellipsoid

$$\frac{x^2}{1 - \beta^2} + y^2 + z^2 = a^2, \tag{11}$$

whose semi-axes in the y and z directions are still equal to a, but whose semi-axis in the x direction is now

$$a\sqrt{1 - \beta^2}. \tag{12}$$

In the system S there is a contraction of the original sphere in the x

direction. This shortening is of course a reciprocal affair. If the surface in question is a sphere in S with equation

$$x^2 + y^2 + z^2 = a^2,$$

in S' it appears again as an ellipsoid with semi-axis in the x direction again given by (12). The same result follows by the use of the method of the preceding paragraph. As u approaches c, the contraction is such that the x dimension approaches zero. A Lorentz electron considered as a sphere in the system in which it is at rest becomes a sheet of zero thickness in a system with respect to which it is moving with the velocity of light, a fact mentioned in the discussion of electron theory in Sec. 7.2.

A correlative to the contraction in length is the behavior of clocks in different inertial systems. Consider a clock C' which is fixed at the origin of S' and for which therefore $x' = 0$ always, and $x = ut$. Consider further a whole set of clocks fixed in S at $x_1, x_2, x_3, \ldots$, properly synchronized. When C' passes the clock at x_1 let the reading on the latter be t_1 and on the former be t_1'. Similarly at the passing of the clock at x_2, let the corresponding times be t_2 and t_2', respectively. Then from the transformation equations

$$t_1' = \frac{1}{\sqrt{1-\beta^2}}\left(t_1 - \frac{\beta}{c}x_1\right)$$

$$t_2' = \frac{1}{\sqrt{1-\beta^2}}\left(t_2 - \frac{\beta}{c}x_2\right)$$

whence

$$t_2' - t_1' = \frac{1}{\sqrt{1-\beta^2}}\left[(t_2 - t_1) - \frac{\beta}{c}(x_2 - x_1)\right]. \qquad (13)$$

But now

$$x_2 - x_1 = u(t_2 - t_1),$$

and therefore eq. (13) becomes

$$t_2' - t_1' = \sqrt{1-\beta^2}\,(t_2 - t_1). \qquad (14)$$

The time interval on the clock in S' (which is the one here considered to be moving) is shorter than that in S between the same events in the ratio $\sqrt{1-\beta^2}:1$. Likewise, if we consider a clock C in S passing in turn the set of clocks at $x_1', x_2', x_3' \ldots$ fixed in S' and running synchronously, and if we let t_1 and t_1' now denote respectively the time on C when it passes the clock at x_1' and that on the latter clock at the moment of passing, etc., we have as above

$$t_2 - t_1 = \frac{1}{\sqrt{1-\beta^2}}\left[(t_2' - t_1') + \frac{\beta}{c}(x_2' - x_1')\right].$$

But now since the velocity of S with respect to S' is oppositely directed from that of S' with respect to S,

$$x_2' - x_1' = -u(t_2' - t_1'),$$

and therefore in this case

$$t_2 - t_1 = \sqrt{1-\beta^2}(t_2' - t_1'). \tag{15}$$

Here again the indicated time interval on the clock that is considered to be moving is shorter than the time interval on the clock considered stationary for the same two events. This is the basis for the usual statement that on the relativity theory a moving clock always runs slow. Such bare statements are apt to be misleading, and doubtless a good deal of the common misunderstanding of the theory of relativity is caused by the failure to examine *exactly* what the theory implies.

The natural tendency of one who feels that he has a deep-seated intuition of the fundamentals of space and time is to ask, at this point: is the moving rod *really* shorter? Does the moving clock actually run more slowly? And does it not seem paradoxical that if a clock in S' appears to run slow from the standpoint of S, the same clock in S appears to run slow from the standpoint of S'? To those who have grasped the view of modern physics with respect to the nature of space and time as discussed in the earlier chapter, such questions will appear distinctly irrelevant. Space and time are indeed fundamental categories in terms of which physical theories are constructed. Nevertheless we cannot accept the view that the properties of these categories are given once and for all to human minds by direct intuition or otherwise. Different types of space and time are logically possible, and if the common lay view proves inconvenient in the development of physics we need have no scruple in adopting a different one. This has all been discussed before. All we need to say here is that the theory of relativity is seen to imply certain properties of space and time. These are logical deductions of the theory, and if we wish to use the theory in physical reasoning we must accept the characteristics it assigns to space and time even if on first sight they seem peculiar. What they really mean physically is that an observer contemplating a rod, say, in a reference system which is moving with constant velocity with respect to the one in which he is at rest *must* assign to it a length *in that system* shorter than the one he gives it in his own system. He must do this to be consistent with the principle of relativity, and there arises no question as to the *real* length of the rod. A similar situation holds with respect to time. Rather than speak in terms of clocks we may say that an observer contemplating the interval of time between two events must assign a shorter interval between these two events in the system which is moving with constant velocity with respect to him. The question of the *real* nature of time does not enter. As a matter of fact, the practical consequences of these results, i.e., their possible application to ordinary phenomena, are negligible, since the contraction can never be observed directly in any ordinary terrestrial measurement and will play a rôle only in experiments in atomic and cosmic domains. The real test of relativity lies in the agreement of its deductions with such experiments and in the way in which it succeeds in unifying apparently disparate phenomena, and in its freedom from internal contradictions. We shall have more to say of this hereafter. It must be emphasized, of course, that we are still discussing only the special relativity which is restricted to inertial systems.

We shall now proceed to obtain briefly other, more important deductions of the Lorentz-Einstein equations. The first of these concerns the relation of velocities in two inertial systems S and S'. This is already available in (5) wherein we need only substitute for γ the value (6) to obtain

$$\frac{dx'}{dt'} = v' = \frac{v - u}{1 - \beta v/c} \qquad (16)$$

and the corresponding reciprocal expression

$$v = \frac{v' + u}{1 + \beta v'/c}. \qquad (17)$$

These formulas give the instantaneous velocity relative to the system S in terms of that relative to S', and vice versa. For $u \ll c$, they reduce to the ordinary classical kinematical relation $v' = v - u$. But for velocities in the neighborhood of c, the departure from the classical result is very considerable. The formulas (16) and (17) apply to velocities along the x axis only, but the generalization to the case of any velocity is simple. See eq. (7.2–9) for the result. The form of eq. (16) suggests the expression for the hyperbolic tangent of the difference of two angles, viz.,

$$\tanh(\alpha - \delta) = \frac{\tanh \alpha - \tanh \delta}{1 - \tanh \alpha \cdot \tanh \delta}.$$

Letting $\tanh \alpha = v/c$ and $\tanh \delta = u/c$, we may write eq. (16) in the form

$$\text{arc tanh } v'/c = \text{arc tanh } v/c - \text{arc tanh } u/c. \tag{18}$$

Thus the ordinary Newtonian composition rule for velocities is obeyed by the arc tanh v/c, which may be called the *rapidity* of the motion. Since arc tanh $1 = \infty$, $v' = c$ if $v = c$, as is also evident from an inspection of eqs. (16) or (17). The composition of any velocity with the velocity of light yields the velocity of light. The principle of special relativity thus leads to the conclusion that the velocity of light is the limiting velocity for material particles. In fact, the same result holds for the propagation of any event as long as one considers only one-way time. Imagine that an event were to propagate itself from x_1 to x_2 in system S with velocity $v > c$. The time interval in S is

$$\Delta t = \frac{x_2 - x_1}{v}.$$

The corresponding time interval in S' is

$$\Delta t' = \frac{1}{\sqrt{1 - \beta^2}} \left[\Delta t - \frac{\beta}{c}(x_2 - x_1) \right]$$

$$= \frac{\Delta t}{\sqrt{1 - \beta^2}} \left(1 - \frac{\beta v}{c} \right).$$

For u sufficiently close to c and $v > c$, we have $\Delta t'$ negative. The only way to insure positive time intervals in all cases of propagation independently of the system in question is to choose $v \leq c$. The situation proves to be different in general relativity, as we shall see in the next chapter.

Minkowski's interpretation. The various applications of relativity to physics can be most beautifully summarized in terms of a geometrical interpretation, due to Minkowski,[1] which also is of great assistance in the understanding of the general theory. Minkowski's leading idea is that the physical world is a world of events, viz., a four-dimensional

---

[1] H. Minkowski. See THE PRINCIPLE OF RELATIVITY by Einstein, Lorentz, Minkowski, Weyl (Dover Publications, N.Y.).

manifold. The coordinates of any point in this manifold are x, y, z, w ($= ct$), i.e., the three rectangular coordinates of three-dimensional space and the time multiplied by the velocity of light as the fourth coordinate. Every point in the four-dimensional space is a *world-point*, i.e., it represents an event in the physical universe. A curve in this four-space represents a succession of events in the physical world and can therefore be taken to represent the history of a material point as it moves in time through physical three-space. Such a curve is a *world-line*. The history of the physical universe is a collection of world-lines. For example, in the case of a particle moving uniformly in a given reference system with velocity v in the x direction, the world-line is a straight line lying in the xw plane and making the angle arc tan $x/w =$ arc tan v/c with the w axis.

We shall confine our description of Minkowski's four-dimensional representation to the xw plane, i.e., consider only those events in the physical world which correspond to one-dimensional motion. This will be sufficient to exemplify the leading ideas of the theory. Let us draw in the xw plane the two conjugate hyperbolas (Fig. 7.2)

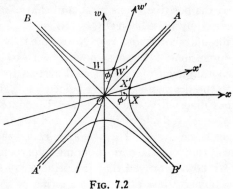

Fig. 7.2

$$\left. \begin{array}{r} w^2 - x^2 = 1 \\ w^2 - x^2 = -1 \end{array} \right\} \quad (19)$$

and their pair of asymptotes

$$w^2 - x^2 = 0. \qquad (20)$$

Let us examine the figure a little. If we denote the asymptotes by AOA' and BOB', any point in the quadrants AOB and $A'OB'$ will represent an event for which $|x| \le |w|$ and hence will correspond to the motions of actual material particles. On the other hand, any point within the quadrants AOB' and $A'OB$ will correspond to $|x| \ge |w|$ and hence will represent events which are not possible for material particles, or more exactly not possible for particles which start their motion at $x = 0$ when $t = 0$. If the point O is regarded as the "here now," points on the negative w axis will correspond to "here in the past," and points on the positive w axis will correspond to "here in the future." Similar designations can be made for the other points of the various quadrants. The asymptotes themselves play an interesting rôle, for $A'OA$ ($x = w$) represents the world-line for light particles traveling in the positive x direction and passing through $x = 0$ at $t = 0$, while $B'OB$ is the world-line for light-particles traveling under the same conditions in the negative x direction. To get from O to any point on an asymptote one must travel with velocity c, while one can get from O to any point in the quadrant AOB exclusive of the asymptotes by traveling with velocity less than c. The world-lines of all particles which are at $x = 0$ at $t = 0$ are included in the quadrants AOB and $A'OB'$.

So far we have paid but little attention to the hyperbolas themselves. Let us denote the intercept of the hyperbola $w^2 - x^2 = -1$ on the positive x axis by X and that of $w^2 - x^2 = +1$ on the positive w axis by W. Then $OX = OW = 1$. That is, these distances provide the units for w and x. (Note that they both have the same physical dimensions.) Suppose now we wish to introduce a transformation of coordinates from one inertial system to another moving with respect to the first with constant velocity u. How may we represent this in our diagram? Rotate the x axis counterclockwise through the angle $\phi = $ arc tan u/c or arc tan β, and rotate the w axis clockwise through the same angle. The new axes, called the x' and w' axes, respectively, will then cut the corresponding hyperbolas at X' and W', respectively. Minkowski proved that the principle of special relativity implies that a physical law expressed in terms of x and w will possess the same form in terms of x' and w'. Thus the geometrical transformation in the xw plane involved in squeezing the x and w axes into x' and w' axes, respectively, is exactly equivalent to the Lorentz-Einstein transformation from the one inertial system to the other. This may be readily demonstrated as follows. The coordinates of X' in the old and the new reference systems are, respectively,

$$x_{X'} = \frac{1}{\sqrt{1 - \beta^2}}, \quad w_{X'} = \frac{\beta}{\sqrt{1 - \beta^2}}$$
$$x_{X'}' = 1, \quad w_{X'}' = 0.$$

This is of course on the assumption that the unit distance along the x' axis is OX' corresponding to the unit distance OX along the x axis. In the same way the coordinates of W' are

$$x_{W'} = \frac{\beta}{\sqrt{1 - \beta^2}}, \quad w_{W'} = \frac{1}{\sqrt{1 - \beta^2}}$$
$$x_{W'}' = 0, \quad w_{W'}' = 1.$$

7.3 THE SPECIAL THEORY OF RELATIVITY

We must now find the linear transformations

$$x' = A_1 x + A_2 w \\ w' = B_1 x + B_2 w \quad (21)$$

which satisfy the above conditions. On substitution and solution for A_1, A_2, B_1, and B_2, eqs. (21) become the regular transformation equations with the use of $w = ct$. In the limiting case where $u = c$, the axes coincide with the asymptote $A'OA$, and the units along both axes become infinite.

The reader may work out the details showing how readily the Minkowski geometrical view yields all the results obtained analytically from the transformation equations. In our two-dimensional representation we have abstracted considerably from the actual situation, in which the hyperbolas $x^2 - w^2 = \pm 1$ must be replaced by the hyper hyperboloids $x^2 + y^2 + z^2 - w^2 = \pm 1$ with the hyper-cone $x^2 + y^2 + z^2 - w^2 = 0$ as asymptotic space. However, in this case the simple geometrical picture has to yield to analysis again, although one familiar with hyper spaces would doubtless find the general form useful.

If in the Minkowski interpretation one replaces the real axis $w = ct$ by the imaginary one $ict = l$, the transformation equations take on an even more familiar aspect. Written in this way the x and t equations assume the form

$$x' = \frac{1}{\sqrt{1-\beta^2}}(x + i\beta l), \quad l' = \frac{1}{\sqrt{1-\beta^2}}(l - i\beta x). \quad (22)$$

Introducing the imaginary angle ψ with

$$\cos \psi = 1/\sqrt{1-\beta^2}, \quad \sin \psi = i\beta/\sqrt{1-\beta^2}, \quad (23)$$

eqs. (22) become

$$x' = x \cos \psi + l \sin \psi, \quad l' = -x \sin \psi + l \cos \psi, \quad (24)$$

and conversely

$$x = x' \cos \psi - l' \sin \psi, \quad l = x' \sin \psi + l' \cos \psi. \quad (25)$$

These are, however, precisely the transformation equations for the rotation of rectangular axes in a plane through the angle $\psi = \arctan i\beta$.

An important feature is the arc element of the world-line in Minkowski's four-dimensional universe. The distance between any two neighboring points in the latter, assuming that it has a Euclidean structure and that the coordinates are Cartesian, is given by

$$ds^2 = dx^2 + dy^2 + dz^2 + dl^2 = dx^2 + dy^2 + dz^2 - c^2 dt^2, \quad (26)$$

which is constructed in a fashion analogous to the arc element in rectangular coordinates in ordinary three-space. Just as $dx^2 + dy^2 + dz^2$ is invariant under a rotation or translation of axes in three-space, so calculation shows that eq. (26) is invariant under the Lorentz-Einstein transformation. As we look upon dx, dy, dz as the components of a 3-vector transforming in accordance with the rule laid down in Sec. 6.2, so we may look upon dx, dy, dz, dl as the components of a 4-vector which transforms in accordance with the equations

$$\left. \begin{aligned} dx' &= \frac{1}{\sqrt{1 - \beta^2}} (dx - u\,dt) \\ dy' &= dy \\ dz' &= dz \\ dt' &= \frac{1}{\sqrt{1 - \beta^2}} \left(dt - \frac{u}{c^2} dx \right). \end{aligned} \right\} \quad (27)$$

One might indeed be inclined to question the ultimate value of such a geometrical presentation. It contains no more than the transformation equations themselves in their original form. Nevertheless it must be admitted that Minkowski's treatment has the great merit of emphasizing the essential equivalence of space and time in relativity theory. We may feel impelled by psychological reasons to make a separation of space and time in our description of physical events, but the way in which we use the corresponding symbols in our equations—if we are to secure the invariance we wish—actually corresponds to a placing of the time and space symbols on an exactly equal footing. In order to describe physical experience symbolically we need only use a four-dimensional manifold wherein logically there is no essential distinction to be drawn between any two of them. The separation into temporal and spatial components occurs only when we try to translate the results of our symbolic reasoning into ordinary language. Whether we shall ever get so far as to be satisfied not to make this translation is a valid question connected with the future of physics. If so, perhaps we may yet take literally Minkowski's famous remark: " From now on space in itself and time in itself dissolve into shadows and only a kind of union of the two retains an individuality."

Relativity dynamics. One of the uses to which we shall proceed to put the Minkowski four-dimensional world is the discussion of special relativity dynamics. The problem here is to find the equation which must replace the fundamental principle of classical mechanics $\mathbf{F} = m\mathbf{a}$ in order to satisfy the invariance requirement of relativity. We shall think of every mechanical phenomenon as given by the four coordinates x, y, z, l. The element of the world-line of any material particle will be the 4-vector $d\mathbf{s}$, the square of whose magnitude is

$$ds^2 = dx^2 + dy^2 + dz^2 + dl^2$$

with $dl = icdt$. Another way of writing this is to express it in terms of the instantaneous velocity components of the particle, dx/dt, etc. Putting $v^2 = v_x^2 + v_y^2 + v_z^2$, we get

$$ds^2 = (v^2 - c^2)dt^2. \tag{28}$$

Consider now another inertial system with coordinates represented by $\xi, \eta, \zeta, \lambda$, where $\lambda = ic\tau$. This system is so chosen that the λ axis lies in the direction of the world-line of the particle at some point while ξ, η, ζ are perpendicular to the world-line at this point. This means that along the world line element near this point there is a change in λ, but no change in ξ, η, ζ. Therefore for the element in the second system we have at this point

$$ds^2 = d\lambda^2 = -c^2 d\tau^2, \tag{29}$$

and it must of course be recalled that the ds^2 here is the same as that in eq. (28) since the latter is invariant under the Lorentz-Einstein transformation. Hence we get on combination the following expression for $d\tau$ in terms of dt

$$d\tau = dt\sqrt{1 - v^2/c^2}. \tag{30}$$

The second inertial system, i.e., that of the $\xi, \eta, \zeta, \lambda$, is called the *proper* system for the particle. It is the system in which the particle is momentarily at rest, since in it instantaneously $\dfrac{d\xi}{d\tau} = \dfrac{d\eta}{d\tau} = \dfrac{d\zeta}{d\tau} = 0$.

For a given particle obviously the proper system will alter from moment to moment unless the particle moves with uniform velocity in some inertial system.

It will now be assumed that in the *proper* system the equation of motion of the particle is in the classical Newtonian form, which we shall write in terms of momentum

$$\mathbf{F} = \frac{d(m\mathbf{v})}{dt}. \tag{31}$$

We look for the general equation which reduces to this form when the transformation is made to the proper system. In seeking for this we introduce the 4-vector $ds/d\tau$ and call it the Minkowski velocity
$$\mathbf{Q} = \left(\frac{dx}{d\tau}, \frac{dy}{d\tau}, \frac{dz}{d\tau}, \frac{dl}{d\tau}\right) = (Q_x, Q_y, Q_z, Q_l)$$
to use the symbolic notation for a 4-vector. Here (using again $\beta = v/c$)

$$\left.\begin{aligned} Q_x &= v_x \frac{dt}{d\tau} = v_x/\sqrt{1-\beta^2} \\ Q_y &= v_y \frac{dt}{d\tau} = v_y/\sqrt{1-\beta^2} \\ Q_z &= v_z \frac{dt}{d\tau} = v_z/\sqrt{1-\beta^2} \\ Q_l &= \frac{dl}{d\tau} = ic/\sqrt{1-\beta^2}. \end{aligned}\right\} \quad (32)$$

To get an equation which is analogous to eq. (31), but is invariant under the relativity transformation, we must use 4-vectors. We introduce the Minkowski 4-vector force $\mathbf{P}$ and write the equation of motion in the 4-vector form which in the proper system is of the form of eq. (31).

$$\mathbf{P} = \frac{d}{d\tau}(m_0 \mathbf{Q}). \tag{33}$$

The quantity m_0 is a scalar, whose significance we shall investigate in a moment. If we let $\mathbf{q}$ be the space part of the 4-vector $\mathbf{Q}$, i.e., $\mathbf{q} = \mathbf{i}q_x + \mathbf{j}q_y + \mathbf{k}q_z$, etc., we can split up $\mathbf{P}$ into a space component

$$\mathbf{p} = \frac{d}{d\tau}(m_0 \mathbf{q}), \tag{34}$$

and a time component

$$p_l = \frac{d}{d\tau}(m_0 q_l). \tag{35}$$

Let us examine the 3-vector $\mathbf{p}$. Since $\mathbf{q} = \mathbf{v}/\sqrt{1-\beta^2}$, we can write

$$\mathbf{p} = \frac{d}{d\tau}\left(\frac{m_0 \mathbf{v}}{\sqrt{1-\beta^2}}\right) = \frac{1}{\sqrt{1-\beta^2}} \frac{d}{dt}\left(\frac{m_0 \mathbf{v}}{\sqrt{1-\beta^2}}\right)$$

by the use of eq. (30). This may then be put into the form

$$\mathbf{p}\sqrt{1-\beta^2} = \frac{d}{dt}\left(\frac{m_0 \mathbf{v}}{\sqrt{1-\beta^2}}\right). \tag{36}$$

Comparing this with the classical form (31) we see that they are identical if

$$p\sqrt{1 - \beta^2} = F \tag{37}$$

and

$$\frac{m_0}{\sqrt{1 - \beta^2}} = m. \tag{38}$$

These then are the general expressions for the classical force and mass in terms of the corresponding relativity quantities in Minkowski's theory. We see that, if m_0 is taken as a scalar constant associated with the particle in its proper system, the mass to be associated with the particle in any other inertial system depends on its velocity in that system, in fact, in precisely the way in which the so-called *transverse* mass of the Lorentz electrodynamics varies with the velocity, eq. (7.1–12). We thus see that the variation of mass with velocity is a general result of the theory of relativity for all particles and is not restricted to electric charges.

The question arises, however, as to why the theory of relativity should provide only one mass, while electrodynamics gives two (i.e., transverse and longitudinal). The answer is found in the fact that there is more than one way to define mass from the standpoint of relativity. It can be shown that if we prefer to write our equation of motion in the commoner classical form $F = ma$ instead of the Newtonian form employing the momentum, we shall encounter the two masses of electrodynamics. For this purpose we need only expand the expression (36) and have

$$F = m_0 v \frac{d}{dt}\left(\frac{1}{\sqrt{1 - \beta^2}}\right) + \frac{m_0}{\sqrt{1 - \beta^2}} \frac{dv}{dt},$$

whence, if we carry out the procedure outlined in Sec. 7.1, we obtain the longitudinal mass $m_0/(1 - \beta^2)^{3/2}$ and the transverse mass $m_0/\sqrt{1 - \beta^2}$. We might call the latter the momentum mass, since it appears as the only mass when we define the relation of mass and force by means of the momentum. In any case the variation of mass with velocity attaches further significance to the limiting case of $v = c$. For the limiting velocity the mass of any particle becomes infinite. This again emphasizes the significance of the velocity of light as the maximum allowed by special relativity.

As a closely related consequence of the relativity theory we shall note the definition of energy. It will be recalled that in classical theory the energy arises from the spatial integration of the equation of motion (see Sec. 4.11). We proceed with the same method here. Writing

$$\mathbf{F} = \frac{d}{dt}(m_0\mathbf{v}/\sqrt{1-\beta^2}), \qquad (39)$$

we obtain, on forming the scalar product with $d\mathbf{r}$ and integrating,

$$\int_{\mathbf{r}_0}^{\mathbf{r}_1} \mathbf{F}\cdot d\mathbf{r} = m_0 \int_{\mathbf{r}_0}^{\mathbf{r}_1} \frac{d}{dt}(\mathbf{v}/\sqrt{1-\beta^2})\cdot d\mathbf{r}. \qquad (40)$$

The right-hand side will then be by definition the change in kinetic energy. We shall write

$$\frac{d}{dt}\left(\frac{\mathbf{v}}{\sqrt{1-\beta^2}}\right)\cdot d\mathbf{r} = \left[\beta^2 c^2 \frac{d}{dt}(1/\sqrt{1-\beta^2}) + c^2\beta\dot{\beta}/\sqrt{1-\beta^2}\right]dt$$

$$= d(c^2/\sqrt{1-\beta^2}), \qquad (41)$$

after some reduction. The right-hand integral in (40) then gives for the kinetic energy

$$T = \frac{m_0 c^2}{\sqrt{1-\beta^2}} + \text{constant}.$$

If we wish T to be 0 for $\beta = 0$, the constant becomes $-m_0 c^2$. Hence

$$T = m_0 c^2 [1/\sqrt{1-\beta^2} - 1]. \qquad (42)$$

This may more conveniently appear as

$$T = (m - m_0)c^2. \qquad (43)$$

As one might expect, it can be shown that the classical laws of the conservation of energy and momentum follow if one defines energy by eq. (42) and momentum by $m_0\mathbf{v}/\sqrt{1-\beta^2}$.

If now we rewrite eq. (43) in the form

$$m = m_0 + T/c^2$$

the mass appears to be made up of an intrinsic quantity m_0, the *rest* mass, plus the contribution T/c^2 from its kinetic energy. When the kinetic energy changes, the mass changes likewise, the change in mass being equal to the change in the energy divided by c^2. If this result is *assumed* to hold in general, i.e., even when the kinetic energy decreases through dissipation associated with radiation, etc., we may naturally connect a loss of energy by radiation with a loss in mass and an absorption of energy through whatever source with a gain in mass.

Einstein has generalized this point of view by associating with every mass m an amount of energy

$$E = mc^2 \tag{44}$$

independently of and in addition to the potential energy it may have in a field of force. If we substitute $m = m_0/\sqrt{1 - v^2/c^2}$, assume that $v \ll c$, and expand the radical, to a first approximation we have

$$E = m_0 c^2 + \tfrac{1}{2} m_0 v^2,$$

wherein $m_0 c^2$ appears as the intrinsic energy and $\tfrac{1}{2} m_0 v^2$ the ordinary classical kinetic energy.

Going still further, Einstein has shown by an elementary argument from classical mechanics that even to the radiation energy E there must be assigned inertia whose mass equivalent is E/c^2. The argument is as follows: we are asked to imagine a long cylindrical tube of length l with two similar bodies A and B at the two ends (Fig. 7.3) which are capable of emitting and absorbing electromagnetic radiation. The tube is initially at rest in some inertial system. We are to suppose that, though the mechanical rest mass of A is the same as that of B, the former possesses an excess of energy E (say in the form of heat) which by some suitable device it is able to emit in the direction of B in the form of electromagnetic radiation. From electrodynamics we know that the emission of the radiation energy E contributes to A a recoil momentum E/c giving to the tube a velocity v toward the left with respect to the system of reference, such that if m is the mass of the tube

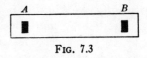

Fig. 7.3

$$v = E/mc.$$

The radiation travels down the tube in the time l/c, and is there absorbed (the absorption time is small and we neglect it altogether along with the emission time) by B, producing a momentum E/c toward the right which brings the tube to rest. In the meantime, however, the tube has traveled to the left a distance x, where

$$x = El/mc^2.$$

We now suppose that by some means *inside the tube* the bodies A and B are interchanged. According to classical mechanics this can bring about no change in the position of the tube—only external forces can do that. But after this interchange the situation will be precisely as in the beginning save that the tube will be a distance x farther to the left. By repeating this process one can produce motion *ad libitum* without the use of external influence, in contradiction to classical mechanics. The only way to avoid this is to associate inertia with the energy E, represented by the mass m_E. The transfer of this mass through the distance l in time t corresponds to a momentum $m_E \frac{l}{t}$, and this must exactly compensate the momentum acquired in the same time by the whole tube, which is mx/t. Hence

$$m_E l = mx = El/c^2,$$

whence

$$m_E = E/c^2$$

is the mass that must be associated with the radiation energy E to bring about the desired result.

The mass energy relationship $E = mc^2$ is considered by Einstein to be the most important result of the special theory of relativity.[1] Strictly speaking, it is not a deduction from the theory, but it is suggested by and consistent with it. Its utility has been widely tested by applications in the domain of atomic theory as well as in astrophysics. In popular language it is widely taken to mean that energy and mass are reduced to the same thing. Leaving aside any epistemological discussion as to how such a process may be logically conceived, we shall merely remark that in relativity theory as in classical theory mass and energy remain precisely defined concepts, although in the relativity theory the mass no longer plays the rôle of an invariant under motion, and this is of course a considerable change. When we talk about the mass of a particle we must now specify the system of reference to which the mass refers. Only in the proper system is the mass a definite characteristic of a particle. And the bold extrapolation of Einstein allows even the proper mass to change provided there appears a corresponding change in the radiation energy in the field surrounding the particle, in accordance with the assumption (44).

Space does not permit us to enter into all the important applications of the special theory of relativity. It will be of interest, however, to discuss with brevity one particular case because of its connection with quantum mechanics. This is the de Broglie wave mechanical theory of matter. On this theory a stationary particle such as an electron is assumed to be represented by a periodic function of space and

---

[1] Einstein has recently given an elementary derivation of the mass energy relationship based on the assumption of conservation of energy and momentum in inertial systems. (*Bull. Amer. Math. Soc.*, **41**, 223, 1935.)

time extending throughout all space and of the same phase throughout. If x_0, y_0, z_0, t_0 denote position and time in the particle's proper system, the analytic representation of the particle is taken to be

$$w(x_0, y_0, z_0, t_0) = f(x_0, y_0, z_0)e^{2\pi i \nu_0 t_0}. \tag{45}$$

This is a simple harmonic vibration with frequency ν_0 and amplitude $f(x_0, y_0, z_0)$, which are assumed to be characteristic of the particle. This at first sight rather incomprehensible assumption gains meaning when one refers the particle to the inertial system with respect to which it has the constant velocity v along the x axis. The mathematical representation of the particle in the new system may then be found from eq. (45) by the use of the Lorentz-Einstein transformation (7) in which the primed coordinates are now to be treated as x_0, y_0, z_0, t_0, respectively, and u becomes v. In making the substitution into eq. (45) we must not forget that ν_0, since it is a frequency, depends on the time and must also be transformed in accordance with

$$dt_0 = \frac{dt}{\sqrt{1 - v^2/c^2}},$$

so that in the new inertial system the frequency appears as

$$\nu = \frac{\nu_0}{\sqrt{1 - v^2/c^2}}. \tag{46}$$

The carrying out of the transformation finally results in the following:

$$w'(x, y, z, t) = f\left(\frac{x - vt}{\sqrt{1 - v^2/c^2}}, y, z\right)e^{2\pi i\nu(t - vx/c^2)}. \tag{47}$$

This is at once recognized as a plane progressive harmonic wave of frequency ν moving in the positive x direction with velocity

$$V = c^2/v. \tag{48}$$

In the inertial system with respect to which it is moving with constant velocity the electron appears as a progressive wave. The wave velocity can clearly never be less than c since $v \leq c$. This might seem to contradict the fundamental result of special relativity in accordance with which c is the maximum possible velocity for the propagation of physical effects. However, de Broglie assumes that the wave representing the particle acts more or less as a pilot for the particle but carries no energy in itself. As a matter of fact, de Broglie later found it more convenient to consider the particle's motion to be represented not by a single wave but by a group of harmonic waves with slightly different frequencies clustered about an average frequency ν and moving in the same direction with slightly different velocities. He was then able to prove that the velocity of the point of phase agreement of the component waves, or *group* velocity as it is called, is equal to the velocity v of the particle. The exact picture is still in question, or perhaps one should say that there is now grave doubt whether such a pictorial representation is at all satisfactory. Nevertheless the de Broglie idea initiated wave mechanics. Its greatest success came in the predicted wave length. De Broglie assumed that the relation between the mechanical energy E of the particle and its frequency considered as a harmonic wave is the famous quantum theory relation

$$E = h\nu. \tag{49}$$

If one combines this with the Einstein mass energy relation for a free particle, eq. (44), one immediately gets

$$h\nu = mc^2 = mvV.$$

But since $V/\nu = \lambda$, the wave length of the wave, it follows that

$$\lambda = \frac{h}{mv}. \tag{50}$$

This relation has been verified by experiments on the diffraction of electrons and other particles at the surfaces of single metallic crystals.[1] Further discussion of wave mechanics and its relation with quantum mechanics will be found in Chapter IX.

We shall close our short account of the special theory of relativity by reiterating that the theory has long since proved its worth in the development of physics during the last three decades. As one examines it dispassionately one will probably be inclined to admit that its chief merit—even greater than the success of its detailed applications—is its insistence on the fundamental nature of the criterion of the invariance of physical principles with respect to transformation of coordinates. To be sure, in its restriction to inertial systems it goes only part of the way, but even to this extent it contains the germ of the general theory of which it was the natural forerunner.

In any discussion of special relativity one inevitably encounters the question: what will be the effect on the theory if an experiment like the Michelson-Morley experiment should finally yield an unimpeachable positive result? Indeed an impartial examination of Miller's data [2] almost forces one to the conclusion that this famous experiment as repeated by him under a variety of conditions actually does not yield by any means a null result, though one much smaller than that predicted by classical pre-relativity electrodynamics. We must consider the matter still unsettled, and it is very desirable that other observers should repeat the experiment with the same care and thoroughness that Miller has bestowed on it. However, if it is substantiated that the experiment gives a positive result, does this mean that the special theory must be discarded? From the attitude which we have taken throughout this book on the nature of physical theories, it should be clear that there is no necessity for this action. It seems that far too much emphasis has been laid on this particular experiment in the establishment of the theory of relativity.

---

[1] See, for example, G. P. Thomson, "The Wave Mechanics of Free Electrons." (New York and London, McGraw-Hill, 1930.)

[2] D. C. Miller, *loc. cit.*

As we have seen, the chief aim of the theory was originally and is now to secure for all physical laws that invariance with respect to transformations of inertial systems which is characteristic of classical mechanics and which appeals so much to the esthetic sense of the physicist. The theory has proved its value in many branches of physics, and it seems an unnecessary recklessness to cast it aside because it proves to be in error by about one part in 10^8 in a single experiment. It would appear to be more sensible to seek for an accounting of the discrepancy within the frame of the theory itself. Suggestions of this sort have indeed already been made. One of these (by Page and Sparrow [1]) points out that a very slight anisotropy of space producing a variation with direction in the velocity of light of only 16.7 cm/sec is sufficient to account for Miller's earlier published results. The principle of relativity is not altered, although the Lorentz-Einstein transformation equations have to be changed slightly to accommodate the variation in c. It might be supposed that this variation will constitute a violation of the relativity principle. As a matter of fact, in the general theory of relativity the velocity of light is considered as a function of the distribution of matter and hence is no longer a constant. In whatever way the question is settled the most reasonable attitude of the physicist would appear to be that the theory of relativity has proved too valuable to be given up unless it encounters much more drastic disagreement with experiment than it has met to date.

[1] L. Page and C. M. Sparrow, *Physical Review*, **28**, 384, 1926.

CHAPTER VIII

THE GENERAL THEORY OF RELATIVITY

8.1. The Fundamental Idea of Relativity. *General invariance.* The special or restricted theory of relativity is based, as we have seen in the previous chapter, on the assumption that the equations expressing physical principles are invariant under the transformation from one inertial system to another. To bring about this invariance certain modifications are necessary in the formulation of physical laws.

On the other hand, we have seen in Chapter VI that invariance is guaranteed if physical laws are stated in terms of 4-vector or tensor equations. Thus, all the results of special relativity can be obtained by recasting physical equations in such a manner that they represent relations between 4-vectors or tensors in a space in which the element of a world-line is given by eq. (7.3–28). In the general theory of relativity we shall free ourselves from the restriction to this particular space, previously called Minkowski space, and require *general invariance*. But before we proceed to this new notion, let us see in a simple instance how we pass from an old law to one which is correct in special relativity by observing the true vector character of the quantities involved.

Newtonian dynamics asserted that, in the absence of forces, the momentum $m \dfrac{d\mathbf{s}}{dt}$ of a particle was to be conserved. m was considered as the invariable mass of the particle. From the standpoint of special relativity, which regards ict as a fourth coordinate, it is at once apparent that differentiation of three components of the vector $\mathbf{s}$ with respect to its fourth component, namely, ict, cannot produce anything like a new vector. The simplest modification which will produce a true 4-vector is the replacement of $m \dfrac{d\mathbf{s}}{dt}$ by $m_0 ic \dfrac{d\mathbf{s}}{ds}$, where $ds = (dx^2 + dy^2 + dz^2 - c^2 dt^2)^{1/2}$, a scalar which reduces to $icdt$ if $v \ll c$. Comparing the two expressions, we have at once that $m = m_0 ic \dfrac{dt}{ds} = \dfrac{m_0}{\sqrt{1 - \beta^2}}$, where $\beta = v/c$. Thus we obtain immediately the variation in mass deduced in the previous chapter. More-

8.1 THE FUNDAMENTAL IDEA OF RELATIVITY

over, if we postulate conservation of momentum, all four components of the vector $m_0 i c \dfrac{d\mathbf{s}}{ds}$ must be conserved. The fourth of these is $m_0 i c \dfrac{d(ict)}{ds} = \dfrac{m_0 i c}{\sqrt{1-\beta^2}}$. But constancy of this implies the conservation of the quantity $\dfrac{m_0 c^2}{\sqrt{1-\beta^2}}$, which in the previous chapter was identified with the total energy of the particle; cf. eq. (7.3–42). Hence, if written in terms of 4-vectors, the statement of the principle of conservation of momentum entails conservation of energy. These few remarks may suffice to illustrate the power of the vector, or in general the tensor, method of writing physical laws.

It is clear that in the general theory of relativity, where we require invariance with regard to unrestricted transformations, not only those from one inertial system to another, we must make use of this method from the outset. In a fundamental sense, then, nothing more need be said of the general theory than that it demands a recasting of all physical laws in tensor form, and a consequent modification of those laws which do not at present conform to it. But the idea of transformability from one system of reference to another has physical consequences which must now be investigated.

Even in pre-relativity physics a law, such as the law of universal gravitation, is true whether it is written in Cartesian or polar or any other kind of *space* coordinates. If, after making the correct transformation, the predictions of the law were found to be in error, the law would certainly be rejected. The fact that the law itself changes its form is insignificant; it would not do so if it were written in tensor form. The same may be said about transformations involving *space and time coordinates*, so long as only inertial systems are considered. The special theory takes care of this situation, as the reader will now clearly see. But what happens if we transform to an accelerated system?

Mathematically, there is hardly any difference; but physically, we *introduce forces*. On passing from a reference system with respect to which the observer is at rest to one which rotates he will suddenly find himself in a field of force, due to the existence of centripetal and Coriolis accelerations. To be sure, he recognizes these as fictitious, for he can remove them by a simple transformation of axes; but they appear as forces none the less, and if he uses the ordinary laws of mechanics he has to introduce extra force terms, not appearing as the result of the transformation, to take account of them. A reverse example is also at hand: in a homogeneous gravitational field every body experiences a force *proportional to its mass*, and therefore all bodies have the same acceleration. But an observer enclosed in a box which is falling freely, and without cognizance of the world around him, would suppose that bodies within the box experience no force at all. In other words, a homogeneous gravitational field can be *removed* by transformation to a proper accelerated system. May it not be that *all* forces can be " transformed away " in a similar manner?

The answer given by Einstein in his general theory of relativity is in the affirmative, at least as regards gravitational forces which are proportional to masses. His stand is that there is some reference system, or in more technical language, some space, in which forces can be ignored. Their occurrence in our physical experience is due only to our failure to use the correct system of reference, i.e., the correct space. If this space is found—and it may not be the crude space of our immediate intuition—then the laws of physics can be stated without reference to forces; the latter make their appearance in other spaces after transformation, and they are then to be identified with transformation parameters. Forces thus assume a geometrical significance.

Spaces are characterized by their metric properties; if we are dealing with four-dimensional spaces, the simplest metric property, namely the interval between events, is the infinitesimal length ds of a world-line which depends on the coordinate elements. We recall from the last chapter that in Minkowski's space, which is not the most general space constructible, every physical event is represented by a point having the four coordinates x, y, z, t or x, y, z, l. This notation, however, stresses too much the separation into space and time components; it will be better to use simply x_1, x_2, x_3, x_4 with $x_4 = ict$, the definition of x_4 being chosen for convenience. We recall that

$$ds^2 = dx_1^2 + dx_2^2 + dx_3^2 + dx_4^2 \tag{1}$$

which follows at once from eq. (7.3–28). Introduce an arbitrary transformation

$$\left. \begin{array}{ll} x_1' = x_1'(x_1, x_2, x_3, x_4); & x_2' = x_2'(x_1, x_2, x_3, x_4); \\ x_3' = x_3'(x_1, x_2, x_3, x_4); & x_4' = x_4'(x_1, x_2, x_3, x_4) \end{array} \right\} \tag{2}$$

or, if this be solved for the x,

$$x_1 = x_1(x_1', x_2', x_3', x_4'); \qquad x_2 = x_2(x_1', x_2', x_3', x_4') \text{ etc.} \tag{3}$$

8.1 THE FUNDAMENTAL IDEA OF RELATIVITY

The square of the resulting arc element may be written

$$ds^2 = \sum_{i,j=1}^{4} g_{ij}' \, dx_i' \, dx_j'. \tag{4}$$

The g_{ij}' are functions of x_1', x_2', x_3', x_4', and form an aggregate of 16 quantities, since i and j both run from 1 to 4. Eq. (4) will not in general have the same form as (1). It is only when (2) is a Lorentz-Einstein transformation that (4) and (1) agree in form, for in that case the g_{ik}' become the matrix

$$\begin{Vmatrix} 1 & 0 & 0 & 0 \\ 0 & 1 & 0 & 0 \\ 0 & 0 & 1 & 0 \\ 0 & 0 & 0 & 1 \end{Vmatrix} \tag{5}$$

In general, the result will be the following. Suppose we start with an arc element

$$ds^2 = \sum_{i,j=1}^{4} g_{ij} dx_i dx_j \tag{6}$$

and introduce an arbitrary transformation (3). Then as in Sec. 6.3, eq. (6.3–11), we have

$$dx_i = \sum_{k=1}^{4} \frac{\partial x_i}{\partial x_k'} \, dx_k', \quad dx_j = \sum_{l=1}^{4} \frac{\partial x_j}{\partial x_l'} \, dx_l'. \tag{7}$$

On carrying through the substitution of (7) into (6) we are led to

$$\sum_{i,j} g_{ij} dx_i dx_j = \sum_{i,j} \sum_{k,l} g_{kl} \frac{\partial x_k}{\partial x_i'} \frac{\partial x_l}{\partial x_j'} dx_i' dx_j'. \tag{8}$$

That is, in terms of the primed coordinates, ds^2 may be written in the form

$$ds^2 = \sum_{i,j} g_{ij}' dx_i' dx_j'$$

if we set

$$g_{ij}' = \sum_{k,l} \frac{\partial x_k}{\partial x_i'} \frac{\partial x_l}{\partial x_j'} g_{kl}. \tag{9}$$

Thus it is seen that we can maintain the invariance of ds^2 under any transformation if we adjust suitably the values of g_{ik} in the new coordinate system. In fact, according to the definitions of Sec. 6.3, it follows that the g_{ij} form a *covariant tensor of the second order*.

The Cartesian coordinates in terms of which (1) is written appear to us to be the most natural to use. Indeed, we feel inclined to regard the particular form of ds expressed by (1) as the world-line element which is typical of "our space-time." That intuitions of this sort may not be trustworthy has already been pointed out in Chapter II. It is important to recognize (1) as a special case, as the metric of Euclidean space, which is essentially different from many other spaces. This distinction would not be so fundamental if every other "metric," as the square of the line element is often called, could be reduced to (1). But this happens not to be the case.

We have already seen that, if we start with the metric (1), many other forms of ds^2, such as (4), can be obtained by taking a transformation like (2). The point we are now making is that, when an arbitrary ds^2 of the form (6) with definite g_{ij} is given, it may not be possible to find a transformation which will reduce the metric to (1), the g_{ij}' to (5). One can see this easily from the following example.

The reader may recall Gauss's expression for the arc element on a surface, i.e., a two-dimensional manifold in which the coordinates of any point are x_1 and x_2:

$$ds^2 = g_{11}dx_1{}^2 + 2g_{12}dx_1dx_2 + g_{22}dx_2{}^2. \qquad (10)$$

This represents, of course, a special two-dimensional example of eq. (6). Here $g_{12} = g_{21}$. If the surface is taken to be a sphere, x_1 and x_2 may be spherical coordinates, e.g., the longitude $\phi = x_1$ and the colatitude $\theta = x_2$. Then, if a is the radius of the sphere,

$$ds^2 = a^2 \sin^2 \theta d\phi^2 + a^2 d\theta^2. \qquad (11)$$

Thus in this case $g_{11} = a^2 \sin^2 \theta$, $g_{22} = a^2$, and $g_{12} = g_{21} = 0$. Can (11) be transformed into

$$ds^2 = dx_1'{}^2 + dx_2'{}^2? \qquad (12)$$

If this were possible through the equations

$$\phi = \phi(x_1', x_2'); \quad \theta = \theta(x_1', x_2'),$$

the following conditions, obtained by differentiation and subsequent substitution, would have to be satisfied for any value of ϕ and θ:

$$a^2 \left[\left(\frac{\partial \phi}{\partial x_1'} \right)^2 \sin^2 \theta + \left(\frac{\partial \theta}{\partial x_1'} \right)^2 \right] = 1$$

8.1 THE FUNDAMENTAL IDEA OF RELATIVITY

$$a^2 \left[\left(\frac{\partial \phi}{\partial x_2'} \right)^2 \sin^2 \theta + \left(\frac{\partial \theta}{\partial x_2'} \right)^2 \right] = 1$$

$$\frac{\partial \phi}{\partial x_1'} \frac{\partial \phi}{\partial x_2'} \sin^2 \theta + \frac{\partial \theta}{\partial x_1'} \frac{\partial \theta}{\partial x_2'} = 0.$$

But this is impossible, for if we put $\theta = 0$, we get

$$\left(\frac{\partial \theta}{\partial x_1'} \right)^2 = \left(\frac{\partial \theta}{\partial x_2'} \right)^2 = \frac{1}{a^2}$$

from the first two conditions, and

$$\frac{\partial \theta}{\partial x_1'} \frac{\partial \theta}{\partial x_2'} = 0$$

from the last, and these values are clearly incompatible.

The meaning of this result is simply that, in a curved space, like the surface of the sphere, the metric is *essentially* different from that in a flat space like a plane. The result may be generalized from two to four dimensions. Suppose now that the laws of motion are expressed as tensor relations in a curved space, without the use of any artificial parameters. They cannot then be transformed into flat, or homaloidal space. If nevertheless we insist on expressing them in our customary Euclidean space we must forego simplicity and introduce artificial parameters, i.e., gravitational forces.

But not all spaces are so radically different from Euclidean space. Since (1) can be transformed to (4), the reverse transformation is also possible. Hence there must be some metrics of the general form (4) which can be reduced to (1). If we chose for coordinates a set of uniformly rotating axes we should obtain such a metric. Of such cases it is characteristic that the force field involved is a *fictitious* one.

We are now ready to summarize. It was seen that forces can appear as the result of a transformation from one 4-space to another. We do not wish forces to appear explicitly in our equations of physics, hence we modify the metric of physical space. If the metric happens to be of the customary Euclidean form, it corresponds to the absence of forces; if the g_{ij} do not have the form (5) but a transformation can be found which will produce this form we say the field of force is a fictitious one; if no such transformation can be found the field is said to be *permanent*. A homogeneous gravitational field is a fictitious one; a varying one must be regarded as permanent, for it cannot be "transformed away" into a Euclidean metric.

If we are permitted to choose the type of space in which the laws of dynamics are to be stated, we can write the equations of motion in the form characteristic of free particles. Free particles, however, describe straight lines, or geodesics, and the equation of a geodesic is $\delta \int ds = 0$. This equation, with ds interpreted as an element of 4-space in which the g_{ij} are in some manner related to what we ordinarily term forces, is the fundamental equation of general relativity. Since our endeavor is to explain the motion of particles in a gravitational field, the g_{ij} must depend on the gravitational masses present in the universe. They must reduce to the form (5) for space free from gravitating matter.

A few remarks with regard to nomenclature are now in order. The space in which the laws of motion have the simple form $\delta \int ds = 0$ is called the *natural* space, its geometry the *natural* geometry. Euclidean space, the space of our customary intuition, in which the laws do not have this ideal form but require the inclusion of forces, is called *abstract* space, its geometry *abstract* geometry. The g_{ij} appearing in the metric of the natural space are frequently referred to as its *potentials*.

The potentials in the natural geometry. We are now confronted with the problem of determining the g_{ij}. The conditions to be imposed upon them are:

1. They must be the components of a covariant tensor.
2. They must reduce to (5) for spaces free from matter.

While these conditions do not fix the potentials with uniqueness, they allow them to be found if certain requirements of mathematical simplicity and elegance are also imposed. We can only sketch the procedure here.

The simplest covariant tensor involving the first and second derivatives of the g's linearly is one of the fourth order, called the Riemann-Christoffel tensor, and defined as

$$B^m_{rst} = \frac{\partial}{\partial x_t}\begin{Bmatrix} rs \\ m \end{Bmatrix} - \frac{\partial}{\partial x_s}\begin{Bmatrix} rt \\ m \end{Bmatrix} + \sum_p \left(\begin{Bmatrix} rs \\ p \end{Bmatrix}\begin{Bmatrix} pt \\ m \end{Bmatrix} - \begin{Bmatrix} rt \\ p \end{Bmatrix}\begin{Bmatrix} ps \\ m \end{Bmatrix} \right), \quad (14)$$

in which the three index symbol $\begin{Bmatrix} rs \\ m \end{Bmatrix}$ is used to denote

$$\begin{Bmatrix} rs \\ m \end{Bmatrix} = \sum_{k=1}^{4} \frac{1}{2} g^{(mk)} \left(\frac{\partial g_{rk}}{\partial x_s} + \frac{\partial g_{sk}}{\partial x_r} - \frac{\partial g_{rs}}{\partial x_k} \right). \quad (15)$$

8.1 THE FUNDAMENTAL IDEA OF RELATIVITY

We further have

$$g^{(mk)} = \gamma_{mk}/\gamma, \qquad (16)$$

where γ is the determinant of the g_{ij}, viz.,

$$\gamma = \begin{vmatrix} g_{11} & g_{12} & g_{13} & g_{14} \\ g_{21} & g_{22} & g_{23} & g_{24} \\ g_{31} & g_{32} & g_{33} & g_{34} \\ g_{41} & g_{42} & g_{43} & g_{44} \end{vmatrix} \qquad (17)$$

and γ_{mk} is the minor of the element g_{mk}. Since each index r, s, t, and m can take the values 1 to 4 independently, it is clear that the tensor B^m_{rst} has $4^4 = 256$ components. However, not all of these are independent, for many are repetitions of others due to the symmetry in the three index symbol, etc. When this is taken into consideration the number of independent components is reduced to twenty. The complete treatment of the construction of the Riemann-Christoffel tensor demands more discussion of tensor analysis than we can indulge in here. Its tensor nature is made clear from the fact that when the coordinate transformation (4) is carried out the transformed tensor becomes

$$(B^m_{rst})' = \sum_{a,b,c,n} \frac{\partial x_a}{\partial x_r'} \frac{\partial x_b}{\partial x_s'} \frac{\partial x_c}{\partial x_t'} \frac{\partial x_m'}{\partial x_n} B^n_{abc}. \qquad (18)$$

It is then evident that B^m_{rst} is a *mixed* tensor, being partly covariant and partly contravariant (Sec. 6.3). The Riemann-Christoffel tensor vanishes when the g's reduce to the set (5) i.e., $g_{11} = g_{22} = g_{33} = g_{44} = 1$ and $g_{ij} = 0$ for $i \neq j$. Hence it vanishes identically when the four-dimensional manifold contains no gravitational field. Since it *is* a tensor it must therefore under the same conditions vanish in all systems of coordinates. It can be proved conversely that the vanishing of the tensor is the condition that by a suitable choice of coordinates it is possible to reduce the general arc element to the form (1) characteristic of the absence of a gravitational field. In other words, the vanishing of the tensor is the necessary and sufficient condition for the absence of a *permanent* gravitational field. This is of course equivalent to the existence of twenty equations obtained by equating to zero the twenty independent components of the tensor.

The simplest relation defining the potentials would be obtained by equating B^m_{rst} to zero. But this is clearly too simple, for it expresses the absence of a gravitational field. It is somewhat reminiscent of Laplace's equation $\nabla^2 V = 0$ which defines the electrostatic potential in a region free from charges.

We then try the next simplest procedure: Einstein showed that it is possible to construct, by linear combinations of the components of B_{rst}^m, a new tensor with the following properties: It vanishes when B_{rst}^m is zero, but it can also be zero under conditions in which B_{rst}^m remains finite. It is called the contracted Riemann-Christoffel tensor

$$G_{rs} = \sum_{m=1}^{4} B_{rsm}^m. \tag{19}$$

The temptation is therefore very great to define the g's by the relation

$$G_{rs} = 0, \tag{20}$$

which is Einstein's law of gravitation. The test of eq. (20) will lie in the correlation of the deductions from it with experiment. It is clear that when these deductions are translated from the four-dimensional manifold back into the ordinary space and time of classical physics the results must agree very closely with those of Newtonian mechanics and differ from them only with respect to small quantities. Otherwise, although we might be charmed by the esthetic elegance of the general theory of relativity, we should be forced to admit that it did not supply a good description of nature, since with all its drawbacks classical mechanics is known to provide a reasonably satisfactory picture. As a matter of fact, this necessity of achieving final numerical results not too different from those of classical mechanics was an important guide to Einstein in his choice of the gravitational field tensor G_{rs}.

In setting up this formalism to insure invariance we have dealt with gravitational fields only. Ultimately it will be necessary, of course, to consider other physical effects such as electric and magnetic ones which must be brought into the picture in order to complete it.

We are now ready to consider the application of the general theory to the motion of a mass particle in a gravitational field. This will afford a clearer view of invariance and will also enable us to understand more precisely the relation between the special and the general theory.

8.2. Motion of a Particle in a Gravitational Field. *Planetary motion.* The description of the motion of any particle in the physical universe on the theory of relativity is based on Einstein's assumption that the world-line of such a particle must be a *geodesic* in the four-dimensional manifold. Thus if we continue to denote the arc element

8.2 MOTION OF A PARTICLE IN A GRAVITATIONAL FIELD

in this space by ds, the equation of motion of the particle is to be written

$$\delta \int ds = 0, \tag{1}$$

where the integral is taken between any two points of the manifold, constituting the termini of a world-line. This equation states that the integral $\int ds$ has a *stationary* value for the actual world-line connecting any two points. Since ds is an invariant under all transformations of coordinates, the relation (1) is independent of the coordinate system used. Its use to describe motion therefore satisfies the fundamental criterion of relativity. The connection with classical mechanics can be very simply shown. Let us suppose that we are dealing entirely with inertial systems—the standpoint of special relativity—and thus take the velocity of light, c, as a universal constant. We can then write Hamilton's principle in the form

$$\delta \int (c^2 - L) dt = 0, \tag{2}$$

where L is the ordinary Lagrangian function. In the present case we shall consider a free particle of unit mass so that L reduces to $v^2/2$, where v is the usual instantaneous particle velocity in 3-space. It is assumed that $v \ll c$. We can now write

$$c^2 - L = c^2(1 - v^2/2c^2)$$

and to the approximation indicated

$$(1 - v^2/2c^2)^2 = 1 - v^2/c^2.$$

Hence the variation principle will appear as

$$\delta \int \sqrt{c^2 - v^2} \, dt = 0. \tag{3}$$

Writing

$$v^2 = \dot{x}^2 + \dot{y}^2 + \dot{z}^2,$$

we see that eq. (3) can be put into the form

$$\delta \int ds = 0$$

if we choose

$$ds^2 = dx^2 + dy^2 + dz^2 - c^2 dt^2$$

which is the arc element in the four-dimensional manifold in the absence of a field, i.e., in the region where the metric is Euclidean and the space-time is "flat" or homaloidal (cf. Sec. 3.4). The above is, of course, not intended as any proof of the condition (1) but merely an illustration of its fundamental connection in a special case with the stationary value principle in classical physics.

It is now necessary to obtain from (1) the corresponding differential equation of motion. The method is similar to that of classical mechanics, but since the resulting equation is of importance the derivation will be briefly sketched here. We have

$$ds^2 = \sum_{i,j} g_{ij} dx_i dx_j,$$

whence applying the variation

$$2ds\,\delta(ds) = \sum_{i,j=1}^{4} [dx_i dx_j \delta g_{ij} + g_{ij} dx_i \delta(dx_j) + g_{ij} dx_j \delta(dx_i)]$$

$$= \sum_{i,j=1}^{4} \left[dx_i dx_j \sum_{k=1}^{4} \frac{\partial g_{ij}}{\partial x_k} \delta x_k + g_{ij} dx_i d(\delta x_j) + g_{ij} dx_j d(\delta x_i) \right].$$

We further get

$$\delta(ds) = \frac{1}{2} \sum_{i,j=1}^{4} \left[\frac{dx_i}{ds} \frac{dx_j}{ds} \sum_{k=1}^{4} \frac{\partial g_{ij}}{\partial x_k} \delta x_k + g_{ij} \frac{dx_i}{ds} \frac{d}{ds} (\delta x_j) + g_{ij} \frac{dx_j}{ds} \frac{d}{ds} (\delta x_i) \right] ds.$$

In writing the variation eq. (1) we shall modify the summation indices somewhat without changing the value of the expression. We get

$$\int \frac{1}{2} \left\{ \sum_{i,j,k=1}^{4} \frac{dx_i}{ds} \frac{dx_j}{ds} \frac{\partial g_{ij}}{\partial x_k} \delta x_k + \left(\sum_{i,k=1}^{4} g_{ik} \frac{dx_i}{ds} + \sum_{j,k=1}^{4} g_{kj} \frac{dx_j}{ds} \right) \frac{d}{ds} (\delta x_k) \right\} ds = 0.$$

Integrating by parts the second part of the integral and recalling that δx_k vanishes at the limits of integration yields

$$\int \frac{1}{2} \sum_{k=1}^{4} \left\{ \sum_{i,j=1}^{4} \frac{dx_i}{ds} \frac{dx_j}{ds} \frac{\partial g_{ij}}{\partial x_k} - \frac{d}{ds} \left(\sum_{i=1}^{4} g_{ik} \frac{dx_i}{ds} + \sum_{j=1}^{4} g_{kj} \frac{dx_j}{ds} \right) \right\} \delta x_k ds = 0.$$

Since this relation must hold for arbitrary δx_k we must have identically

$$\sum_{i,j=1}^{4} \frac{dx_i}{ds} \frac{dx_j}{ds} \frac{\partial g_{ij}}{\partial x_k} - \sum_{i=1}^{4} \frac{dg_{ik}}{ds} \frac{dx_i}{ds} - \sum_{j=1}^{4} \frac{dg_{kj}}{ds} \frac{dx_j}{ds}$$

$$- \sum_{i=1}^{4} g_{ik} \frac{d^2 x_i}{ds^2} - \sum_{j=1}^{4} g_{kj} \frac{d^2 x_j}{ds^2} = 0. \quad (4)$$

We now recall, however, that

$$\frac{dg_{ik}}{ds} = \sum_{j=1}^{4} \frac{\partial g_{ik}}{\partial x_j} \frac{dx_j}{ds}, \quad \frac{dg_{kj}}{ds} = \sum_{i=1}^{4} \frac{\partial g_{kj}}{\partial x_i} \frac{dx_i}{ds}.$$

8.2 MOTION OF A PARTICLE IN A GRAVITATIONAL FIELD

Using these relations, and noting that the last two terms in eq. (4) are equal since $g_{kj} = g_{jk}$, we get

$$-\frac{1}{2}\sum_{i,j=1}^{4}\frac{dx_i}{ds}\frac{dx_j}{ds}\left(\frac{\partial g_{ik}}{\partial x_j} + \frac{\partial g_{kj}}{\partial x_i} - \frac{\partial g_{ij}}{\partial x_k}\right) - \sum_{n=1}^{4} g_{kn}\frac{d^2x_n}{ds^2} = 0, \qquad (5)$$

where the index n has been introduced to replace i and j. We can get rid of g_{nk} by introducing $g^{(k\alpha)}$ defined as in eq. (8.1-16). Thus we have from the latter equation

$$g^{(k\alpha)}g_{kn} = \frac{\gamma_{k\alpha}g_{kn}}{\gamma}.$$

Now by definition $\gamma_{k\alpha}$ is the minor of the element $g_{k\alpha}$ in the determinant, γ, of the g's; cf. (8.1-17). Hence

$$\sum_{k=1}^{4} g_{kn}\gamma_{k\alpha} = \begin{cases} 0, & n \neq \alpha \\ \gamma, & n = \alpha, \end{cases}$$

and therefore

$$\sum_{k=1}^{4} g^{(k\alpha)}g_{kn} = \begin{cases} 0, & n \neq \alpha \\ 1, & n = \alpha. \end{cases}$$

We now multiply (4) by $g^{(k\alpha)}$ and sum with respect to k. Clearly

$$\sum_{k,n=1}^{4} g^{(k\alpha)}g_{kn}\frac{d^2x_n}{ds^2} = \frac{d^2x_\alpha}{ds^2},$$

and our equation of motion becomes

$$\frac{d^2x_\alpha}{ds^2} + \frac{1}{2}\sum_{i,j,k=1}^{4}\frac{dx_i}{ds}\frac{dx_j}{ds}g^{(k\alpha)}\left(\frac{\partial g_{ik}}{\partial x_j} + \frac{\partial g_{kj}}{\partial x_i} - \frac{\partial g_{ij}}{\partial x_k}\right) = 0.$$

By using the three-index symbol as defined previously in eq. (8.1-15) we may simplify the above to the form

$$\frac{d^2x_\alpha}{ds^2} + \sum_{i,j=1}^{4}\begin{Bmatrix} ij \\ \alpha \end{Bmatrix}\frac{dx_i}{ds}\frac{dx_j}{ds} = 0. \qquad (6)$$

Four equations are involved in (6) as α takes the values 1 to 4. They are, so to speak, parametric equations of motion for the particle, which one may ultimately solve to find x_1, x_2, x_3, x_4 in terms of the parameter s, which is itself later eliminated in the process of expressing x_1, x_2, x_3 in terms of x_4. The latter may be translated back into ict for purposes of comparison with classical physics. For simplicity we shall, however, refer to (6) as a single equation. Since it has come from the invariant relation (1) its invariance with respect to all transformations of coordinates is assured. It is a tensor equation. But it is well to note the exact nature of its invariance. If we introduce a general transformation of coordinates of the form (8.1-2), eq. (6) will have precisely the same form in the primed as in the unprimed coordinates, provided for the primed g's one uses the values given in (8.1-9). This can also be verified by direct trial. It is important to recognize the condition with respect to the g's in order to avoid confusion. Thus one might at first imagine that one could pick out a particular set of g's (functions of x_1, x_2, x_3, x_4) which seem suitable for the solution of a certain problem, then substitute them into (6), getting differential equations in which only the x's and s appear. One might then expect that these equations would be invariant under all transformations, but this expectation would be disappointed, for that is not the meaning of invariance as understood in general relativity. In fact, we can easily see this by the very simple example of the case where the g tensor reduces to the form (8.1-5). Then (6) becomes simply

$$\frac{d^2 x_\alpha}{ds^2} = 0, \qquad (7)$$

which does *not* have the same form in all systems of coordinates. In order to preserve the invariance of (6) the g's must be left in place without specific assignment of their functional values in terms of x_1, x_2, x_3, x_4.

The question then arises as to how we are going to use (6) in the actual evaluation of the orbit of a particle in a gravitational field. At first sight it would seem that we have paid too high a price for the property of invariance, for the equation looks entirely too general It says in fact that it will stand in its present form no matter what g's are chosen. Nevertheless, before we can use it we must get somewhere a set of g's. Where do we get them? We have decided that the g's must satisfy the tensor equation $G_{rs} = 0$ for a gravitational field. However, since this is made up of ten partial differential equations there is still a great deal of arbitrariness left in the quantities which satisfy it. What is actually done is to *choose* a certain set of g's satisfying $G_{rs} = 0$ and such that when they are substituted into (6) they yield equations which on solution give results agreeing with experiment, i.e., agreeing to the first order of small quantities with the Newtonian theory. This may seem at first blush an arbitrary procedure, yet on closer consideration it is a remarkable fact that it is *possible* to choose successfully quantities which have the proper invariance properties and which at the same time cause the fundamental

8.2 MOTION OF A PARTICLE IN A GRAVITATIONAL FIELD

equations to assume a form not too different from that of the classical Newtonian theory.

We proceed as follows. As we know that in the classical planetary motion problem spherical coordinates are most useful, we shall interpret x_1, x_2, x_3, x_4 in terms of coordinates which reduce to ordinary spherical coordinates far away from all gravitating matter, plus the time as a fourth coordinate. We also use units such that the velocity of light c is unity far away from all gravitating matter. Thus

$$x_1 = r, \quad x_2 = \theta, \quad x_3 = \phi, \quad x_4 = t, \tag{8}$$

and we shall assume that the arc element characterizing the space-time metric in the presence of the gravitational field can be written in the form

$$ds^2 = f_1 dt^2 - f_2 dr^2 - r^2(d\theta^2 + \sin^2\theta d\phi^2), \tag{9}$$

where f_1 and f_2 are functions of r only. Our problem is to substitute the values of the g's as thus assumed into the equation $G_{rs} = 0$ and determine the f_1 and f_2 so that the equation is satisfied. Our hope is, of course, that a possible choice of these parameters can be made. The assumption of spherical symmetry for the field, the center of which is supposed to be placed at the origin of the coordinate system, leads to the particular choice of $r^2 d\theta^2 + r^2 \sin^2\theta d\phi^2$ as the angular part of ds^2, for this is of course the arc element on a sphere of radius r. The absence of the product terms $drd\theta$, $d\theta d\phi$, etc., is dictated by the fact that owing to symmetry, ds^2 should be independent of the sign of $d\theta$ and $d\phi$. A similar situation prevails with respect to product terms in dt, since the arc element should be independent of the sign of dt. The evaluation of f_1 and f_2 by substitution into $G_{rs} = 0$ finally yields after algebra too tedious to be repeated here

$$ds^2 = \mu dt^2 - \mu^{-1} dr^2 - r^2(d\theta^2 + \sin^2\theta d\phi^2), \tag{10}$$

where

$$\mu = 1 - 2m/r, \tag{11}$$

and m is a constant of integration. If it vanishes, $\mu = 1$ and the resulting arc element is that of a field-free region. The metric is then Euclidean and characterizes a homaloidal space-time. For $m \neq 0$, however, the metric is no longer Euclidean. Not only is the dt^2 term modified, but also the dr^2 term is changed. The radial part of ds is no longer dr but $dr/\sqrt{\mu}$.

The next step is the writing of the eqs. (6) for the general case.

This is a straightforward algebraic process, and we need write only the result in the form of the following four equations

$$\frac{d^2r}{ds^2} - \frac{1}{2\mu}\frac{d\mu}{dr}\left(\frac{dr}{ds}\right)^2 - \mu r\left(\frac{d\theta}{ds}\right)^2 - \mu r \sin^2\theta\left(\frac{d\phi}{ds}\right)^2 + \frac{\mu}{2}\frac{d\mu}{dr}\left(\frac{dt}{ds}\right)^2 = 0 \quad (12)$$

$$\frac{d^2\theta}{ds^2} + \frac{2}{r}\frac{dr}{ds}\frac{d\theta}{ds} - \sin\theta\cos\theta\left(\frac{d\phi}{ds}\right)^2 = 0 \quad (13)$$

$$\frac{d^2\phi}{ds^2} + \frac{2}{r}\frac{dr}{ds}\frac{d\phi}{ds} + 2\cot\theta\frac{d\theta}{ds}\frac{d\phi}{ds} = 0 \quad (14)$$

$$\frac{d^2t}{ds^2} + \frac{1}{\mu}\frac{d\mu}{dr}\frac{dt}{ds}\frac{dr}{ds} = 0. \quad (15)$$

To simplify the treatment without material loss of generality we shall assume that the particle begins to move from rest in the plane $\theta = \pi/2$. This value of θ satisfies identically eq. (13), indicating that the motion once started in this plane will continue there and we need consider only the other three equations. Integrating eqs. (14) and (15) under this assumption we get

$$r^2\frac{d\phi}{ds} = h \quad (16)$$

and

$$\mu\frac{dt}{ds} = k, \quad (17)$$

where h and k are constants. We now substitute the results into eq. (12) and at the same time make use of the relation

$$\frac{1}{\mu}\left(\frac{dr}{ds}\right)^2 = \frac{k^2}{\mu} - \frac{h^2}{r^2} - 1$$

which follows from (10), (16), and (17).

Thus we get finally

$$\frac{d^2r}{ds^2} + \frac{m}{r^2} = \frac{h^2}{r^3}\left(1 - \frac{3m}{r}\right). \quad (18)$$

Writing

$$\frac{d^2r}{ds^2} = \left(\frac{d\phi}{ds}\right)^2\frac{d^2r}{d\phi^2} + \frac{d^2\phi}{ds^2}\frac{dr}{d\phi}$$

and substituting from (16) and (17) there results

$$\frac{d^2r}{ds^2} = \frac{h^2}{r^4}\frac{d^2r}{d\phi^2} - \frac{2h^2}{r^5}\left(\frac{dr}{d\phi}\right)^2,$$

8.2 MOTION OF A PARTICLE IN A GRAVITATIONAL FIELD

and further, upon changing to a new coordinate $u = 1/r$, eq. (18) reduces to

$$\frac{d^2u}{d\phi^2} + u = \frac{m}{h^2} + 3mu^2, \tag{19}$$

which together with eq. (16), viz., $\frac{d\phi}{ds} = hu^2$, suffices to determine the geodesic corresponding to the path of the particle. It is interesting to compare these equations with the corresponding ones in classical planetary theory, namely [1]

$$\frac{d^2u}{d\phi^2} + u = \frac{m}{h^2}, \quad \frac{d\phi}{dt} = hu^2. \tag{20}$$

In the case of the gravitational field produced by an assumed fixed mass M (like the sun) m in eq. (20) is the quantity GM, where G is the constant of gravitation. The relativity equation in u differs from the Newtonian one only in the addition of the term $3mu^2$, if it is assumed that m in the relativity case has the same meaning as in the classical case. In the former it entered the problem merely as a constant of integration. We ought to estimate the value of the term $3mu^2$. If it can become large grave doubt might thus be cast on the theory. The ratio of $3mu^2$ to $\frac{m}{h^2}$ is $3h^2u^2$, which becomes $3\left(\frac{d\phi}{ds}\right)^2 / u^2$. If we let $\frac{d\phi}{ds} = \frac{d\phi}{dt}\frac{dt}{ds}$ and take $-\left|\frac{dt}{ds}\right| = \frac{1}{c}$ for a first approximation (cf. arc element (8.1-1)), where c is the velocity of light in free space far from gravitating matter, the dimensions of the ratio $3\left(\frac{d\phi}{ds}\right)^2 / u^2$ are seen to be correct. We have been using units such that c = unity. Keeping this in mind, and recalling that the diameter of the earth's orbit is about 1.5×10^{13} cm and ω = angular velocity in radians per second = 2×10^{-7} sec^{-1} approximately, we get for the case of the earth as an illustration of the motion of a particle in a gravitational field

$$3\left(\frac{d\phi}{ds}\right)^2 / u^2 \sim 10^{-8}.$$

Consequently in the ordinary planetary case the correction term in the relativity equation is practically negligible. However, the influence on the shape of the orbit is worth noticing. In classical mechanics

---

[1] Cf., for example, Lindsay, "Physical Mechanics" (Van Nostrand, New York, 1933), p. 63.

the orbit if closed is an ellipse. Solution of eq. (19) (we shall omit the the details of the analysis) yields an orbit which is approximately elliptical in form but in which the perihelion rotates through an angle $6\pi m^2/h^2$ during each revolution. This value proves to be very small for all the planets in the solar system save Mercury, for which it corresponds to about 43 seconds of arc per century. This serves to remove an outstanding discrepancy in the agreement between Newtonian mechanics and the observed behavior of the planet.

An interesting question that arises at this point concerns the rôle played by the velocity of light in the general theory. In the special theory it appears as a fundamental constant which has the same value in all inertial systems. Can we expect it to manifest itself in the same way in the general theory? In our present discussion of the motion of a particle we have indeed concealed the presence of c rather thoroughly by choosing our units so that its value is unity. Suppose, however, we had chosen to write the arc element in the form

$$ds^2 = \mu c^2 dt^2 - \mu^{-1} dr^2 - r^2(d\theta^2 + \sin^2\theta d\phi^2) \tag{21}$$

where c^2 is here considered merely as an extra parameter and has at first no connection with the velocity of light. The usual analysis then gives eq. (19), but in place of $\dfrac{d\phi}{ds} = hu^2$ we now have

$$r^2 \frac{d\phi}{dt} = h\sqrt{c^2\mu - \mu^{-1}\left(\frac{dr}{dt}\right)^2 - r^2\left(\frac{d\phi}{dt}\right)^2} \tag{22}$$

from (16) and (21). The equation of the orbit comes out in the same form as before and contains the same perihelion rotation $6\pi m^2/h^2$. We now note that the observation of the orbit alone, in so far as it is an approximate ellipse, cannot give us the value of h, but only the value of m/h^2. However, the perihelion rotation term gives the value of of m^2/h^2 and consequently leads to a definite value of h. This could never have been obtained from the exact elliptical orbit of classical theory—for that the area equation $r^2 \dfrac{d\phi}{dt} = h$ is necessary. Now, however, that h is fully determined by the perihelion rotation, we can use (22) to determine c. Solving for $\dfrac{d\phi}{dt}$, we get

$$r^2 \frac{d\phi}{dt} = h\sqrt{\frac{c^2\mu - \mu^{-1}\left(\dfrac{dr}{dt}\right)^2}{1 + h^2/r^2}}.$$

8.2 MOTION OF A PARTICLE IN A GRAVITATIONAL FIELD

However, it turns out that μ in planetary problems is very close to unity, $h^2/r^2 \ll 1$ and $c \gg \dfrac{dr}{dt}$. Hence the above relation takes the approximate form

$$r^2 \frac{d\phi}{dt} = hc, \qquad (23)$$

and from this we can determine c. (The calculation assures us that our assumption $c \gg \dfrac{dr}{dt}$ is justified.) Now if $\dfrac{ds}{dt}$ in the above is interpreted as having the dimensions of velocity, c must also be a velocity. Suppose that we have a particle whose velocity in the ordinary space and time of classical mechanics is given by

$$v^2 = \left(\frac{dr}{dt}\right)^2 + r^2 \left(\frac{d\theta}{dt}\right)^2 + r^2 \sin^2\theta \left(\frac{d\phi}{dt}\right)^2.$$

Then the fact that ds^2 is positive in eq. (21) indicates that v can never be greater than c. That is, in c we have a velocity which can never be exceeded by that of any material particle. It is natural then to treat this velocity as that of light, which on the special theory of relativity plays exactly this rôle of limiting velocity for the propagation of physical effects. In this way the velocity of light enters the equations of general relativity merely as a parameter whose value is ultimately determined by the deviation of the actual motion of a particle from that predicted by the classical theory of gravitation.

The path of a light ray. As a special case of the particle motion treated above we shall now discuss the path of a light ray in a gravitational field. The result has achieved considerable publicity owing to the attempts to verify it by observing the deflection of starlight passing close to the sun. In so far as light has inertia it must be affected by a gravitational field. This is easily visualizable in terms of the light particle or photon theory now generally accepted, but is actually independent of the precise nature of the theory of light. For the purpose of the brief analysis which will be sketched here, it will be most convenient to speak in terms of photons. It is seen from the arc element (21) that the geodesic along which a photon moves is given by

$$ds = 0. \qquad (24)$$

In other words, the space-time interval between *any* two points along a photon path is zero. The condition for a stationary value is thus satisfied in the extreme sense (we do not attach significance to negative ds

and we can use the results of our previous analysis, merely setting $\frac{1}{h} = 0$ —from eq. (16)—in the differential equation (19). The equation for the photon then is

$$\frac{d^2u}{d\phi^2} + u = 3mu^2, \qquad (25)$$

where once more the term $3mu^2$ is considered small compared with the rest, the reason being that, except for very large fields, the path of the photon will be a straight line. This is precisely what we get by neglecting the $3mu^2$. For the solution of $\frac{d^2u}{d\phi^2} + u = 0$ is

$$u = \frac{1}{r} = a \cos(\phi - \phi_0), \qquad (26)$$

which represents a straight line perpendicular to the line making the constant angle ϕ_0 with the polar axis and distant $1/a$ from the origin. The maximum value of u is a. Suppose that the ray of light in question is to graze a sphere with center at the origin and radius R, e.g., the sun, if we assume that the gravitational field is of this origin. Then $a = 1/R$. To get the next approximation to the path of the photon we substitute the solution (26) into eq. (25) and obtain finally

$$u = a \cos(\phi - \phi_0) + ma^2[1 + \sin^2(\phi - \phi_0)], \qquad (27)$$

if we neglect higher powers of ma^2, etc. Transforming to rectangular coordinates by setting

$$x = r \cos(\phi - \phi_0), \quad y = r \sin(\phi - \phi_0),$$

we obtain for the approximate equation of the path

$$x = \frac{1}{a} - ma\left[\sqrt{x^2 + y^2} + \frac{y^2}{\sqrt{x^2 + y^2}}\right]. \qquad (28)$$

For $r = \sqrt{x^2 + y^2}$ very large in the positive y direction the curve approaches the straight line

$$x = 1/a - 2may,$$

and for r very large in the negative y direction, the straight line

$$x = 1/a + 2may.$$

These two straight lines represent respectively the limiting positions of the light ray before and after its passage by the limb of the sun.

8.2 MOTION OF A PARTICLE IN A GRAVITATIONAL FIELD

The angle between them is then the deflection of the light ray due to the gravitational field. This angle is

$$\Delta\theta = 2 \arctan 2ma \doteq 4ma$$

to a first approximation. In ordinary units we have $m = GM/c^2$ (the value quoted previously, viz., GM, was in units in which c has the value unity), and hence

$$\Delta\theta = \frac{4GM}{c^2 R}. \tag{29}$$

It is of interest to observe that, if one calculates on *classical* mechanics the path of a photon moving in a gravitational field from a very great distance with initial velocity 3×10^{10} cm/sec, the resulting deflection comes out to be one-half of the relativity value (29). In seconds of arc the latter is 1.7. Many attempts have been made to observe the deflection during total solar eclipses. The quantity in question is so small that the results have of necessity been somewhat uncertain, but up to within the past few years they have been generally quoted as in better agreement with the relativity value than with the classical value. The latest examination of the problem by Freundlich,[1] including a summary of the results obtained by the Potsdam expedition for the observation of the total solar eclipse of May 9, 1929 in Sumatra indicates that the observed value is nearer 2.2 sec of arc. Freundlich is also of the opinion that if the previous eclipse measurements are properly reduced they will agree with his new value more closely than with the relativity value. One can hardly say that the problem is settled as yet, either from the theoretical or the experimental standpoints. At any rate, it is of interest to observe that in no case do the experimental findings agree better with the Newtonian point of view than with that of general relativity.

Shift of spectral lines. A third illustration of the general problem of this section is provided by the vibration of an atomic particle in a gravitational field. Such a particle may be regarded as a clock located at a definite point in the field. If it remains at rest there in a certain system of reference we have in the arc element

$$dr = d\theta = d\phi = 0,$$

while

$$ds^2 = \mu dt^2. \tag{30}$$

---

[1] E. Freundlich, H. v. Klüber, A. v. Brunn, *Zs. f. Astrophys.*, 3, 171, 1931.

Now the arc element is invariant in all systems of coordinates and hence is the same in all gravitational fields. Therefore if we denote the fields at the surface of the sun and the earth by μ_S and μ_E, respectively, we shall have

$$\frac{dt_S}{dt_E} = \frac{\sqrt{\mu_E}}{\sqrt{\mu_S}}. \tag{31}$$

But this means that the times of vibration of the particle in the solar and terrestrial field respectively are in the inverse ratio of the μ values associated with the two fields. The field of the earth is so weak compared with that of the sun that $\mu_E = 1$ very closely, while

$$\mu_S = 1 - 2GM/c^2R.$$

Hence approximately

$$\frac{dt_S}{dt_E} = 1 + \frac{GM}{c^2R} = 1.00000212. \tag{32}$$

The result is that the frequency of vibration of a solar particle is less than that of a corresponding terrestrial particle. If this frequency may be associated with that of emitted light, the lines in the solar spectrum should be shifted slightly toward the red (long wave) end of the spectrum as compared with the corresponding lines in terrestrial spectra. Although here once more the expected effect is a very small one and extremely difficult to measure with much exactness, it is now generally conceded that the relativity prediction is verified to a sufficient degree of approximation.

As must necessarily be the case, all three " practical " predictions of general relativity connected with motion in a gravitational field correspond to second order effects in which agreement of theory and experiment may long remain dubious on account of the experimental difficulties, the masking effects of larger order, etc. It is therefore not surprising that many people who have found the fundamental idea of relativity repugnant to their physical taste have been inclined to minimize the significance of the reported verifications, preferring to seek for explanations of the observed effects in other directions—a not wholly hopeless or onerous task considering their small magnitude. This attitude is perfectly justifiable. It must be confessed that the theory of relativity is by no means complete, its greatest shortcomings being its failure to take account of electromagnetic properties and of the quantum theory in the atomic domain. It is quite probable that when the necessary modifications are made the precise results in the three examples examined will also be changed. It is, however, very unlikely that the change will be material. However, it may be appropriate to raise again the question discussed in connection with the experimental findings of D. C. Miller and their bearing on the special theory: if the predicted observations of an eventually modified general relativity theory do not agree with precise experimental results, what attitude should one take toward the theory? In attempting to answer this it seems that we must place emphasis on the continual striving of physics to achieve more general, all-embracing points of view. These not only provide the general structure of physical theory with greater esthetic elegance but also open up new lines of thought and fields of application. This has been to a very eminent degree true of the theory of relativity in both its special and general forms. It is therefore unlikely that physicists will be willing to discard the concept if some of its special applications are not in precise agreement with experiment. Rather it is likely that continual attempts will be made at minor modification to which the theory readily lends itself (e.g., alterations in the metric of space-time, etc.). Its real realm of usefulness in the future will in all probability lie in the field of cosmological speculation.

8.3. General Relativistic Mechanics. We have so far discussed the fundamental idea of general relativity and its application to certain special cases. It is now in order to treat in a more general way the revision of the structure of mechanics made necessary by it. Strictly speaking, this might more logically have preceded the specific applications; however, one usually learns more readily about a theory when one examines first some concrete illustrations.

In facing the general problem we have to remember that by its very nature the theory of relativity is a *medium* theory: it seeks to reduce the study of all physical phenomena to the properties of a four-dimensional manifold whose points represent those coincidences in space and time which alone have physical significance. In modifying classical mechanics to satisfy the criterion of relativity it thus seems most natural to consider the equations descriptive of a continuous medium. We have already had occasion to refer to these in earlier parts of this book in connection with the principle of elementary abstraction (Chapter I) and later (Chapter VI) in connection with the classical mechanics of continua.

The behavior of a continuous medium on the basis of classical mechanics (electromagnetic phenomena are excluded from consideration) is completely described by the vector equation of motion and the equation of continuity. Denoting medium density by ρ, and medium velocity by $\mathbf{v}$, we can write the latter in the form

$$\dot{\rho} + div(\rho \mathbf{v}) = 0. \tag{1}$$

The equation of motion can be written

$$\rho \frac{d\mathbf{v}}{dt} = \mathbf{F}_e + \mathbf{F}, \tag{2}$$

where $\mathbf{F}_e$ denotes the external force on the medium and $\mathbf{F}$ the internal force. We must recall that in eq. (2) the time derivative is the acceleration for a particle moving from place to place. Thus denoting component velocities by v_1, v_2, v_3, respectively, we have

$$\frac{dv_i}{dt} = \frac{\partial v_i}{\partial t} + \sum_{j=1}^{3} v_j \frac{\partial v_i}{\partial x_j} \quad (i = 1, 2, 3). \tag{3}$$

The internal force components are in the nature of stresses which form a nine-component tensor X_{ij} (cf. Chapter VI). It is shown in the theory of deformable media that the force component in the x_i direction on an element of unit volume due to these stresses is

$$F_i = \sum_{j=1}^{3} \frac{\partial X_{ij}}{\partial x_j}. \tag{4}$$

If then we either omit reference to the external forces or consider them somehow absorbed in the stress forces, we may write the equation of motion in the form of the three equations

$$\frac{\partial (\rho v_i)}{\partial t} + \sum_{j=1}^{3} \frac{\partial}{\partial x_j} (\rho v_i v_j - X_{ij}) = 0 \quad (i = 1, 2, 3). \tag{5}$$

These presuppose the equation of continuity

$$\dot{\rho} + \sum_{j=1}^{3} \frac{\partial}{\partial x_j} (\rho v_j) = 0, \tag{6}$$

yielding four equations in all for the description of the medium. The temptation is strong to look upon eqs. (5) and (6) as the components of a tensor equation in which, moreover, the time is no longer separated in character from the space components. This would at first appear to be possible by introducing the aggregate T_{ij} with

$$T_{ij} = T_{ji} = \rho v_i v_j - X_{ij} \quad (i, j = 1, 2, 3). \tag{7}$$

8.3 GENERAL RELATIVISTIC MECHANICS

In addition we assume that

$$\left.\begin{array}{c} T_{i4} = T_{4i} = c\rho v_i \\ T_{44} = c^2\rho \end{array}\right\} \tag{8}$$

with $x_4 = ct$. We can then write the four equations above in the form

$$\sum_{j=1}^{4} \frac{\partial T_{ij}}{\partial x_j} = 0 \quad (i, j = 1, 2, 3, 4). \tag{9}$$

Unfortunately the aggregate T_{ij} is not actually a tensor, for while, as we have proved, X_{ij} is a quantity of this character, it is so only for the three coordinates x_1, x_2, x_3. Moreover, the terms $\rho v_i v_j$ are also of non-tensorial character. We must then search for a possible alteration which will not change the numerical statement made by the equations —for we know that they *do* form an accurate description of phenomena —but which will enable them to assume the tensor form. To retain analogy with Sec. 8.1 we shall here let $x_4 = ict$ (i.e., follow the Minkowski notation) and write the arc element in the region where the metric is Euclidean in the form (8.1-1). We introduce the quantities

$$\lambda^i = \frac{dx_i}{ds} \quad (i = 1, 2, 3, 4), \tag{10}$$

which, as we know from the discussion in Sec. 6.3, form a *contravariant* tensor of the *first* order. In general the superscript is used to denote tensors of this type. Defining

$$v_i = c \frac{dx_i}{dx_4}, \tag{11}$$

we are then led to

$$\lambda^i = \frac{v_i}{c\sqrt{1-\beta^2}}, \tag{12}$$

where

$$\beta^2 = -\frac{dx_1^2 + dx_2^2 + dx_3^2}{dx_4^2}. \tag{13}$$

From these contravariant components it is possible to form a *covariant* tensor with components

$$\lambda_i = \sum_k g_{ik} \lambda^k. \tag{14}$$

In the present case all the g's vanish save $g_{11} = g_{22} = g_{33} = g_{44} = 1$, and hence

$$\lambda_i = \lambda^i. \tag{15}$$

We shall now define a new aggregate T_{ij} with the components

$$\left. \begin{aligned} T_{ij} = T_{ji} &= c^2\rho\lambda_i\lambda_j - X_{ij} \quad (i,j = 1, 2, 3) \\ T_{i4} = T_{4i} &= c^2\rho\lambda_i\lambda_4 \\ T_{44} &= c^2\rho\lambda_4{}^2. \end{aligned} \right\} \quad (16)$$

Numerically the new T_{ij} differ from those defined in eqs. (7), (8), and (9) only through quantities of the order β^2. If now we suppose that ρ is an invariant, from the tensorial nature of the λ_i, the aggregate T_{ij} would form a covariant tensor were it not for the components X_{ij}, which we have just seen do not form a tensor in the four coordinates x_1, x_2, x_3, x_4. A possible procedure is to search for a new set of quantities χ_{ij} which *are* the components of a tensor. Then we shall have

$$T_{ij} = T_{ji} = c^2\rho\lambda_i\lambda_j - \chi_{ij} \quad (i,j = 1, 2, 3, 4) \quad (17)$$

as the components of an actual tensor and eqs. (5) and (6) will be replaced by a genuine tensor equation with the desirable invariance character. Clearly χ_{ij} will have to be numerically equivalent to X_{ij} to within small quantities of the order $1/c$. If such a set can be found we can then write the equations describing the continuous medium in the form

$$\sum_{j=1}^{4} \frac{\partial T_{ij}}{\partial x_j} = 0 \quad (i = 1, 2, 3, 4), \quad (18)$$

or in the alternative slightly more general fashion

$$\sum_{j,k=1}^{4} g^{(jk)} \frac{\partial T_{ij}}{\partial x_k} = 0 \quad (i = 1, 2, 3, 4), \quad (18a)$$

where $g^{(jk)}$ has the usual meaning of the minor of g_{jk} in the determinant $\gamma(8.1$–$17)$. From the form of the g's in the present case it is clear that $g^{(jk)} = 1$ for $j = k$ and $= 0$ for $j \neq k$. Hence eq. (18) reduces to eq. (17). It is shown in discussions of tensor analysis that corresponding to the *divergence* of a vector it is possible to define analogously the divergence of a tensor. The latter is rather more complicated than the former. Hence without going into details we shall merely note the definition. The divergence of a symmetrical tensor T_{ij} of the second order is defined to be the set of four quantities D_i where

$$D_i = \sum_{j,k=1}^{4} g^{(jk)} \left[\frac{\partial T_{ij}}{\partial x_k} - \sum_{l=1}^{4} \begin{Bmatrix} ik \\ l \end{Bmatrix} T_{lj} - \sum_{l=1}^{4} \begin{Bmatrix} jk \\ l \end{Bmatrix} T_{il} \right] \quad (i = 1, 2, 3..4). \quad (19)$$

8.3 GENERAL RELATIVISTIC MECHANICS

In the coordinates we are using (viz., Galilean coordinates, i.e., those in which the space-time metric is Euclidean) the three-index expressions $\begin{Bmatrix} ik \\ l \end{Bmatrix}$, etc., all vanish and the divergence reduces to the derivative terms. Hence we see that in this case eq. (18) says that the divergence of T_{ij} vanishes. We write therefore

$$\operatorname{div} T_{ij} = 0 \tag{20}$$

as the final tensor equation descriptive of the behavior of a continuous material medium. It has the same form in all systems of coordinates. The tensor T_{ij} is known as the stress tensor or more elaborately as the energy-momentum-stress tensor. From its formation it is seen to involve all three concepts. The vanishing of its divergence may be interpreted as including the statements of the conservation of energy and the conservation of momentum. It appears rather remarkable that by the alteration of the classical equations to an extent which is numerically negligible for all large-scale physical phenomena it is found possible to put the classical equations into a form which satisfies the fundamental criterion of relativity. Of course, all this is contingent on the construction of the tensor χ_{ij}. Its existence in general cannot be proved without at the same time postulating the relativity principle: we see of course that it must exist if the principle is to be applicable to mechanical phenomena. It will be sufficient for our present purposes if we indicate its construction for the special case of a perfect fluid.

A perfect fluid cannot support shearing stresses. Hence we have in this special case

$$\left. \begin{array}{l} X_{ij} = 0 \text{ for } i \neq j \\ X_{ij} = -p \text{ for } i = j \ (i, j = 1, 2, 3) \\ X_{44} = 0, \end{array} \right\} \tag{21}$$

where p denotes the pressure. Define

$$\chi_{ij} \equiv p(\lambda_i \lambda_j - g_{ij}). \tag{22}$$

This is a tensor. Moreover, for Galilean coordinates in which all the g's are zero save $g_{ii} = 1 (i = 1, 2, 3, 4)$ it differs very little from χ_{ij}. Specifically for $i \neq j$

$$\chi_{ij} = p\lambda_i \lambda_j,$$

which from the definition of the λ's is a small quantity of the order of v^2/c^2. For $i = j = 1, 2, 3$ χ_{ij} differs from $-p$ by quantities of the order of v^2/c^2, and for $i = j = 4$, since $\lambda_4 = 1/\sqrt{1 - \beta^2}$, $\chi_{44} = 0$ to within v^2/c^2. Hence χ_{ij} satisfies all the conditions laid upon it by (21). Consequently the stress tensor for a perfect fluid is

$$T_{ij} = c^2\rho\lambda_i\lambda_j - p(\lambda_i\lambda_j - g_{ij}). \tag{23}$$

Corresponding to this tensor we can form the invariant quantity

$$T = \sum_{i,j=1}^{4} g^{(ij)} T_{ij} \tag{24}$$

$$= (c^2\rho - p) \sum_{i,j=1}^{4} g^{(ij)} \lambda_i \lambda_j + p \sum_{i,j=1}^{4} g^{(ij)} g_{ij}.$$

We can show very readily that $\sum_{i,j=1}^{4} g^{(ij)} \lambda_i \lambda_j = 1$. Moreover, $\sum_{i,j=1}^{4} g^{(ij)} g_{ij} = 4$, so that

$$T = c^2\rho + 3p. \tag{25}$$

Let us next imagine that the continuous medium is really made up of a large number of discrete particles, i.e., has an atomic constitution. Then if the particles are at rest with respect to each other we may set $p = 0$, for there is no longer anything which will give rise to *fluid* pressure. In this case the stress tensor reduces to the form

$$T_{ij} = c^2\rho\lambda_i\lambda_j,$$

and the stress invariant to

$$T = c^2\rho.$$

Here ρ must, however, be regarded as the *static* invariant density of the mass distribution, i.e., the proper density corresponding to the *rest* mass m'_0 of the special theory (cf. Sec. 7.3). We shall call it therefore ρ_0 and write for a discrete collection of particles

$$\left. \begin{array}{c} T_{ij} = c^2\rho_0\lambda_i\lambda_j, \\ T = c^2\rho_0. \end{array} \right\} \tag{26}$$

But T is a genuine invariant and hence should be the same whether the aggregate be at rest or in motion. Hence by combining eq. (25) with eq. (26) we obtain the general relation

$$c^2\rho = c^2\rho_0 - 3p. \tag{27}$$

The actual density ρ is therefore not the same as the static or rest density, but is obtained from it by incorporating the pressure term

8.3 GENERAL RELATIVISTIC MECHANICS

which comes from the motions in the medium. An extremely interesting application of eq. (27) may be made to the case in which the medium consists of photons, i.e., is a distribution of radiation. If the photons were to possess any rest mass, since they travel with the velocity c in any inertial system, their actual mass would be infinite. The only way to avoid this is to take their rest mass equal to zero. But this makes the rest density of a radiation continuum equal to zero, or $\rho_0 = 0$ in eq. (27). The actual density ρ need not vanish, however. In fact, we recognize in $c^2\rho$ the *energy* density of the radiation from the Einstein law $E = mc^2$ giving the mass associated with radiation of energy E. This enables us at once to obtain the pressure in the radiation distribution. The latter is indeed just $-c^2\rho/3$, and the pressure exerted *by* the radiation is the equal and opposite quantity

$$p = c^2\rho/3, \tag{28}$$

a well-known result in radiation theory.

The foregoing considerations have shown the possibility of modifying the fundamental equations of mechanics as applied to a continuous medium and, by extension, to a discrete aggregate so as to secure for them a four-dimensional tensor form. This insures the invariance demanded by the relativity concept. However, Einstein wished to go a step further. All physical phenomena, on his view, should appear as ultimately based on the structure of the space-time manifold. But this structure is exhibited by the tensor G_{ij} which is a function of the quantities g_{ij} that in turn characterize the space-time metric. The thought immediately arises that there should be some connection between the stress tensor T_{ij} and the gravitational or world tensor G_{ij}. Now it is an interesting fact (the proof of which need not be given here) that if one forms the tensor quantity

$$G_{ij} - \tfrac{1}{2}Gg_{ij} + \lambda g_{ij} \tag{29}$$

its divergence vanishes for any constant λ. In this equation the quantity G is the so-called Einstein invariant which is formed from the gravitational tensor G_{ij} in precisely the same way that the stress invariant T is formed from the stress tensor T_{ij}; in other words,

$$G = \sum_{i,j=1}^{4} g^{(ij)} G_{ij}. \tag{30}$$

We have already indicated that div $T_{ij} = 0$. There is thus a strong temptation to assume a simple connection between the stress tensor and the tensor (29). The simplest is of course a relation of the form

$$G_{ij} - \tfrac{1}{2}Gg_{ij} + \lambda g_{ij} + \epsilon T_{ij} = 0 \tag{31}$$

where ϵ is a fixed constant. This is the assumption which Einstein made, and the tensor equation just written was taken as the fundamental equation of gravitation. This tensor equation is much more general than $G_{ij} = 0$ which we used in Sec. 8.2 in our elementary discussion. We shall soon see the connection between the two. We note that in any case the number of independent component differential equations involved in (31) is ten as is the situation in $G_{ij} = 0$. However, the former are to be supplemented by the four differential equations expressing the vanishing of the divergence of $G_{ij} - \tfrac{1}{2}Gg_{ij} + \lambda g_{ij}$. This leaves but six independent relations among the g's.

Let us examine eq. (31) more closely. We shall multiply through by $g^{(ij)}$ and sum over i and j, obtaining (see eqs. (24) and (30))

$$G - 4\lambda = \epsilon T, \tag{32}$$

wherein we have utilized the fact that

$$\sum_{i,j=1}^{4} g^{(ij)} g_{ij} = 4.$$

Now (32) is a very interesting equation, for it says that, if $T = 0$, we shall have

$$G = 4\lambda. \tag{33}$$

But if there is no matter (meaning no manifestation of inertia) anywhere in space-time, we should expect to have $T_{ij} = 0$, for there can in this case no longer be any stresses present. This in turn entails $T = 0$, so that eq. (33) can be taken to represent the state of affairs in empty space-time. But under these conditions eq. (31) becomes

$$G_{ij} = \lambda g_{ij}. \tag{34}$$

This tensor equation is the gravitational field equation for space-time free of matter (including, of course, radiation so far as it is characterized by inertia). We must again emphasize that purely electromagnetic effects are excluded from consideration.

An interesting geometrical interpretation of eq. (34) is possible. It can readily be shown that for the case of a two-dimensional surface embedded in a three-dimensional Euclidean manifold the Einstein invariant G reduces precisely to the Gaussian curvature $G = 1/R_1R_2$, defined in (2.4–2), where R_1 and R_2 are the principal radii of curvature of the surface. Moreover, if the four-dimensional manifold representing space-time is assumed to be embedded in a five-dimensional Euclidean manifold, G becomes [1]

$$G = \sum_{i,j=1}^{4}{}' 1/R_i R_j, \qquad (35)$$

where the prime indicates that, in the summation of i and j from 1 to 4, terms for which $i = j$ are omitted. Here $R_i (i = 1, 2, 3, 4)$ are the four principal "radii of curvature" of the manifold, defined analogously to the quantities R_1 and R_2 in the simpler and more obvious two-dimensional example. It thus appears that G in eq. (35) is the natural generalization of the Gaussian curvature. The difficulty arises, however, as Eddington points out (*loc. cit.*) that in general a four-dimensional Riemannian continuum like that demanded by relativity cannot be represented as a surface embedded in a five-dimensional Euclidean manifold. Actually in general ten dimensions are required for the latter. This makes the geometrical interpretation more intricate. However, it is customary to continue to call G the Gaussian curvature and to introduce in addition the radius of the hypersphere—a two-, three-, or four-dimensional sphere embedded in five dimensions, to which eq. (35) applies with the appropriate summation of i, j—that has the same Gaussian curvature. Thus if the radius of the four-dimensional hypersphere is R, we have from eq. (35), $G = 12/R^2$ (with $R_1 = R_2 = R_3 = R_4 = R$). If the hypersphere is of three dimensions instead of four, we get $G = 6/R^2$, again using eq. (35) with $R_1 = R_2 = R_3 = R$ and $R_4 = \infty$. For the two-dimensional hypersphere, finally, $G = 2/R^2$. Now consider the generalized quadric surface

$$\sum_{i,j=1}^{4} (G_{ij} - \tfrac{1}{2} g_{ij} G) x_i x_j = -3. \qquad (36)$$

The *radius* of the quadric in the direction from the origin to the point (x_1, x_2, x_3, x_4) will be given by

$$R_q = x_i \bigg/ \frac{dx_i}{ds}. \quad (i = 1, 2, 3, 4). \qquad (37)$$

More extended analysis now shows (cf. Eddington, *loc. cit.*) that this same R_q is equal to the radius R of spherical curvature of the corresponding three-dimensional section of space-time (i.e., is the radius of the hypersphere of three dimensions). We can then at once ascer-

[1] Cf. Eddington, "Mathematical Theory of Relativity" (Cambridge, 1923), pp. 149 ff.

tain the relation between R and λ in the case in which there is no matter present in the world, i.e., when $G_{ij} = \lambda g_{ij}$, for then

$$G_{ij} - \tfrac{1}{2}Gg_{ij} = -\lambda g_{ij}$$

and the quadric (36) becomes

$$-\lambda \sum_{i,j=1}^{4} g_{ij}x_ix_j = -\frac{\lambda R^2}{ds^2}\sum_{i,j=1}^{4} g_{ij}dx_idx_j = -\lambda R^2 = -3. \quad (38)$$

But from this we have at once

$$R = \sqrt{3/\lambda}. \quad (39)$$

In words this says that at any point in empty space the radius of curvature of any three-dimensional section of space-time is $\sqrt{3/\lambda}$. It is true that space-time is here considered as a four-dimensional continuum embedded in a five-dimensional Euclidean manifold. However, the analysis of Eddington indicates that this restriction is really unnecessary and the result remains the same even if the four-dimensional space-time is embedded in the ten-dimensional manifold which is actually required. The quantity λ in the law of gravitation, which for space-time free of matter takes the form (34), thus plays an important rôle in providing a measure of the curvature of space-time, that is, our universe. It has received the name of *cosmical constant*. If it vanishes, $R = \infty$ and space-time is homaloidal throughout with a Euclidean metric. In this case the Einstein law of gravitation in empty space-time reduces to the form

$$G_{ij} = 0 \quad (40)$$

which has been used in our discussion in Sec. 8.2 of the motion of a particle in a symmetrical gravitational field. The success with which eq. (40) accounts for such motion must be taken as an indication that the value of λ in our universe is extremely small. The assignment of a non-vanishing value to it leads to the various brands of relativistic cosmology for which reference should be made to special treatises on the subject.[1]

---

[1] See, for example, the excellent report by H. P. Robertson, "Relativistic Cosmology," in *Reviews of Modern Physics*, 5, 1, 1933. Reference should also be made to R. C. Tolman, "Relativity, Thermodynamics, and Cosmology (Oxford, 1934).

CHAPTER IX

QUANTUM MECHANICS

9.1. Introduction. Before we embark on a study of the latest form of the quantum theory it seems well to outline the course of events which has led, almost compellingly, to the formalism discussed in the subsequent sections of this chapter. The present remarks must of necessity be brief and for that reason wholly inadequate to place the older facts and theories in their proper setting. Each of them marked a signal achievement in its day; some of them, e.g., the Bohr-Sommerfeld theory of the atom, enjoyed successes more brilliant and striking than those which have attended the slower evolution of our present notions; our hasty despatch of these important matters has no reason other than the rather accidental fact that they are not, in a logical sense, the foundation of our latest quantum mechanical ideas, although historically the connection is extremely close.

What we must set down here is the story of the failures of classical physics. The reader who desires information beyond this brief review—and such information will definitely aid in an understanding of the following sections—may be referred to Ruark and Urey's comprehensive book, "Atoms, Molecules, and Quanta."[1] The trouble started in connection with the problem of black-body radiation. A black body in the technical sense, i.e., one which absorbs all the radiation incident on it, emits, at any given temperature, a characteristic spectrum of frequencies. This spectrum contains all frequencies (i.e., is continuous), but the intensity associated with these frequencies differs in a manner which was the subject of extensive experimental investigations. This spectral distribution of intensities, while varying with the temperature of the emitting black body, proved to be independent of its material composition. It was this distribution which theorists found difficult to explain. The physicist recalls in this connection the regularities known as the Stefan-Boltzmann law, Wien's displacement law, and the Rayleigh-Jeans law.

The method of calculating the intensity distribution seemed perfectly clear: emission of radiation must be due to the vibrating charges composing the black body. These charges vibrate with many different

[1] McGraw-Hill Book Company, New York, 1930.

frequencies, and it is the variation in the relative intensity of these different frequencies which causes the intensity distribution within the black-body spectrum. To be sure, there are many different types of vibrator, such as the harmonic oscillator, the anharmonic oscillator, the circulating charge, etc.; and the mathematical physicist, in his quest for an explanation of this phenomenon, does not know at once which type to choose. But he reflects immediately upon the circumstance that the material constitution of a black body has nothing to do with the characteristics of its spectrum, which is nature's way of telling him that the choice of vibrator will not affect the result. Consequently, he chooses the simplest type, the linear harmonic oscillator. He takes an unlimited collection of them with all possible frequencies, imposes the condition that they shall be in thermal equilibrium (cf. Chapter V), and sees what the distribution of intensities among the different frequencies will be.

This method leads to a unique result, but not one in agreement with the observed facts. It predicts that the very high frequencies in the black-body spectrum should be extremely intense, whereas the actual intensity of these frequencies goes to zero. No modification of the method of deriving the result is possible that would not do violence to some fundamental fact of classical physics. Hence, Max Planck concluded, there must be something radically wrong with classical notions.

Planck combined the courage to break loose from current theories with the genius to construct a new one. In 1900 he showed in what manner the old calculation could be made to yield the correct result: it was by assuming that, while the oscillators are still allowed all possible frequencies of vibration, *each may not possess any energy it pleases.* In classical physics, the reader recalls, the energy of a given oscillator depends merely on its amplitude, and this amplitude is subject to no restriction. Planck assumed that the energy of each oscillator must be an integral multiple of the quantity $h\nu$, called a *quantum* of energy, where h is a universal constant and ν the frequency of the oscillator in question. Thus energy was quantized; it appeared in packages, not of equal size, but of sizes proportional to the frequencies of the oscillator. The constant h is sometimes called the "quantum of action." The formula thus produced is known as Planck's radiation law. It was derived in Sec. 5.7. Experience has shown it to be correct.

Quantization, established in this manner, affects both the energy of material systems, i.e., the oscillators constituting the black body, and that of the light composing the black-body spectrum. But now, if the energy of light is to be quantized, can light still be a wave? The classical conception of a wave carries with it the inevitable assumption of continuous energies. If, however, we were wrong in supposing radiation to be an undulatory disturbance, and the advocates of a corpuscular theory were right, then quantization of light energy could be easily conceived. For we should merely have to associate with every corpuscle a certain discrete energy to satisfy the demands of quantum theory. Curiously enough, the interpretation of several new phenomena supported strongly the existence of light quanta as discrete entities in space.

The first is the photoelectric effect. Light falls, for example, on a metal plate, and electrons are ejected from the metal surface. It is possible to measure both the number, N, of electrons liberated per second and their maximum kinetic energy, E. This energy must, of course, come entirely from the incident light, if the principle of conservation is to be valid. Now this effect presents two striking features, the first suggesting, the second very nearly forcing, the quantum hypothesis.

In the first place, if we increase the intensity of the incident light, the maximum kinetic energy of the photoelectrons is unaltered; it is merely N which changes. But if we increase the frequency, ν, of the incident light, the energy changes (essentially) according to the law $E = h\nu -$ constant. Clearly, the simplest explanation of this fact is to suppose that energy is converted in quanta $h\nu$, as pointed out by Einstein in 1905.

Secondly, assume the light of frequency ν to be made very feeble and consider a very small portion of the metal surface, so small indeed as to contain but one electron to be ejected. If the incident light is undulatory in character, the energy which the small portion of area absorbs, and which it is presumably capable of imparting to the electron, will increase uniformly in time. In particular, since the light is very feeble, there will be an interval of time, following the instant at which the light first hits the surface, during which the energy absorbed is insufficient to eject the electron with the requisite energy $h\nu -$ constant. Hence photoelectric emission should exhibit a time lag depending on the intensity of the ejecting radiation. This lag should, in fact, be measurable. But none is ever observed. The electrons appear to leave any portion of the metal surface with a random distribution in time, the event of emerging immediately after exposure being as likely as a finite lag. This observation points almost conclusively to the supposition that light, instead of being composed of waves, consists of a swarm of particle-like quanta, each of which may liberate an electron upon impact, and each carrying an energy $h\nu$.

The famous Compton effect is another instance which leads to this conclusion. A light wave, on encountering a material obstacle, may have its intensity changed or be retarded, but its frequency should not be altered. Nevertheless, it was found by Arthur H. Compton (in 1923) that just such a change in frequency occurred when x-rays fell upon matter. In fact, the change was precisely as though an x-ray quantum of energy $h\nu$ collided elastically with an electron within the material, imparting a fraction of its energy to the latter and retaining a smaller energy $h\nu'$, and hence a smaller frequency ν'. Moreover, nature made this phenomenon particularly beautiful and impressive by allowing the "recoil electron" to be observed (by means of the Wilson cloud chamber). A more compelling argument for the corpuscular nature of radiation could hardly be desired.

But how are we to reconcile this fact with the experiments of interference and polarization, which demand an explanation in terms of light waves? This was indeed a perplexing question, and one that remains perplexing as long as we adhere to the classical way of describing physical phenomena. The incongruity of waves and quanta haunted all the early theories which we are here reviewing and found its resolution only in the later formalism which is discussed at length in this chapter, a formalism which definitely renounces the attempt of visual characterization and which grants no ultimate significance to the terms "wave" and "particle." We merely state at this time that there is no longer any logical difficulty as a result of this abstraction, and that physics is not forced to admit an essential dualism in its manner of describing phenomena, as is often asserted. This will be clear later.

In the fields of spectroscopy and atomic structure, similar departures from classical physics took place. There had been accumulated an overwhelming mass of evidence showing the atom to consist of a heavy, positively charged nucleus surrounded by negative, particle-like electrons. According to Coulomb's law of attraction between electric charges, such a system will collapse at once unless the electrons revolve about the nucleus. But a revolving charge will, by virtue of its acceleration, emit radiation. A mechanism for the emission of light is thereby at once provided.

However, this mechanism is completely at odds with experimental data. The two major difficulties are easily seen. First, the atom in which the electrons revolve continually should emit light all the time. Experimentally, however, the atom radiates only when it is in a special, "excited" condition. Second, it is impossible by means of this model to account for the occurrence of spectral lines of a single frequency (more correctly, of a narrow range of frequencies). The radiating electron of our model would lose energy; as a result it would no longer be able to maintain itself at the initial distance from the nucleus, but fall in toward the attracting center, changing its frequency of revolution as it falls. Its orbit would be a spiral ending in the nucleus. By electrodynamic theory, the frequency of the radiation emitted by a revolving charge is the same as the frequency of revolution, and since the latter changes, the former should also change. Thus our model is incapable of explaining the sharpness of spectral lines.[1]

Bohr saw that the classical laws of electrodynamics must be abandoned. His great achievement was to perceive a connection between the anomalies in the atomic-structure problem and those in the black-body problem. Planck had solved the latter by supposing his oscillators to vibrate with quantized energies; was it not possible to introduce quantization into the motion of the atomic electrons? Was not the existence of sharp spectral lines an indication that only certain states of motion were realized to the exclusion of all others? What Bohr did when, in 1913, he published his famous theory of the hydrogen atom, was not merely to transcribe Planck's quantum theory into terms applicable to the atom, but also to generalize it in a natural and beautiful manner so as to include Planck's form as a special case. He created, in fact, a new system of dynamics, in which classical laws were augmented by the following two postulates:

1. Among the classically possible states of motion of a charged particle there exist certain "stationary states," distinguished by the fact that the particle in these states does not radiate, even though it be accelerated. In consequence of this postulate, the particle, unable to lose or gain energy, will remain in a stationary state as long as it is undisturbed (except for "spontaneous" changes causing light emission). To determine these stationary states of motion Bohr and Sommerfeld gave rules commonly known as quantum conditions, which in their most compact form simply require that the amount of action (cf. Sec. 3.12) gained by the particle in one period of its "stationary" motion be an integral multiple of Planck's constant h. In symbols,

$$\oint p\, dq = nh,$$

---

[1] It is also apparent that, after some time, the atom would collapse. Indeed this time happens to be a very small fraction of a second. This model, therefore, makes the continued existence of the universe difficult to understand.

where q is a generalized coordinate, p its conjugate momentum, and n an integer. With each stationary state of " quantum number " n there is associated a (constant) energy E_n.

2. When, for any reason, a transition occurs from one stationary state (n) to another (m) the difference in energy $E_n - E_m$ enters or leaves the atom as monochromatic radiation of frequency ν, such that

$$|E_n - E_m| = h\nu.$$

The radiation is emitted by the atom if $E_n > E_m$, absorbed if $E_n < E_m$. Bohr's theory makes no attempt to explain the dynamical situation within the atom during the acts of emission or absorption, which are considered to require a very short time.

The enormous success of Bohr's theory in explaining the spectra of one-electron systems (H, He^+, Li^{++}, Na, etc.), its application to the calculation of critical and ionizing potentials, and its confirmation by the observations on the spectra of molecules, are now well known. The further introduction, by Bohr, of his " correspondence principle " permitted the calculation of intensity and polarization of the spectral lines emitted, rounding off the theoretical structure in a temporarily pleasing manner.

Yet the unsatisfactory features of this set of postulates soon began to be felt. In contemplating the first one, physicists experienced a slight metaphysical uneasiness, for, after all, if we are to accept blindly a postulate, it might well be a bit less specific. Moreover, while to postulate a certain fundamental fact about nature is a scientific procedure that might pass muster, to require of her the establishment of a " correspondence " between two theories in a postulational sense is another matter. Criticism of this type, however, rarely makes an impact upon a physical theory, unless it is weighted down by the force of practical arguments. In the case of Bohr's theory, this force was soon supplied by the theory's failures.

For our purpose a brief list of the shortcomings will suffice.

1. Calculations yielded definitely incorrect answers for the stationary energies of atoms with more than one valence electron. They appeared to fail whenever the charged particle was subject, not only to the static field of a nucleus, but also to the rapidly varying force of another moving charge. For He and H_2, for instance, there was no possible stationary arrangement of orbits of the two electrons that would produce the energies observed spectroscopically.

2. Probably for a similar reason, the theory failed to explain the phenomenon of dispersion, for here again the electron is subject to a very rapidly varying force, namely, that of the light wave which the

atom disperses. According to Bohr's theory, anomalous dispersion should take place at the frequencies of revolution of the electron, whereas it actually occurs at the frequencies which the atom is capable of absorbing; these are not, in general, the same as the former.

3. The anomalous Zeeman effect, i.e., the magnetic splitting of spectral lines into more than three components, was beyond the capacity of Bohr's theory to explain.

4. In band spectroscopy, the number n appearing in the definition of stationary states was soon discovered to be not always integral, but sometimes half-integral.

5. Less directly related, but quite definitely traceable, to Bohr's postulates is the failure of the earlier quantum theory to explain the Ramsauer effect, i.e., the ease with which electrons having certain velocities will pass through gases.

The efforts to avoid these difficulties have led to the system of quantum mechanics to which the remainder of this chapter is devoted. In introducing it, we feel that the historical course thus far pursued is an ill-suited method of approach. Its full meaning, in our opinion, is more easily grasped if the acquaintance is made on thoroughly logical lines. But before we abandon this historical note, let us mention another experimental discovery which has materially strengthened our confidence in the subsequent more abstract formalism.

In 1927 experiments by Davisson and Germer disclosed, and further observations by G. P. Thomson, Rupp, and others confirmed, the wave nature of material particles. Electrons, on being reflected from or passing through metal crystals, showed diffraction patterns similar to those displayed by x-rays under analogous conditions. The wave length of the electrons was given correctly by the wave theory advanced by Louis de Broglie in 1924 (cf. Sec. 7.3). The peculiar paradox which beclouded the physical nature of light now found its counterpart in the ambiguity regarding the nature of the electron. The appearance of this dilemma in both cases indicates strongly that the old question as to the physical nature of the ultimate constituents of the universe requires more than an empirical treatment; it demands a radical modification of the concepts involved. We shall see how the difficulty resolves itself in a more refined form of theory, in which both the concepts of wave and particle lose their fundamental distinction.

This new theory is known as quantum mechanics, a term meant to set it off from those parts of physics with which we have dealt in the previous chapters and which are collectively referred to as classical physics. In the remainder of this section we shall discuss the main points of difference between classical and quantum mechanical reasoning, a difference which is indeed incisive, for quantum mechanics is not merely a limited discipline apart from others—it is a new point of view from which all other disciplines are to be treated.

To grasp the distinction more clearly, it is well to survey the field of classical physics and look for its salient features. But these are not easy to find, for classical physics is a quite heterogeneous mixture of conflicting elements. Indeed, many of the supposedly profound philosophical implications of quantum mechanical reasoning are inherent in classical physics and become apparent if one desires to look for them. This observation is of some importance in forming a correct perspective and in rating the philosophical value of modern theories. Hence it is well to begin by reviewing some of the various logical attitudes hidden under the guise of classicism. Much of this has already been presented in the previous chapters, but there is a certain advantage in recalling it here in order to point out as sharply as possible the relation between classical and quantum mechanics.

First there is the fundamental difference, already discussed in Chapter V, between dynamical and statistical theories, the one implying unvarying regularity, reversibility, and uniqueness; the other mere predominance, irreversibility, and chance. It would be easy to make out a case for determinism or indeterminism depending on the type of theory chosen as the basis of discussion. Next we note the several different answers to the question regarding the nature of the ultimate physical entities. In mechanics they are considered as mass points, or approximations to mass points, connected by central forces. This point of view, though successful in a limited realm, has proved useless in electrodynamics which calls for the introduction of more complicated forces. In the case of light, classical physics makes use neither of particles nor of forces, but feels called upon to assume the new phenomenon of immaterial waves. It is evident that the initial monistic trend which dominated Helmholtz's famous mechanistic program has been lost, even in classical analysis, and will probably never appear again.

Not even the ancient antithesis of continuity and discreteness is held sacred in the theories leading up to quantum mechanics. They embrace both of these contradictory elements as the occasion may require, often with a shameful ignorance of Greek paradoxes. The atom, or the electron, is clearly a particle separable and distinct from its surroundings, while an electromagnetic field is thought of as possessing continuity, as discussed in Chapter VI.

Upon examination of the various theories of classical physics it will be seen that different *methods of describing physical experience* are employed. We have emphasized in Chapter I that no physical theory merely portrays facts, but that it involves postulates and ideas which are not directly derived from experience. The reader who has been accustomed to taking literally, and as statements about ultimate reality, all physical propositions regarding nature may be slightly shocked by this statement. He might look upon such things as the electron, or the more recently discovered positron, as entities of the same empirical status as the tree he sees, or the table at which he works. This, we feel, would be erroneous; the electron should be regarded as a construct rather than a fact, for it is not an immediate datum of our experience, but the result of an elaborate sequence of inferences which, incidentally, is not unique. It is in this sense that we wish to refer to atoms, photons, electromagnetic fields, and all other objects of indirect experimentation as physical constructs, reserving for every individual the right to ascribe to them any epistemological rank he pleases.

In classical mechanics we deal with macroscopic bodies. Here, and here only, do we use descriptional constructs which experience immediately suggests: we define the state of the body, or group of bodies, as the aggregate of the velocities (or momenta) and positions of all bodies at any given time. It will appear later that the concept of "physical state" is most directly related to the type of constructs used in the description; hence our present inquiry can be conducted most profitably by focusing attention upon the different definitions of state in classical physics. The mechanical definition just mentioned has been introduced into the description of phenomena, like those of thermodynamics, which do not suggest it in a very direct manner. Here positions and velocities of particles become very definitely constructs which have lost their immediate reference to experience. This particular concept of state, while it is not forced upon us, has turned out to be adequate because it permits a complete description of the regularities which it is designed to explain. A large part of atomic theory, astronomy, electromagnetism, and acoustics yields to the tools of analysis if positions and velocities of particles are considered as defining a physical state. In fact, this definition has come to be regarded *par excellence* as the only correct one.

Nevertheless there are fields even in classical physics in which this particular formalism is known to be not the only one that can be applied. The state of a thermodynamic substance is much more conveniently defined by using the variables pressure and volume, for the functional relation between the variables of state then becomes simpler. But here we encounter a very remarkable situation, since two different symbolisms can be used to explain the same group of facts with different ease, but equal final success. The adoption of the definition of state becomes a matter of preference and is usually decided on the basis of simplicity considerations. As far as the physical description of phenomena is concerned this choice is unimportant, but it has an important bearing upon the question of determinacy. If the pressure-volume definition of state is adopted in connection with thermodynamic processes, then one perfectly determinate state is followed by another, while the use of the other definition in terms of velocities and positions of innumerable particles would render the succession of states indeterminate or at least incompletely determinate. Another way of expressing this situation would be to say that the theory is encumbered with details that have no counterpart in experience.

Finally, classical physics presents instances in which the mechanical definition fails entirely. For a long time this definition was forced upon the facts of electromagnetism which were explained as waves and strains in a hypothetical ether. But this procedure led to absurdities that have finally overthrown this view. It then became customary to regard as the state of an electromagnetic field either all simultaneous values of the electric and magnetic field strengths at different points of space, or what amounts essentially to the same thing, the four-potential of the electromagnetic field (Chapter VI). Hence we are confronted in classical physics with at least three different and unrelated methods of reducing the facts of experience to scientific formulation.

Despite these differences there are a few traits which all classical theories share. We shall be better prepared for an appreciation of the spirit of quantum mechanics if we devote some attention to these common features.

First is the assumption, implicitly contained in every pronouncement of classical physics, that the data of experience are of all imaginable degrees of difference and of infinite detail. The bodies of our experience move visually with any speed, any energy, and may have any position.[1] In principle, any change whatever may occur in their state of motion. Moreover, if one state of motion is observed now and another later, it is perfectly plausible and justifiable to ask: in what detailed manner did the change take place? Certainly there is no principle of reasoning which makes it necessary that nature should realize all imaginable changes in the state of motion of a body, nor even that she should behave in such a way as to permit our chosen

---

[1] As an illustration we recall the method of elementary abstraction, discussed in 1.7.

formalism to portray her changes. The point is, however, that she does so in our immediate experience with large-scale bodies. Whether or not the physical constructs by which we analyze more indirect experiments partake of this simplicity is an open question for which an answer can be found only by trying both alternatives and seeing which is more successful. The circumstance that our symbols are fashioned largely after the elements of large-scale experience merely reflects anthropomorphic limitations and should not predispose us in any way. Classical physics assumes that the symbols of description (atoms, electrons, etc.) conform to the suggestion of crude experience; quantum mechanics, as we shall see, abandons this hypothesis.

The second feature which all classical theories possess in common concerns the character of analysis. Limited use is made of the great variety of mathematical notions because of the excessive homage which is paid the arithmetical concept of *number*. It is true that every observation provides a number, and that numbers are the only objects with which scientific observations deal. Hence one might plausibly suppose that all analysis should also deal with numbers or those more general mathematical schemes which synthesize numbers (vectors, tensors, matrices). This assumption may be conveniently expressed by the statement that the elements of calculation are identical with the elements of observation. This, as we have said, is a characteristic hypothesis underlying classical physics. Again, its logical foundation is found to be precarious if we realize that the constructs in terms of which we attempt to explain nature are not immediate objects of observation. There is indeed no compelling reason why they should be combined in the manner necessary for the elements of observation. It will be seen that quantum mechanics allows itself much greater freedom in the use of abstract combinatory mechanisms. There appears at this point the fact, paradoxical but true, that quantum mechanics, though it was initially founded with the purpose of connecting more closely the facts of experiment with theory, has actually severed this bond and drifted away into a formalism that belies its initial promise.

Any theory which operates exclusively with the elements of observation naturally carries itself in close alignment with experimental facts. If, on the basis of an observed situation, it predicts the outcome of another experiment, that prediction, being in terms of numbers, will designate a very definite observation. This precise determination of experience is inherent in classical theory. It might seem, at first glance, that the statistical theories should be classified separately in this connection, for statistical laws do admit exceptions, as was discussed in Chapter V. Although this is true in principle, it should be remembered that statistical predictions, though slightly uncertain in an absolute sense, are always definite; they are unique, but not always correct. When the pressure of a given mass of gas is calculated from its volume and its temperature the equation of state gives but one answer, and if we should have the good fortune of observing one of the extremely rare violations of the laws of thermodynamics, these laws tell us nothing about what to expect. No classical law which is complete in its formulation predicts alternatives; if they appear they are due to ignorance of boundary conditions. Of course the ultimate precision of its pronouncements exposes classical theory to tests that are far too severe. But here the theory of errors comes conveniently to our aid and mitigates the requirements for agreement with experience so that they are no longer burdensome.

These three hypotheses: (1) possibility of an infinitely detailed experience, (2) identity of the elements of calculation with those of observation, (3) sharp prediction of the data of observation, are dismissed in quantum mechanics. In place of the first it selects from the manifold of all possible observations only a certain class. And it does more than give a vague characterization of this permissible class inasmuch as it provides us with definite mathematical rules for its selection. These will be set forth in the next section. The second hypothesis is abandoned for the sake of greater freedom of mathematical operation. We shall see that the notion of an "operator," no longer that of a number, dominates the new theory. Finally, sharp prediction of observations is removed as the result of a redefinition of the concept of state which is in conformity with the operator notion. Before we proceed to discuss the new postulates in detail we wish to outline two important aspects which the removal of hypothesis (3) forces upon the new description of experience—aspects which are so radical and unfamiliar that the uninitiated reader will profit by their explicit discussion.

The general principle by which (3) is replaced may be termed the *principle of elementary diffuseness*. We use this phrase here merely for the purpose of qualitative description; its exact nature will appear more clearly later. We will first illustrate it by the special example of indeterminacy of position and momentum. When a particle has a definite position, quantum mechanics does not allow us to specify its exact momentum, or velocity; and conversely, when the exact momentum is given the new theory is incompetent to predict its position (except with an infinitely large error). This peculiar interrelatedness of errors affects all quantities connected by Hamilton's canonical equa-

9.1 INTRODUCTION

tions (cf. 3.14–17), such as coordinate and momentum, energy and time. Just how this relation arises will be shown later; at present we will state the relation without proof. If Δx is the uncertainty in the position x of the particle (the term uncertainty requires definition; at present the reader may think of it as the probable error of a measurement) and Δp the uncertainty in its momentum, then

$$\Delta x \cdot \Delta p \geqq \frac{h}{4\pi}, \qquad (1)$$

where h is Planck's constant.[1] Similarly, if ΔE is the uncertainty in the energy of a physical system and Δt the time interval during which it is known to possess approximately the energy E, then $\Delta E \cdot \Delta t \geqq \frac{h}{4\pi}$. Although both these relations are usually thought of as special examples of Heisenberg's "uncertainty principle," it is only metaphorically that one can speak of Δt as an uncertainty. To cite an application of the last-mentioned example of "elementary diffuseness" one may refer to the sharpness, or lack of sharpness, of spectral lines. If the energy states involved have a long lifetime (like the lowest state of any atom), Δt is large, hence ΔE is small and the spectral line sharp, and

---

[1] It happens that this relation follows from the point of view, originally advanced by Louis de Broglie and tested by the famous experiments of Davisson and Germer, that the motion of a particle can be described as a group of waves whose (group) velocity coincides with that of the particle, and whose wave length λ is given by the de Broglie formula (Sec. 7.3).

$$\lambda = \frac{h}{p},$$

p being the momentum of the particle. If the particle is known to be somewhere in the interval Δx, the waves must be confined to this range and disappear outside. But the only way of causing waves filling a certain portion of space to disappear is to let them interfere destructively where they are not wanted. Thus the waves representing the particle must interfere at the end points of Δx and everywhere outside. Suppose now that two waves of wave lengths λ_1 and λ_2 re-enforce each other at the midpoint of the range Δx. Then, if they are to interfere destructively at the end points, the interval $\Delta x/2$ must contain at least half a wave length more of one wave than of the other. Hence $\frac{\Delta x}{\lambda_1} - \frac{\Delta x}{\lambda_2} \geqq 1$, or $\Delta x \cdot \Delta \left(\frac{1}{\lambda}\right) \geqq 1$. Now by de Broglie's formula, $\Delta \left(\frac{1}{\lambda}\right) = \frac{\Delta p}{h}$, so that we find $\Delta x \cdot \Delta p \geqq h$. This, to be sure, is a derivation of eq. (1), but not a proof, for it is now clear that the axioms of quantum mechanics are far more general than the hypothesis of de Broglie. It shows, however, that "wave mechanics" conforms well to these axioms. (The reason for the absence of the factor $1/4\pi$ in the last relation lies in our arbitrary identification of uncertainty with the total range Δx.)

vice versa. In fact, its mean breadth can be fairly accurately estimated from the relation $\Delta E \cdot \Delta t \sim h$.

Another interesting glimpse into the workings of the principle of elementary diffuseness is afforded by a study of what is known as a "pure state." By this we mean simply any physical situation in which all individuals concerned are as much alike as possible in some one respect. For instance, if a number of atoms are all in the same energy state they present a pure state with respect to the energy; or if a number of similar projectiles move in the same direction with the same speed, they form a pure state with respect to momentum. There are many instances in which the quantum mechanical picture of a pure state agrees with the classical one, but there are others where this agreement fails. To anticipate matters, let us think of a swarm of electrons, all moving in the same direction. Each electron is a little magnet. There is, in general, no reason why these little magnets should have their magnetic moments in the same direction; in other words, they will not represent a pure state with respect to the magnetic moment. We shall now perform an experiment (known as the Stern-Gerlach experiment) and discuss its results first from the classical then from the quantum mechanical point of view. We impress a uniform magnetic field upon the stream of electrons, thereby causing alignment of the magnetic moments either parallel or anti-parallel to the field. But a separation of the two component streams has not been effected thereby. To do so the electrons are subjected to a strongly inhomogeneous field where they experience not only torques, but also forces which separate the two components. Hence, when they emerge from the field, they form two distinct beams, each containing, in classical parlance, electrons with their magnetic moments oriented in one direction and therefore representing a pure state.

This interpretation is inconsistent with quantum mechanical notions. They require us to think of these two beams as not entirely homogeneous, but as having merely a preponderance of magnetic axes in one direction, with a definite probability assigned to the observation that, in one of these beams of maximum homogeneity, an electron shall be found with its moment at right angles to the aligning field. Nevertheless, this is the greatest homogeneity attainable. It is seen, therefore, that a pure state in quantum mechanics may not require ideally perfect homogeneity. The principle of elementary diffuseness may operate in such a manner as to make this impossible, as it does in the case here discussed.

One may be tempted to ask the questions: which interpretation corresponds to the facts? Is there no way of deciding? Unfortunately we are here dealing with one of the many instances where experimental decisions are inconclusive, instances which remind us forcibly of the previous inference that the concepts with which we work are not immediate facts but constructs. The only way in which the matter could be decided would be to conduct another Stern-Gerlach experiment wherein each " pure " beam is subjected to a field in a direction different from that which effected the first separation. But here we observe at once that a field which separates will also cause orientation of the axes in the beam and therefore twist some of the electrons into the new direction. We cannot discern the situation without at the same time interfering with it. In a great number of similar instances, this peculiar feature of many experiments prevents decisions on the basis of facts, leaving the alternatives of classical and quantum mechanical interpretation entirely open. Indeed, the frequent occurrence of examples of experimental indecision is regarded by many advocates as a proof of the validity of quantum mechanical interpretation. However, it is difficult to see the logical force of such arguments beyond establishing that the principle of elementary diffuseness is not experimentally precluded.

9.2. Axiomatic Foundation of Quantum Mechanics. *Definitions.* Any object of physical inquiry which, in classical physics, is thought of as an entity as distinct from a mere property, will be termed a *physical system*. Such objects may be capable of direct observation (large-scale bodies, measuring devices, light sources, etc.), or may indicate their presumed existence by a sequence of inferences based upon direct observations (molecules, atoms, electrons, photons, etc.). This notion of a system is largely carried over from previous physical experience; however, for our present purposes it would be well to strip the classical conceptions as much as possible of their inherent picturesqueness and regard them as abstract characterizations of the things which we are going to discuss. Whether or not it is necessary to prohibit imagination completely, and to reject all attempts of visualization, is difficult to say; this is certainly a psychological rather than a logical question. It seems safe to suppose, however, that the least amount of imaginative impediments is desirable if we wish to proceed most easily, for it may always turn out in the light of later developments that qualities with which systems are commonly endowed are untenable. We know, for instance, that Anaxagoras imputed color to his atoms, whereas it is now clear that an atom is about a thousand times smaller than any visible wave length. Let us, then, think of a physical system in the most abstract manner possible.

With a system are associated a great number of properties, or quantities, such as position, energy, momentum, and the like, capable of measurement. These will be called *observables*, although it is not intended to imply that they are observable directly. Every measurement of an observable yields a number. In classical physics this number was assigned to the observable as its only characterization and referred to as its magnitude. Hence the notion of an observable was relatively unimportant because it was synonymous with result of observation. In quantum mechanics we accord to this notion a more independent character inasmuch as we shall associate with it a more general construct (operator) which, to be sure, determines numbers, but not a single number. The relation between these numbers and the results of measurements performed on an observable will form the subject of a later discussion.

We must now define the *state* of a physical system. Classical physics used several unrelated definitions of state, as has been shown, the principal one being the designation of all momenta and all coordinates of the particles composing the system, amounting to an enumeration of $2s$ numbers, if the system has s degrees of freedom. In our present definition we renounce all attempts of specifying the configuration of parts and take a function ϕ of the s coordinates as representing the state. The reader will ask at once: what does this function mean? The answer would be: nothing, if he were trying to associate the function with a distribution of things in space. But we will show later how we can derive, from a knowledge of ϕ, a great amount of useful information regarding the behavior of the system when it is subjected to observation. The point of importance is this: In quantum mechanics the state of a system is no longer defined by means of a number of variables having an immediate intuitive appeal and recalling exact configurations of constituent parts. In fact, it is not defined in terms of observables at all; it is simply a function in configuration space. To use a very crude analogy we might say that in classical physics we had given the biography of a man by describing minutely all his actions and reviewing all his utterances, but that we are now giving a story of the succession of all his thoughts, sentiments, and volitions. If we knew the precise manner of the man's reactions —supposing them to be governed by regularity—we could restore from them all that was contained in the former account of his life. In the case of physical systems we do possess the clue for translating the story of ϕ functions into the story of observations.

This abstract definition of states has an effect upon the character of physical laws. The latter in general connect *states*. In classical physics they connected positions and momenta at different times.

9.2 AXIOMATIC FOUNDATION OF QUANTUM MECHANICS

To be consistent, we must not expect laws of this kind in quantum mechanics, but simply laws connecting ϕ functions at different instants of time. This statement will have an important bearing on the question of causality which will be discussed later.

The independent variables which ϕ contains as arguments are the s coordinates defining the degrees of freedom of the system. This raises the question: if the system is to be so abstract and not to be thought of in the classical sense, how can we know its degrees of freedom? Is it not necessary, for this purpose, to picture the system as a composite and to count the number of constituents together with their modes of interactions? If this has to be affirmed, then quantum mechanics is a logical absurdity, for it will forever be obliged to draw upon the heterogeneous (and alien) store of knowledge of classical physics which it pretends to supersede. But there is no necessity for this affirmation. As in many other problems we can here take the attitude of trial and error with regard to the number of degrees of freedom. Beginning with the smallest number of coordinates which gives any promise, we can start building up ϕ functions, and if then, in applying the rules subsequently to be developed, we find our states incompetent to describe observations, we try a different set of coordinates or a greater number of them. But it is rather premature to go into details regarding this matter before we have discussed our states more fully. It is certainly true, on the other hand, that in its historical development the new theory has leaned heavily upon the old; that even now in setting up a problem an investigator will choose the variables in his ϕ functions in conformity with the classical picture; but from the logical standpoint which we are adopting this is a fortunate coincidence rather than a basic similarity.

Our definition of a state is not yet complete. There are definite limitations to be imposed on the state functions if they are to have physical meaning. These restrictions are purely postulational, although they possess plausibility in view of the axioms of the next paragraph. Two of these restrictions are most fundamental; hence we discuss them here. ϕ, which may in general be a complex function (ϕ^* denotes its complex conjugate), must be quadratically integrable:

$$\int \phi^* \phi \, d\tau \quad \text{exists.} \qquad (A)$$

This integration is to be extended over the fundamental domain of all the variables of which ϕ is a function. $d\tau$, therefore, corresponds to the element of configuration space in classical physics. If the fundamental domain of any of the variables extends to infinity, then

(A) implies in general that ϕ should vanish at infinity in a suitable manner.[1] Condition (A), though necessary, is not always sufficient. The second condition which must be satisfied by every ϕ is that of single-valuedness.

$$\phi \text{ is single-valued.} \qquad (B)$$

Cases in which this condition is important arise whenever one of the variables in ϕ is an angle, for then (B) requires that $\phi(2\pi + \alpha) = \phi(\alpha)$. We conclude our somewhat extensive definition of states by noting that the state functions ϕ need not satisfy any particular known differential equation.

A notion which proves very useful in quantum mechanical discussions is that of an *operator*. Any mathematical operation, like addition, multiplication, differentiation, integration, and so forth, can be represented by a characteristic symbol which is then called an operator. If such a symbol appears in front of a mathematical expression it indicates that the operation in question is to be carried out upon the expression. For instance, if we let D stand for the operation d/dx then Dy means simply dy/dx. Sometimes it makes sense to write these operators independently and without the expression to which they are applied. For instance, if D means d/dx and X means multiplication by x, then $DXf - XDf = f$ is true of a large class of functions f of x and might be written $DX - XD = 1$, where the f is, crudely speaking, "canceled." It is possible to formulate a large portion of quantum mechanics in terms of differential operators (operators signifying differentiations together with multiplications), although it is often convenient to use other operators as well.

Relation between states and experience. After having defined the state of a physical system without any reference to experience, or, more particularly, measurements, we must now construct a bridge relating states to measurements, thereby assigning to the states a physical meaning. This involves a formalism which can be expressed in three fundamental axioms forming the basis of quantum mechanical reasoning.

I. *To every observable p there corresponds an operator P.* In order to ascertain the correct operator which is to be associated with a given observable we must rely on trial. One of the principal concerns of the new discipline has therefore been, and will be in the future, the development of operators for all physical observables. The table which

---

[1] Examples of functions which do not vanish at infinity but have an integrable square have hitherto played no rôle in quantum mechanics.

follows gives an assignment which is sufficient for the solution of many physical problems.

p	P
q_k	q_k.
p_k	$\dfrac{h}{2\pi i}\dfrac{\partial}{\partial q_k}$
m_k	$\dfrac{h}{2\pi i}\left(q_l\dfrac{\partial}{\partial q_m} - q_m\dfrac{\partial}{\partial q_l}\right)$
$H(p, q)$	$H\left(\dfrac{h}{2\pi i}\dfrac{\partial}{\partial q}, q\right)$.

By q_k is meant any Cartesian coordinate; by p_k, the associated momentum. When it is necessary to express a momentum operator or a combination of momentum operators in some other system of coordinates (e.g., polar coordinates) it is best to write the combination first in its Cartesian form and then to transform the differential expression in the usual manner. m_k denotes the observable "angular momentum." It is evident that the corresponding operator is constructed by first expressing m in terms of coordinates and linear momenta ($m_z = xp_y - yp_x$) and then replacing each of the latter by its operator. The same statement refers to $H(p, q)$, the energy in its Hamiltonian form (cf. Sec. 3.14) whose operator is constructed by replacing all Cartesian momenta in the classical Hamiltonian by their operators. This rule seems to be a very general one and may be expected to guide us in the construction of operators not hitherto employed; but it has serious limitations inasmuch as it may become ambiguous, as will be seen later. One further remark should be made regarding the observable "energy." $H(p, q)$ in the usual simple form: kinetic energy + potential energy, is not Lorentz-invariant. In classical physics it is not difficult to find a modified expression which satisfies relativity requirements. How are we to proceed in order to find the invariant operator? We shall answer this question later; at present we merely wish to warn the reader that the golden rule of replacing p's by $h/2\pi i \cdot \partial/\partial q$'s is inapplicable in this case.

In classical mechanics q_k and p_k are known as canonically conjugate variables. The corresponding operators Q_k and P_k which obey the relation

$$P_k Q_k - Q_k P_k = \frac{h}{2\pi i},$$

are said to be canonically conjugate operators, and the same term is applied to all operators with this property.

It is not true that every observable has its own unique operator. The form of the operator will in general also depend on the nature of the system in question. Thus the energy operator for a single point of mass M not subject to forces is

$$-\frac{h^2}{8\pi^2 M}\left[\frac{\partial^2}{\partial x^2} + \frac{\partial^2}{\partial y^2} + \frac{\partial^2}{\partial z^2}\right] \equiv -\frac{h^2}{8\pi^2 M}\nabla^2; \tag{1}$$

if it has a potential energy of the form $V(x, y, z)$, this term has to be added (unmodified because the operator for a coordinate is simply the coordinate itself); the energy operator for two mass points M_1 and M_2 is

$$-\frac{h^2}{8\pi^2}\left(\frac{1}{M_1}\nabla_1^2 + \frac{1}{M_2}\nabla_2^2\right) + V(x_1, y_1, z_1, x_2, y_2, z_2), \tag{2}$$

where ∇_1^2 is the Laplacian involving the coordinates of the first mass point and ∇_2^2 contains those of the second.

We are now ready to state the second fundamental postulate.

II. *The only possible values which a measurement of the observable p can yield are the eigenvalues p_λ of the equation*

$$P\psi_\lambda = p_\lambda \psi_\lambda, \tag{II}$$

where ψ_λ satisfies conditions (A) and (B).

This statement calls for explanatory comment. If P is a differential operator, then (II) is a differential equation. ψ_λ is clearly a function of the variables upon which P operates; p_λ is a number. Now it may occur to the reader that such a differential equation has solutions for *any* value of p_λ, so that it cannot be instrumental in selecting only a certain set. This would indeed be true if no boundary conditions were to be satisfied. The solutions consistent with these boundary conditions do select (in general) only certain values of p_λ. These are called the eigenvalues of the operator P. To clarify this matter let us discuss a simple classical example illustrating the way in which boundary conditions select eigenvalues.

The differential equation of a vibrating string is well known to be

$$\frac{\partial^2 y}{\partial x^2} - \frac{1}{v^2}\frac{\partial^2 y}{\partial t^2} = 0,$$

Here y is the transverse displacement; x, the coordinate of a point along the string; v (=square root of the tension divided by that of the linear density) is the velocity of propagation of the waves along the string; and t, the time. If we are interested in

9.2 AXIOMATIC FOUNDATION OF QUANTUM MECHANICS

harmonic solutions only, we can put $y = f(x)e^{2\pi i \nu t}$ and obtain the ordinary differential equation

$$\frac{d^2 f}{dx^2} + \frac{4\pi^2 \nu^2}{v^2} f = 0,$$

This is precisely of the form of the operator equation in which we are interested, the combination of constants $\frac{4\pi^2 \nu^2}{v^2}$, for which we shall now write p, corresponding to the eigenvalue. There is a solution for every value of p of the familiar type

$$f = A \sin(\sqrt{p}\, x + \gamma),$$

A and γ being arbitrary constants. But most of these solutions are without physical interest, for they represent modes of vibration in which the ends of the string flop about in any manner. We desire solutions satisfying boundary conditions amounting in this case to the requirement that the ends of the string be fixed at all times. Let us, then, seek solutions that conform to this condition. If the end points of the string have the coordinates 0 and l, we require that $f(0) = 0, f(l) = 0$. The first of these relations can be satisfied only by putting $\gamma = 0$, the second by equating $\sqrt{p}\, l$ to $n\pi$, where n is any integer. Hence the only permissible values of p are those given by $p_n = \frac{n^2 \pi^2}{l^2}$; they are the eigenvalues p_n of the operator $P = \frac{d^2}{dx^2}$ for the boundary condition $f(0) = f(l) = 0$. The p's of our example are, of course, related to the wave lengths of the standing waves in the string, for we have seen that $p = \frac{4\pi^2 \nu^2}{v^2}$, which is the same as $\frac{4\pi^2}{\lambda^2}$, if λ is the wave length. If p is restricted to certain values, λ is subject to the analogous restriction, i.e.,

$$\lambda_n = \frac{2l}{n}.$$

This is a familiar result of acoustics. For a more thorough discussion of this matter see Sec. 1.8. Let us now return to the more general operator equation in postulate II.

Whether or not this postulate corresponds to the facts can be decided only on the basis of experiment. Suffice it to say here that the "spectrum" of eigenvalues generated by II has always been found to be the one observed.

What meaning is to be attached to the functions ψ_λ? They are called eigenfunctions of the operator P and define *states* in the sense outlined previously. The states to which they refer are termed *eigenstates*. We do not wish to give the answer to the question as to their meaning here, for it will present itself more naturally when we discuss the final postulate. But it is well to anticipate two mathematical properties of these ψ_λ's, to be proved for special cases later on. They are: (1) their orthogonality, defined by

$$\int \psi_\lambda^* \psi_\mu \, d\tau = 0, \text{ if } \lambda \neq \mu; \tag{3}$$

and (2) their completeness, which can be expressed somewhat crudely by stating that any function of state ϕ referring to the system in question (which therefore contains the same arguments as ψ_λ and satisfies the same boundary conditions) can be expressed as an expansion of the form

$$\phi = \sum_\lambda a_\lambda \psi_\lambda, \tag{4}$$

where the a_λ's are numerical coefficients. To be precise, we should add that relation (3) may break down where there is more than one eigenfunction ψ_λ belonging to the same eigenvalue (degeneracy), but even then it is possible to make these functions orthogonal by choosing linear combinations, so that the generality of (3) is not impaired. The exact meaning of completeness will be examined later.

In concluding the discussion of the second postulate we give a few examples of physically useful operator equations and their solutions.

(a) Let P in eq. (II) represent the operator: x-coordinate. This, as was pointed out, is always simply "multiplication by x," regardless of the nature of the system. Hence the resulting operator equation defines the results of possible x-coordinate measurements performed on any system we wish to imagine, for instance, any particle, charged or neutral. It is

$$x\psi_\lambda(x) = \xi_\lambda \psi_\lambda(x). \tag{5}$$

x is variable, ξ_λ a fixed number. This is an algebraic equation of which the solution is at once evident if we write it in the form: $(x - \xi_\lambda)\psi_\lambda(x) = 0$. Here the first factor on the left is *different* from 0 except when $x = \xi_\lambda$, hence $\psi_\lambda(x)$ must be a function *equal* to 0 unless $x = \xi_\lambda$. At this point it may have any value and still be consistent with (5). Such a function is unknown to mathematicians; it possesses indeed some paradoxical properties.[1] Nevertheless, we shall refer to the construct in question as Dirac's δ function, for while it is an improper function, we can define accurately what we mean by it, viz.:

$$\delta(q) = 0 \text{ if } q \gtrless 0, \quad \delta(q) = \delta(-q), \quad \text{and} \quad \int_{-\infty}^{\infty} \delta(q)dq = 1. \tag{6}$$

The last relation amounts to a definition of the value of the "function" at $q = 0$. If we identify the argument q of δ with $x - \xi_\lambda$ we have what is equivalent to the eigenfunction of the operator x, corresponding to the eigenvalue ξ_λ; i.e.,

$$\psi_\lambda = \delta(x - \xi_\lambda). \tag{7}$$

[1] Cf. Von Neumann, "Math. Grundlagen der Quantenmechanik" (Julius Springer, Berlin, 1932).

9.2   AXIOMATIC FOUNDATION OF QUANTUM MECHANICS

We can also verify this by substituting eq. (7) in eq. (5) and then integrating. Before doing this we remark that one can prove on the basis of (6) the following identity:

$$\int_{-\infty}^{\infty} f(q)\delta(q)dq = f(0), \text{ or more generally, } \int_{-\infty}^{\infty} f(q)\delta(q-a)dq = f(a).$$

With the use of this relation, eq. (5) is seen to be satisfied.

But now, what are the eigenvalues, ξ_λ? ψ_λ already satisfies conditions (A) and (B),[1] and we can form a ψ_λ for every ξ_λ. Hence ξ_λ is not restricted at all in this case; any value of x may be observed, *the coordinate operator has a continuous spectrum*. We shall see that this is not true for all operators. In the remaining examples we shall meet with more familiar functions.

(*b*) What are the eigenvalues of the linear momentum operator P_x? From the table on p. 405 we see that the operator equation reads

$$\frac{h}{2\pi i}\frac{d\psi_\lambda}{dx} = p_\lambda \psi_\lambda. \tag{8}$$

It has the solution

$$\psi_\lambda = \text{constant} \cdot e^{\frac{2\pi i}{h}p_\lambda x}. \tag{9}$$

To determine the possible values of p_λ it is necessary to consider the fundamental range of x. Let this be the domain $-l \leq x \leq l$. Conditions (A) and (B) then merely require that

$$\int_{-l}^{l} \psi_\lambda^* \psi_\lambda dx = 2|\text{constant}|^2 l$$

shall exist, and this is evidently true if the constant is finite.[2] Hence here again we find that all values of p_λ may occur; λ does not specify a discrete set of momenta.

(*c*) Let us now consider the eigenvalues of the angular momentum operator. If the reader desires a classical picture he may, for sim-

---

[1] In this connection as well as many others the "δ function" has to be handled with more than mathematical tact, for if we calculate $\int_{-\infty}^{+\infty}[\delta(x-\xi_\lambda)]^2 dx$ the result is $\delta(0)$, which is of doubtful existence. In such circumstances one must replace $\delta(x-\xi_\lambda)$ by a true analytic function with somewhat similar behavior, such as $\frac{a}{\sqrt{2\pi}}e^{-a^2(x-\xi_\lambda)^2}$, with a large but not infinite.

[2] l may be very large, but to avoid mathematical difficulties, we shall not assume it to be infinite; cf. Sec. 9.6.

plicity, think of a mass point revolving in the $x\,y$ plane. The operator equation is

$$\frac{h}{2\pi i}\left(x\frac{\partial}{\partial y} - y\frac{\partial}{\partial x}\right)\psi_\lambda = m_\lambda \psi_\lambda. \qquad (10)$$

If we introduce polar coordinates, putting $x = r\cos\theta$, $y = r\sin\theta$, we have $\dfrac{d}{d\theta} = -r\sin\theta\,\dfrac{\partial}{\partial x} + r\cos\theta\,\dfrac{\partial}{\partial y} = x\,\dfrac{\partial}{\partial y} - y\,\dfrac{\partial}{\partial x}$. Therefore the equation in polar coordinates is simply

$$\frac{h}{2\pi i}\frac{d}{d\theta}\psi_\lambda = m_\lambda \psi_\lambda. \qquad (10')$$

It has the solution

$$\psi_\lambda = \text{constant}\cdot e^{\frac{2\pi i}{h}m_\lambda \theta}. \qquad (11)$$

Any value of m_λ will cause ψ_λ to satisfy condition (A), but not condition (B). For we require that

$$\psi_\lambda(\theta) = \psi_\lambda(\theta + 2\pi),$$

hence

$$e^{\frac{2\pi i}{h}m_\lambda \theta} = e^{\frac{2\pi i}{h}m_\lambda(\theta+2\pi)}, \quad \therefore\ e^{\frac{4\pi^2 i}{h}m_\lambda} = 1.$$

This can be true only if $m_\lambda = \lambda h/2\pi$, where λ is an integer. This, the reader will observe, is simply Bohr's condition for the angular momentum of an electron. In the present theory it results most naturally from the condition that the function describing the state shall be single-valued.

(d) Lastly, let us consider the differential equation arising from the energy operator:

$$H\psi_\lambda = E_\lambda \psi_\lambda. \qquad (12)$$

Owing to the difference in the structure of the operator H this equation takes on a variety of forms depending on the nature of the system to which H refers. Usually, however, it will be a second order equation in which the Laplacian ∇^2 figures prominently. The most important forms of eq.(12) were strikingly discovered by Schrödinger even before the point of view here presented was known. Eq. (12) is therefore generally known as Schrödinger's equation. Its importance for the solution of physical problems is great, for we are usually concerned with the calculation of energies. For instance, the values of E_λ in eq. (12) are the energy levels encountered in spectroscopy if H refers to the atom or molecule in question. No generally valid statements

9.2 AXIOMATIC FOUNDATION OF QUANTUM MECHANICS

can be made about the E_λ; for some systems they form a discrete set, for others a continuous one, corresponding to the experimental existence of discrete and continuous spectra. In most examples both discrete and continuous energies occur. We shall postpone the detailed solution of special forms of Schrödinger's equation to a more suitable time. Let us now conclude the presentation of fundamental postulates by stating the remaining one.

III. *When a given system is in a state ϕ, the expected mean of a sequence of measurements on the observable p is*

$$\bar{p} = \frac{\int \phi^* P \phi d\tau}{\int \phi^* \phi d\tau}, \qquad \text{(III)}$$

where P is the operator corresponding to p.

The term "expected mean" is here used in the sense defined in Chapter IV. It implies that measurements form a probability aggregate. The multiple repetition of the measurement involved in the idea of a probability aggregate may occur either in time or in space: in the former case $\bar{p}$ is the mean of the numerical results of a great number of experiments on the same system in the state in question; in the second case it is the mean of the numerical results of simultaneous observations on a great number of similar systems in the same state. Most physical measurements explained by quantum mechanics are of the second type; hence $\bar{p}$ is practically always understood in the sense of an average over the results of simultaneous observations.

But there comes to mind at this point a troublesome question: how can we know that the systems are in the same state if states are not defined in terms of observable things? Are we not simply begging the question, or at best, defining states by this postulate? We are certainly not begging the question, for the next paragraph will show that our statement is a positive, indeed a synthetic, proposition capable of producing non-tautological theorems. The question can be answered correctly in two ways.

First: Postulate III defines states. One must then regard our previous definition of "states" in terms of ϕ functions as a definition of the *symbol* of state only, and as requiring a complement in the form of the present postulate. This is a possible standpoint, and will probably be taken by all those who insist that every good physical definition must be in terms of operations or measurements.[1] The authors of this book incline in the other direction, believing that this

---

[1] This point of view has been very aptly set forth by P. W. Bridgman in his book: "The Logic of Modern Physics" (Macmillan, 1927).

restriction would involve a short-sighted curtailment of the realm of physics, and that such definitions are not always feasible, although many of them appear to be so at the present time. Moreover, this restriction would separate physics sharply from many other fields of human interest where definitions in terms of operations are often impossible.[1]

From this point of view one must give the second answer: Mathematical concepts, such as ϕ functions, are perfectly proper tools for defining physical states, although their meaning in terms of operations is obscure. Postulate III, from this angle, is not a definition, but *forms the necessary link between states and experience.* In order to be able to identify systems which are in the same state we need merely assume that equal experimental processes of preparation produce identical states—a regularity in the behavior of nature which forms the basis of all science. If anyone still insists that, after all, we do not know that the systems were in the same state before the process of preparation was applied, we can only reply that this is a working hypothesis with which no theory, not even classical, can dispense. At any rate, if the reader feels safer in taking the former stand he may well do so; the formulation of quantum mechanics as we are presenting it is independent of this metaphysical detail. The question is important, however, from a philosophical point of view.

We shall henceforth refer to postulate III somewhat loosely as the "mean value relation." This term is suggested by the formal analogy of the expression $\dfrac{\int \phi^* P \phi d\tau}{\int \phi^* \phi d\tau}$ with the usual expression of an expected mean. If, for instance, the intrinsically positive expression $\phi^*\phi$ were interpreted as a probability density, and P as a function of configuration space, III would be an identity. But this is much too specific an interpretation, since P is an operator.

III implies further conditions on state functions. We have already observed that conditions (A) and (B) are necessary but not sufficient. We now have a criterion for examinining ϕ functions as to their suit-

---

[1] We are aware that these remarks are open to criticism on the score that definitions in terms of operations are characteristic of a science, and that we should not compare scientific definitions with unscientific ones. But this leads to the more serious difficulty regarding logical coherence of scientific concepts. On the basis of purely operational definitions, all concepts are strictly empirical and *isolated*; in fact, a length measured by a meter stick would be a concept generically different from an astronomical length, and there would be no justification for reckoning both in the same units. For a detailed discussion cf. H. Margenau, "Methodology of Modern Physics," *Philosophy of Science*, 2, 48; 2, 164, 1935.

ability for any particular physical problem at hand. If we want to represent a state with a finite energy, or a finite momentum, we must choose state-functions for which not only $\int \phi^*\phi d\tau$, but also $\int \phi^* H \phi d\tau$, or $\int \phi^* \dfrac{h}{2\pi i} \dfrac{\partial}{\partial q} \phi d\tau$, exist.

Postulate III, which alone forms the connection between observable quantities and state-functions, allows an interesting conclusion. If ϕ were multiplied by any constant, the value of $\bar{p}$ would remain unchanged. Hence ϕ is arbitrary to within a constant multiplier. One can take advantage of this fact in simplifying the form of the theory. Choosing the constant so that condition (A) becomes

$$\int \phi^* \phi d\tau = 1, \tag{13}$$

the mean value relation takes the form

$$\bar{p} = \int \phi^* P \phi d\tau. \tag{III'}$$

Henceforth every ϕ function will be assumed to satisfy (13), i.e., to be *normalized*. Thus, condition (A) may now be combined with the orthogonality relation (3) in the form

$$\int \psi_\lambda^* \psi_\mu d\tau = \delta_{\lambda\mu} \tag{14}$$

where $\delta_{\lambda\mu}$, known as the "Kronecker symbol," has the value 1 when $\lambda = \mu$, zero otherwise. It will be observed that this procedure has not yet removed all arbitrariness from the choice of ϕ's. We are still entitled to multiply ϕ by any constant factor $e^{i\theta}$, without disturbing III' or (13). State-functions differing by a constant of absolute value unity are therefore empirically indistinguishable.

9.3. Some Useful Theorems. The three postulates of the preceding section are pregnant with implications which will now be worked out in the form of propositions and proofs. In connection with the proofs it is interesting to realize that no positive information except that contained in the foregoing section is to be used. This circumstance appears to us as a favorable commentary on the logical simplicity of quantum mechanics.

Prop. 1. The formal "mean" of the operator P, viz., $\int \phi^* P \phi d\tau$, is equivalent to an average over all possible eigenvalues of P with weights depending upon the function ϕ (state of the system).

The proof is simple. Expand ϕ by (9.2-4): $\phi = \sum_\lambda a_\lambda \psi_\lambda$, and

choose ψ_λ to be an eigenfunction of the operator in question, P; in symbols:
$$P\psi_\lambda = p_\lambda \psi_\lambda. \tag{1}$$
Then
$$\int \phi^* P \phi d\tau = \int \sum_\lambda a_\lambda^* \psi_\lambda^* P \sum_\mu a_\mu \psi_\mu d\tau = \sum_{\lambda,\mu} a_\lambda^* a_\mu \int \psi_\lambda^* P \psi_\mu d\tau.$$

Now because of (1) and (9.2-13) the integral reduces to p_λ when $\mu = \lambda$; otherwise it is 0. Hence there remains a single sum instead of a double one, and the result is
$$\int \phi^* P \phi d\tau = \sum_\lambda |a_\lambda|^2 p_\lambda, \tag{2}$$
whence the proposition is proved.

This result, considered in connection with III′, is indeed suggestive. Can we regard the positive coefficients $|a_\lambda|^2$ as the frequencies with which the various possible values p_λ occur in our measurements? This is of course not an obvious consequence of (2) and III′, for there are many different ways of choosing coefficients $|a_\lambda|^2$ which satisfy the equation
$$\bar{p} = \sum_\lambda |a_\lambda|^2 p_\lambda,$$
but only one set can define the true probability distribution (cf. Chapter IV).

Prop. 2. The $|a_\lambda|^2$ define the probability distribution of the p_λ; i.e., the $|a_\lambda|^2$ are the probabilities in the probability aggregate of which measurements of the observable p are the elements, p_λ the properties.

Proof: Suppose the set of numbers $w_\lambda (w_\lambda \geq 0)$ defines the true probability distribution of the p_λ's. Then, by definition,

or in general
$$\begin{array}{ll} \sum_\lambda w_\lambda = 1, & \sum_\lambda w_\lambda p_\lambda = \bar{p}, \\ \sum_\lambda w_\lambda p_\lambda^r = \bar{p^r}. & r = 1, 2, \ldots. \end{array} \tag{3}$$

On the other hand, we know that $\sum_\lambda |a_\lambda|^2 = 1$, and that in general
$$\bar{p^r} = \int \phi^* P^r \phi d\tau = \sum_\lambda |a_\lambda|^2 p_\lambda^r, \quad r = 1, 2, \ldots. \tag{4}$$

This follows at once if we choose for the operator in III′ that corresponding to the observable "rth power of p." P^r designates then simply an r fold iteration of the operator P. But as a result of II,
$$P^r \psi_\lambda = P^{r-1}(P\psi_\lambda) = P^{r-1}(p_\lambda \psi_\lambda) = p_\lambda P^{r-2}(P\psi_\lambda)$$
$$= p_\lambda^2 P^{r-2} \psi_\lambda = \ldots = p_\lambda^r \psi_\lambda.$$

Hence by expanding ϕ in (4) in terms of the ψ_λ's and then making use of this last identity one verifies (4) at once. The fact that the sum of the squares of the a_λ equals 1 follows from the normalization of the function to which the a_λ's refer:

$$1 = \int \phi^* \phi d\tau = \sum_{\lambda,\mu} \int a_\lambda^* \psi_\lambda^* a_\mu \psi_\mu d\tau = \sum_{\lambda,\mu} a_\lambda^* a_\mu \delta_{\lambda\mu} = \sum_\lambda |a_\lambda|^2.$$

On combining eqs. (3) and (4) we obtain the system of equations

$$\sum_\lambda (|a_\lambda|^2 - w_\lambda) p_\lambda^r = 0, \quad r = 1, 2, 3 \ldots$$

The p_λ here are given real numbers, all different. If these equations are to be true, either all the differences $|a_\lambda|^2 - w_\lambda$ vanish, or the determinant

$$\begin{vmatrix} 1 & 1 & 1 & 1 & \cdots \\ p_1 & p_2 & p_3 & p_4 & \cdots \\ p_1^2 & p_2^2 & p_3^2 & p_4^2 & \cdots \\ \cdot & \cdot & \cdot & \cdot & \\ \cdot & \cdot & \cdot & \cdot & \end{vmatrix} = 0.$$

The latter is certainly not in general the case; hence

$$|a_\lambda|^2 = w_\lambda. \tag{5}$$

Corollary:

$$w_\lambda = |a_\lambda|^2 = \left| \int \psi_\lambda^* \phi d\tau \right|^2. \tag{6}$$

For writing $\phi = \sum_i a_i \psi_i$, which defines the a_i (called coefficients in the development of ϕ in the system of ψ's), we have

$$\int \psi_\lambda^* \phi d\tau = \sum_i \int \psi_\lambda^* a_i \psi_i d\tau = \sum_i a_i \delta_{i\lambda} = a_\lambda.$$

We are now in a position to calculate the relative frequency with which a given eigenvalue of any observable will be measured when a great number of measurements on the system is made, provided we are given its state ϕ. For this purpose we either expand ϕ in terms of the eigenstates of the observable, pick out the coefficient belonging to the eigenvalue in question, and square its absolute value; or we perform the integration in eq. (6) (which the reader will probably recognize as the customary formula for the development coefficients of a function ϕ in an orthogonal and normal system ψ_λ).

Prop. 3. The probability (relative frequency) of finding the system, which is in the state $\phi(q_1, q_2, \ldots q_s)$, at the point $(\xi_1, \xi_2, \ldots \xi_s)$ is given by $\phi^*(\xi_1 \ldots \xi_s)\phi(\xi_1 \ldots \xi_s)$.

Proof: We have shown that the eigenstates belonging to the coordinate operator x have the form $\delta(x - \xi_\lambda)$, where the eigenvalues ξ_λ have a continuous distribution; cf. eq. (9.2-7). By an obvious generalization of the argument leading to eq. (9.2-7) we find that the eigenfunctions belonging to the operator $(q_1, q_2, \ldots q_s)$ corresponding to the observable: "simultaneous occupation of the points $q_1, q_2, \ldots q_s$" (s being the number of degrees of freedom) are simply products of δ functions, viz.,

$$\delta(q_1 - \xi_{1\lambda}) \cdot \delta(q_2 - \xi_{2\lambda}) \ldots \delta(q_s - \xi_{s\lambda}).$$

The eigenvalues $\xi_{i\lambda}$ form again a continuous distribution. Let us now refer to some one particular λ, so that we can omit the index λ, meaning by ξ_1 some fixed value of q_1 and by ξ_2 some fixed value of q_2, etc. The probability sought is then, by eqs. (5) and (6),

$$\left| \int \phi(q_1, q_2 \ldots q_s) \delta(q_1 - \xi_1) \ldots \delta(q_s - \xi_s) dq_1 dq_2 \ldots dq_s \right|^2$$
$$= \phi^*(\xi_1 \ldots \xi_s) \phi(\xi_1 \ldots \xi_s),$$

which proves our proposition.

If, in particular, ϕ is the state of a single particle and therefore a function only of x, y, z, $|\phi(x, y, z)|^2$ is the probability that the particle be found at the point x, y, z of ordinary space. Hence it is permissible, in this simple case, to think of the square of the state function as a distribution of mass, or, if the particle be an electron, as a distribution of charge in space. But this simple interpretation is not always applicable and is, in view of III, a very special notion which does not exhaust the meaning of a ϕ function.

Prop. 4. Certainty of measuring the eigenvalue p_k exists if, and only if, the state function

$$\phi = e^{i\alpha} \psi_k, \tag{7}$$

where α is a real constant. This statement is almost obvious, for if certainty is to exist, there can be only one a_λ different from 0, namely a_k, and this must have an absolute value unity. Eq. (7) is the only ϕ function which satisfies this requirement. This fact throws an interesting light upon eigenfunctions: they characterize states which yield certainty with regard to the outcome of at least one kind of measurement, that of the observable of which the state is an eigenstate. In the preceding discussion the latter have been denoted by ψ, while any state in general was designated by the symbol ϕ. We shall continue this usage of letters wherever feasible, but it should be borne in mind that there is no real distinction between states and eigenstates. Presumably every ϕ is an eigenstate of some operator, although it may be difficult (or perhaps impossible) to find it, or the operator may not be of physical interest. When we say that a state is an eigenstate of some operator we mean that we know, and are interested in, this operator.

Suppose that a physical system is in an eigenstate belonging to one observable, say the energy; will it then be true, also, that some other observable, like the position of the system, will yield one definite value without spread upon measurement? In other words, can one state be a simultaneous eigenstate of several observables? This is indeed possible, but only under a rather special condition:

Prop. 5. If P and Q are permutable operators, the eigenstates belonging to P and Q are simultaneous eigenstates; that is, if the state of the system is such that a value p_i will be measured with certainty, then one value q_i will also be measured with certainty. Two operators are said to permute when their order of operation may be reversed without changing the result.[1] The proof of the proposition is as follows:

Let the eigenstates of P be denoted by ψ_λ, those of Q by χ_μ. We then know that the following equations are satisfied by hypothesis:

(a) $P\psi_\lambda = p_\lambda \psi_\lambda$, (b) $Q\chi_\mu = q_\mu \chi_\mu$, (c) $PQ = QP$,

The state function is one of the ψ_λ's, say ψ_i. Then, because of (c) and (a),

$$QP\psi_i = PQ\psi_i = p_i Q\psi_i.$$

If we write the last two members of this equation in the form

$$P(Q\psi_i) = p_i(Q\psi_i)$$

it simply tells us that $Q\psi_i$ is an eigenfunction of the operator P, namely, that belonging to the ith eigenstate. In other words, $Q\psi_i = $ constant$\cdot \psi_i$. Comparing this with (b) we are forced to identify ψ_i with one of the χ_μ's, say χ_j, and the constant with q_j. Hence

$$\psi_i = \chi_j.$$

If now we apply the mean value relation to find the observed value of the q observable,

$$\bar{q} = \int \psi_i^* Q\psi_i d\tau = \int \chi_j^* Q\chi_j d\tau = q_j,$$

i.e., the value q_j is measured with certainty.

The opposite case, in which P and Q are not permutable, is also of interest. Then the above proof cannot be conducted, and we infer

---

[1] Thus coordinate operators are permutable because the order in which multiplications are performed is immaterial; but the operator $\partial/\partial x$ does not permute with x.

that certainty of measurement for one observable will in general imply uncertainty for another. In fact, the present formalism permits an answer to the more general question: for any given state ϕ, how is the precision in the measurements of one observable related to the precision in the measurements of another? (By precision we mean, in keeping with Chapter IV, a measure of the smallness of the range over which the various measurements are scattered.) We shall here discuss this question as it regards two canonically conjugate observables, i.e., those whose corresponding operators are canonically conjugate.

But what quantity can we take for precision? If the distribution of the aggregate of measurements were a normal one ("Gauss error law") we could choose the conventional definition. This is not, however, the case in general. Therefore, in the absence of a more suitable alternative, we shall take the statistical dispersion of the measurements, in our terminology $\overline{(p - \bar{p})^2}$ or for short $\overline{(\Delta p)^2}$, if p is the observable, as a measure of the reciprocal precision, i.e., "uncertainty." But in view of the many abuses to which this last term has been subjected we shall continue to refer to dispersion or mean square deviation in its well-established mathematical sense. On the basis of these conventions we shall now prove the following proposition.

Prop. 6. If P and Q are canonically conjugate operators, then

$$\overline{(\Delta p)^2} \cdot \overline{(\Delta q)^2} \geq \frac{h^2}{16\pi^2}. \qquad (8)$$

The present demonstration requires that P and Q are operators satisfying the relation

$$\int \phi^* P \phi d\tau = \int (P^*\phi^*) \phi d\tau; \quad \int \phi^* Q \phi d\tau = \int (Q^*\phi^*) \phi d\tau, \qquad (a)$$

if ϕ has properties (A) and (B), (as for instance all state-functions do). (a) will be assumed at present; in Sec. 9.5 we shall show that all quantum mechanical operators conform to this relation.

We first proceed to establish the inequality

$$\int u^* u d\tau \cdot \int v^* v d\tau \geq \tfrac{1}{4}[\int (u^*v + v^*u) d\tau]^2, \qquad (b)$$

where u and v are any integrable functions of those coordinates on which the state-function for the system in question depends. This is done as follows: let λ be a real variable not depending on coordinates. Then

$$\int (\lambda u + v)(\lambda u^* + v^*) d\tau$$

9.3 SOME USEFUL THEOREMS

is always positive or zero, the latter only for the trivial case where u is directly proportional to v which we are here excluding. Hence

$$\lambda^2 \int u^* u d\tau + \lambda \int (u^* v + u v^*) d\tau + \int v^* v d\tau > 0 \text{ for every real } \lambda. \quad (c)$$

Consider the left-hand side as a quadratic form in λ. We know that it can have no real roots. But the roots of $(a\lambda^2 + b\lambda + c)$ are $-\dfrac{b}{2a} \pm \dfrac{1}{2a}\sqrt{b^2 - 4ac}$; they are real unless

$$4ac > b^2.$$

Identifying the coefficients a, b, and c with the integrals in (c) we arrive at once at the inequality (b).

Now suppose that the system is in a state ϕ. If then we define

$$u = (P - \bar{p})\phi; \quad v = i(Q - \bar{q})\phi,$$

the inequality (b) reads

$$\int (P - \bar{p})^* \phi^* (P - \bar{p}) \phi d\tau \cdot \int (Q - \bar{q})^* \phi^* (Q - \bar{q}) \phi d\tau$$
$$\geq \tfrac{1}{4} [i \int (P - \bar{p})^* \phi^* (Q - \bar{q}) \phi d\tau - i \int (Q - \bar{q})^* \phi^* (P - \bar{p}) \phi d\tau]^2,$$

or, upon using (a) ($\bar{p}$ and $\bar{q}$ are simply numbers!) and canceling terms on the right,

$$\int \phi^* (P - \bar{p})^2 \phi d\tau \cdot \int \phi^* (Q - \bar{q})^2 \phi d\tau \geq -\tfrac{1}{4} [\int \phi^* (PQ - QP) \phi d\tau]^2. \quad (d)$$

But what is the meaning of a quantity like $\int \phi^*(P - \bar{p})^2 \phi d\tau$? By prop. 1 (eq. 2) it is equivalent to $\sum_\lambda |a_\lambda|^2 (p_\lambda - \bar{p})^2$, and since prop. 2 tells us that the $|a_\lambda|^2$ are probabilities, the quantity is simply $\overline{(\Delta p)^2}$, the dispersion of the measured values of the observable p about their expected mean $\bar{p}$ when the system is in the state ϕ. Hence (d) may be written

$$\overline{(\Delta p)^2} \cdot \overline{(\Delta q)^2} \geq -\tfrac{1}{4} [\int \phi^* (PQ - QP) \phi d\tau]^2. \quad (e)$$

This is a theorem of great importance, more general indeed than the one we set out to derive. It contains prop. 5 as a special case, for if P and Q are permutable the right-hand side of (e) vanishes, and it is possible for $\overline{(\Delta p)^2}$ and $\overline{(\Delta q)^2}$ to be 0. This means, of course, that there is but one outcome of the measurement.

If P and Q are operators belonging to canonically conjugate observables, like momentum and position, we have

$$PQ\phi - QP\phi = \frac{h}{2\pi i} \frac{\partial}{\partial q}(q\phi) - q \frac{h}{2\pi i} \frac{\partial}{\partial q} \phi = \frac{h}{2\pi i} \phi,$$

and (e) becomes
$$\overline{(\Delta p)^2} \cdot \overline{(\Delta q)^2} \geqq \frac{h^2}{16\pi^2}$$
which is identical with (8).

Eq. (9.1–1) which was derived (cf. footnote on p. 399) on the basis of entirely different assumptions, is now seen to be a special example of the more general proposition 6. The reason for the appearance of the factor h in the simple relation $\Delta x \cdot \Delta p \geqq h$, in place of $h/4\pi$ in (9.3–8) lies in the use of different definitions of "uncertainty."

9.4. Remarks on Prop. 6. Uncertainty. An inequality similar to (9.3–8) was first discovered by Heisenberg, and has been made the basis of a far-reaching and fruitful system of analogies, known as indeterminacy relations, which the reader will find discussed in most treatments of quantum mechanics. To sum up the situation: If the state of a physical system is such that measurements on a given observable p have a standard deviation ($= \sqrt{\overline{(\Delta p)^2}}$ equal to δp, then the results of measurements on the canonically conjugate observable q have a standard deviation δq of *at least* $h/4\pi \delta p$. In particular if the state is an eigenstate of the observable p, so that $\delta p = 0$, then the "spread" of the measurements on q is infinite: the latter observable cannot be measured with any precision at all. It is tempting to interpret this result as follows. Suppose the system is, at present, in an eigenstate of p. In order to ascertain this fact it is necessary to make an observation. But every observation amounts to an interaction between the system and a measuring device and will therefore alter its "state" in such a way that the canonically conjugate observable can no longer be measured with any precision. To illustrate the destruction of a sharply defined "state" by observation, many thought experiments have been devised which support this point of view. Let it be required, for instance, to determine the coordinates and the momentum of an electron. Because of its smallness, only radiation of very short wave length can be used to measure the electron's position (radiation of wave length larger than the electron's diameter would not be reflected by it and fail to betray its presence). But if this is done, the electron will, by virtue of the great momentum of short-wave-length radiation, experience a recoil which changes completely its initial state of motion and therefore precludes every possibility of determining it. The difficulty arises from the fact that the momentum p of the measuring agency (light ray) is related to its size (wave length) by the relation $p = h/\lambda$, (since $p = h\nu/c$, and $\nu/c = 1/\lambda$). Many similar experiments lead to the same dilemma, always because canonically conjugate quantities, pertaining to ele-

9.4 REMARKS ON PROP. 6. UNCERTAINTY 421

mentary particles, if multiplied together, give a product of the dimensions and order of magnitude h. These observations are certainly correct, but are they the equivalent to prop. 6? Is the uncertainty attending the measurement of q conditioned by the destruction of the "state" through a measurement of p?

In the foregoing statements we have placed the word state in quotation marks, for clearly we were using the term in its classical sense. Quantum mechanically, coordinates and momentum of an electron do not define its state. Hence we have been guilty, in the immediately preceding remarks, of carrying classical terminology into the discussion of a problem which is foreign to classical concepts. We feel that it is essential to guard against such confusion. Nevertheless it is possible to restate the matter using the correct definition of states in terms of ϕ functions. We shall then have to answer the question: will measurements change the state-function of the system?

The answer cannot be derived from the postulates given so far, for we have not yet considered the manner in which ϕ functions change in time. Hitherto we have been concerned with a classification of possible states and the behavior of systems when one of these states is realized. Later (Sec. 9.11) we shall learn how, in interactions where the energy operator involves the time, ϕ functions are modified in time. The answer to our query will then be in the affirmative: quantum mechanical states will change upon interaction with measuring devices.

But proposition 6 is evidently not concerned with such interactions, for it follows from purely static hypotheses involving no time changes. When a given state is realized, the experimental correlate of this state is a number of probability aggregates, one for every observable, with definitely assignable properties and probabilities. This assignment does not depend on the circumstance whether or not observations have been made, or what observation has been made first. Experimental indeterminacy as to the outcome of a given measurement is inherent in the fundamental manner of describing states. Moreover, it is by no means true that predictions regarding experience must always appear in the form of an inequality like prop. 6; this indefiniteness is incidental to the particular form of the proposition. We can, in general, calculate accurately the entire distribution within every probability aggregate, not merely a lower limit of its dispersion. The mean value relation III provides all this information.

Finally, if the scattering of measurements were caused by the destruction of states, this scattering would necessarily be influenced by the choice of measuring apparatus, and by the skill of the observer. If the postulates of quantum mechanics permitted only the calculation of the greatest precision, i.e., the smallest standard deviation attainable, thereby affording a margin for experimental inaccuracy, the situation could well be explained by the assumption of destruction of states. But, as we have seen, this is not the case. The rules for an assignment of probabilities take no account of the choice of apparatus. When an electron is in a given quantum mechanical state the probabilities of its detection at the various points of space are fixed, and the probabilities of momentum measurements are definite; these assignments remain the same independently of the choice of wave length used in measuring positions and momenta. We therefore conclude that the scattering of measurements has its roots in a fact more fundamental than the destruction of states by interaction with measuring devices, namely, in the definition of states peculiar to quantum mechanics.

The point of view regarding uncertainty here presented is not one predominantly held by physicists who appear somewhat loath to abandon the classical description of states. But we feel that an attempt at logical clarity and simplicity leads to it compellingly. At any rate, it seems to avoid a great deal of confusion.

9.5. Hermitean Operators. Having acquainted ourselves with the fundamental facts of quantum mechanics, let us now examine some of the details of its mathematical apparatus. There is an important class of operators, called Hermitean operators, which play a special rôle in physics. If u and v are any two functions satisfying given boundary conditions—in our case the latter are conditions (A) and (B)—then P, an operator affecting the variables of u and v, is said to be Hermitean if

$$\int u^*(Pv)d\tau = \int (P^*u^*)v d\tau. \tag{1}$$

(A *real* operator is therefore Hermitean if its position inside the integral is immaterial.) The importance of these operators lies in the fact that they have real eigenvalues. To prove this, let u be one of the eigenfunctions of P, so that

$$u = \psi_k, \quad P\psi_k = p_k\psi_k, \quad P^*\psi_k^* = p_k^*\psi_k^*.$$

Then we have, on expanding v in terms of the ψ's,

$$\int u^*(Pv)d\tau = \int \psi_k^* P \sum_\lambda a_\lambda \psi_\lambda d\tau = a_k p_k$$

$$\int (P^*u^*)v d\tau = \int (P^*\psi_k^*) \sum_\lambda a_\lambda \psi_\lambda d\tau = a_k p_k^*.$$

If the two results are to be equal, a fact which follows from the Hermitean character of P, then the eigenvalues p_λ must all be real, for k

may be taken to be any one of the λ's. Since eigenvalues, representing possible values of measurements, must be real, we are safe in physics if we admit only Hermitean operators. It is now necessary to show that the choices already made correspond to this restriction. Let us prove, therefore, that the coordinate, momentum, and energy operators are Hermitean.

(a) Since q is real, $q^* = q$. Hence $\int u^* q v d\tau = \int q^* u^* v d\tau$.

(b) If q is a Cartesian coordinate [for simplicity we shall suppose its fundamental range to be $(-\infty, \infty)$] the momentum operator is $\dfrac{h}{2\pi i}\dfrac{\partial}{\partial q}$.

Now
$$\int u^* \left(\frac{h}{2\pi i}\frac{\partial v}{\partial q}\right) d\tau = \frac{h}{2\pi i}\int_\sigma u^* v d\sigma - \frac{h}{2\pi i}\int \frac{\partial u^*}{\partial q} v d\tau.$$

But the surface integral, to be extended over the infinite boundary at $q = \pm\infty$, vanishes, and the last integral is simply

$$\int \frac{h}{2\pi(-i)}\frac{\partial u^*}{\partial q} \cdot v d\tau.$$

This proves the momentum operator to be Hermitean.

(c) The usual form of the energy operator for a system of f degrees of freedom is

$$H = \sum_{\lambda=1}^{f} H_\lambda + V(q_1, q_2 \ldots q_f), \tag{2}$$

where
$$H_\lambda = -\frac{h^2}{8\pi^2 m_\lambda}\frac{\partial^2}{\partial q_\lambda^2}$$

and V is a function of the coordinates alone. Then

$$\int u^* H v d\tau = \sum_\lambda \int u^* H_\lambda v d\tau + \int u^* V v d\tau. \tag{3}$$

Writing each term of the sum explicitly,

$$\int u^* H_\lambda v d\tau = -\frac{h^2}{8\pi^2 m_\lambda}\int\cdots\int^{(f)} u^* \frac{\partial^2}{\partial q_\lambda^2} v dq_1 dq_2 \ldots dq_f.$$

On performing a partial integration over q_λ this becomes

$$-\frac{h^2}{8\pi^2 m_\lambda}\int\cdots\int^{(f-1)} \left[u^*\frac{\partial v}{\partial q_\lambda}\bigg|_{q_\lambda=\infty} - u^*\frac{\partial v}{\partial q_\lambda}\bigg|_{q_\lambda=-\infty}\right] dq_1 \ldots dq_{\lambda-1} dq_{\lambda+1} \ldots dq_f$$

$$+ \frac{h^2}{8\pi^2 m_\lambda}\int\cdots\int^{(f)} \frac{\partial u^*}{\partial q_\lambda}\frac{\partial v}{\partial q_\lambda} d\tau.$$

The first integral vanishes. If we perform another, similar partial integration with respect to q_λ in the second, the remaining term takes the form

$$-\frac{h^2}{8\pi^2 m_\lambda}\int \overset{(s)}{\cdots}\int \frac{\partial^2 u^*}{\partial q_\lambda^2} v \, dq_1 \ldots dq_s,$$

which is the same as $\int (H_\lambda^* u^*) v \, d\tau$. Hence we have demonstrated the Hermitean character of each term, in the sum of (3). Since V is real and contains no differentiations, it is certainly Hermitean. Therefore $\int u^* H v \, d\tau = \int (H^* u^*) v \, d\tau \, (= \int (Hu^*) v \, d\tau)$.

The class of operators under discussion has another interesting property. The eigenfunctions belonging to different eigenvalues of an Hermitean operator are orthogonal. Let p_l and p_k be two different eigenvalues of the operator P in question. Then

$$P\psi_l = p_l \psi_l,$$

hence

$$\int \psi_k^* P \psi_l \, d\tau = p_l \int \psi_k^* \psi_l \, d\tau.$$

Also

$$P\psi_k = p_k \psi_k, \quad P^* \psi_k^* = p_k \psi_k^*,$$

since p_k is real; hence

$$\int \psi_l P^* \psi_k^* \, d\tau = p_k \int \psi_k^* \psi_l \, d\tau.$$

But P being Hermitean,

$$\int \psi_l P^* \psi_k^* \, d\tau = \int \psi_k^* P \psi_l \, d\tau = p_l \int \psi_k^* \psi_l \, d\tau,$$

so that

$$p_l \int \psi_k^* \psi_l \, d\tau = p_k \int \psi_k^* \psi_l \, d\tau.$$

Since, by hypothesis,

$$p_l \neq p_k,$$

it is evident that

$$\int \psi_k^* \psi_l \, d\tau = 0. \tag{4}$$

Clearly, the proof breaks down if two eigenvalues are equal, i.e., in the case of degeneracy. That, however, is not a serious matter, for it is then possible to take linear combinations of the non-orthogonal functions, thereby producing orthogonal ones. These new orthogonal functions are eigenfunctions in as proper a sense as the old, for they satisfy the operator equation with the eigenvalue to which the old ones referred, the equation being linear. A process by which the orthogonalization can be carried out conveniently will now be described; it is due to the mathematician Schmidt.

9.5 HERMITEAN OPERATORS

We have previously spoken of degenerate eigenvalues as eigenvalues for which the operator equation has more solutions than one. This statement was not one of perfect precision, for we have not, so far, set up a criterion by which solutions are to be distinguished. Clearly, a solution which is a constant multiple of another is not to be regarded as different from it since both are the same after normalization, except for a physically insignificant factor of absolute value 1. In general, we consider solutions of an operator equation as different only when they are *linearly independent*. Perhaps this term requires explanation. n functions $u_1, u_2 \ldots u_n$ are said to be linearly independent if it is impossible to find a relation

$$\sum_{i=1}^{n} a_i u_i = 0 \tag{5}$$

with at least one of the a_i's different from zero, which is true for every value of the independent variables of the u's. Our reason for adopting this convention is obvious. If eq. (5) could be satisfied we should be able to express at least one of the n functions in terms of the remaining ones by forming a linear combination, merely by solving eq. (5) for one of the functions in question.

Let us suppose, then, that for a given eigenvalue the operator equation has n linearly independent solutions $u_1, u_2 \ldots u_n$. They are not in general orthogonal. We seek a set of orthogonal and normal functions $w_1, w_2, \ldots w_n$, which are linear combinations of the u's and orthogonal. Let

$$w_1 = \frac{u_1}{\sqrt{\int |u_1|^2 d\tau}}.$$

Then put

$$v_2 = a_{21} w_1 + u_2$$

and determine a_{21} so that

$$\int w_1^* v_2 d\tau = a_{21} \int |w_1|^2 d\tau + \int w_1^* u_2 d\tau = 0.$$

Thus

$$a_{21} = -\int w_1^* u_2 d\tau.$$

v_2 is orthogonal to w_1, but not normalized. We therefore take

$$w_2 = \frac{v_2}{\sqrt{\int |v_2|^2 d\tau}}$$

as the second function of the desired set. Next we put

$$v_3 = a_{31}w_1 + a_{32}w_2 + u_3.$$

The coefficients a_{31} and a_{32} are again determined by the orthogonality conditions

$$\int w_1^* v_3 d\tau = \int w_2^* v_3 d\tau = 0$$

which lead to

$$a_{31} = -\int w_1^* u_3 d\tau \quad \text{and} \quad a_{32} = -\int w_2^* u_3 d\tau.$$

Finally we normalize v_3 and let

$$w_3 = \frac{v_3}{\sqrt{\int |v_3|^2 d\tau}}.$$

This process can be continued, leading to

$$w_i = \frac{v_i}{\sqrt{\int |v_i|^2 d\tau}}, \quad v_i = a_{i1}w_1 + a_{i2}w_2 + \ldots a_{i,i-1}w_{i-1} + u_i,$$

where

$$a_{ij} = -\int w_j^* u_i d\tau. \tag{6}$$

In defining w_i in this manner we are safe, for none of the v_i can vanish because eq. (5) can never be true.

Now it will be recalled that each one of the n u functions is already orthogonal with respect to the eigenfunctions belonging to any other eigenvalue by virtue of eq. (4). If, therefore, all degenerate eigenfunctions of the Hermitean operator P are orthogonalized by means of the process here described, all eigenfunctions of P may be said to be orthogonal.

It is important to observe in this connection that Schmidt's process is not unique in the sense of fixing a set of w's when the u's are given, for if we take the u's in a different order a different set of w's will result. It is clear, therefore, that there are in general many different orthogonal linear combinations, all of which can be regarded as proper eigenfunctions belonging to a degenerate eigenvalue of P.

9.6. Special Forms of Schrödinger's Equation. Since the energy of a physical system is a matter of great concern in many problems, we shall do well to consider in some detail the various common types of differential equation which the energy operator generates, all of which are of the standard form

$$H\psi_\lambda = E_\lambda \psi_\lambda,$$

9.6 SPECIAL FORMS OF SCHRÖDINGER'S EQUATION

where the E_λ's are the possible values of the energy. H is given by (9.5-2).

(a) The simplest example is that of a freely moving mass point. Let its constant potential energy in the region in which it moves be V. The point has three degrees of freedom, hence the Schrödinger equation becomes

$$\frac{-h^2}{8\pi^2 M}\left(\frac{\partial^2 \psi}{\partial x^2} + \frac{\partial^2 \psi}{\partial y^2} + \frac{\partial^2 \psi}{\partial z^2}\right) + V\psi = E\psi,$$

or

$$\frac{\partial^2 \psi}{\partial x^2} + \frac{\partial^2 \psi}{\partial y^2} + \frac{\partial^2 \psi}{\partial z^2} + k^2 \psi = 0, \text{ where } k^2 \equiv (E-V)\frac{8\pi^2 M}{h^2}. \quad (1)$$

We are omitting the subscripts on ψ and E for convenience. It will be observed that k, according to its definition, corresponds to $2\pi/h$ times the classical momentum of the particle. Eq. (1) can be solved at once by writing ψ as a product of three functions: $X(x)$, $Y(y)$, and $Z(z)$. On substitution of $\psi = X \cdot Y \cdot Z$ and division by ψ, eq. (1) reduces to

$$\frac{X''}{X} + \frac{Y''}{Y} + \frac{Z''}{Z} + k^2 = 0.$$

Primes denote differentiations with respect to the variable of which the primed quantity is a function. Since, except for the constant, each of the terms of this equation depends on one independent variable only which does not occur in any other term, each term must separately equal a constant. Hence, denoting these constants by $-k_1^2$, $-k_2^2$, $-k_3^2$, respectively,

$$X'' + k_1^2 X = 0; \quad Y'' + k_2^2 Y = 0; \quad Z'' + k_3^2 Z = 0; \quad k_1^2 + k_2^2 + k_3^2 = k^2.$$

These have the well-known solutions

$$X = A_1 e^{ik_1 x} + B_1 e^{-ik_1 x}; \quad Y = A_2 e^{ik_2 y} + B_2 e^{-ik_2 y};$$

$$Z = A_3 e^{ik_3 z} + B_3 e^{-ik_3 z}. \quad (2)$$

The A's and B's are any constants compatible with the boundary conditions. The next question is: what are the proper boundary conditions? Let us first assume that the fundamental range of x, y, and z covers all space. Condition (B), Sec. 9.2, is automatically satisfied. It is then necessary to satisfy condition (A). But it will at once be seen that this is impossible without making all A's and B's zero. This case corresponds to an eigenfunction which vanishes everywhere. According to prop. 3, Sec. 9.3, the particle would then be non-existent or, more conservatively, would have no probability of being detected anywhere. We shall return to this situation presently.

Another possibility is that the particle is, classically speaking, somewhere in the range $0 \leq x \leq l_1$, $0 \leq y \leq l_2$, $0 \leq z \leq l_3$, and that it has a zero probability of being found at the boundaries of this parallelepiped. In view of prop. 3, Sec. 9.3, $\psi^*\psi$ must then vanish at the boundaries, which is possible only if ψ vanishes there. If we substitute eq. (2) into $\psi = X \cdot Y \cdot Z$ and impose this condition we find

$$A_s + B_s = 0; A_s e^{ik_s l_s} + B_s e^{-ik_s l_s} = 0, s = 1, 2, 3.$$

Hence $B_s = -A_s$, $\sin k_s l_s = 0$. This last relation is equivalent to

$$k_s = \frac{n_s \pi}{l_s},$$

where the three numbers n_1, n_2, n_3 are integers. But in view of eq. (1) this fixes the energies of the problem. They are

$$E_{(n_1, n_2, n_3)} - V = \frac{h^2}{8\pi^2 M}(k_1{}^2 + k_2{}^2 + k_3{}^2)$$
$$= \frac{h^2}{8M}\left[\left(\frac{n_1}{l_1}\right)^2 + \left(\frac{n_2}{l_2}\right)^2 + \left(\frac{n_3}{l_3}\right)^2\right]. \quad (3)$$

To every choice of three integers (n_1, n_2, n_3) there corresponds an eigenfunction

$$\psi_{n_1 n_2 n_3} = c \sin\left(\frac{n_1 \pi}{l_1}x\right) \sin\left(\frac{n_2 \pi}{l_2}y\right) \sin\left(\frac{n_3 \pi}{l_3}z\right), \quad (4)$$

as is seen if we put $B = -A$ in eq. (2). The constant c in eq. (4) is determined by integrating $\psi^*\psi$ over all space and setting the result equal to unity. The result is

$$c = \sqrt{\frac{8}{l_1 l_2 l_3}} = \sqrt{\frac{8}{v}}, \quad (5)$$

where v is the volume of the parallelepiped. From eq. (3) we conclude that the energy values form an infinite but denumerable set; they are discrete. To every triple of numbers there corresponds an "energy level" of the system. The spacing of these "levels," i.e., the separation between successive possible values of the energy, is not uniform. A geometrical picture of their distribution can be obtained by imagining a point lattice filling all space, with a parallelepiped of sides $1/l_1$, $1/l_2$, $1/l_3$ as its crystallographic unit. (Cf. Fig. 9.1, which represents such a lattice in two dimensions.) The energies are then represented, except for a constant factor, by the squares of the lengths of the lines

9.6 SPECIAL FORMS OF SCHRÖDINGER'S EQUATION

connecting one lattice point with all the others. We notice that some of the arrows are equally long. This means that to different ψ functions there correspond equal energies: these states are degenerate.

If we increase l_1, l_2, l_3, that is, the volume of the space in which the particle is confined, two things happen: (1) our lattice points move closer together, hence the separation between the possible energies decreases; (2) the constant c [cf. eq. (5)] becomes smaller. In the limit of an infinite volume, the lattice points fill all space (any given point will be approached by a lattice point as closely as we please) and c approaches zero. This limiting case is evidently identical with the one discussed at the beginning of this example, but we view it now in a different perspective. As the spatial confinement of the particle is removed the energy spectrum becomes more and more nearly continuous, but the state-function loses the ability of being normalized. This situation is a very general one in quantum mechanical problems. All continuous spectra present this inherent difficulty.

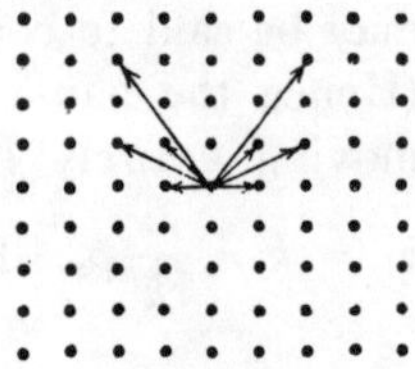

Fig. 9.1

How, then, are we to deal with continuous spectra, or more generally, with continuous ranges of eigenvalues? The first possibility is to make condition A less stringent or replace it by a different one in the case of continuous eigenvalues. We can, for instance, normalize the eigenfunction (known in the literature as "eigendifferential") belonging to an infinitesimal range of eigenvalues instead of normalizing the eigenfunction belonging to a spectral "point." This is the usual procedure and the only one applicable in the numerical solution of physical problems. The other, logically perhaps more satisfactory, possibility is to treat the problem as though the system were confined. All energy eigenvalues will then be discrete. The continuous spectrum will appear as the result of a limiting process in which the range of the fundamental variables approaches infinity. In the next section (9.7) we shall deal with some general properties of Schrödinger's equation. Because the analysis in that connection is very much simpler if we restrict ourselves to discrete eigenvalues we shall ignore the existence of continuous spectra and proceed as though the system were confined. This assumption does no violence to the physical situation since, as we have shown, a progressive removal of the confinement diminishes the difference between the discrete energies and approximates the physical condition as closely as is desired. Moreover, no physical system moves, strictly speaking, in an infinite space.

As a final observation on the problem of the free mass point we add the following remarks: eqs. (2) define a sinusoidal variation. If we wish to complete the time picture of a wave we are at liberty to multiply these equations by $e^{2\pi i \nu t}$, for, as we have seen (cf. 9.2), multiplication of a state-function by a constant (in space!) of absolute value unity does not alter the state. From this point of view the free mass point may be said to correspond to a sinusoidal wave in three dimensions. (Hence the name "wave mechanics.") Suppose that the wave is moving entirely in the x direction, so that $k_2 = k_3 = 0$. Then $k_1 = k = \frac{2\pi}{h} p_x$. But the wave length of this wave is clearly given by $k_1 \lambda = 2\pi$. Therefore $\lambda = \frac{2\pi}{k_1} = \frac{h}{p_x}$ which is de Broglie's equation.

(b) As our next example we take the problem of finding the possible energy levels of a linear harmonic oscillator. Since we are here interested in fundamentals we shall not go deeply into mathematical detail.[1] The classical Hamiltonian of the oscillator, which has but one degree of freedom, is the sum of the kinetic energy $p^2/2M$ and the potential energy $M/2 \cdot \omega^2 x^2$, where ω is 2π times the classical frequency, a constant. Hence the operator H has the form $-\frac{h^2}{8\pi^2 M} \frac{d^2}{dx^2} + \frac{M}{2} \omega^2 x^2$. Schrödinger's equation therefore takes the form:

$$\frac{d^2\psi}{dx^2} + \left(\frac{8\pi^2 M}{h^2} E - \frac{4\pi^2 M^2}{h^2} \omega^2 x^2\right)\psi = 0.$$

Using the abbreviations

$$\mathcal{E} = \frac{8\pi^2 ME}{h^2}, \alpha = \frac{2\pi M\omega}{h}, \qquad (6)$$

this takes the form:

$$\frac{d^2\psi}{dx^2} + (\mathcal{E} - \alpha^2 x^2)\psi = 0. \qquad (7)$$

We shall find in this case that a continuous spectrum will not arise even if we take the fundamental range of x to be $(-\infty, +\infty)$, the physical reason being that the potential energy increases so strongly as $|x| \to \infty$ as to provide effective confinement of the mass point. The solution of eq. (7) for large values of x, found by neglecting $\mathcal{E}$ against $\alpha^2 x^2$, is easily seen to be $e^{\pm \frac{\alpha}{2} x^2}$. This is verified by substituting the expres-

---

[1] The reader will find this and the following examples worked out more fully in the following books: A. Sommerfeld, "Ergänzungsband" (Vieweg und Sohn, 1929); E. U. Condon and P. M. Morse, "Quantum Mechanics" (McGraw-Hill, 1929).

9.6 SPECIAL FORMS OF SCHRÖDINGER'S EQUATION

sion into $\psi'' - \alpha^2 x^2 \psi = 0$. We can choose either the plus or the minus sign and still satisfy the asymptotic equation. But in order to satisfy condition (A) we must choose the minus sign. For suppose that the form of the asymptotic solution is appreciably correct from $x = a$ on outward. Then the integral

$$\int_a^\infty \psi^* \psi \, dx = \int_a^\infty e^{\alpha x^2} \, dx$$

certainly diverges, and this divergence cannot be offset by contributions from other parts of the total range, which are all positive

We now assume ψ to have the form

$$\psi = v(x) e^{-\frac{\alpha}{2} x^2} \tag{8}$$

On substituting this into eq. (7), we get for v the differential equation:

$$v'' - 2\alpha x v' + (\varepsilon - \alpha) v = 0. \tag{9}$$

This reduces to

$$\frac{d^2 v}{d\xi^2} - 2\xi \frac{dv}{d\xi} + \left(\frac{\varepsilon}{\alpha} - 1\right) v = 0 \tag{10}$$

by the substitution $\xi = \sqrt{\alpha} x$. Eq. (10) can be conveniently solved by expanding $v(\xi)$ as a power series. Put $v = \sum_\lambda a_\lambda \xi^\lambda$ and substitute in eq. (10). If eq. (10) is to be satisfied for all values of ξ, the coefficient of every power of ξ must vanish. But the coefficient of the jth power of ξ is:

$$(j+2)(j+1) a_{j+2} - 2j a_j + \left(\frac{\varepsilon}{\alpha} - 1\right) a_j.$$

If this is to be zero, the coefficients of the power series must be related as follows:

$$a_{j+2} = -\frac{\frac{\varepsilon}{\alpha} - 1 - 2j}{(j+2)(j+1)} a_j. \tag{11}$$

Any power series in which the coefficients satisfy eq. (11) is a solution of eq. (10). Let us investigate the behavior of such series for large values of ξ, or x. It will be determined solely by the high powers, that is, by the terms of large j. But the coefficients of these high powers are given by

$$a_{j+2} = \frac{2}{j} a_j,$$

obtained from eq. (11) by neglecting finite numbers against j. From a certain j onward, the series will therefore have the form

$$a_j \xi^j + \frac{a_j}{\frac{j}{2}} \xi^{j+2} + \frac{a_j}{\frac{j}{2}\left(\frac{j}{2}+1\right)} \xi^{j+4} + \frac{a_j}{\frac{j}{2}\left(\frac{j}{2}+1\right)\left(\frac{j}{2}+2\right)} \xi^{j+6} + \cdots$$

If j is even we can put $j = 2\lambda$ and write these terms

$$a_{2\lambda} \left(\xi^{2\lambda} + \frac{\xi^{2\lambda+2}}{\lambda} + \frac{\xi^{2\lambda+4}}{\lambda(\lambda+1)} + \cdots \right).$$

If this is multiplied and divided by $(\lambda - 1)!$ and ξ^2 is factored out, it takes the form

$$(\lambda - 1)!\, a_{2\lambda}\xi^2 \left(\frac{\xi^{2\lambda-2}}{(\lambda-1)!} + \frac{\xi^{2\lambda}}{\lambda!} + \frac{\xi^{2\lambda+2}}{(\lambda+1)!} + \frac{\xi^{2\lambda+4}}{(\lambda+2)!} + \cdots \right),$$

which is identical with the higher terms in the expansion: constant $\cdot \xi^2 e^{\xi^2}$. A similar result is obtained if j is odd. Therefore, if we take v to be an infinite series, with coefficients given by eq. (11), the ψ function for large values of x will be

$$e^{\xi^2} e^{-\frac{\alpha}{2}x^2} = e^{\frac{\alpha}{2}x^2}$$

which, as pointed out before, is inadmissible in view of condition (A).

The only other possibility is that the series for v should terminate, i.e., be a polynomial. In that case we have an infinite number of possibilities for breaking off the series. If we wish the polynomial to be a constant, the series must have only one term, a_0, and a_2 must be 0. But according to eq. (11), this means that $\varepsilon/\alpha = 1$. In general, if n is the degree of the polynomial, the condition for termination is

$$\frac{\varepsilon}{\alpha} = 2n + 1. \tag{12}$$

The polynomials obtained in this manner are known as Hermite's polynomials. From eq. (12) we can find at once the eigenvalues of our problem: by substituting eq. (6) it is seen that

$$E_n = (n + \tfrac{1}{2}) \frac{h\omega}{2\pi}. \tag{13}$$

The energies are all discrete; they are equally spaced, and their spacing increases with the "stiffness" (or the classical frequency $\omega/2\pi$) of the oscillator.[1]

This result has an important application to the problem of the diatomic molecule, which to some degree of approximation can be considered as a simple linear oscillator. When it vibrates, the possible values of its energy are nearly given by the sequence (13) as may be verified by spectroscopic measurements. But in general such a molecule has other degrees of freedom and therefore a more complex spectrum. It will be of interest, for instance, to find the possible energy levels of such a system if we ascribe to it what corresponds to

---

[1] The reader who is familiar with standard types of differential equations may be somewhat annoyed by the lengthy analysis leading to eq. (13). He will observe that eq. (7), upon the substitution $x = \xi/\sqrt{\alpha}$, goes into the form $\psi'' + (1 - \xi^2)\psi + \left(\frac{\varepsilon}{\alpha} - 1\right)\psi = 0$. This he may recognize as the differential equation for Hermite's orthogonal functions:

$$\psi'' + (1 - \xi^2)\psi + \lambda\psi = 0$$

where λ is to be identified with $\varepsilon/\alpha - 1$. It is known to have the eigenvalues $\lambda = 0, 2, 4, 6 \ldots$, so that we are led at once to eq. (13). (Cf. R. Courant and D. Hilbert, "Meth. d. Math. Phys." [Springer, Berlin, 1924], p. 261.)

9.6 SPECIAL FORMS OF SCHRÖDINGER'S EQUATION

classical motion of rotation, i.e., if we describe its state in terms of a function with an angular variable. This will be done in (c).

The energy of a simple harmonic oscillator can never be 0; classically speaking, the vibrating point can never be at rest. For even if n in eq. (13) is zero, E_0 is not zero but $\frac{1}{2}h\omega/2\pi$. This quantity is called the "zero point energy" of the oscillator, for it represents an irreducible minimum of energy which the oscillator cannot lose even at the absolute zero point of temperature. Spectroscopic and thermal observations support this conclusion.

(c) Next we consider the problem of finding the allowed energies of a system classically described as a rotator. The Hamiltonian of such a system, e.g., a dumb-bell having a moment of inertia I and rotating with angular momentum m_z about the z axis, has the form

$$H = \frac{m_z^2}{2I}$$

since there is no potential energy. If in this expression we replace m_z by its operator $\frac{h}{2\pi i}\left(x\frac{\partial}{\partial y} - y\frac{\partial}{\partial x}\right) = \frac{h}{2\pi i}\frac{\partial}{\partial \phi}$, where here ϕ is the usual longitude angle about the z axis, the energy operator is seen to be $H = -\frac{h^2}{8\pi^2 I}\frac{\partial^2}{\partial \phi^2}$. We are ascribing to the system only one degree of freedom, as seen from the fact that the energy operator affects only one variable, ϕ. Schrödinger's equation, if multiplied through by $-h^2/8\pi^2 I$, is:

$$\frac{d^2\psi}{d\phi^2} + k^2\psi = 0, \text{ with } k^2 = \frac{8\pi^2 I}{h^2} E.$$

It has the solution:

$$\psi = Ae^{ik\phi} + Be^{-ik\phi}.$$

Condition (B) now becomes operative; it demands that $k = n$, an integer. Hence the possible energies are

$$E_n = \frac{n^2 h^2}{8\pi^2 I}. \tag{14}$$

This prediction *is not* verified! The system which should exhibit these energy levels, namely the diatomic molecule, does not do so. Instead, it has its energies arranged according to the law

$$E_l = \frac{l(l+1)h^2}{8\pi^2 I}, \tag{15}$$

where l is an integer, as can be definitely ascertained from measurements on band spectra. Why this discrepancy? The reason is simply that we have not assigned to the system a sufficient number of degrees of freedom. Classical mechanics does not always guide us in the choice of degrees of freedom; this is, in principle, a matter of trial.

Let us then work the problem by assigning to the dumb-bell two degrees of freedom instead of one. We shall now do well first to write the complete Hamiltonian in Cartesian form, so that Schrödinger's equation becomes

$$\nabla^2 \psi + \frac{8\pi^2 M}{h^2} E \psi = 0.$$

On transforming to spherical coordinates by using the well-known spherical form for ∇^2 and then putting $r = a = $ constant, there results:

$$\frac{1}{a^2 \sin^2 \theta} \frac{\partial^2 \psi}{\partial \phi^2} + \frac{1}{a^2} \frac{\partial^2 \psi}{\partial \theta^2} + \frac{\cot \theta}{a^2} \frac{\partial \psi}{\partial \theta} + \frac{8\pi^2 M}{h^2} E \psi = 0. \quad (16)$$

We now repeat essentially the procedure used in example (a): Assuming ψ to be a product of two functions $\Theta(\theta)$ and $\Phi(\phi)$ with but a single independent variable each, we substitute in eq. (16) and divide through by ψ. Then, upon multiplying by $a^2 \sin^2 \theta$, the equation assumes the form

$$\frac{\Phi''}{\Phi} + \sin^2 \theta \frac{\Theta''}{\Theta} + \sin \theta \cos \theta \frac{\Theta'}{\Theta} + \mathcal{E} \sin^2 \theta = 0, \quad (17)$$

where

$$\mathcal{E} \equiv \frac{8\pi^2 M a^2}{h^2} E.$$

The first term is independent of θ, the remainder independent of ϕ. Hence the first term is equal to a constant. If this constant, c, were positive, Φ would be of the form $e^{+\sqrt{c}\phi}$ and hence not periodic in ϕ. This case must be dismissed as being contradictory to condition (B). The only other alternative is to choose a negative constant, say $-m^2$. But the solution of $\frac{\Phi''}{\Phi} = -m^2$ is:

$$\Phi = A e^{im\phi} + B e^{-im\phi} \quad (18)$$

Single-valuedness requires that m be an integer.

If this solution is substituted in eq. (17) the equation reads, after simple transpositions:

$$\Theta'' + \cot \theta \, \Theta' + \left(\mathcal{E} - \frac{m^2}{\sin^2 \theta} \right) \Theta = 0. \quad (19)$$

9.6 SPECIAL FORMS OF SCHRÖDINGER'S EQUATION

Let us now change the independent variable from θ to $x \equiv \cos \theta$. The result is

$$(1 - x^2) \frac{d^2 \Theta}{dx^2} - 2x \frac{d\Theta}{dx} + \left[\mathcal{E} - \frac{m^2}{(1 - x^2)} \right] \Theta = 0. \tag{19'}$$

This is the well-known differential equation for the "associated Legendre function," $P_l^m(x)$, defined as follows:

$$P_l^m(x) \equiv \left(\sqrt{1 - x^2}\right)^m \frac{d^m}{dx^m} P_l(x), \tag{20}$$

where $P_l(x)$ is the ordinary Legendre polynomial of degree l. The eigenvalues of eq. (19) are known to be

$$\mathcal{E} = l(l + 1), \tag{21}$$

where l is any positive integer or zero. These eigenvalues turn out to be independent of m. Because of the definition of $\mathcal{E}$, eq. (17), this result means that

$$E_l = \frac{l(l + 1)h^2}{8\pi^2 I},$$

since $I = Ma^2$. The energy levels for the rotator with two degrees of freedom are therefore correctly given by formula (15). The complete eigenfunctions are

$$\psi = \text{constant} \cdot P_l^m(\cos \theta)(A e^{im\phi} + B e^{-im\phi}). \tag{22}$$

The definition of $P_l^m(\cos \theta)$, eq. (20), shows that the value of m cannot be greater than l, nor can it be negative, since otherwise the differentiation would lose its meaning. It is very convenient, however, to admit negative m's. In order to make this possible we must replace m by its absolute value $|m|$ in eq. (20). We can then regard eq. (18) as the sum of two functions, one with a positive and one with an equal but negative value of m, so that, to every l, there correspond not $(l + 1)$ (that is $m = 0, 1, 2, 3, \ldots$), but $(2l + 1)$ values of m (i.e., $m = -l, -l + 1, \ldots 0, 1, \ldots l$).

As we have noted, the energy of the rotator depends only on l, not on m. It is thus clear that every energy value shows a $(2l + 1)$ fold degeneracy, for there are $2l + 1$ different state-functions all of which belong to the same eigenvalue E_l. These functions are, aside from constant factors:

$$P_l^l (\cos \theta) e^{il\phi}, \ P_l^{l-1} (\cos \theta) e^{i(l-1)\phi}, \ldots, P_l(\cos \theta),$$

$$P_l^1(\cos \theta) e^{-i\phi} \ldots, P_l^l(\cos \theta) e^{-il\phi}. \tag{23}$$

Every one of these satisfies eq. (16) with $E = E_l$, and since the equation is linear, every linear combination of them will be an eigenfunction. It happens that the various functions of the set (23) are orthogonal among each other, but this is not generally true for degenerate state functions. These considerations are interesting in themselves because of their applicability to the problem of the rotating molecule, but their importance is more basic, for these same solutions are encountered in any problem where the potential energy is a function only of the distance from a point, and where the field is central. We will now proceed to show this.

(*d*) The classical picture to which the present example refers is that of a mass point moving under the influence of an attracting or repelling stationary center of force. If the mass point is supposed to be an electron and the force a Coulomb attraction excited by a proton, the example describes the H atom. We are not, however, including the case of more complicated atoms, for the forces to which the electrons are subject are not in general central ones. In some instances only may they be considered as central forces without too great an error; and then the present analysis is applicable if a suitable form of the potential energy is known. Such a case is the motion of the valence electron of the alkalis.

If $V(r)$ is the potential energy of the particle, and M its mass, Schrödinger's equation has the form

$$\nabla^2 \psi + \frac{8\pi^2 M}{h^2}(E - V)\psi = 0. \tag{24}$$

We transform this to spherical coordinates, put

$$\psi = R(r) \cdot \Theta(\theta) \cdot \Phi(\phi), \tag{25}$$

and divide by ψ. The result is

$$\frac{R''}{R} + \frac{2}{r}\frac{R'}{R} + \frac{1}{r^2 \sin^2\theta}\frac{\Phi''}{\Phi} + \frac{1}{r^2}\frac{\Theta''}{\Theta} + \frac{1}{r^2}\cot\theta\frac{\Theta'}{\Theta}$$

$$+ \frac{8\pi^2 M}{h^2}(E - V(r)) = 0. \tag{26}$$

As before, primes denote differentiations with respect to the different variables upon which the functions depend. This equation separates at once into two parts if it is multiplied through by r^2. The first part is

$$\frac{r^2 R''}{R} + 2r\frac{R'}{R} + \frac{8\pi^2 M r^2}{h^2}(E - V);$$

9.6 SPECIAL FORMS OF SCHRÖDINGER'S EQUATION

the second:

$$\frac{1}{\sin^2\theta}\frac{\Phi''}{\Phi} + \frac{\Theta''}{\Theta} + \cot\theta\frac{\Theta'}{\Theta};$$

their sum is to be zero. Each of these parts must therefore be constant, for the first is independent of the angles, the second is independent of r. Hence if the first equals $\mathcal{E}$, the second must equal $-\mathcal{E}$. But then the second part gives the equation

$$\frac{\Phi''}{\Phi} + \sin^2\theta\cdot\frac{\Theta''}{\Theta} + \sin\theta\cos\theta\frac{\Theta'}{\Theta} + \mathcal{E}\sin^2\theta = 0,$$

which is formally identical with eq. (17). Hence we can make use of the results of the preceding example and write the angular part of ψ in accordance with eqs. (22) and (23):

$$\Theta(\theta)\cdot\Phi(\phi) = \text{constant}\cdot P_l^{|m|}(\cos\theta)e^{im\phi}. \tag{27}$$

The permissible values of $\mathcal{E}$ are given by eq. (21). The radial part of the equation then reads, after multiplication by R/r^2:

$$R'' + \frac{2}{r}R' + \left[\frac{8\pi^2 M}{h^2}(E - V) - \frac{l(l+1)}{r^2}\right]R = 0. \tag{28}$$

We are here principally interested in the eigenvalues E of the equation and shall not investigate closely the behavior of R. If we use, instead of R, the dependent variable $P = rR$, eq. (28) takes the simpler form

$$P'' + \left[\frac{8\pi^2 M}{h^2}(E - V) - \frac{l(l+1)}{r^2}\right]P = 0. \tag{29}$$

This equation permits an interesting conclusion: For large r, V approaches zero whatever its exact form may be. Hence, as $r \to \infty$, P satisfies:

$$P'' + \frac{8\pi^2 ME}{h^2}P = 0. \tag{30}$$

If $E > 0$, P is oscillatory; it partakes of all the properties of the eigenfunctions of the free electron discussed in example (a). In particular, it cannot be normalized in the usual way, and the probability density P^*P of finding the particle a great distance from the attracting center is finite. This case corresponds in every detail to the continuous energy states of an atom which are spectroscopically known to lie beyond the ionization energy $E = 0$. If $E < 0$, the particle is said to be bound.

To deal with intrinsically positive constants, let us put $E = -E'$. According to eq. (30)

$$\lim_{r \to \infty} P = \text{constant} \cdot e^{\pm \sqrt{8\pi^2 ME'/h^2} r}. \tag{31}$$

Condition (A) requires that we choose the negative sign in the exponent. Put

$$\frac{8\pi^2 ME'}{h^2} = \alpha^2.$$

Eq. (31) suggests that in solving eq. (29) we try the substitution

$$P = rR = e^{-\alpha r} \cdot v,$$

which transforms (29) into

$$v'' - 2\alpha v' - \left(\frac{8\pi^2 MV}{h^2} + \frac{l(l+1)}{r^2}\right) v = 0. \tag{32}$$

Before going on it is necessary to fix the form of V. In the hydrogen problem, $V = -e^2/r$, since the force is a Coulomb attraction acting between a proton and an electron. As in the previous examples, we now develop v as a series in positive powers of r about the origin. The lowest power, γ, of this series, is evidently determined by the behavior of eq. (32) at the origin, i.e., by the equation

$$v'' = \frac{l(l+1)}{r^2} v,$$

and if $\lim_{r \to 0} v = \text{constant} \cdot r^\gamma$, the only possible values of γ, found by substitution, are

$$\gamma = l+1 \quad \text{and} \quad \gamma = -l.$$

Closer investigation shows the second possibility to be inadmissible, for it will lead to contradiction with condition (A). We must, therefore, put

$$v = r^{l+1} \sum_\lambda a_\lambda r^\lambda, \tag{33}$$

substitute in eq. (32), and equate the coefficient of every power to zero. Picking out the coefficient of r^{l+j} we find

$$[(l+j+2)(l+j+1) - l(l+1)] a_{j+1} = \left[2\alpha(l+j+1) - \frac{8\pi^2 Me^2}{h^2}\right] a_j. \tag{34}$$

9.6 SPECIAL FORMS OF SCHRÖDINGER'S EQUATION

Every power series in which the coefficients a_λ obey this relation is a solution of eq. (32). But we shall find, as we did previously, that the radial function R cannot be normalized unless the series for v terminates. This is easily seen by examining the behavior for large values of r of a non-terminating series with coefficients given by eq. (34). Neglecting finite numbers against j, this relation reads

$$a_{j+1} = \frac{2\alpha}{j} a_j,$$

a recurrence formula which, for large j, defines the function

$$\text{constant} \cdot r e^{2\alpha r}.$$

Multiplication by r^{l+1} gives v. If we choose this function for v and multiply by $e^{-\alpha r}$ to get P, the resulting function is essentially $e^{\alpha r}$. This tends to infinity so strongly that $\int_0^\infty P^2 dr$ does not exist. Hence the series for v must terminate.

If n_r is to be the exponent of the highest power of r in $\sum_\lambda a_\lambda r^\lambda$, then a_{n_r+1} must vanish. This can happen only if the right-hand side of eq. (34) vanishes, hence if

$$2\alpha(n_r + l + 1) = \frac{8\pi^2 M e^2}{h^2}.$$

Substituting for α, we see that

$$E' = \frac{2\pi^2 M e^4}{h^2(n_r + l + 1)^2}. \tag{35}$$

We note first that the total energy depends, not on n_r and l singly, but on the sum $(n_r + l + 1)$. According to eq. (33) this is simply the degree of the v polynomial. $(n_r + l + 1)$ is often denoted by n and called the "principal quantum number." n_r has the name "radial quantum number," and l is referred to as "azimuthal quantum number." Clearly, $n \geq l + 1$. Result (35) is identical with that of Bohr's theory if $l + 1$ is used in place of Bohr's k. The true total energies, it will be recalled, are given by $-E'$. Relation (35) has the best of experimental support.

The functions R are now easily constructed. They are related in a simple way to the better-known Laguerre polynomials, and this relation, though not of interest at present, is very useful in making numerical calculations. For future reference let us recall that

$$\psi = cR(r)P_l^{|m|}(\cos\theta)e^{im\phi}, \tag{35a}$$

where c is so chosen as to make $\int \psi^*\psi r^2 \sin\theta\, dr\, d\theta\, d\phi = 1$. It is customary to split the normalizing factor c into three factors $c_1 c_2 c_3$ in such a way that $c_1 R$, $c_2 P_l^{|m|}$, and $c_3 e^{im\phi}$ are individually normalized.

These examples will suffice to illustrate the detailed manner of solving Schrödinger's equation. They have been chosen partly because of their physical importance, partly because of their mathematical simplicity. For the sake of fairness we should remark that in many cases of great interest exact solutions are not available and recourse must be taken to approximate methods of solution. These approximate methods, usually comprised under the general heading of "perturbation theory," form in themselves a subject of considerable physical significance and justify a closer study from a fundamental point of view, not so much because of their mathematical nicety, but because of the insight which they afford into the workings of physical disturbances upon systems in an eigenstate. But in a discussion of this matter we shall be aided by a deepened knowledge of the basic properties of solutions and eigenvalues of Schrödinger's equation. These we propose to study now. The reader may remember that, in the general theory of the preceding paragraphs, we have often assumed the "completeness" of the eigenfunctions. This property, among others, will now be established.

9.7. General Properties of Solutions and Eigenvalues of Schrödinger's Equation. *Preliminary considerations.* The various component forms of Schrödinger's equation encountered in the last paragraph, and all others, belong to a type of differential equation known to mathematicians as the Sturm-Liouville equation. It is usually written:

$$L(u) + \lambda \rho u = 0, \tag{1}$$

where $L(u)$ is a differential operator defined by:

$$L(u) = (pu')' - qu. \tag{2}$$

Primes denote differentiations with respect to the (single) independent variable for which we shall often use the letter x; p, q, and ρ are functions of x. We shall suppose that $\rho(x)$, which is of the nature of a weighting function, satisfies $\rho \geq 0$ in the entire fundamental domain of x. The latter is taken to be *finite*.[1] (Cf. remarks in Sec. 9.6(a).) λ is a constant, the eigenvalue of the Sturm-Liouville equation. We first show that the most common forms of Schrödinger's equation, or rather the component equations into which it can be separated, are of the form (1).

Using the abbreviation

$$k^2 \equiv \frac{8\pi^2 M}{h^2},$$

[1] Cf. E. C. Kemble, *Proc. Nat. Acad. Sci.*, 19, 710, 1933, for the case of an infinite domain.

9.7 GENERAL PROPERTIES OF SOLUTIONS

we can write every one-dimensional Schrödinger equation as follows:

$$\psi'' - k^2 V(x)\psi + k^2 E\psi = 0. \tag{3}$$

Comparison with eqs. (1) and (2) shows that

$$p = 1, \quad q = k^2 V(x), \quad \rho = 1, \quad \lambda = k^2 E. \tag{4}$$

The equation of the simple harmonic oscillator, (9.6–7), where $V = \tfrac{1}{2} M\omega^2 x^2$, is a special example of eq. (3). The same is true for each of the three equations into which the Schrödinger equation for the free electron ($V = 0$) resolves itself; cf. (9.6–1).

The three equations encountered in connection with the central field problem are also of the type (1). The equation of which (9.6–18) is the solution can be written

$$(\Phi')' + m^2 \Phi = 0. \tag{5}$$

Comparison shows that

$$p = 1, \quad q = 0, \quad \rho = 1, \quad \lambda = m^2. \tag{6}$$

(9.6–19) may be stated

$$(\sin\theta\,\Theta')' - \frac{m^2}{\sin\theta}\Theta + \varepsilon \sin\theta\,\Theta = 0; \tag{7}$$

$$p = \sin\theta, \quad q = \frac{m^2}{\sin\theta}, \quad \rho = \sin\theta, \quad \varepsilon = \lambda. \tag{8}$$

Finally, (9.6–28) is equivalent to the equation

$$(r^2 R')' - [l(l+1) + k^2 r^2 V(r)]R + k^2 r^2 E R = 0. \tag{9}$$

Here

$$p = r^2, \quad q = l(l+1) + k^2 r^2 V(r), \quad \rho = r^2, \quad \lambda = k^2 E. \tag{10}$$

The fundamental ranges of the independent variables are all different. In eq. (3) both limits are finite but arbitrary; in (9) the lower limit is 0 and the upper is any large value of r; in (5) the range is from 0 to 2π; and in (7) it goes from 0 to π. Nor do the boundary conditions agree! Aside from the condition that $|\psi|^2$ be integrable we must impose in (3) the restriction that ψ shall vanish at both end points of the range, in (5) and (7) that ψ be periodic (single valued!), and in (9) that it shall vanish at the upper limit. It would be difficult to phrase all these conditions in a uniform mathematical way. Fortunately this is not necessary. But we observe, upon examining these boundary conditions together with eqs. (4), (6), (8), (10), that if u and v are any two admissible ψ functions, then

$$vpu'\big|_b = vpu'\big|_a \tag{11}$$

if a and b are the end points of the range in any one problem. We can develop our theory on the basis of this property of the functions u in eq. (1). The solutions of eqs. (3), (5), (7), (9), and in general of (1), are real functions. When complex functions are used, as we have done in example (c), this was a matter of convenience; sines and cosines would have done as well. In the subsequent discussion of this paragraph we shall, for the sake of simplicity, assume all u's and v's to be real.

The operator L in eq. (2) is Hermitean with respect to functions satisfying eq.(11). If we work out $\int vL(u)dx$ we obtain

$$\int v(pu')'dx - \int vqudx = vpu' \Big|_b^a - \int v'pu'dx - \int vqudx.$$

The integrated part vanishes because of eq. (11). Performing another partial integration on the last result we find

$$\int vL(u)dx = -v'pu \Big|_a^b + \int u(pv')'dx - \int uqvdx = \int uL(v)dx.$$

Since L is a real operator, its Hermitean character is proved. The relation

$$\int vL(u)dx = \int uL(v)dx \tag{12}$$

is a special case of a theorem known as Green's theorem.

The stationary character of the eigenvalues. As will now be shown, eq. (1) represents the condition that the integral

$$D(u) = \int (pu'^2 + qu^2)dx \tag{13}$$

shall be stationary, i.e., either a maximum or a minimum, provided that

$$\int \rho u^2 dx = 1. \tag{14}$$

(14) is the ordinary normalizing condition as will at once be seen upon identification of the function ρ in eqs. (4–10). To prove the theorem, let us recall the following facts from the calculus of variations:[1] If u is an unknown function of x, the form of which is to be determined, and $I(u, u', x)$ is a given function of u, du/dx, and x, then the u's which make $\int I dx$ stationary, i.e., $\delta \int I dx$ zero, must satisfy Euler's equation:

$$\frac{\partial I}{\partial u} - \frac{d}{dx} \frac{\partial I}{\partial u'} = 0. \tag{15}$$

But if there is an accessory condition

$$\int G(u)dx = \text{constant}$$

to be obeyed, the I in eq. (15) has to be replaced by $I - \lambda G$, where λ is an undetermined but constant multiplier. Therefore, the condition that δD be 0 subject to eq. (14) is simply eq. (15) if we write:

$$I = pu'^2 + qu^2 - \lambda \rho u^2.$$

If this is substituted in eq. (15), eq. (1) results. We see, then, that the process of solving the Sturm-Liouville equation is equivalent to a search for the stationary values of $D(u)$. (This statement will *not* pass closest scrutiny, since no proof for the sufficiency of Euler's equation as a condition for stationarity can be given. But the statement is true for all physical problems.) D has in general many stationary values, just as a curve may have many maxima and minima.

Let us now order the eigenfunctions of eq. (1) in the following way. Suppose that $u = u_1(x)$ satisfies eqs. (1) and (14), and that it makes eq. (13) a *minimum* (which is usually the case in connection with Schrödinger's equation). We shall call

---

[1] Cf., for instance, Courant-Hilbert.

the corresponding eigenvalue λ_1. We now seek a minimum of D subject not only to condition (14) but also to

$$\int \rho u u_1 dx = 0. \qquad (16)$$

Let $u = u_2(x)$ be the function which produces this minimum. It is at once clear that the minimum in question cannot lie lower than the first, because the requirement upon the admissible u's has been made more stringent: u_2 must satisfy two accessory conditions, u_1 only one. What is the differential equation for u_2?

The interesting answer is that u_2 is also subject to eq. (1). To prove this, we form the minimizing condition (15) with

$$I = pu_2'^2 + qu_2^2 - \lambda_2 \rho u_2^2 - \mu \rho u_1 u_2,$$

λ_2 and μ being undetermined constants, which have to be introduced because of the existence of two accessory conditions, (14) and (16). We thus obtain:

$$2qu_2 - 2\lambda_2 \rho u_2 - \mu \rho u_1 - 2(pu_2')' = 0,$$

which is the same as:
$$L(u_2) + \lambda_2 \rho u_2 + \tfrac{1}{2}\mu \rho u_1 = 0. \qquad (17)$$

To determine μ we multiply by u_1 and integrate over x. In doing this we observe that

$$\int u_1 L(u_2) dx = \int u_2 L(u_1) dx = -\lambda_1 \int \rho u_1 u_2 dx = 0.$$

The first step is justified by (12), the second because u_1 satisfies (1), the third because u_2 is subject to (16). Eq. (17) will then lead to

$$\lambda_2 \int \rho u_1 u_2 dx = -\tfrac{1}{2}\mu \int \rho u_1^2 dx.$$

Here the left-hand side vanishes—cf. eq. (16)—and the integral on the right is 1. Hence
$$\mu = 0.$$

But then (17) reduces to the Sturm-Liouville equation (1). In other words, we have shown that the function u_2, which produces another minimum in $D(u)$ but under the additional condition (16), is simply another solution of (1). λ_2, the multiplier appearing in the minimizing problem, is a second eigenvalue of (1).

This process can be continued. We seek next a function u_3 which will minimize $D(u)$, but subject to the three conditions

$$\int \rho u_3^2 dx = 1, \quad \int \rho u_1 u_3 dx = 0, \quad \int \rho u_2 u_3 dx = 0.$$

The minimum thus obtained will lie at least as high as that due to u_2, for the choice of admissible functions has been further restricted. If we write down Euler's equations, there will now appear three undetermined constants: λ_3, μ, and ν. On multiplying first by u_1 and integrating, then by u_2 and integrating, μ and ν can be shown to be zero. The resulting equation is again identical with (1), and λ_3 turns out to be another eigenvalue. In this manner we obtain an ordered sequence of eigenfunctions $u_1, u_2, u_3 \ldots$ and corresponding eigenvalues $\lambda_1, \lambda_2, \lambda_3$, the arrangement being such that the minimum of D corresponding to u_1 is the lowest, that due to u_2 the second lowest, etc., except for the possibility of coincidence of these minima.

The last and most important step is to show that the minima of $D(u)$ are identical with the successive λ's. This is done by substitution, as follows

$$D(u_n) = \int (pu_n'^2 + qu_n^2) dx = u_n p u_n' \Big|_a^b - \int [u_n(pu_n')' - u_n q u_n] dx$$
$$= -\int u_n L(u_n) dx = \lambda_n \int \rho u_n^2 dx = \lambda_n.$$

The steps involved are fairly obvious. We have proved that the eigenvalues of the Sturm-Liouville equation are the successive minima of $D(u)$ as u is subjected to increasingly restrictive conditions.

These results have an immediate physical consequence. They permit us, in some cases, to determine approximately the lowest energy state, and to a poorer approximation, even the eigenfunction corresponding to this state, without solving Schrödinger's equation at all. We know that the lowest eigenvalue is the lowest possible value of D, with u subject only to the condition of normalization. Now in some cases a good guess can be made at the general form of the eigenfunction u_1, which is left undetermined by inserting several variable parameters. One can then easily calculate $\lambda_1' = -\int \overline{u_1} L(\overline{u_1}) dx$, where $\overline{u_1}$ is the trial function which includes the parameters. λ_1' will then also depend on these parameters. If now we minimize λ_1' with respect to the parameters, its lowest possible value will be λ_1, and $\overline{u_1}$ corresponding to this choice of parameters will be u_1. The accuracy of this approximation depends, of course, on the adequacy of the initial selection of $\overline{u_1}$. In general, the λ_1 thus determined will be too high.

On comparing eq. (1) with the one-dimensional form of Schrödinger's equation, (3), it is seen that the operator L is the same as $-H$, aside from multiplicative constants. Hence $\lambda_1 = \text{constant} \cdot \int u_1 H(u_1) dx$, and the method just outlined amounts to a minimization of the energy. The process is frequently referred to by this name.

Another conclusion may be of some interest. If p and q in eq. (2) are both greater than zero, $D(u_n) = \int (p u_n'^2 + q u_n^2) dx > 0$, and therefore every $\lambda_n > 0$. This observation is often useful. For instance, it allows us to make the prediction—somewhat trivial from a classical point of view but in need of proof in quantum mechanics—that there can be no energy states below the minimum of the potential energy. Let us, for simplicity, consider the one-dimensional case, to which eq. (3) applies. If V_0 is the minimum of V, this equation may be written:

$$\psi'' - k^2(V(x) - V_0)\psi + k^2(E - V_0)\psi = 0.$$

Here p and $q \geqq 0$; hence $\lambda = k^2(E - V_0) \geqq 0$, so that

$$E \geqq V_0.$$

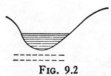

FIG. 9.2

Energy levels like the dotted ones in Fig. 9.2 cannot exist.[1]

Distribution of high energies. Although the sequence of eigenvalues of the Sturm-Liouville equation follows no uniform law, depending, as it does, on the form of p, q, and ρ, it is nevertheless true that the high eigenvalues of all Sturm-Liouville equations are distributed according to the same law, viz.:

$$\lim_{n \to \infty} \lambda_n = \text{constant} \cdot n^2. \tag{18}$$

To prove this, we transform eq. (1) by the substitution

$$u = p^n \rho^k z, \quad t = \int_a^x p^l \rho^m dx, \tag{19}$$

---

[1] In Dirac's equation (cf. Sec. 9.14), which is the relativistic generalization of Schrödinger's, this is no longer true.

and use t as independent variable. If the reader performs the algebra involved in this transformation and then solves for the values of l, m, k, n which will cause: (1) the coefficient of $\dot{z}$ to vanish (dots denote differentiation with respect to t), and (2) the coefficient of $\ddot{z}$ to be the same as that of λz, he will find that

$$k = n = -\tfrac{1}{4}, \quad l = -\tfrac{1}{2}, \quad m = \tfrac{1}{2}.$$

Eq. (1) then reduces to the simple form:

$$\ddot{z} - r(t)z + \lambda z = 0. \tag{20}$$

Here r is a function of t which in general is bounded, but its maximum depends on p, q, and ρ. Let the upper bound of its absolute value be M. Then, if $\lambda \gg M$, the eigenvalues of eq. (20) are the same as those of the equation

$$\ddot{z} + \lambda z = 0,$$

which has the general solution $z = A \cos \sqrt{\lambda}\, t + B \sin \sqrt{\lambda}\, t$. Now, whatever the boundary condition on u required by the physical problem may be, z must vanish at the end points of the t range. This is easily verified by examining the boundary conditions for eqs. (3)–(9)[1] and observing that $z = (p\rho)^{\frac{1}{4}} u$, as follows from (19). But the end points of the t range are 0 and $\tau = \int_a^b \sqrt{\dfrac{\rho}{p}}\, dx$. This allows for z only the solution $B \sin \sqrt{\lambda}\, t$, with λ given by $\dfrac{n^2 \pi^2}{\tau^2}$. We have completed the proof that

$$\lim_{n \to \infty} \lambda_n = \frac{\pi^2}{\left(\int_a^b \sqrt{\dfrac{\rho}{p}}\, dx \right)^2} \cdot n^2, \tag{21}$$

which agrees with (18). The result obtained reminds us immediately of the free electron case (9.6–3), where the energy levels are also arranged according to the law constant $\cdot n^2$. The spectrum (21) is discrete if (a, b) is a finite range. However, if $\int_a^b \sqrt{\dfrac{\rho}{p}}\, dx$ tends to ∞ in the same manner as n it becomes continuous, for then $\lambda_{n+1} - \lambda_n$ is found to be zero. We encounter here all the characteristics of the free electron problem, and may therefore express the result of this paragraph by saying that the high energies of any physical system are distributed like those of a free electron.

The simple harmonic oscillator, (9.6–13), apparently violates this rule. The analytical reason for this behavior is that $r(t)$ in eq. (20) is not bounded. This, in turn, results from the form of V which goes to infinity like x^2. Clearly such a case is over-idealized, and if V is given a finite upper limit our result holds for the oscillator as well.

In the next section we shall prove the completeness of the solutions of Schrödinger's equation. For the purpose of that proof a relation like (18) is required, but it can be used in a milder form:

$$\lambda_n \to \infty, \text{ as } n \to \infty. \tag{22}$$

---

[1] Eq. (5) is no exception; if we take the cosine function for Φ, the fundamental range may be taken from $-\pi/2$ to $3/2\,\pi$.

Even the simple harmonic oscillator, with its idealized potential function, obeys this rule and will therefore require no separate treatment.

The completeness of the solutions of Schrödinger's equation. When we say that a set of eigenfunctions $u_1, u_2, u_3, \ldots, u_n \ldots$ is complete we mean the following. Take any function $f(x)$ which satisfies the same conditions as the functions $u(x)$. Then define coefficients c by the rule

$$c_n = \int \rho f u_n dx \qquad (23)$$

and differences

$$\Delta_n = f - \sum_{1}^{n} c_i u_i. \qquad (24)$$

The system of u's is complete if

$$\lim_{n \to \infty} \overline{\Delta_n^2} = 0, \qquad (25)$$

where

$$\overline{\Delta_n^2} = \int \Delta_n^2 \rho dx. \qquad (26)$$

Verbally, completeness means that any function f subject to the same conditions as the u's can be "approximated in the mean" by a series $\sum_{1}^{\infty} c_i u_i$. It does not imply in general that

$$f = \sum_{1}^{\infty} c_i u_i, \qquad (27)$$

because the left and the right of this equation may not be equal at certain points although (25) is satisfied. Nevertheless uniform convergence of $\Sigma c_i u_i$, and hence the validity of (27), are always assumed in physical problems, usually without proof.

Eq. (25) is the relation we wish to prove for the solutions of the Sturm-Liouville equation, i.e., for the u_i's discussed in the preceding section. For simplicity let us denote $\sqrt{\overline{\Delta_n^2}}$ by a_n. The quantity Δ_n/a_n is evidently normalized in the sense that

$$\int \left(\frac{\Delta_n}{a_n}\right)^2 \rho dx = 1.$$

We also observe that Δ_n/a_n is orthogonal to every u_i up to and including u_n, the order of the u's being arranged as described above. To show this we simply calculate:

$$\int \left(\frac{\Delta_n}{a_n}\right) u_i \rho dx = \frac{1}{a_n} \left\{ \int f u_i \rho dx - \sum_{j=1}^{n} c_j \int u_j u_i \rho dx \right\}$$

$$= \begin{cases} \dfrac{1}{a_n}(c_i - c_i) = 0 \text{ if } i \leq n, \\ \dfrac{1}{a_n}(c_i - 0) = \dfrac{c_i}{a_n} \text{ if } i > n. \end{cases}$$

These results follow at once from the fact that u_n is orthogonal upon all preceding u's, and from the definition of c_n, eq. (23).

We can say, therefore, that *the function* Δ_n/a_n *satisfies all accessory conditions of* u_{n+1}, except that of minimizing D. Consider now the class of functions which satisfy these conditions. Of all these, u_{n+1} produces the smallest $D(u)$, namely, λ_{n+1}. Therefore

$$D\left(\frac{\Delta_n}{a_n}\right) \geq D(u_{n+1}),$$

or

$$\int \left[p\left(\frac{\Delta_n}{a_n}\right)'^2 + q\left(\frac{\Delta_n}{a_n}\right)^2\right]dx \geq \lambda_{n+1}.$$

Upon substitution of eq. (24) for Δ_n this inequality takes the form

$$\frac{1}{a_n^2}\left\{\int (pf'^2 + qf^2)dx - 2\sum_{i=1}^{n} c_i \int \alpha_i dx + \sum_{i,j=1}^{n} c_i c_j \int \beta_{ij} dx\right\} \geq \lambda_{n+1} \qquad (28)$$

with the abbreviations

$$\alpha_i = pu_i' f' + qfu_i$$

$$\beta_{ij} = pu_i' u_j' + qu_i u_j.$$

But

$$\int \alpha_i dx = pu_i' f \Big|_a^b - \int [f(pu_i')' - fqu_i]dx = \int f\lambda_i \rho u_i dx = c_i \lambda_i.$$

The first step involves a partial integration; the second involves use of eqs. (11) and (1), the last follows from eq. (23). Similarly,

$$\int \beta_{ij} dx = u_j p u_i' \Big|_a^b - \int [u_j(pu_i')' - u_j qu_i]dx = \int u_j \lambda_i \rho u_i dx = \lambda_i \delta_{ij}.$$

Substituting these values in eq. (28) we find immediately that

$$\frac{1}{a_n^2}\left\{\int (pf'^2 + qf^2)dx - \sum_{1}^{n} c_i^2 \lambda_i\right\} \geq \lambda_{n+1}. \qquad (29)$$

Since f satisfies the same continuity and boundary conditions as the u's, which are essentially the state functions of the physical problem, the integral in eq. (29) exists. (If f violates this rule, an "approximation in the mean" by the u functions may be impossible.) Let its value be A. Also, because the lowest eigenvalue is the minimum of D, the very existence of such an eigenvalue insures that it will lie above some finite lower bound. If we reckon all λ's from this lower bound, i.e., measure energies taking the lowest state as the zero level, which is clearly permissible, they are all positive. Hence the sum in eq. (29) is certainly not smaller than 0. Therefore

$$A \geq \lambda_{n+1} \cdot a_n^2.$$

But we know from (22) that λ_{n+1} approaches ∞ as $n \to \infty$. Therefore, since $a_n^2 = \overline{\Delta_n^2}$,

$$\lim_{n\to\infty} \overline{\Delta_n^2} = 0.$$

We have proved eq. (25).

It is important to observe that, if the system of u's is to be complete, no solution, or set of solutions, may be excluded from consideration. Supposing, for instance, that u_l were omitted; we could certainly not approximate u_l itself, which is a perfectly admissible f, by the series $\sum_{i \neq l} c_i u_i$, since every c_i would vanish. Nor is it permissible in cases like that of the hydrogen problem (6d) where, in the absence of spatial confinement, a continuous spectrum appears, to neglect the eigenfunctions belonging to this continuous range of eigenvalues ($E > 0$ in problem 6d), for it is precisely this set of solutions which has the property (22), a property indispensable in our completeness proof.

Considerations like those in this paragraph, being of limited practical use, are rarely presented in textbooks on quantum mechanics. Nevertheless they are the mathematical basis upon which the formalism of the new discipline stands. It is for this reason that we have included them in our discussion of the foundations of this subject. The result here proved has been anticipated in Secs. 9.2 and 9.3, even with wider generality than our present analysis permits. (We have limited ourselves to the energy operator.) Further use of it will be made in the development of perturbation theory.

9.8. Matrices. The notion of a matrix arises most naturally in many fields of mathematics, where it presents itself as a useful generalization of the concept of number. It would be well from a pedagogical point of view to introduce it in connection with a definite mathematical problem, for instance that of linear transformations of vectors, and to derive all the properties of a matrix on the basis of the requirements of such a problem. To save space, however, we shall proceed differently. Instead of following the inductive course of presentation, we will take the concept out of its natural setting and discuss its properties in terms of definitions and rules of operations, limiting ourselves to those aspects which are of immediate importance in quantum mechanics. It should be recalled at every step, of course, that the categorical rules to be discussed are by no means arbitrary, but spring from a connected group of mathematical facts to which we are not referring.

A matrix is an array of numbers, ordered in rows and columns, this aggregate being usually symbolized by a single letter. The numbers are known as the elements of the matrix; in order to designate them fully it is necessary to designate the row and the column in which each element stands. This is done by adding two subscripts, the first indicating the ordinal number of the row, the second that of the column to which the element belongs. Thus, if α is a matrix, we can define it fully by the following scheme:

$$\alpha = \left\{ \begin{array}{c} \alpha_{11}\ \alpha_{12}\ \alpha_{13}\ \ldots\ \alpha_{1m} \\ \alpha_{21}\ \alpha_{22}\ \alpha_{23}\ \ldots\ \alpha_{2m} \\ \cdot\ \cdot\ \cdot\ \cdot\ \cdot\ \cdot\ \cdot\ \cdot \\ \cdot\ \cdot\ \cdot\ \cdot\ \cdot\ \cdot\ \cdot\ \cdot \\ \alpha_{n1}\ \alpha_{n2}\ \alpha_{n3}\ \ldots\ \alpha_{nm} \end{array} \right\}.$$

As pointed out, the α_{ij} are ordinary numbers. α has n rows and m columns. If $n = m$ the matrix is said to be a square matrix.

In order to be able to use matrices in calculations one must know the rules according to which they are combined in addition, subtraction, multiplication, and division. Subtraction is merely the addition of the negative matrix, and division means multiplication by the reciprocal. Hence if we define the negative and the reciprocal of a matrix we shall have to state only two rules of operation: addition and multiplication. The results of these operations upon matrices must again be matrices.

1. Matrices are added by adding the corresponding elements individually. Thus, if α and β are two matrices

$$(\alpha + \beta)_{ij} = \alpha_{ij} + \beta_{ij}. \tag{1}$$

Knowing all the elements we can of course construct the sum-matrix. It is clear that addition is possible only if the matrices to be added have equal numbers of rows and columns. Also, matrices cannot be added to ordinary numbers. Addition of matrices is both commutative and associative, commutative because

$$\alpha + \beta = \beta + \alpha,$$

and associative because

$$(\alpha + \beta) + \gamma = \alpha + (\beta + \gamma)$$

if γ is another matrix. Both of these relations follow at once from eq. (1).

2. Matrices are multiplied by the rule:

$$(\alpha\beta)_{ij} = \sum_\lambda \alpha_{i\lambda}\beta_{\lambda j}. \tag{2}$$

To give an example, if

$$\alpha = \begin{Bmatrix} 1 & 0 & 2 & 1 & 3 \\ 2 & 1 & 3 & 2 & 1 \\ 0 & 4 & 2 & 0 & 3 \\ 2 & 1 & 0 & 1 & 2 \end{Bmatrix}, \quad \beta = \begin{Bmatrix} 2 & 1 & 2 & 0 \\ 1 & 3 & 2 & 1 \\ 2 & 0 & 4 & 3 \\ 2 & 1 & 3 & 0 \\ 1 & 2 & 3 & 2 \end{Bmatrix},$$

then

$$\alpha\beta = \begin{Bmatrix} 11 & 8 & 22 & 12 \\ 16 & 9 & 27 & 12 \\ 11 & 18 & 25 & 16 \\ 9 & 10 & 15 & 5 \end{Bmatrix}.$$

It will be observed that this is formally the same as the rule for multiplication of determinants. Multiplication of matrices is possible only when the first has as many columns as the second has rows. If α has l rows and m columns, and β has m rows and n columns, then $\alpha\beta$ has l rows and n columns. It is possible to multiply a matrix by a number; this is done by multiplying each element individually. We can therefore define the negative of a matrix as the result of multiplying it by -1, i.e., as the matrix with the signs of all its elements reversed.

When more than two matrices are to be multiplied, rule 2 must be carried out several times in succession. In doing so, the order of operations is indifferent since the associative law holds for multiplication. This is easily seen by carrying out the operations:

$$[(\alpha\beta)\gamma]_{ij} = \sum_\lambda (\alpha\beta)_{i\lambda}\gamma_{\lambda j} = \sum_{\lambda,\mu} \alpha_{i\mu}\beta_{\mu\lambda}\gamma_{\lambda j} = \sum_\mu \alpha_{i\mu}(\beta\gamma)_{\mu j} = [\alpha(\beta\gamma)]_{ij}.$$

The commutative law, however, does not generally hold; that is, in general,

$$\alpha\beta \neq \beta\alpha.$$

For if we carry out the multiplication on the left-hand side of this inequality we get, as the ijth element, $\sum_\lambda \alpha_{i\lambda}\beta_{\lambda j}$, while the right side gives us $\sum_\lambda \beta_{i\lambda}\alpha_{\lambda j}$, and these two sums are not equal, except in very special cases.

There are some simple types of matrices which, because of their frequent occurrence, deserve special consideration. First we have the so-called unit matrix, designated by $\mathbf{1}$, which has the form

$$\mathbf{1} = \begin{Bmatrix} 1 & 0 & 0 & 0 & \ldots & 0 \\ 0 & 1 & 0 & 0 & \ldots & 0 \\ 0 & 0 & 1 & 0 & \ldots & 0 \\ \cdot & \cdot & \cdot & \cdot & & \cdot \\ 0 & 0 & 0 & 0 & 0 & \ldots 1 \end{Bmatrix}$$

Ones appear in the "principal diagonal," while all other elements are zero ($\mathbf{1}_{ij} = \delta_{ij}$). The reader will easily verify that the unit matrix, if multiplying any other matrix, leaves the latter unchanged. Hence it is commutative with any matrix.

Next we shall define the reciprocal of a matrix α, denoted by α^{-1}. It is fixed by postulating that $\alpha^{-1}\alpha = \mathbf{1}$. Not every matrix has a reciprocal. For if we try to determine α^{-1} when α is given by using the relation $\alpha^{-1}\alpha = \mathbf{1}$ (which is equivalent to n^2 ordinary equations if

n is the number of columns of α) we find this to be possible only when α is a *square matrix* whose determinant $|\alpha_{ij}| \neq 0$. Hence division by a matrix can be defined only if the latter satisfies these conditions. In our subsequent discussions we shall be concerned with square matrices only.

Every matrix commutes with its own reciprocal. To prove this multiply both sides of the equation $\alpha^{-1}\alpha = \mathbf{1}$ by α in front and by α^{-1} behind. The result is

$$\alpha\alpha^{-1}\alpha\alpha^{-1} = \alpha\mathbf{1}\alpha^{-1} = \alpha\alpha^{-1}.$$

If we now denote the product $\alpha\alpha^{-1}$ by β and make use of the associative law, we have from this:

$$\beta\beta = \beta,$$

which can only mean that β is the unit matrix. Therefore $\alpha\alpha^{-1} = \mathbf{1}$, while $\alpha^{-1}\alpha = \mathbf{1}$ by definition: α and α^{-1} commute.

We are now enabled to define the *function of a matrix*, $f(\alpha)$. Let $f(x)$ be a function of the variable x defined in some domain by the series

$$f(x) = \ldots c_{-2}x^{-2} + c_{-1}x^{-1} + c_0 + c_1 x + c_2 x^2 + \ldots$$

We shall then agree to mean by $f(\alpha)$ the series of matrices

$$f(\alpha) = \ldots c_2\alpha^{-2} + c_1\alpha^{-1} + c_0 1 + c_1\alpha + c_2\alpha^2 + \ldots,$$

where $\alpha^{-2} = \alpha^{-1}\alpha^{-1}$, $\alpha^2 = \alpha\alpha$, etc. Such a function has as many rows and columns as does α itself; it has meaning only if α is square and possesses a non-vanishing determinant. Since α and α^{-1} commute it is evident that one function of α commutes with α itself and with any other function of α.

Another important type of matrix is that which has finite elements in the principal diagonal only while all others vanish. It is known as a diagonal matrix. If α and β are diagonal and have elements $\alpha_{ij} = a_i\delta_{ij}$, $\beta_{ij} = b_i\delta_{ij}$, then

$$(\alpha\beta)_{ij} = \sum_\lambda a_i\delta_{i\lambda}b_\lambda\delta_{\lambda j} = a_i b_i \delta_{ij} = a_j b_j \delta_{ij}.$$

This shows that diagonal matrices commute with one another, and that their product is again a diagonal matrix.

So far we have assumed the number of rows and columns to be finite. This restriction is not essential if there is certainty that all sums encountered in the manipulation of matrices converge. Provided that the matrices themselves, although being infinite, have elements which are large in certain areas of the array but decrease sufficiently rapidly toward the infinite border, this will always be the case. In quantum mechanics, where we are dealing principally with infinite matrices, the latter will be assumed to satisfy this requirement.

Having now acquainted ourselves with the rules of operation respecting matrices we are confronted with the problem of how to construct them conveniently. Among the numerous ways of forming square arrays of numbers one is of particular interest, for it permits an immediate connection between the present subject and the theory of operators discussed in the preceding paragraphs. Let P be an operator generating a complete system of orthogonal functions by means of the equation

$$P\phi_i = p_i\phi_i.$$

The number of different ϕ functions will in general be infinite. If Q is some other operator, defined with respect to the same variables as P, we can form a doubly infinite array of numbers

$$Q_{ij} = \int \phi_i^* Q \phi_j d\tau. \tag{3}$$

This fact in itself is trivial, but it is noteworthy that the rules for combining several operators according to (eq. 3), for the purpose of finding the resulting array, are precisely the same as the rules for combining matrices. In other words, if M and N are two operators, then

$$M_{ij} + N_{ij} = (M + N)_{ij} \tag{4}$$

and

$$(MN)_{ij} = \sum_\lambda M_{i\lambda} N_{\lambda j}. \tag{5}$$

This will now be proved.

Eq. (4) is obvious, because

$$M_{ij} + N_{ij} = \int \phi_i^* M \phi_j d\tau + \int \phi_i^* N \phi_j d\tau = \int \phi_i^*(M + N)\phi_j d\tau$$

$$= (M + N)_{ij}.$$

To demonstrate the rule for multiplication, we write

$$(MN)_{ij} = \int \phi_i^*(MN\phi_j) d\tau \tag{6}$$

and expand the function $N\phi_i$ as follows:

$$N\phi_j = \sum_\lambda a_{\lambda j} \phi_\lambda. \tag{7}$$

The $a_{\lambda j}$ are constant coefficients which we now wish to identify. According to eq. (7)

$$\int \phi_i^* N \phi_j d\tau = \sum_\lambda a_{\lambda j} \int \phi_i^* \phi_\lambda d\tau = a_{ij}$$

because of the orthogonality of the ϕ's. But according to eq. (3), $\int \phi_i{}^* N \phi_j d\tau$ is simply N_{ij}, so that

$$a_{ij} = N_{ij}.$$

Hence eq. (7) may be written:

$$N\phi_j = \sum_\lambda N_{\lambda j} \phi_\lambda.$$

This is to be substituted in eq. (6), which then reads:

$$(MN)_{ij} = \int \phi_i{}^* (M \sum_\lambda N_{\lambda j} \phi_\lambda) d\tau = \sum_\lambda N_{\lambda j} \int \phi_i{}^* M \phi_\lambda d\tau = \sum_\lambda M_{i\lambda} N_{\lambda j}.$$

Thus the proof of eq. (5) is completed.

To form a matrix by our rule (3), two things are required: first, an operator such as Q; second, a complete orthogonal system of functions ϕ_i. This system need not be the system of eigenfunctions of Q; it may belong to some other operator P. If, however, the system of functions is that belonging to the operator itself, we have another consequence: the resulting matrix will be diagonal. In forming the matrix elements of P we observe that

$$P_{ij} = \int \phi_i{}^* P \phi_j d\tau = \int \phi_i{}^* p_j \phi_j d\tau = p_j \delta_{ij}.$$

The diagonal elements of the P matrix are simply the eigenvalues of P. This theorem provides the link between Schrödinger's theory and the matrix theory of Heisenberg, Born, and Jordan, as will be seen presently. But before turning to this subject, let us consider briefly another special type of matrix.

It has been observed that quantum mechanics deals exclusively with Hermitean operators (cf. Sec. 9.5). One would like to know the properties of the matrix corresponding to such an operator. If H is Hermitean, then $\int \phi_i{}^* H \phi_j d\tau = \int (H^* \phi_i{}^*) \phi_j d\tau$. Hence

$$H_{ij} = \int (H^* \phi_i{}^*) \phi_j d\tau = \int \phi_j H^* \phi_i{}^* d\tau = H_{ji}{}^*.$$

A matrix whose elements satisfy the relation $H_{ij} = H_{ji}{}^*$ is also called Hermitean. A real Hermitean matrix is known as a symmetrical one. Its elements consist of equal pairs distributed symmetrically about the principal diagonal: $H_{ij} = H_{ji}$.

9.9. Formal Structure of Matrix Mechanics and Its Relation to the Operator Theory. In 1925 even before the appearance of Schrödinger's famous papers on the subject of wave mechanics, Heisenberg founded a most remarkable calculus to which the name quantum mechanics was originally given. During the succeeding year this calculus was greatly amplified by the work of Heisenberg, Born, and Jordan. The Bohr theory of the atom, which had itself enjoyed splendid successes, was at that time encountering difficulties that appeared insurmountable. Not only was it compelled to yield the clarity of its fundamental axioms by making implausible modifications (introduction of half quantum numbers to account for the observed spectrum of the rotator), but it was definitely incapable of dealing with atoms containing more than one electron (helium problem). Since it operated with the classical notions of electron orbits, electron speeds, and relative phases of electronic orbits, it was constantly encumbered with a mass of difficulties for which there was no experimental counterpart. Heisenberg recognized the trouble: Bohr's theory had grown too complicated; it had woven too close a fabric of (classical) hypotheses to be a useful filter for observable facts. His proposal was to start anew, to build a theory containing nothing but the elements of observation.

The theory was to explain atomic structure. But the main experimental avenue to the knowledge of atomic structure was spectroscopy. Hence the theory had to be built upon the facts of spectroscopy. What, then, are the immediate data of spectroscopy? In their crudest form they are spectral lines of definite frequencies and definite intensities. Upon unraveling the spectrum all frequencies appear as transitions between energy levels the positions of which can be completely determined. The intensities, on the other hand, are simply related to the probabilities of transition between the various levels. In general, therefore, atomic spectroscopy provides us with two sets of numbers: possible energies $E_1, E_2, \ldots E_i \ldots$, and transition probabilities, $T_{11}, T_{12} \ldots T_{ij} \ldots$, T_{ij} being the probability of transition between the ith and the jth energy state, measuring the intensity of the line of frequency $(E_j - E_i)/h$. The latter form a square array reminding us of a matrix, the former a single sequence. If we feel tempted to represent the energies as a matrix also, we should plausibly look for a diagonal matrix which has the energy values strung along the principal diagonal. The problem of Heisenberg, Born, and Jordan was to devise a method of calculating these two matrices. The following ingenious scheme was developed with utter abandon of classical postulates, and its success was striking.

For the present we shall merely state the directions for solving the problem without giving any reasons for them. Afterwards we will show how this heuristic method is related to, and indeed follows from, the operator theory of the preceding sections. Let it be desired to find the observable properties of a physical system which has a classical Hamiltonian function $H(q_1 \ldots q_k, p_1 \ldots p_k)$, where the q's

9.9 FORMAL STRUCTURE OF MATRIX MECHANICS

are coordinates and the p's the conjugate generalized momenta (cf. Chapter III). We then seek a system of $2k$ matrices $Q_1, Q_2 \ldots Q_k$, $P_1, P_2, \ldots P_k$ which satisfy the following conditions:

$$\left. \begin{array}{l} Q_m Q_n - Q_n Q_m = P_m P_n - P_n P_m = 0 \\[6pt] P_m Q_n - Q_n P_m = \dfrac{h}{2\pi i} \delta_{nm} \cdot \mathbf{1} \end{array} \right\} \quad (1)$$

and

$$H(Q_1 \ldots Q_k, P_1 \ldots P_k) \text{ is diagonal.} \quad (2)$$

If such a system can be found, *the diagonal elements of H are the energies of the problem, and the squares of the elements of the Q matrices are the transition probabilities.*

A few words of explanation seem in order. First it is to be noted that the subscripts on Q_1, Q_2, P_1, P_2, etc., distinguish not elements but entire matrices whose elements would be written $(Q_\lambda)_{ij}$, etc. The first part of (1) states simply that all the P and the Q matrices commute among one another, while the second part indicates that a given coordinate matrix does not commute with the matrix assigned to its conjugate momentum. It is interesting to observe that relations (1) are essentially the same as (3.14–21, 24) if only we replace $PQ - QP$ by the *Poisson bracket* (PQ). The formal similarity between the axioms of the present theory and the transformation theory of mechanics, discussed in Sec. 3.14, is indeed very thoroughgoing and has formed a fruitful basis for the development of quantum mechanics (cf. Dirac, "The Principles of Quantum Mechanics"). H in (2) is to be regarded as the function of a matrix, and therefore, of course, as a matrix itself. It is to be constructed by expanding the classical H as a series in the p's and q's and then replacing each of the latter by its associated matrix. We state without proof that conditions (1) and (2) define a set of matrices unique with respect to the diagonal elements of H and the squares of the elements of Q. (The elements of Q are indeterminate to within a constant multiplier of absolute value one. By squares we mean squares of absolute values.) The remark that the squares of the elements of the Q matrices are the transition probabilities T_{ij} was somewhat loose. To explain precisely what is meant let us abandon the generality of the formalism and think of a system with only three degrees of freedom (for instance, the electron in the hydrogen atom). Let us then regard q_1 as x, q_2 as y, and q_3 as z. The directions then state this: $|(Q_1)_{ij}|^2$ is the probability of a transition from the ith to the jth energy level *resulting in the emission (or absorption) of light having its electric vector along* x, $|(Q_2)_{ij}|^2$ is the similar probability for emission with the electric vector along y, etc. Hence the theory is competent to inform us not only about the total probability of transition, which would clearly be

$$T_{ij} = |(Q_1)_{ij}|^2 + |(Q_2)_{ij}|^2 + |(Q_3)_{ij}|^2,$$

but even about the state of polarization of the light emitted or absorbed.

These directions for guidance in the construction of the desired matrices cannot fail to appear to the uninitiated like an alchemist's formula for producing gold. The significant difference, however, lies in the fact that these directions work. Moreover, they are not too difficult to carry out.

Before proceeding to consider the details it is well to discuss a few useful mathematical theorems. The first has to do with *contact transformations*. Such a transformation amounts to a multiplication on the right by some matrix S, and on the left by its reciprocal S^{-1}. A contact transformation with S applied to the matrix H thus converts H into $S^{-1}HS$.

Theorem 1: If H' is Hermitean, there exists a matrix S which, by a contact transformation, will make H' diagonal, i.e., $S^{-1}H'S = H$ is diagonal.

We shall not trouble to prove this theorem.[1]

Theorem 2: A contact transformation applied to a function of a number of matrices has the same effect as its application to every matrix individually; i.e., if α, β, γ, etc., are matrices,

$$S^{-1}f(\alpha, \beta, \gamma, \ldots)S = f(S^{-1}\alpha S, S^{-1}\beta S, S^{-1}\gamma S, \ldots). \tag{3}$$

To prove it, we remember the meaning of a function of matrices. $f(\alpha, \beta, \gamma \ldots)$ is a sum of matrix products each of the type

$$\alpha\gamma\beta\gamma\gamma\alpha\beta\alpha\beta\beta \ldots,$$

and each with a different numerical coefficient. Now we see, for instance, that $S^{-1}\alpha\gamma\beta \ldots S = S^{-1}\alpha SS^{-1}\gamma SS^{-1}\beta SS^{-1} \ldots S$. If this identity is applied to every term of the sum which constitutes $f(\alpha, \beta, \gamma \ldots)$, eq. (3) results immediately.

Let us now see how the directions for finding the matrices in question can be carried out. It is usually fairly simple to find a set which will satisfy conditions (1). These may be called $Q_1' \ldots Q_k'$, $P_1' \ldots P_k'$. But they will not in general make $H(Q_1' \ldots Q_k',$

---

[1] Cf. E. Wigner, "Gruppentheorie und ihre Anwendung auf die Atomspektren" (Vieweg und Sohn, Braunschweig, 1931), p. 29.

9.9 FORMAL STRUCTURE OF MATRIX MECHANICS

$P_1' \ldots P_k'$) diagonal. By means of a contact transformation with some, at present undetermined, matrix S, however, $H(Q_1' \ldots P_1' \ldots)$ can be rendered diagonal. Hence if we write H' for $H(Q_1' \ldots Q_k', P'_1 \ldots P_k')$ and H for the diagonal matrix which we are seeking,

$$S^{-1}H'S = H \tag{4}$$

S is called the *transformation matrix*. If it is known, the energies, i.e., the diagonal elements of H, are given by eq. (4).

Not only that; we can then calculate likewise the elements of the correct Q and P matrices. Eq. (3) shows that, if the correct H is related to H' by (4), then

$$Q_i = S^{-1}Q_i'S, \text{ and } P_i = S^{-1}P_i'S.$$

The solution of our problem is therefore equivalent to a determination of the transformation matrix S.

To complete our discussion we indicate briefly how this is done. Upon multiplication on the left by S, eq. (4) takes the form

$$H'S = SH.$$

The elements of H' in this equation are known. In terms of them, the matrix equation above may be written as a sequence of ordinary equations:

$$\sum_\lambda H_{i\lambda}' S_{\lambda j} = \sum_\lambda S_{i\lambda} H_{\lambda j} = S_{ij} H_{jj}, \tag{5}$$

because H is a diagonal matrix. For any one fixed value of j (which will now be omitted) we find, therefore, that

$$\sum_\lambda (H_{i\lambda}' - H\delta_{i\lambda})S_\lambda = 0, \; i = 1, 2, 3, \ldots \tag{6}$$

(Here H is simply a number.) This is a set of linear equations to be solved for the unknown quantities S_λ. But if it is to have solutions, the determinant of the coefficients of the S_λ must vanish. Let us suppose the maximum value of i to be n (n may be infinite), so that (6) consists of n equations. The vanishing of the determinant, as the reader will easily verify by writing it down, is equivalent to an algebraic equation of the nth degree in H and has therefore n solutions for H, of which some may coincide (case of degeneracy). By introducing any one of the values of H thus determined into (6), n values of the S_λ can be calculated. We therefore obtain n^2 S-values in all, and these form, as they should, the square array composing the transformation matrix.

We have illustrated in principle the method of matrix mechanics; the technique will not be discussed for want of space. The solution of an infinite number of linear equations may appear to present difficulties, but can usually be obtained by fairly simple means. The important point is that this method leads in all cases to which it has been applied to results identical with those attained by solving the operator equations of the preceding paragraphs. Such a coincidence is all the more surprising because of the difference in the methods of calculation—analytical on the one hand and purely algebraic on the other—nevertheless, we hardly expect it to be accidental.

The fundamental equivalence of the two theories was proved by Schrödinger and by Eckart; it is very easily demonstrated on the basis of the developments in the preceding paragraph. Let us think of the matrices which are to satisfy (1) and (2) as being formed from a set of operators according to the rule (9.8–3). The matrices Q' and P' will obey (1) if the operators from which they are formed obey these relations, as follows at once from (9.8–5). Interpreting (1) as operator equations one sees that they are satisfied by the operators:

$$Q_m = q_m, \quad P_m = \frac{h}{2\pi i}\frac{\partial}{\partial q_m}.$$

But this is precisely the assignment which has already been made. (Cf. p. 405.) The elements of the matrices Q_m' and P_m' are then formed by choosing any complete orthogonal system of functions, ϕ_i, and using (9.8–3).

These matrices will not make H diagonal. The condition under which

$$[H(Q_1', \ldots Q_k', P_1' \ldots P_k')]_{ij}$$
$$= \int \phi_i^* H\left(q_1 \ldots q_k, \frac{h}{2\pi i}\frac{\partial}{\partial q_1}, \ldots \frac{h}{2\pi i}\frac{\partial}{\partial q_k}\right)\phi_j d\tau \quad (7)$$

will be diagonal is simply that the functions ϕ be chosen to satisfy the operator equation

$$H\psi_i = E_i\psi_i, \quad (\phi = \psi_i), \quad (8)$$

which is none other than Schrödinger's. Thus the problem of making H in (2) a diagonal matrix *is mathematically identical with solving Schrödinger's equation*. The ϕ's satisfying (8) furnish the diagonal elements $\int \psi_i^* H \psi_i d\tau = E_i$; the equivalence of the two theories is evident. The correct choice of matrices in Heisenberg's matrix mechanics is the exact equivalent of the proper selection of eigenfunctions in the operator method. When, in the former theory, we determine the elements of S by solving (6) we are doing the same as when, in the operator theory, we pass from one system of state functions ϕ which are not solutions of (8) to the ψ_i's satisfying (8). This, indeed, becomes entirely apparent if we try to find the counterpart of the elements of S in the operator theory.

For this purpose we suppose that we have satisfied (1) by matrices with elements $\int \phi_i{}^* q\phi_j d\tau,\ \int \phi_i{}^* \dfrac{h}{2\pi i}\dfrac{\partial}{\partial q}\phi_j d\tau$, which do not make H diagonal because the ϕ's are not solutions of (8).[1] These elements compose the matrix H' of (5), which now reads

$$\sum_\lambda \int \phi_i{}^* H \phi_\lambda d\tau\, S_{\lambda j} = S_{ij} E_j. \tag{9}$$

It is solved if we put

$$\sum_\lambda \phi_\lambda S_{\lambda j} = \psi_j, \tag{10}$$

where ψ is an eigenfunction of (8), for upon substituting this in (9) there results:

$$E_j \int \phi_i{}^* \psi_j d\tau = E_j S_{ij},$$

whence

$$S_{ij} = \int \phi_i{}^* \psi_j d\tau, \tag{11}$$

and this same result is obtained if we multiply (10) by $\phi_i{}^*$ on both sides and integrate. By closer examination (11) can be shown to be the only solution of (9), although we have here merely shown that it is a solution.

What is the physical meaning of S_{ij}? According to (10), S_{ij} is the ith coefficient in the development of the jth energy eigenfunction in terms of the state-functions ϕ. But in view of prop. 2, Sec. 9.3, this implies that $|S_{ij}|^2$ is the probability of measuring the ith eigenvalue of the observable Q for which the ϕ are the eigenstates, when the physical system is in its jth energy eigenstate. On this basis of physical interpretation it is possible to develop the theory of matrix mechanics further, to apply it, for example, to the problem of finding the possible values, not of the energy as we have done in this section, but of any observable. The only modification necessary for that purpose is to demand that, instead of H in (9.9-2), the matrix corresponding to the observable in question be diagonal. But the results of this development, as we now see, cannot be different from those of the operator calculus, and nothing is gained in the way of basic information by further pursuit in this direction. The matrix point of view does, however, simplify many problems; matrix terminology appears to

---

[1] They will in general be eigenstates of some other operator Q.

many investigators to be more concise and direct and therefore enjoys considerable popularity. Thus, instead of saying: a system is in an eigenstate with respect to the energy, many prefer to state briefly that the energy is diagonal. The reader will not fail to see the essential equivalence of these phrases.

Historically, the two theories have greatly fertilized each other. Although each is self-sufficient, an investigator will nowadays rarely limit himself to the use of matrices alone. If the latter are required, they are usually determined by finding eigenfunctions first and then constructing the matrices according to (9.8–3), which is indeed a simpler problem than the solution of an infinite number of linear equations (6).

We have not as yet shown why the squares of the Q_{ij} should be related to the transition probabilities, as Heisenberg's theory postulates. For the present we shall take this for an experimental fact. Later, when we discuss the process of light absorption, we shall present a proof for the remarkable coincidence.

Both formulations of quantum mechanics, operator as well as matrix calculus, are linked with classical physics through the Hamiltonian in a manner which is fairly definite but not always unambiguous. Classical theory does not require the q's and p's appearing in $H(q_1 \ldots q_k, p_1 \ldots p_k)$ to be written in any particular order since they are permutable. In quantum mechanics the order is not indifferent. Ambiguities regarding the arrangement of coordinates and momenta will not arise in problems dealing with physical systems whose classical motions are described by means of scalar potentials. The reason is that in all such cases H contains no products of p's and q's. (The q's are Cartesian coordinates!) A different situation arises when the classical forces depend on velocities. In the problem of a charged particle moving in a magnetic field (e.g., Zeeman effect) the Hamiltonian involves products of the vector potential A, which is in general a function of the q's, and the momenta p. Here the quantum mechanical solution is different depending on whether the combination $A_x p_x$, say, or $p_x A_x$ is chosen. As to the proper choice, classical physics gives us no clue; the correct order has to be ascertained by trial.

9.10. Perturbation Theory. In many physical problems an exact solution of the operator equation is difficult to obtain. It may then become necessary to use approximation methods in order to find the answer. The details of one such method which has proved highly fruitful will now be discussed in connection with Schrödinger's equation. Its application to any operator equation is immediate, but limitation to the energy equation is appropriate, partly for the sake of definiteness, partly because the method is rarely applied in practice to any other case. Let us distinguish at the outset between two possibilities: first the one that the eigenfunction of the energy operator is non-degenerate for the state in question, second that we are dealing with a degenerate state. The second possibility involves complications not appearing in the first, and requiring a somewhat different procedure.

Non-degenerate case. To provide a specific physical basis for our discussion we think of a physical system, like an atom, subject to a "perturbation," that is, to forces which, although they shift the energy levels slightly, do not change the general arrangement of the levels appreciably. Mathematically, the effect of this perturbation is to introduce added terms into the Hamiltonian H of the unperturbed system. These added terms may be constants, or functions of the q's, or functions of the p's and q's. In the last case they appear as differential operators in the Schrödinger equation, otherwise as ordinary functions. Whatever they are, we shall denote them by the symbol V, without implying, of course, that V is anything of the nature of a scalar potential. If then we suppose that the unperturbed equation

$$(H - E_k^0)\psi_k = 0 \tag{1}$$

is solved, our problem is to find solutions of the perturbed equation

$$(H + V - E_k)\phi_k = 0. \tag{2}$$

In particular we wish to know the ϕ_k's, i.e., the eigenfunctions of the perturbed problem, and the E_k's, which we suppose to be slightly, but not greatly, different from the known E_k^0's of eq.(1). V is considered to be a "small" operator, and by this we mean that the matrix elements of V, formed with the use of the complete system ψ_k, are small compared with the diagonal elements of H in this system, i.e., with the E_k^0. We approximate the ϕ's by the series

$$\phi_k = \sum_\lambda a_{k\lambda}\psi_\lambda + \sum_\lambda b_{k\lambda}\psi_\lambda + \sum_\lambda c_{k\lambda}\psi_\lambda + \ldots \tag{3}$$

and the energies by

$$E_k = E_k^{(0)} + E_k^{(1)} + E_k^{(2)} + \ldots \tag{4}$$

and determine the various a's, b's, c's, $E^{(1)}$'s, and $E^{(2)}$'s. To effect the calculation we assume that the terms on the right of (3) and (4) are arranged in descending orders of magnitude, i.e., that

$$a \gg b \gg c, \quad \text{and} \quad E_k^{(0)} \gg E_k^{(1)} \gg E_k^{(2)}$$

unless b and $E_k^{(1)}$ vanish, in which case we simply suppose that

$$a \gg c, \quad \text{and} \quad E_k^{(0)} \gg E_k^{(2)}.$$

We shall not be interested in the higher approximations. If the results of an application of the method to be outlined are inconsistent with these conditions, our scheme fails and the results are not significant. To avoid circumlocutions we shall speak of the a's and $E^{(0)}$'s as quantities of the zeroth order, of the b's and $E^{(1)}$'s as quantities of the first order, etc. The matrix elements of V are considered to be of the first order.

Upon substitution of (3) and (4) in (2) we obtain

$$\sum_\lambda (H - E_k^{(0)}) a_{k\lambda} \psi_\lambda$$
$$+ \sum_\lambda (H - E_k^{(0)}) b_{k\lambda} \psi_\lambda + \sum_\lambda (V - E_k^{(1)}) a_{k\lambda} \psi_\lambda$$
$$+ \sum_\lambda (H - E_k^{(0)}) c_{k\lambda} \psi_\lambda + \sum_\lambda (V - E_k^{(1)}) b_{k\lambda} \psi_\lambda - E_k^{(2)} \sum_\lambda a_{k\lambda} \psi_\lambda$$
$$+ \quad . \quad . \quad . \quad . \quad . \quad . \quad . \quad . \quad . \quad . \quad . \quad . \quad . \quad = 0.$$

The terms are arranged in rows according to their orders of smallness. Let us multiply this equation by ψ_j^* and integrate over all coordinates of the problem. Then, if $V_{jk} = \int \psi_j^* V \psi_k d\tau$ in accordance with (9.8-3), the result is

$$(E_j^{(0)} - E_k^{(0)}) a_{kj}$$
$$+ (E_j^{(0)} - E_k^{(0)}) b_{kj} + \sum_\lambda a_{k\lambda} V_{j\lambda} - E_k^{(1)} a_{kj}$$
$$+ (E_j^{(0)} - E_k^{(0)}) c_{kj} + \sum_\lambda b_{k\lambda} V_{j\lambda} - E_k^{(1)} b_{kj} - E_k^{(2)} a_{kj} \quad (5)$$
$$+ \quad . \quad . \quad . \quad . \quad . \quad . \quad . \quad . \quad . \quad . \quad . \quad . \quad . \quad = 0.$$

In the zeroth approximation we can neglect all but the first row of this equation, which is to be satisfied for any value of j. Putting the first row equal to 0 for $j \neq k$, we find that $a_{kj} = 0$ because the energy difference is then finite. If $j = k$ the equation is identically satisfied and does not yield a value for a_{kk}. We can determine it, however, if we recall that ϕ_k is to be normalized. To this approximation, (3) tells us that $\phi_k = a_{kk} \psi_k$; hence, because ψ_k is already normalized, $a_{kk} = 1$. Our first result is therefore

$$a_{kj} = \delta_{kj}. \quad (6)$$

In accordance with the usual process of making successive approximations we substitute this answer back into (5) and solve again, this time retaining the next order of terms, i.e., the second row. Eq. (6) causes the first row to vanish; we must therefore equate the second row to zero. If $j \neq k$ we see, using (6), that

$$b_{kj} = \frac{V_{jk}}{E_k^{(0)} - E_j^{(0)}}, \; j \neq k; \tag{7}$$

and if $j = k$,

$$E_k^{(1)} = V_{kk}. \tag{8}$$

This constitutes the first order approximation. To get the second, we substitute these results back into (5) and solve again, this time retaining all the terms written down. Now the first two rows vanish, and to make the third equal to zero we proceed as before. If $j = k$ we get

$$E_k^{(2)} = \sum_\lambda b_{k\lambda} V_{k\lambda} - b_{kk} E_k^{(1)},$$

and this, by virtue of (7) and (8), reduces to

$$E_k^{(2)} = \sum_\lambda{}' \frac{V_{k\lambda} V_{\lambda k}}{E_k^{(0)} - E_\lambda^{(0)}}. \tag{9}$$

The prime attached to the summation sign is to indicate that the term for which the two indices are equal must be omitted. On putting $j \neq k$ we get an expression for c_{jk} in which, however, we are not interested since we are limiting ourselves to two approximations. By continuing this process we can obtain higher approximations without difficulty. We are particularly interested in the energy changes, given by (8) and (9). To the first approximation, the increments of the energies due to the perturbation are simply the diagonal elements of the "perturbation matrix"; to the second, they involve all unperturbed frequencies $E_k^{(0)} - E_\lambda^{(0)} = h\nu_{k\lambda}$, where $\nu_{k\lambda}$ is the frequency emitted during the transition from $E_k^{(0)}$ to $E_\lambda^{(0)}$ of the system. It would be of interest to compare these results with the classical formulas which astronomers use to calculate, for instance, the effect of the moon's perturbation upon the earth's orbit. The reader who investigates this matter will find the formulas essentially identical.

Let us illustrate our results by applying them to a simple physical example: the Stark effect. (1) is then the Schrödinger equation for an atom (with one electron, for simplicity); V in (2) is the additional potential energy of this electron in a constant electric field F. If this field is along the x axis

$$V = -eFx.$$

According to (8) the first approximation in the energy is

$$E_k{}^{(1)} = - eF \int \psi_k{}^* x \psi_k d\tau.$$

This energy change represents what is known as the first order Stark effect. Let us suppose that k designates the lowest energy state, and that the atom is hydrogen. (It will be recalled that this state is non-degenerate.) Then ψ_k is given by (9.6–25) and (9.6–27) with $l = m = 0$. In other words, ψ_k is a function of r alone. But if a product of x with a function of r is integrated over all space, the result is 0. The ground state of H is therefore unaffected by the field to this order of approximation; it exhibits no "first order Stark effect." The same is true for the normal states of other atoms and in all other states which have a spherically symmetrical energy state-function.

It is not true, however, that $\int \psi_k{}^* x \psi_\lambda d\tau$ vanishes if $k \neq \lambda$, as the reader can easily verify by substituting the hydrogen eigenfunctions discussed in 9.6d. Hence (9), which yields the second order Stark effect, will not in general vanish. In fact,

$$E_k{}^{(2)} = e^2 F^2 {\sum_\lambda}' \frac{x_{k\lambda} x_{\lambda k}}{E_k{}^{(0)} - E_\lambda{}^{(0)}}, \qquad (10)$$

which can be calculated for any atom if the energy eigenfunctions and the energy levels are known. In classical physics, the increment in energy of an atom due to a static electric field in the absence of a dipole moment is usually written in the form

$$E_k{}^{(2)} = - \frac{\alpha_k}{2} F^2,$$

and α_k is called the polarizability of the kth state. Comparing this expression with (10) we see that the polarizability in quantum mechanics is given by

$$\alpha_k = 2e^2 {\sum_\lambda}' \frac{x_{k\lambda} x_{\lambda k}}{E_\lambda{}^{(0)} - E_k{}^{(0)}},$$

an expression which has been tested experimentally for many instances and found to be correct.

Degenerate case. Eq. (9) is useless when some of the energies $E_\lambda{}^{(0)}$, over which the sum is to be extended, coincide with $E_k{}^{(0)}$, i.e., if the state ψ_k for which the perturbation is to be computed is a degenerate one, since then some of the denominators vanish and we can no longer be sure that the condition $E_k{}^{(0)} >> E_k{}^{(2)}$ is satisfied. In that case we shall have to modify our procedure. It is a somewhat fortunate circumstance that under these conditions we may usually content ourselves with a first approximation, for it is found in practice that this suffices for most purposes. We shall anticipate this fact and work out only first order perturbations.

Let us suppose that the state, the perturbations of which we desire to investigate, possesses an s fold degeneracy, so that there are s eigenfunctions for the eigenvalue $E_k{}^0$. We know these s functions and label them $\psi_{k1}, \psi_{k2}, \ldots \psi_{ks}$. They are considered to be orthogonalized by the method summarized in eq. (9.5–6), so that not only

$\int \psi_{k\lambda}{}^*\psi_l d\tau = 0$, but also $\int \psi_{k\lambda}{}^*\psi_{k\mu} d\tau = 0$ if $\lambda \neq \mu$.

The index l here stands for any state belonging to some other energy $E_l{}^{(0)}$. The problem can then be set up as follows. The solutions of the "unperturbed equations"

$$(H - E_k{}^{(0)})\psi_{k\lambda} = 0, \lambda = 1, 2, \ldots s \\ (H - E_l{}^{(0)})\psi_l = 0 \qquad\qquad \Biggr\} \quad (11)$$

are known. The states ψ_l may, or may not, be degenerate.

Solutions of

$$(H - E_{k\nu} + V)\phi_{k\nu} = 0 \qquad (12)$$

are required. It is now necessary to add another subscript ν to E_k, for the energies will in general not be degenerate after the perturbation. Let us suppose that

$$\phi_{k\nu} = \sum_{\lambda=1}^{s} \alpha_{\nu\lambda}\psi_{k\lambda} + \sum_{l\neq k} a_l \psi_l + \sum_{\lambda=1}^{s} \beta_{\nu\lambda}\psi_{k\lambda} + \sum_{l\neq k} b_l \psi_l + \ldots \qquad (13)$$

and

$$E_{k\nu} = E_k{}^{(0)} + E_{k\nu}{}^{(1)} + \ldots \qquad \bullet(14)$$

The α's and a's in (13) are considered to be of zeroth order, the β's and b's of the first order of smallness. It is merely for convenience and clarity that we have separated the sums over the degenerate members composing the kth state from the others. Substitution of (13) and (14) in (12) yields

$$(H - E_k{}^{(0)})\left(\sum_{\lambda=1}^{s}\alpha_{\nu\lambda}\psi_{k\lambda} + \sum_{l\neq k} a_l\psi_l\right) \\ + (H - E_k{}^{(0)})\left(\sum_{\lambda=1}^{s}\beta_{\nu\lambda}\psi_{k\lambda} + \sum_{l\neq k} b_l\psi_l\right) \\ + (V - E_{k\nu}{}^{(1)})\left(\sum_{\lambda=1}^{s}\alpha_{\nu\lambda}\psi_{k\lambda} + \sum_{l\neq k} a_l\psi_l\right) \qquad (15) \\ + \quad . \quad . \quad . \quad . \quad . \quad . \quad . \quad . \quad = 0.$$

Terms in the first row are of zeroth order, those in the second and third of the first order. If we multiply (15) on the left by ψ_m^*, where $m \neq k$, and integrate, neglecting the second and third lines we find that every $a_l = 0$. Thus we may, to the approximation here sought, neglect all the a's. Now let us multiply (15) by $\psi_{k\mu}^*$, any one of the s degenerate functions, and integrate. Because of (11), nothing will then be left of the first line, while the second and third give

$$\sum_{\lambda=1}^{s} \alpha_{\nu\lambda}(V_{\mu\lambda} - E_{k\nu}^{(1)}\delta_{\mu\lambda}) = 0.$$

Here we have written $V_{\mu\lambda}$ for $\int \psi_{k\mu}^* V \psi_{k\lambda} d\tau$. This equation must be true for every value of μ from 1 to s. It is therefore a system of s linear equations from which the coefficients $\alpha_{\nu\lambda}$ are to be determined. We can simplify the writing if we omit the fixed indices k (designating the group of originally degenerate levels) and ν (numbering the members of this group). Then these equations are

$$\left.\begin{array}{l} (V_{11}-E^{(1)})\alpha_1 + V_{12}\alpha_2 \quad\quad + V_{13}\alpha_3 + \ldots + V_{1s}\alpha_s = 0 \\ V_{21}\alpha_1 \quad\quad + (V_{22}-E^{(1)})\alpha_2 + V_{23}\alpha_3 + \ldots + V_{2s}\alpha_s = 0 \\ \cdot \cdot \cdot \cdot \cdot \cdot \cdot \cdot \cdot \cdot \cdot \cdot \cdot \cdot \cdot \cdot \cdot \cdot \cdot \cdot \cdot \cdot \cdot \cdot \cdot \\ V_{s1}\alpha_1 \quad\quad + V_{s2}\alpha_2 \quad\quad + V_{s3}\alpha_3 + \ldots + (V_{ss}-E^{(1)})\alpha_s = 0. \end{array}\right\} (16)$$

They can be solved for the α's only if the determinant

$$\begin{vmatrix} (V_{11}-E^{(1)}) & V_{12} & V_{13} \ldots V_{1s} \\ V_{21} & (V_{22}-E^{(1)}) & V_{23} \ldots V_{2s} \\ \cdot & \cdot & \cdot \\ V_{s1} & V_{s2} & V_{s3} \ldots (V_{ss}-E^{(1)}) \end{vmatrix} = 0. \quad (17)$$

The expansion of this determinant produces an equation of the sth degree in $E^{(1)}$, and this has at most s different roots. Hence there are at most s different values of $E^{(1)}$ for which (16) can be satisfied. They are the new energies $E_k^{(1)}$ which we set out to find. If some of the roots coincide, some of the $E_k^{(1)}$ coincide: the perturbation has not removed the degeneracy completely. Eq. (17) is often called the "secular" equation for the determination of the perturbed energies. If V is zero, (17) has the s equal solutions $E^{(1)} = 0$; as the $V_{\mu\lambda}$ increase in magnitude, the energies move apart. Figuratively speaking, the perturbation forces the s initially coincident energies apart.

For every $E_k^{(1)}$, introduced in (16), there is one set of α's, s in number. Altogether, this method allows therefore the calculation of s^2 coefficients α, which is, of course, just the required number. We shall not attempt to calculate the remaining coefficients in (13) since they will not be needed. Since the a's are zero, we have already found the full zeroth approximation for ϕ_k.

A study of the Zeeman effect provides a very simple illustration of the method discussed. Let us again think of an atom with one electron in a central field. The reader will recall from (9.6–35a), that, in the absence of a perturbation, the eigenfunctions are

$$\psi_{k\lambda} = c_1 R(r) \cdot c_2 P_l^{|m|}(\cos\theta) \cdot c_3 e^{im\phi}. \tag{18}$$

They possess a $(2l+1)$ fold degeneracy corresponding to the fact that the energy does not depend on m, and there are $2l+1$ possible values of m for a given integer l. Hence λ runs from 1 to $s = 2l+1$, and the various degenerate eigenfunctions can be numbered in such a way that $\lambda = 1$ corresponds to $m = -l$, $\lambda = 2$ to $m = -l+1$, etc., and finally $\lambda = 2l+1$ corresponds to $m = +l$.[1] Now it can be shown that, if a uniform magnetic field $\mathcal{H}$ along the z axis is present, the added term [2] in the Hamiltonian operator is

$$V = \frac{he}{4\pi i M c} \mathcal{H} \frac{\partial}{\partial \phi},$$

where M is the mass of an electron. In forming the matrix elements of V between the degenerate states (18) every integral will contain the part:

$$c_1{}^2 c_2{}^2 \int [R(r) P_l^{|m|}(\cos\theta)]^2 r^2 \sin\theta \, dr \, d\theta,$$

which is 1 since each of the three factors of (18) is normalized individually. The remainder of each integral will be

$$\frac{he\mathcal{H}}{4\pi i M c} \cdot \frac{1}{2\pi} \int_0^{2\pi} e^{-im_1\phi} \frac{\partial}{\partial \phi} e^{im_2\phi} d\phi = \frac{he\mathcal{H}}{4\pi M c} m_2 \cdot \frac{1}{2\pi} \int_0^{2\pi} e^{i(m_2 - m_1)\phi} d\phi.$$

m_1 and m_2 are the two m values for the states between which the matrix element is taken, and $1/\sqrt{2\pi}$ is the normalization factor c_3. But this integral is 0 unless $m_2 = m_1$. We see that all non-diagonal elements of V vanish and that the diagonal ones are

$$V_{11} = -Al, \; V_{22} = -A(l-1), \; V_{33} = -A(l-2) \ldots V_{ss} = Al$$

---

[1] In the case of a Coulomb field the degeneracy is even greater!
[2] Cf., for instance, Condon and Morse, p. 128.

where A has been written for $he/4\pi Mc \cdot \mathcal{H}$. Thus the secular equation (17) reads:

$$\begin{vmatrix} -lA - E^{(1)} & 0 & 0\ldots 0 & 0\ldots 0 \\ 0 & -(l-1)A - E^{(1)} & 0\ldots 0 & 0\ldots 0 \\ \cdot & \cdot & \cdot & \cdot \\ 0 & 0 & 0\ldots -E^{(1)} & 0\ldots 0 \\ \cdot & \cdot & \cdot & \cdot \\ 0 & 0 & 0\ldots 0 & 0\ldots lA - E^{(1)} \end{vmatrix} = 0.$$

It is evident that the solutions are $E^{(1)} = mA = m \cdot \dfrac{he\mathcal{H}}{4\pi Mc}$, where m can take any integral value from $-l$ to $+l$. Thus every degenerate energy state with azimuthal quantum number l will, under the influence of the field, "split up" into $2l + 1$ sublevels, all equally spaced and grouped symmetrically about the unperturbed level $E^{(1)} = 0$, the energy difference being $he\mathcal{H}/4\pi Mc$. What effect will this have upon the appearance of spectral lines emitted in the presence of a field?

Let us consider a spectral line which is due to the transition of an atom from an energy state in which $l = 2$ to another state in which $l = 1$. If the horizontal lines in Fig. 9.3 (a) represent the two energy levels in question, the frequency of the line will be proportional to the length of the arrow between them. When a magnetic field is applied, the levels split up as shown in (b). The two sets of levels can be connected by fifteen different arrows (not those drawn!) representing seven different lengths. We might expect, therefore, that instead of one line we should see seven lines when the atom is in a field. But we have overlooked an important fact. In Sec. 9.9 we saw that the probabilities of transition are given by the squares of the matrix elements of the coordinates. If we calculate these we find that they vanish whenever the two states between which a transition is considered have values of m different by more than 1. Transitions which violate this rule have a zero probability, and hence do not take place.

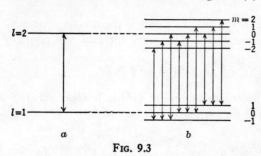

FIG. 9.3

We should, therefore, connect by arrows only those levels of (b) whose m's differ by 0 or 1. In that way we obtain the arrows drawn. It will be seen that, although there are nine arrows, there occur only three different lengths, one equal to that of the arrow in (a), one shorter, and one longer by an equal amount. Hence the line should appear as a symmetrical triplet under the influence of the magnetic field, and the separation between its components, on a frequency scale, should be $1/h$ times the energy difference between the levels in (b); i.e., it should be

$$\Delta \nu = \frac{e \mathcal{H}}{4\pi M c}.$$

This quantity is known as the Larmor frequency.

The theory here presented explains the "normal" Zeeman effect and is in perfect agreement with experiment. However, we have not told the entire story. In general, the situation is not as simple as that here considered, for the electron has magnetic properties not considered in our expression for V. It possesses indeed a magnetic moment (spin) which we have left out of account. Our theory, therefore, does not describe all features of the Zeeman effect but only its simplest aspects.

9.11. Quantum Dynamics. *Time-dependent states in general.* Thus far it has been our principal endeavor to *classify* the possible states of physical systems. We have been confronted with the problems of finding eigenstates and eigenvalues, of interpreting the meaning of eigenstates in terms of observations. But we have always taken these states to be unchangeable, for the state-functions did not involve the time. Indeed we have been careful to consider only operators whose form was independent of the time, so that their eigenfunctions could be completely formed without the time coordinates. In brief, our considerations have been limited to stationary states.

It is important to observe that a stationary state in quantum mechanics is altogether different from the corresponding notion in classical physics. A stationary state is one whose description in terms of the variables of state is possible without the explicit use of the time. In classical physics, where the state variables are the p's and q's, a system is said to be stationary when neither the p's nor the q's vary with the time (i.e., the p's are zero in some inertial system). The branch of mechanics dealing with such systems is called statics. In quantum mechanics, however, the variables describing states are the state-functions, and a system is in a stationary state if its state-function does not involve the time. Thus the electron in a central field, treated in 9.6(*d*), or the simple harmonic oscillator of 9.6(*b*), are in stationary states from the quantum mechanical point of view, while the classical descriptions of these motions would clearly exhibit them as instances of non-stationary, i.e., dynamical, phenomena. Hence, if we were to impose the customary division into statics and dynamics upon quantum mechanics, quantum statics would be a much larger field than its classical counterpart. Let us now turn to quantum dynamics.

The fundamental axioms of Sec. 9.2 afford no clue as to the treatment of cases which change in time. It will be necessary, therefore, to introduce a new and additional postulate regulating the behavior of state-functions in time. The only safe demand to be made upon this new principle is that, beside being simple and plausible, it should not violate the previous axioms. All further guidance in its formulation must be derived from pragmatical considerations of its utility and freedom of contradiction with experiment. The principle has been found by Schrödinger to be this: If $u(q_1 \ldots q_k, t)$ is a time-dependent state-function and H the energy operator, then u obeys the differential equation

$$Hu = -\frac{h}{2\pi i}\frac{\partial}{\partial t}u. \qquad (1)$$

The relation between this u and the previous time-free state-functions, and our reason for using a new symbol, will be made clear presently. Let us first see how one could be led to the formulation of (1).

The principal operators of quantum statics are the coordinate q and the momentum operator $p = \frac{h}{2\pi i}\frac{\partial}{\partial q}$. Their importance arises partly from the fact that p and q are canonically conjugate quantities. E, the energy, and the time t are also canonically conjugate, and this suggests that their operators may be similarly related. If, then, we treat t like q and assign to it the operator: multiplication by t, E should be represented by $\frac{h}{2\pi i}\frac{\partial}{\partial t}$. Hence the operator equation should read $\frac{h}{2\pi i}\frac{\partial}{\partial t}u = Eu$. But in conformity with previous axioms, $Eu = Hu$; if we observe these facts there results eq.(1) with the wrong sign for the H operator. As far as the physical significance of the equation is concerned the sign of i must, however, be immaterial and its choice arbitrary, because our interpretation in terms of experience is concerned with absolute values alone. Custom and convenience have standardized the use of the sign as given in (1), and we shall adhere to this custom.[1] The reader must not feel that these remarks are intended as a proof of eq. (1), which should rather be considered as an axiom to be added to the list of Sec. 9.2.

To be sure, the postulate of quantum dynamics is simple, and we have shown it to be plausible, but is it consistent with those of quantum statics? This matter can easily be tested by supposing the operator H in (1) to be stationary in the sense that it does not contain the time. It then has the form considered previously, and we should in some manner be led back to the stationary energy eigenfunctions.

Let us write u as a product of one function ψ depending only on the coordinates of configuration space, and another function $f(t)$. Since H acts only on the coordinates, eq. (1) becomes

$$f(t)(H\psi) = -\frac{h}{2\pi i}\dot{f}(t)\psi.$$

On dividing through by $\psi \cdot f$ we have

$$\frac{H\psi}{\psi} = -\frac{h}{2\pi i}\frac{\dot{f}}{f}.$$

But here, the left-hand side depends on the space coordinates, the right on t. Hence each side must equal a constant. We therefore have two equations:

$$H\psi = \text{constant} \cdot \psi \quad \text{and} \quad \dot{f} = -\frac{2\pi i}{h} f \cdot \text{constant}.$$

The first of these permits us at once to identify both ψ and the constant: ψ must be one of the former eigenfunctions of H, and the constant can only be one of the eigenvalues of H, say E_k. This means that the second equation has the solution $f = e^{-\frac{2\pi i}{h}E_k t}$, so that

$$u_k = \psi_k e^{-\frac{2\pi i}{h}E_k t}. \tag{2}$$

Since (1) contains no explicit eigenvalue, it is also satisfied by the more general expression

$$u = \sum_\lambda a_\lambda u_\lambda, \tag{3}$$

where the a's are independent of coordinates and time. It is to be remembered that both eqs. (2) and (3) were derived on the assumption that H does not contain t. If this is not true they are not solutions of (1).

---

[1] If the sign is thus chosen, the u function of a free electron with momentum p in the x direction becomes the function of a wave progressing in the x direction; if it is reversed the wave would progress in a direction opposite to that of p.

So far the present theory is a harmless extension of the previous postulates, for u_k and ψ_k differ only by a factor of absolute value 1 and constant in configuration space, which, as we have seen before, is of no physical consequence. Indeed we could restate all theorems affecting the eigenfunctions of H, with the ψ_k's replaced by the u_k's. In particular the mean value relation [cf. (9.2–III) and (9.3-2)], in the form

$$\int \phi^* H \phi \, d\tau = \sum_\lambda |a_\lambda|^2 E_\lambda,$$

is true even if we replace $\phi = \sum_\lambda a_\lambda \psi_\lambda$ by the u of eq. (3). The reader can easily verify this statement by substitution.

Having assured ourselves that (1) leads to the old results when H does not involve t we may now apply this equation to truly dynamical problems where

$$H = H(q_1 \ldots q_k, \frac{h}{2\pi i}\frac{\partial}{\partial q_1}, \ldots \frac{h}{2\pi i}\frac{\partial}{\partial q_k}, t).$$

Theoretically, our equation determines $u(t)$ completely if its value u_0 at any given time t_0 is known. However, $u(t)$ will not be related to u_0 in the trivial fashion of (2), that is, $u(t) \neq u_0 e^{-\frac{2\pi i}{h}Et}$. The meaning of $u(t)$ is of course precisely that of the state-functions discussed earlier. $u^*(q_1 \ldots q_k, t)u(q_1 \ldots q_k, t)$, for instance, is the probability density of finding the system in question at the point $q_1 \ldots q_k$ at time t. Also, if $w_i(q_1 \ldots q_k, t)$ is a complete set of eigenfunctions belonging to some operator $R(t)$ with observable eigenvalues $r_i(t)$, then the probability of measuring the value $r_i(t)$ when the system is in the state $u(q_1 \ldots q_k, t)$ is given by $|a_i|^2$, where $u = \sum_\lambda a_\lambda w_\lambda$. (Cf. Sec. 9.3, prop. 2.)

We shall not consider exact solutions of eq. (1) for special forms of H, although it is possible to obtain such solutions. Our interest at present is in an approximate method of solving the equation, but a method applicable regardless of the form of H. It is known as the *method of variation of constants* and was originally applied to quantum dynamics by Dirac.

Suppose that a physical system has been isolated from its surroundings for a long time. Its Hamiltonian function H, during this time, will be independent of t. If it happens to be in an eigenstate of the energy, its state function will be ψ_k, satisfying

$$H\psi_k = E_k \psi_k,$$

or if we wish to state explicitly the trivial time dependence of the state function, we write

$$Hu_k = -\frac{h}{2\pi i}\frac{\partial}{\partial t}u_k, \quad u_k = \psi_k e^{-\frac{2\pi i}{h}E_k t}. \quad (4)$$

If it is not in an eigenstate of the energy, we can expand its state function ϕ, or u, in the two equivalent ways

$$\phi = \sum_\lambda a_\lambda^0 \psi_\lambda$$

or

$$u = \sum_\lambda a_\lambda^0 u_\lambda, \quad (5)$$

where the a^0's are constants (independent of the time). Let us now suppose that at the time $t = 0$ the system is suddenly exposed to a perturbation varying with the time, and adding the function or operator $V(t)$ to H. To describe this perturbation we must abandon the method of time-free state-functions and use expansions of the second type. In a short time t the initial state-function u will have transformed itself into a different function v, but v can still be expanded as a series of the type

$$v = \sum_\lambda a_\lambda(t) u_\lambda \quad (6)$$

if we take the a's to be functions of t. The reason is that each u_λ, aside from a time factor, is a member of the complete set ψ_λ. But v satisfies the equation

$$[H + V(t)]v = -\frac{h}{2\pi i}\dot{v}$$

which, if (6) is substituted for v, takes the form

$$\sum_\lambda a_\lambda(Hu_\lambda + Vu_\lambda) = -\frac{h}{2\pi i}\sum_\lambda (\dot{a}_\lambda u_\lambda + a_\lambda \dot{u}_\lambda).$$

The first sum on the left cancels the second on the right because of (4). Let us now multiply the remainder of this equation by u_j^* on the left and integrate over configuration space. The result is

$$\dot{a}_j = -\frac{2\pi i}{h}\sum_\lambda a_\lambda \int u_j^* V u_\lambda d\tau, \quad j = 1, 2, 3, \ldots \quad (7)$$

The integral on the right is a known function of t. There will in general be an infinite number of quantities a_λ, all unknown functions of t, and an infinite number of equations. Hence (7) is a linear set of first order differential equations from which the a_λ can, in principle, be determined if their values at any given time are known. According to (5), however, their values at $t = 0$ are $a_\lambda{}^0$, and these coefficients are assumed to be known. The problem is therefore perfectly determinate.

But the exact solution of (7) is in practice impossible, so that we have to look for approximate methods of solution. Fortunately we are not often interested in the fate of the perturbed system after a long time, but in its immediate response to the perturbing agency. That is, we desire mainly to know the a's after a very short time. Now for that case a good approximation is obtained by identifying the a's on the right of (7) with the a^0's. When this is permissible, each equation can be integrated individually and the result is at once obtained.

Absorption of light by atoms. An application of this method to the case of an atom exposed to radiation provides a most interesting example. We shall suppose that at the instant $t = 0$ the atom is in its lowest energy state. Let the states be numbered with increasing energy. Then, in terms of our previous notation,

$$a_1{}^0 = 1, \text{ all other } a^0\text{'s are } 0. \tag{8}$$

For the sake of simplicity let us assume that the incident light wave is monochromatic (of frequency ν) and plane polarized (electric vector along x). It will then constitute an electric field of magnitude

$$F = F_0 \sin 2\pi \left(\nu t - \frac{z}{\lambda} \right)$$

(z is distance in direction of propagation) and a magnetic field at right angles. The electron in the atom is subjected to both of these, and its energy changed thereby. But the magnetic vector of the light wave will hardly affect the electron because the latter has a small average momentum. The only appreciable interaction arises from the electric vector. The expression for this vector, however, can be simplified if we consider radiation of fairly large wave length, such as visible light. The atom itself (quantum mechanically, its state-function) extends over a region of a few Å units, whereas λ is several thousand Å units. Hence if z is measured from any point within the atom, z/λ is very small in the whole region in which interaction can take place. The term νt, however, is of appreciable magnitude, for ν is about 10^{15}/sec, and although we shall assume t to be small, our method permits it to be at least about 10^{-10} sec.[1] We can, therefore,

---

[1] t must be much smaller than the "mean life" of the atomic states.

take F to be $F_0 \sin 2\pi\nu t$, without sensible error, so that the perturbing term in the Hamiltonian becomes

$$V = -eF_0 x \sin 2\pi\nu t. \tag{9}$$

If (8) and (9) are substituted in (7) the result is

$$\dot{a}_j = \frac{2\pi i}{h} eF_0 x_{j1} \sin(2\pi\nu t) e^{\frac{2\pi i}{h}(E_j - E_1)t}$$

x_{j1} is the usual matrix element (cf. 9.8-3) formed between the jth and the ground state. Let us put $\dfrac{E_j - E_1}{h} = \nu_{j1}$, a positive quantity since $E_j > E_1$. Then, on expanding the sine and integrating we have

$$a_j = \frac{eF_0 x_{j1}}{2ih} \left\{ \frac{e^{2\pi i(\nu_{j1} + \nu)t}}{\nu_{j1} + \nu} - \frac{e^{2\pi i(\nu_{j1} - \nu)t}}{\nu_{j1} - \nu} \right\} + C.$$

C, the constant of integration, has to be determined in such a way that $a_j = 0$ when $t = 0$ for $j \neq 1$. But let us first compare the two terms in brackets. Both numerators are of absolute value 1. ν in the denominator is of the order 10^{15} sec^{-1}. If we consider in a qualitative manner the variation of a_j with ν, the frequency of the incident light, we see that a_j is very small unless ν lies in the neighborhood of ν_{j1}. It is to this case that we shall confine our attention. We can thus safely omit the first term in brackets, and write our result, on proper adjustment of C, for $j \neq 1$

$$a_j = i \frac{eF_0 x_{j1}}{2h} \left(\frac{e^{2\pi i(\nu_{j1} - \nu)t} - 1}{\nu_{j1} - \nu} \right). \tag{10}$$

But now, what is the physical meaning of a_j? The reader will recall that $|a_j|^2$ is the probability of finding the system in the state ψ_j (or u_j) which has an energy E_j, at the time t. In other words, it is the probability that, t seconds after the atom has been exposed to the radiation, the atom has made a transition to the jth energy state. The energy necessary for this excitation is naturally withdrawn from the radiation field. The chance that the atom will make a transition is therefore the same as the chance that radiation will be absorbed. Hence the importance of the quantity $|a_j|^2$ is quite fundamental in physical problems. Indeed it is greater than might appear from this argument, for it may be shown by statistical reasoning that the probability of emission of a light wave of frequency ν_{j1}, and hence its intensity, is very simply related to $|a_j|^2$. Let us then compute this quantity.

According to (10)

$$|a_j|^2 = \frac{e^2 F_0^2 |x_{j1}|^2}{2h^2} \cdot \frac{1 - \cos 2\pi(\nu_{j1} - \nu)t}{(\nu_{j1} - \nu)^2}. \tag{11}$$

This result yields a great deal of useful information. First, the transition probability is proportional to the square of the matrix element x_{j1}. This was a postulate in Heisenberg's matrix theory (cf. Sec. 9.9); we have here deduced it from other, more fundamental axioms. Second, the transition probability is proportional to the square of the amplitude of the electric vector, and therefore to the intensity of the incident radiation, a result which is expected on the basis of the classical theory of radiation. Third, we have essentially deduced Bohr's frequency requirement which claims that the frequency absorbed by an atom during a transition from E_1 to E_j is given by $(E_j - E_1)/h$. To realize this, imagine $|a_j|^2$ plotted against ν. The graph will have a steep maximum at $\nu = (E_j - E_1)/h = \nu_{j1}$, the ordinates being small everywhere else. This means that the atom will not absorb light with appreciable intensity unless ν is very nearly equal to ν_{j1}. The stringency of Bohr's condition according to which no light of other frequencies should be absorbed is mitigated by the present theory, as experiment indeed requires. Finally (11) gives us some information about the manner in which the absorption occurs in time. In the plot of $|a_j|^2$ against ν, the principal maximum at $\nu = \nu_{j1}$ (which corresponds to the intensity plot of an absorption line) will be broad if t is small, becoming narrower as t increases. If the incident radiation acts only for a short time, the absorption band will be broad, and vice versa.[1]

Eq. (11) represents the transition probability when the light wave has its electric vector along x. Had we chosen it along y or z, we would have obtained the same expression with x_{j1} replaced by y_{j1} or z_{j1}. If, for a given transition, x_{j1} and y_{j1} were found to vanish, this would mean that in the transition only light of a definite polarization, i.e., having its electric vector along z, could be absorbed; and if emission took place we could similarly conclude that only light of this state of polarization could originate in the reverse transition. Hence (11),

---

[1] This fact has an interesting physical consequence. One can easily prove that the "half width" of the principal maximum of (11), that is the frequency range over which $|a_j|^2$ is greater than half of its maximum, is approximately $1/t$. Now one feature which can limit the time of application of the light is the finite lifetime of the atomic energy states. Clearly, if one of the states involved in the transition can exist only for a time Δt we cannot properly inquire, using our present method, what is the chance of finding the atom in the jth state after a time longer than Δt. Hence the half width of a line should be at least $1/\Delta t$. Now in some cases Δt can be determined by independent experiments and is found to be, in general, of the order of magnitude 10^{-8} sec, except for the ground state. The half width should therefore be of the order 10^8 sec^{-1}, which is about the limit of sharpness of spectral lines actually observed.

if applied to spectroscopy, will not only "select" possible transitions (those for which coordinate matrices are different from zero) and determine their probability of occurrence, but will also function as a "polarization rule."

9.12. Review and Orientation. Our discussion of the fundamental principles of quantum mechanics has now reached a point at which it seems well to pause for a moment and to determine our bearings. The first ten sections dealt entirely with the development and expansion of a set of very general axioms regarding the interpretation of nature, and with their application to simple cases of physical interest. These cases were all characterized by stationarity. Putting it another way, the systems whose behavior was studied were regarded as closed. Aside from incidental mathematical difficulties, no important obstacles were encountered in these applications, and the results were in reasonable harmony with experiment. Only at times did we find it necessary to admit that our treatment was idealized, particularly in connection with the Zeeman effect. But if we do not carry our scrutiny too far the theory provides an excellent background for physical observations. Its principal beauty, however, is on the logical side; for it is of extreme simplicity,[1] and its texture is perfectly uniform.

There appears to be a slight break in both the conceptual and the structural continuity of the theory when temporal changes are incorporated in it. We have seen in the last paragraph that the required extension of the basic posulates is indeed suggested by these postulates themselves; nevertheless, one cannot help feeling that time is introduced in the manner of an afterthought. It does indeed not appear on the same footing as the space coordinates. As a consequence of this fact the dynamical law (9.11-1) cannot, of course, be Lorentz-invariant. But we have not at present attempted to formulate a relativistic theory and should not hold this feature against our method. Another consequence of this lack of symmetry between time and space coordinates is apparent in the structure of the general solution of (9.11-1), that is, the expansion (9.11-6). If we write this in the form $u = \sum_\lambda b_\lambda(t)\psi_\lambda$, so that the time factor of u_λ is contained in b_λ and ψ_λ is a function of space coordinates alone, the latter form a complete orthogonal set in the fundamental range of the space coor-

---

[1] The common complaint that quantum mechanics is complicated and difficult to understand arises mainly as the result of two errors: (1) confusion of unfamiliarity with difficulty, (2) failure to realize that many of the problems solved by quantum mechanics are precisely those which classical physics, even with the use of its "most difficult" devices, was unable to handle.

dinates (cf. Sec. 9.7). This completeness does not exist for the b_λ, considered as functions of the time.

The same situation can be viewed from another angle. Eq. (9.11-1) is not of the Sturm-Liouville type as are the operator equations studied previously. As a result, there are no eigenvalues corresponding to the time coordinate. Thus, while a physical problem is in general described by as many quantum numbers as it requires space coordinates (cf. examples of Sec. 9.6), the time coordinate produces no quantum number of its own. These peculiarities have not been entirely removed even by the very recent innovations in quantum theory of which we shall speak later. Whether their elimination would open the way to further progress is unknown, but seems possible.

In spite of these slight formal flaws the theory is self-consistent up to this point. If we wish to go further we shall have to introduce elements of an extraneous character, things which, although they fit nicely into the general scheme, do not naturally evolve from it. Notable among these are the electron spin [1] and Pauli's exclusion principle. It is impossible to avoid them because experimental evidence undeniably demands their consideration. In the subsequent paragraphs we shall be mainly concerned with their study. Since we feel, however, that these things belong to a field in which considerable changes of approach are likely to occur, or at least are possible, it may be well to proceed from now on with somewhat less detail, even if it should be necessary for the reader occasionally to accept mathematical results without detailed proof.

9.13. The Electron Spin. *Preliminary considerations.* A careful comparison with experiment of the verifiable results so far obtained shows the following major discrepancies. In Sec. 9.6(d) we concluded that the energy levels of an atom whose electron is subject to a central field should depend only on the two quantum numbers n and l, unless it is placed in an external field. Now the hydrogen atom certainly belongs to this category and should show this characteristic.[2] As long as experimental accuracy was not too great this was believed to be the case; but when the spectral lines of hydrogen were finally investigated with instruments of great resolution they were found to possess a

---

[1] It is only fair to say that the spin results quite naturally from Dirac's generalization of the Schrödinger theory, as we shall see later. But this generalization leads to other consequences which are not yet entirely convincing as to their finality.

[2] Reference should here be made to Sommerfeld's relativistic treatment of the Bohr atom to which the present remarks do not apply. This was singularly successful in explaining the "fine structure" of the H lines but has now been superseded by a new interpretation. (Cf. Sec. 9.17.)

structure incompatible with the above conclusion. All energy levels except those for which $l = 0$ were split into two; hence they depended on something besides n and l, and this something was not the quantum number m since all levels were double regardless of the variety of possible m's.

The spectra of the alkalis were long known to possess features contradictory to our theoretical conclusion. The two components of the well-known sodium doublet of wave lengths 5890 Å and 5896 Å both originate in a transition from an energy state for which $n = 4$, $l = 1$ to another state for which $n = 3$, $l = 0$. Why, then, should there be two lines? Again we see that one of the states must be double.

Finally, there are but few cases in which the Zeeman effect exhibits the simple features deduced in Sec. 9.10. The line in the absence of a field usually resolves itself into more than three components when the magnetic field is applied. There is evidently some feature which so far we have failed to bring into the analysis, another degree of freedom which we have neglected.

As early as 1925 Uhlenbeck and Goudsmit suggested an explanation which brought all these discrepancies into line and appeared to eliminate the difficulties. They ascribed to the electron a motion of rotation about its axis and a consequent magnetic moment, the rotation being now known as the electron spin. They found that, if the additional angular momentum thus produced is $h/4\pi$ and the corresponding magnetic moment has the value $he/4\pi mc$ (known as the Bohr magneton), where m is the mass of the electron, the experimental results can be explained. Their method of reasoning was semi-classical, as indeed the early date of their discovery necessitated. Describing the electron as a sphere of small but finite radius they could account correctly for the ratio of its mechanical to its magnetic moment. In a magnetic field, however, the little gyroscope would forget its classical manners and set its axis of rotation either parallel or anti-parallel to the lines of force of the field with complete disregard of other possibilities. If one calculates classically the increments in energy in these two modes of alignment of the spinning electron and adds them to the energies previously computed the result is in agreement with experiment.

In a fundamental sense this can be but an interesting coincidence pointing the way to further discoveries. The first question which arises is this: can this explanation be transcribed into consistent quantum mechanical terminology? Is it possible to endow the electron with an additional degree of freedom which, when introduced into the Hamiltonian, leads to the results observed? Among the results to be explained are these: the operator corresponding to this new degree of freedom, which we shall metaphorically continue to call "spin," must have *only two eigenvalues* corresponding to the values $\pm \frac{h}{4\pi}$ of the angular momentum in any one direction in which the external field is imagined to be placed. The magnetic moment along any axis should have the two eigenvalues $\pm \mu$, where

$$\mu = \frac{he}{4\pi mc}; \qquad (1)$$

and the energy in a magnetic field $\mathcal{H}$, in the absence of all other forces, should be $\pm\mu\mathcal{H}$. The first possibility which comes to mind is to introduce into the usual Hamiltonian a term expressing the energy of a spinning electron, and then to replace the angular momentum by its quantum mechanical operator. This procedure fails, mainly for two reasons: it makes the eigenvalues for both energy and angular momentum twice too large, and it produces too many eigenvalues. One might then try some entirely new differential operator for the spin, unhampered by classical analogies, and attempt to produce the desired results. This procedure, also, is without success. To make a long story short, it is found very difficult to obtain a differential operator (of the type used hitherto in our examples) which, acting upon a function of a continuous variable, has only two eigenvalues.

Pauli's spin theory. The first successful solution of this perplexing problem was given by Pauli. His success was due to the use of a method not previously employed in quantum mechanics, but essentially consistent with its postulates. Instead of continuing the fruitless search for an operator with two eigenvalues Pauli forced the issue by choosing a variable whose range consists only of two points. We shall illustrate his theory by considering first an electron which has no energy other than that due to its spin.

Let s be the variable which measures the spin along some direction, say z, so that the state-function ϕ is $\phi(s)$ in the same sense as the state-functions previously encountered were functions of coordinates. These coordinates had definite ranges which extended usually from a lower to an upper limit, but none of the points between these limits were excluded. We shall now suppose that the range of s consists of two points: $+1$ and -1. If the reader wishes he may think of s as the cosine of the angle between the electron's spin axis and some direction in space, so that our limitation of the range amounts to the double possibility of alignment: $s = +1$ means the spin is parallel to the chosen direction, and $s = -1$ means it is opposite to it. A state-function $\phi(s)$ is then capable of assuming two values only, one, say a, at the point $s = +1$, and another one, b, at the point $s = -1$. The most general form of such a function is clearly

$$\phi(s) = a\delta_{s,1} + b\delta_{s,-1}. \tag{2}$$

If $\phi(s)$ is to be normalized to 1, then

$$\int \phi^*(s)\phi(s)ds = \int (a^*a\delta_{s,1}\delta_{s,1} + a^*b\delta_{s,1}\delta_{s,-1}$$
$$+ b^*a\delta_{s,1}\delta_{s,-1} + b^*b\delta_{s,-1}\delta_{s,-1})ds = a^*a + b^*b = 1.$$

We are using the integral here to bring out the analogy of our present operations with the previous ones. What we mean is a *summation over the values of the function at the two points* $s = \pm 1$, not an integral (which would be zero). We have seen that if x is an ordinary variable, $\phi^*(x)\phi(x)$ is the probability that x be the measured value of the coordinate (Sec. 9.3, prop. 3). Similarly, if the general axioms are to be applicable, $\phi^*(s)\phi(s)$ must now be the probability of finding the system in the condition characterized by s. This observation permits us at once to find the eigenfunctions of the spin operator, even before we have determined the nature of this operator itself. The eigenfunction describing the state in which s is certainly $+1$, which we shall denote by $\psi_+(s)$, must be such that $\psi_+^*(s)\psi_+(s) = \delta_{s,1}$. If now we take for $\psi_+(s)$ the most general function (2) and determine the constants a and b so that this relation is satisfied, we find immediately $a = 1$, $b = 0$ (aside from insignificant factors of absolute value unity). Hence

$$\psi_+(s) = \delta_{s,1}. \tag{3}$$

Similarly, the eigenfunction describing the state in which s is certainly -1 (let us call it $\psi_-(s)$) must be such that $\psi_-^*(s)\psi_-(s) = \delta_{s,-1}$, and this leads in the same way to

$$\psi_-(s) = \delta_{s,-1}. \tag{4}$$

We see that (2) is already an expansion in terms of the eigenfunctions of the spin operator, and these eigenfunctions (3) and (4) form, in a very trivial sense, a complete and orthogonal set.

Next, let us find the form of the spin operator, or rather *operators*. To understand the reason for this plural it is necessary to recall a few facts about angular momenta. To be sure, there are three different and in general independent angular momenta, those along three mutually perpendicular axes. One might think at first, however, that they ought to be equivalent and should not require separate treatment.

We might thus be led to suppose that the results obtained for the angular momentum operator σ_z could simply be taken over and applied to σ_y and σ_x. But this procedure would be erroneous, for it fails to take account of the fact of elementary diffuseness: *an eigenstate of the operator σ_z may not be an eigenstate of σ_y or σ_x*, and this would make a simple transfer of the results from σ_z to σ_y or σ_x impossible. The test of this matter, according to Sec. 9.3, prop. 5, is to determine whether the operators are permutable. Not having as yet determined the form of these operators we cannot apply this test literally. But let us look back at the previous angular momentum operators we have encountered. In Sec. 9.2, II, c (p. 409) we were interested in the angular momentum m_z about the z axis of a particle which, indeed, had no spin, but nevertheless m_z was an angular momentum. We saw there that the corresponding operator is given by

$$m_z = \frac{h}{2\pi i}\left(x\frac{\partial}{\partial y} - y\frac{\partial}{\partial x}\right).$$

In fact we might have added

$$m_x = \frac{h}{2\pi i}\left(y\frac{\partial}{\partial z} - z\frac{\partial}{\partial y}\right) \quad \text{and} \quad m_y = \frac{h}{2\pi i}\left(z\frac{\partial}{\partial x} - x\frac{\partial}{\partial z}\right). \quad (5)$$

The reader will easily verify that these operators are not permutable.

Let us, then, determine σ_z, the *spin* angular momentum about z, first, and then see what is to be done about σ_y and σ_x. By hypothesis, σ_z has the eigenvalues $\pm h/4\pi$ and the eigenfunctions (3) and (4). To save writing let us put

$$\sigma_{x,y,z} = \frac{h}{4\pi} S_{x,y,z},$$

so that $S_{x,y,z}$ has the eigenvalues ± 1. S_z must satisfy the operator equations

$$\left.\begin{array}{rl} S_z\psi_+ =& \psi_+ \\ S_z\psi_- =& -\psi_- \end{array}\right\} \quad (6)$$

S_z cannot be a differential operator because the functions ψ are not differentiable. There is really no necessity to determine its form further, for it may be regarded as *defined* by (6). It is merely a symbol which, whenever it appears before ψ_+, leaves it unchanged; but whenever it appears before ψ_-, it changes the sign of this quantity. Since, by (2), the most general " spin function " is a linear combination of these two, we know what is meant by the application of S_z to any spin function.

9.13 THE ELECTRON SPIN

It is often convenient, however, to regard S_z as a matrix. This makes for convenience in writing, if not for greater clarity. Let us think of ψ_+ and ψ_- as the elements of a matrix with but one column:

$$\psi = \begin{pmatrix} \psi_+ \\ \psi_- \end{pmatrix}.$$

Then we can combine both equations (6) in one matrix equation

$$S_z \psi = s_z \psi \qquad (7)$$

if we take

$$S_z = \begin{pmatrix} 1 & 0 \\ 0 & -1 \end{pmatrix}. \qquad (8)$$

s_z denotes the eigenvalues of the operator S_z, that is $+1$ and -1. Equating the (11) element of the matrix product on the left of (7) to the (11) element of the matrix on the right we obtain the first of eqs. (6); the (21) elements give the second.

We have no exact knowledge as to the relation between S_z and the other components of the spin. We expect, however, that the commutation rules valid for other angular momenta will also apply to the components of spin. The reader can easily verify on the basis of (5) that $m_y m_z - m_z m_y = \dfrac{h}{2\pi i} m_x$, and three similar relations are obtained by cyclical permutation of the subscripts. Hence we shall assume similar relations between the spin operators σ. If then we replace σ by $\dfrac{h}{4\pi} S$, the following operator equations must be true:

$$\left. \begin{array}{l} S_x S_y - S_y S_x = 2i S_z \\ S_y S_z - S_z S_y = 2i S_x \\ S_z S_x - S_x S_z = 2i S_y. \end{array} \right\} \qquad (9)$$

Since we already know S_z, these equations are sufficient to determine the other two components. In doing so it is convenient to use the matrix form of the operators. According to (8), S_z is diagonal; S_y and S_x cannot be diagonal, for diagonal matrices commute, and we have just shown that S_z, S_y, and S_x do not commute. The solution of (9) is seen to be

$$S_z = \begin{pmatrix} 1 & 0 \\ 0 & -1 \end{pmatrix}, \quad S_y = \begin{pmatrix} 0 & -i \\ i & 0 \end{pmatrix}, \quad S_x = \begin{pmatrix} 0 & 1 \\ 1 & 0 \end{pmatrix}. \qquad (10)$$

(This can be verified by substitution.) These matrices are known as Pauli's spin matrices. They are merely a shorthand way of writing the following set of substitutions:

$$\left.\begin{array}{l}S_x\psi_+ = \psi_+ \\ S_x\psi_- = -\psi_-\end{array}\right\} \left.\begin{array}{l}S_y\psi_+ = -i\psi_- \\ S_y\psi_- = i\psi_+\end{array}\right\} \left.\begin{array}{l}S_z\psi_+ = \psi_- \\ S_z\psi_- = \psi_+\end{array}\right\} \quad (11)$$

It is evident that ψ_+ and ψ_- are eigenstates only for the operator S_x; the other operators " mix " the ψ's.

A simple spin problem. In applying this theory we proceed exactly as we did in previous examples: we write down the energy equation $H\psi = E\psi$, replace the spin components in H by their operators, and solve the equation, which now is not a differential, but an ordinary equation. Let us find, for instance, the energy values of a stationary spinning electron in a uniform magnetic field. $\mathcal{H}_x, \mathcal{H}_y, \mathcal{H}_z$ are the components of the field along the three axes. The classical energy then becomes

$$\mu_x \mathcal{H}_x + \mu_y \mathcal{H}_y + \mu_z \mathcal{H}_z.$$

If we substitute for μ_x [cf. eq. (1)] the operator $\dfrac{he}{4\pi mc} S_x$ and make similar replacements for μ_y and μ_z, we find the operator equation for the energy:

$$\frac{he}{4\pi mc}(S_x\mathcal{H}_x + S_y\mathcal{H}_y + S_z\mathcal{H}_z)\psi = E\psi. \quad (12)$$

In solving this equation we distinguish two different cases.

I. The field $\mathcal{H}$ is along z.
(12) reduces to

$$|\mu| \mathcal{H} S_x \psi = E\psi. \quad (12')$$

The energy operator has here the simple form $|\mu| \mathcal{H} S_x$; it is permutable with S_x and therefore has the same eigenstates as S_x, that is, ψ_+ and ψ_-. If the electron is in the state ψ_+, it has its spin directed along $\mathcal{H}$ and the energy $|\mu| \mathcal{H}$; if it is in the state ψ_- it has its spin directed oppositely to $\mathcal{H}$ and has the energy $-|\mu| \mathcal{H}$.

II. The field is in any direction.

We shall now find that the energy eigenstates are not the same as those of S_x, because the operator in (12) is not permutable with S_x. We assume ψ to have the form

$$\psi_j = a_j\psi_+ + b_j\psi_-. \quad (13)$$

The index j is necessary since we expect to find several eigenstates, i.e., several ψ's satisfying (12). Our problem is to determine the various a_j and b_j. Let us substitute (13) into (12). The result is

$$|\mu| \{\mathcal{H}_x S_x(a_j\psi_+ + b_j\psi_-) + \mathcal{H}_y S_y(a_j\psi_+ + b_j\psi_-) + \mathcal{H}_z S_z(a_j\psi_+ + b_j\psi_-)\}$$
$$= E_j(a_j\psi_+ + b_j\psi_-).$$

Now let the various S operators be applied to the ψ's according to (11). We then have

$$|\mu| \{\mathcal{H}_x(a_j\psi_- + b_j\psi_+) - i\mathcal{H}_y(a_j\psi_- - b_j\psi_+) + \mathcal{H}_z(a_j\psi_+ - b_j\psi_-)\}$$
$$= E_j(a_j\psi_+ + b_j\psi_-).$$

If we multiply this equation by ψ_+, only the coefficients of ψ_+ remain; if we multiply it by ψ_-, we are left only with the coefficients of ψ_-. We can therefore equate these two sets of coefficients separately and obtain

$$\left.\begin{array}{l} |\mu|(\mathcal{H}_x a_j - i\mathcal{H}_y a_j - \mathcal{H}_z b_j) = E_j b_j \\ |\mu|(\mathcal{H}_x b_j + i\mathcal{H}_y b_j + \mathcal{H}_z a_j) = E_j a_j \end{array}\right\} \quad (14)$$

The condition that these equations shall have solutions different from zero is:

$$\begin{vmatrix} |\mu|(\mathcal{H}_x - i\mathcal{H}_y) & -|\mu|\mathcal{H}_z - E_j \\ |\mu|\mathcal{H}_z - E_j & |\mu|(\mathcal{H}_x + i\mathcal{H}_y) \end{vmatrix} = 0.$$

But this is equivalent to

$$|\mu|^2(\mathcal{H}_x^2 + \mathcal{H}_y^2 + \mathcal{H}_z^2) - E_j^2 = 0.$$

Hence there are two possible values for E_j:

$$E_1 = |\mu|\mathcal{H} \quad \text{and} \quad E_2 = -|\mu|\mathcal{H}. \tag{15}$$

Consequently, there must be two sets of coefficients a and b. They are determined by substituting (15) successively into (14). Putting $E_j = E_1$ we find from the first of eqs. (14), after squaring,

$$(\mathcal{H}_x^2 + \mathcal{H}_y^2)|a_1|^2 = (\mathcal{H} + \mathcal{H}_z)^2|b_1|^2.$$

But $|b_1|^2 = 1 - |a_1|^2$. Let us call the angle between $\mathcal{H}$ and the z axis, θ. This equation then reduces to

$$\sin^2\theta |a_1|^2 = (1 + \cos\theta)^2(1 - |a_1|^2),$$

whence, after simple manipulation,

$$\left.\begin{array}{l}|a_1|^2 = \cos^2\dfrac{\theta}{2}\\[6pt]|b_1|^2 = \sin^2\dfrac{\theta}{2}.\end{array}\right\} \qquad (15')$$

We have found that

$$\psi_1 = \cos\frac{\theta}{2}\psi_+ + \sin\frac{\theta}{2}\psi_-. \qquad (16)$$

What does this mean? ψ_1 is the eigenstate corresponding to the eigenvalue $E_1 = |\mu|\mathcal{H}$. That is, if the electron is in state ψ_1, its spin is certainly along $\mathcal{H}$ (which is taken to make an angle θ with z).

But (16) is at the same time an expansion in terms of eigenstates, ψ_+ and ψ_-, in which the spin moment is certainly along z or opposite to z, respectively. If now we recall the meaning of a_1 and b_1, this fact is tantamount to the statement:

If the electron's spin makes an angle θ with the z axis, then the probability of finding the spin directed along z is nevertheless finite and equal to $\cos^2\dfrac{\theta}{2}$; the probability of finding it directed oppositely to z is $\sin^2\dfrac{\theta}{2}$. This statement, interpreted classically, is paradoxical; its quantum mechanical meaning has already been discussed in Sec. 9.1 in connection with the significance of quantum mechanical pure states. We have now actually proved the results which we then anticipated.

To complete the problem, let us find ψ_2. We substitute E_2 of (15) into (14) and find in a similar manner

$$|a_2|^2 = \sin^2\frac{\theta}{2}; \quad |b_2|^2 = \cos^2\frac{\theta}{2}.$$

Hence

$$\psi_2 = \sin\frac{\theta}{2}\psi_+ + \cos\frac{\theta}{2}\psi_-.$$

The details are simple and may be left to the reader. He will also verify that the interpretation of ψ_2 is consistent with that of ψ_1.

The spin in atomic problems. The question now arises as to how we shall deal with the spin in general, that is, when the total energy of an electron is due, classically speaking, to translatory motion and spin. To have a definite example, let us think of the electron in a hydrogen atom, discussed in Sec. 9.6(d). If we wish to find the permissible energies we write as before

$$H\psi = E\psi, \tag{17}$$

but include in H a term due to the spin. To make this procedure more plausible we imagine a field $\mathcal{H}$ applied along z. If we neglect the effect of this field on the "orbital" part of the Hamiltonian the operator H will then be $H_0 + |\mu| \mathcal{H} S_z$, where H_0 is the ordinary Schrödinger operator. Correspondingly we represent ψ as a product $\psi(q)\psi(s)$ where q stands for the three space coordinates and s for the spin coordinate. H_0 operates only upon $\psi(q)$, S_z only upon $\psi(s)$. These substitutions split (17) into the two equations

$$H_0\psi(q) = E_q\psi(q) \text{ and } |\mu| \mathcal{H} S_z\psi(s) = E_s\psi(s),$$

where $E_q + E_s = E$. The first of these equations is identical with the ordinary Schrödinger equation, the second with (12'). Hence $\psi(s)$ is either ψ_+ or ψ_-, and E_s is either $+|\mu|\mathcal{H}$ or $-|\mu|\mathcal{H}$. The introduction of the spin has the effect of either increasing or decreasing the energy of the Schrödinger problem by $|\mu|\mathcal{H}$. If the eigenfunction without spin is $\psi(q)$, a solution of Schrödinger's equation, the spin will transform it either into $\psi(q)\psi_+$ or $\psi(q)\psi_-$. Let us now suppose that $\mathcal{H}$ is very small. The energy values will then practically coincide, and the two eigenfunctions will be degenerate. We see, therefore, that all energy states calculated by the ordinary Schrödinger method possess a twofold "spin degeneracy." To take account of this fact it is necessary to write the complete ψ function of an atomic state either $\psi(q)\delta_{s,1}$ or $\psi(q)\delta_{s,-1}$.

If we wish to characterize an atomic state by quantum numbers, the effect of the spin gives rise to a new quantum number. s, a variable with a domain of two points, is indeed equivalent to a quantum number. It is capable of the two values $+1$ and -1. The complete specification of an atomic state thus requires four numbers: n, l, m, and s; cf. Sec. 9.6(d).

Our discussion of spin degeneracy was based on an idealization. In reality, $\mathcal{H}$ is never zero. Although there is no external field, the constituents of the atom itself produce a field, although not a constant one. Thus the two "spin energies" are really never coincident. Their difference accounts for the doublet structure of the alkali spectra and many other phenomena.

Pauli's theory does not explain the origin of the spin, nor does it give any reason for its magnitude. It merely provides a method for incorporating it into quantum mechanics. The questions which this theory leaves open are most satisfactorily answered by Dirac's relativistic generalization of the Schrödinger equation. The latter actually solves the difficulties and informs us that the electron's spin arises from its efforts of satisfying the exacting requirements of relativistic invariance. But Dirac's theory is important enough to be postponed and reserved for closer examination.

9.14. The Exclusion Principle. Up to the present, all physical objects under discussion have been single systems. We have learned to calculate the possible eigenvalues of one electron, one oscillator, or one atom. Logically there is nothing to prevent us, however, from regarding several isolated systems as a single one with possible energies equal to the sums of the possible energies of the individual systems. Quantum mechanics would be incomplete if it did not lead to this result. In proving that it does we shall at once answer another interesting question: how is the state-function of the combination related to the state-functions of the individuals? This, in turn, will raise a complex of questions leading at once to the famous exclusion principle of Pauli and to quantum statistics. Although our analysis can be carried through for any operator equation, we shall base it on Schrödinger's equation, for it is the eigenstates of the energy in which we are primarily interested.

(a) Let us consider two different physical systems, one with a Hamiltonian operator H_1 and the other with a Hamiltonian H_2. The first has energies $\mathcal{E}_i^{(1)}$ and eigenfunctions $\psi_i^{(1)}$ if considered alone. The second, if isolated, has energies $\mathcal{E}_j^{(2)}$ and eigenfunctions $\psi_j^{(2)}$. If all coordinates (including the spin coordinate s) of the first system are symbolized by the single letter q_1, and those of the second by q_2, these quantities are related by the equations

$$H_1\psi_i^{(1)}(q_1) = \mathcal{E}_i^{(1)}\psi_i^{(1)}(q_1) \quad \text{and} \quad H_2\psi_j^{(2)}(q_2) = \mathcal{E}_j^{(2)}\psi_j^{(2)}(q_2).$$

We shall now think of the two as one system, supposing, however, that they do not influence each other. Consequently the Hamiltonian of the combined system will be $H = H_1 + H_2$, where H_1 acts only upon the coordinates q_1 and H_2 only on q_2. In concrete physical cases this will not be true, because when systems are combined they will always interact (unless they are kept infinitely far apart). The mathematical equivalent of an interaction is an additional term H_{12}, operating on both q_1 and q_2, in the Hamiltonian. The case which we are considering is therefore idealized.

If the eigenstate of the combined system is $\psi(q_1, q_2)$, it must be a solution of the equation

$$(H_1 + H_2)\psi(q_1, q_2) = E\psi(q_1, q_2).$$

But this is found to be satisfied by

$$\psi(q_1, q_2) = \psi_i^{(1)}(q_1)\psi_j^{(2)}(q_2),$$

and on substitution we see that $E = \mathcal{E}_i^{(1)} + \mathcal{E}_j^{(2)}$. Hence our expectation regarding the energies is verified, and we have also learned that, while the energies are the sums of the individual ones, the ψ functions are the products of the ψ's for the individual systems.

(b) Let us now consider two *similar* physical systems. We then have the following simplifications: the form of H_1 is identical with that of H_2, except that the first acts on q_1 and the second on q_2. As a result, the $\psi^{(1)}$ functions have the same form as the $\psi^{(2)}$, and the sequence $\mathcal{E}^{(1)}$ becomes identical with the sequence $\mathcal{E}^{(2)}$. We may, therefore, omit all superscripts and write our result:

$$\Psi(q_1, q_2) = \psi_i(q_1)\psi_j(q_2); E = \mathcal{E}_i + \mathcal{E}_j. \tag{1}$$

It expresses the fact that system 1 is in a state with energy $\mathcal{E}_i$ and system 2 in a state with energy $\mathcal{E}_j$. Suppose we interchange the systems, assuming 1 to have energy $\mathcal{E}_j$ and 2 to have energy $\mathcal{E}_i$. This leaves E unchanged, but it converts Ψ into $\psi_i(q_2)\psi_j(q_1)$. This is not the same function as that of (1)! To see this clearly one might think of $\psi_i(q_1)$ as $\sin x$ and $\psi_j(q_2)$ as $\cos y$. It is then evident that $\sin x \cos y$, if mapped in space, defines a different surface from $\cos x \sin y$. This means, however, that the state of the combined system represented by (1) is a degenerate one, although there may be no degeneracy in either ψ_i or ψ_j alone. If there is, the degeneracy of the combined system is greater.

According to previous remarks (cf. end of Sec. 9.5) the two functions

$$\Psi_1 = \psi_i(q_1)\psi_j(q_2), \quad \Psi_2 = \psi_j(q_1)\psi_i(q_2),$$

both of which belong to $E = \mathcal{E}_i + \mathcal{E}_j$, are by no means the only possible eigenfunctions for this energy. Any linear combination:

$$\Psi = a\Psi_1 + b\Psi_2 \tag{2}$$

is equally good. To be sure, a and b must be so adjusted that Ψ is normalized, but otherwise they are not subject to restriction on the basis of any principle introduced so far. We observe, of course, that only two combinations like (2) can be linearly independent.

Two sets of coefficients a and b are of particular interest. If, aside from a normalization constant ($\sqrt{1/2}$), $a = 1$ and $b = 1$,

$$\Psi = c(\Psi_1 + \Psi_2). \tag{3}$$

We observe that, when q_1 and q_2 are interchanged, $\Psi_1 \to \Psi_2$ and $\Psi_2 \to \Psi_1$. Physically, an interchange of all coordinates of the systems means an interchange of all quantum numbers, including that of the spin. Eq. (3) is entirely unaltered when such an interchange is made. It is said to be symmetrical with respect to an interchange of all coordinates of the systems.

If we choose $a = 1$ and $b = -1$, we have another interesting result:

$$\Psi = c(\Psi_1 - \Psi_2); \tag{4}$$

an interchange produces a change in the sign of Ψ. This function is said to be anti-symmetrical with respect to an interchange of all coordinates of the systems. But a change in the sign of Ψ amounts to multiplication by a factor of absolute value unity, and is physically insignificant. Hence both the symmetrical and the anti-symmetrical function preserve quantum mechanical states in an interchange of the coordinates of the systems. Moreover, they are the only ones which possess this property.

So far, our interest in the two combinations (3) and (4) arises merely from this simplicity which they alone exhibit. We shall soon see that they alone are realized in nature.

(c) Let us see what happens when we combine ν similar systems which are individually nondegenerate (each $\mathcal{E}_i$ has but one ψ function). By the same reasoning as in (a) we can show that the eigenfunction belonging to the state in which the first system has an energy $\mathcal{E}_i$, the second $\mathcal{E}_j$, ... the νth $\mathcal{E}_s$, has the form

$$\Psi = \psi_i(q_1)\psi_j(q_2) \ldots \psi_s(q_\nu), \tag{5}$$

and the state has a total energy

$$E = \mathcal{E}_i + \mathcal{E}_j + \ldots \mathcal{E}_s. \tag{6}$$

But any state whose Ψ function has the same form as (5) but with the q's permuted (distributed differently over the ν functions) has the same energy. Now there are $\nu!$ different ways of distributing the ν sets of coordinates, and hence there are altogether $\nu!$ different Ψ functions belonging to the energy (6). Let Ψ_P be any function obtained from (5) by a particular permutation, P, of the q's. Then any linear combination of the type

$$\Psi = \sum_P a_P \psi_P \tag{7}$$

will be a proper state-function corresponding to the energy (6) if the numerical coefficients a_P are so chosen that Ψ is normalized.

Here the theory grows unwieldy. If we actually had $\nu!$ linear combinations like (7), all with different sets of constants a_P, we should

9.14 THE EXCLUSION PRINCIPLE

indeed be confronted with a difficult problem whenever we had to employ the ψ functions in calculation.[1] Fortunately, however, we can introduce at this point a very great simplification in the form of a new postulate, known as Pauli's exclusion principle. It states:

If the individual systems in question are elementary charged particles (in particular electrons or protons) then the only combinations realized in nature are anti-symmetrical with respect to an interchange of the coordinates of two systems.

There is no way of deducing Pauli's principle; its validity has to be inferred from its results. The reason for its restriction to elementary charged particles (photons do not obey it!) is not completely understood.

As a result of this exclusion principle the multitude of functions (7) is reduced to one, for it is easily seen that only one of the linear combinations possesses the property of changing its sign when any pair of q's is interchanged. This combination has a simple form; it can best be written as a determinant:

$$\Psi = c \begin{vmatrix} \psi_i(q_1)\psi_i(q_2) \ldots \psi_i(q_\nu) \\ \psi_j(q_1)\psi_j(q_2) \ldots \psi_j(q_\nu) \\ \cdot \quad \cdot \quad \cdot \quad \cdot \quad \cdot \quad \cdot \\ \cdot \quad \cdot \quad \cdot \quad \cdot \quad \cdot \quad \cdot \\ \psi_s(q_1)\psi_s(q_2) \ldots \psi_s(q_\nu) \end{vmatrix} \quad (8)$$

where c is a constant. It is evident that this function is anti-symmetrical, for an interchange of two q's is equivalent to an interchange of two columns of the determinant, and this means a change in sign. Logically there is an interesting difference between a single function of the type (5) and the anti-symmetrical combination (8). In (5) a unique assignment of the individual systems to the various energy states can be made; in (8), where all possible permutations appear, no such assignment is possible. We cannot specify, therefore, in what state the jth system is to be found; all systems may be said to occupy all states indiscriminately. Thus it is seen that the exclusion principle deprives elementary particles of their discernibility. Physicists have occasionally interpreted this as an actual identity; at any rate it is often spoken of as an identity of the elementary particles. It is to be doubted, however, whether this usage reflects any intent on the part of physicists to voice their decision on the philosophical question regarding the relation between indistinguishability and identity.

[1] In the case of a perturbation, the secular determinant (cf. Sec. 9.10) would lead to an equation of the ν!th degree.

Our postulate of anti-symmetry has a very important consequence. Suppose that two electrons are in the same state, such that, for instance, $i = j$. Then two rows of our determinant are equal, and the determinant vanishes. Thus Ψ, and hence $\int \Psi^* \Psi d\tau$, are 0 whenever two electrons have identical quantum numbers. *It is impossible for two charged elementary particles to be in identical states.* This postulate is sometimes regarded as the essence of Pauli's principle. It is indeed an immediate result of the former statement, as we have shown; it may be used whenever it is convenient. But the first form is undoubtedly more fundamental.

What happens if we exchange two pairs of q's in (8)? If one exchange alters the sign of Ψ, two exchanges leave it unchanged. Hence the interchange of two pairs of elementary particles can have no effect upon the wave function; the latter must be symmetrical with regard to a double interchange of elementary systems. Thus, if our individual systems already contain two identical elementary particles or, by simple generalization, any even number of such entities, so that $\psi_i(q_1)$, etc., become the state-functions of these composite systems, we must choose symmetrical combinations in place of (8). Atoms with even mass number are composite systems of this character. Their states are therefore to be described by symmetrical state-functions. To form such functions from the aggregate of individual ψ_i's is not difficult. In expanding the determinant (8) alternate signs must be chosen for alternate permutations of the elements. If we violate this rule and choose equal signs for all permutations we obtain a symmetrical combination. Indeed, one can show that the Ψ function resulting from this procedure is the only symmetrical one.

A symmetrical Ψ function reproduces itself when all coordinates of two individual systems are interchanged. Hence it is *not* impossible for two such systems to have identical quantum numbers. But the state which results upon interchange is *not a different state* and must not be counted twice in statistical considerations. This is also an important consequence of Pauli's principle which would evidently not hold if the principle were inoperative, for if all the $\nu!$ linear combinations (7) were realized in nature an interchange would convert one degenerate state into another. Later, when we consider quantum statistics, we shall make use of this fact.

Finally we shall have to insure that the exclusion principle is consistent with the rest of quantum mechanical postulates. Is the limitation which it implies at all possible? In other words, will a symmetrical [1] state, once realized, remain symmetrical? The answer

---

[1] Symmetrical in this connection always means symmetrical with respect to an interchange of all coordinates of two individual systems.

comes from a study of (9.11–1). The reader will verify that, if a symmetrical operator is applied to a symmetrical function, the result will be again symmetrical. Hence if Ψ is symmetrical, $H\Psi$ will be symmetrical, provided H is a symmetrical operator. But one can prove from Newton's third law that every physical energy function must be symmetrical. Now $H\Psi dt$, in view of the dynamical law $H\Psi = -\dfrac{h}{2\pi i}\dfrac{\partial}{\partial t}\Psi$, is simply the increment of Ψ in time dt, aside from constant factor. If Ψ, however, always receives symmetrical increments, it must forever remain symmetrical. The same argument may be used to establish the continued anti-symmetry of initially anti-symmetrical states.

It will now be of interest to investigate a few of the major results of Pauli's principle in order to provide at least a partial verification of this very powerful postulate. To exhibit its beauty completely would, unfortunately, require more space than we have at our disposal.

9.15. Atomic Structure. We have seen that it is impossible for electrons to be in identical states. This implies that no two electrons whose ψ functions are referred to the same system of coordinates can have the same set of quantum numbers. (This rule evidently does not apply to electrons in different atoms.) In this form, the principle should be applicable to the electrons which make up an atom; it should help to explain the regularities of atomic structure.

Briefly, the regularities are these: all the elements of the periodic table arrange themselves in groups (rows of the periodic table) in such a way that corresponding elements within each group have similar chemical properties. In particular, the first element of each group (H, Li, Na, etc.) appears to have one electron outside a group of rather inactive electrons (the evidence comes from chemistry and spectroscopy), while the last element of each group is a rare gas which appears to have only a group of inactive electrons. All the evidence points to the conclusion that the electrons are arranged in " shells," and that any given shell has a limited capacity for electrons. In fact, it is seen that the capacities of the various shells are 2, 8, 18, 32, ...[1] Let us see what we should expect theoretically.

In passing from one element to the next the nuclear charge increases by one unit. Hence, to neutralize the structure, one electron must be added to the outside. Two rules will be operative in this addition of electrons: (1) the new electron will seek the lowest possible energy state; (2) the exclusion principle will be obeyed. If the

[1] For reasons to be discussed immediately the successive members of a group in the periodic table are not necessarily successive steps in the building up of one shell.

many-electron atom had the same energy states as a one-electron structure, the energies could all be classified in terms of the quantum number n, as was done in 9.6(d), and we should know that the energies are arranged in the order of the quantum numbers n, in the sense that high energies correspond to high values of n. This arrangement of the energies would be correct for a many-electron problem if there were no interactions between the electrons. However, the Coulomb forces between the electrons, which become more and more important as the number of electrons increases, will eventually modify this arrangement so that we can no longer say, for instance, that the energy of a state with $n = 4$ is always greater than that of a state with $n = 3$. We shall first, however, neglect this interaction and proceed as though the electrons did not influence one another.

In that case the states with $n = 1$ are filled first. We recall that the quantum number l is always smaller than n; hence if $n = 1$, l can only be 0. Now for every l there can be $2l + 1$ m values, that is, only one in the present case. But s can be either $+1$ or -1. Hence there are two states with different sets of quantum numbers (n, l, m, s): 1, 0, 0, $+1$ and 1, 0, 0, -1, if $n = 1$.

The next higher energy is realized when $n = 2$. l may now have the two values 0 and 1. If $l = 0$, there are two possible states with different quantum numbers: 2, 0, 0, $+1$, and 2, 0, 0, -1. If $l = 1$, however, there are six, for m can now be $-1, 0,$ and $+1$. This makes altogether eight different sets of quantum numbers.

In general for any value of l there are $2(2l + 1)$ different sets. These are said to form a subshell, which can be labeled by the value of n and l. For a given n, there are $\sum_{l=0}^{n-1} 2(2l + 1) = 2n^2$ different sets of quantum numbers. The electrons having these sets are said to form a shell, characterized by the number n alone.

The rule that a subshell can at most contain $2(2l + 1)$ electrons is known as *Stoner's rule*. It is at once explained by Pauli's principle. On the other hand, we have shown the reason for the progression 2, 8, 18, 32 ... indicating successive numbers of electrons in closed shells, this progression being simply $2n^2$.

If nature conformed to these principles, the atoms ought to be built up according to the following scheme, in which the upper row gives the quantum numbers n and l, the lower the number of possible states in each sublevel:

(10),	(20),	(21),	(30),	(31),	(32),	(40),	(41),	(42),	(43)
2	2	6	2	6	10	2	6	10	14

This scheme is actually followed in the periodic table up to the element A where the vertical line is drawn in the table. H and He represent the first shell of two. Next, Li and Be fill the (20) subshell, while the elements from B to Ne constitute the (21) subshell. This closes the second shell ($n = 2$). The elements from Na to A fall into the (30) and (31) subshells in perfect accord with our table, but there the regularity of sequence stops. Instead of continuing to place electrons into the third shell, nature now begins building up the (40) subshell (K, Ca) and then (Sc, Ti, etc.) goes back to fill in (32). The reason for this irregularity is simply the fact that more energy is required, after completion of (31), to place an electron in (32) than is necessary to place it in (40) because of the repulsive forces between the now highly condensed electrons. The truth of this assumption can actually be verified by an examination of the spectral terms of K. From Sc to Ni electrons are mainly added to (32), and Cu finally emerges with the (32) subshell completed and one electron in (40). The next seven elements Zn to Kr are then constructed in regular order, the added electrons occupying the subshells (40) and (41). Thus the process goes on until the addition of a further electron with a simultaneous increase of the nuclear charge will no longer liberate energy, but require it for its completion.

9.16. Quantum Statistics.[1] Pauli's principle has a most profound effect upon statistical considerations. Let us reconsider from the enlarged point of view of quantum mechanical axioms the results of the statistical mechanics of Darwin and Fowler. We were there confronted with the problem of finding average values over all possible macroscopic states of an assembly, when these states satisfied the conditions

$$\sum_\lambda a_\lambda = \nu, \quad \sum_\lambda a_\lambda \mathcal{E}_\lambda = E. \tag{1}$$

To review the situation briefly: ν similar systems are distributed over a great number of energy states $\mathcal{E}_\lambda$; a_λ is the number of systems in the λth state. A given macroscopic state was defined by the assignment of numbers:

$$a_0, a_1, a_2, \ldots . \tag{2}$$

Each one of these macroscopic states had a certain weight which was determined by counting the number of microscopic states which yield this assignment. If each energy state has a g_λ-fold degeneracy,

---

[1] Before studying this paragraph the reader is advised to review the theory of Sec. 5.7, for it is partly on the basis of the developments there presented that the details of quantum statistics will here be derived.

then, according to eq. (5.7-8), the weight of the macroscopic state corresponding to the assignment (2) has a weight

$$w = \frac{\nu!\, g_0{}^{a_0} g_1{}^{a_1} g_2{}^{a_2} \cdots}{a_0!\, a_1!\, a_2! \cdots}. \tag{3}$$

In most experiments, however, we cannot observe how many systems are in state $\mathcal{E}_0$, how many in state $\mathcal{E}_1$, etc. We are interested in finding the *average number* in any one state when the total number of systems is ν and the total energy a constant, E. Thus we arrived at the formula for $\overline{a_r}$, the average number of systems in the rth state subject to conditions (1), given by eq. (5.7-9), which we now rewrite:

$$\overline{a_r} = \frac{\sum_a \dfrac{a_r g_0{}^{a_0} g_1{}^{a_1} g_2{}^{a_2} \cdots}{a_0!\, a_1!\, a_2! \cdots}}{\sum_a \dfrac{g_0{}^{a_0} g_1{}^{a_1} g_2{}^{a_2} \cdots}{a_0!\, a_1!\, a_2! \cdots}}. \tag{4}$$

The evaluation of this expression leads to the Maxwell-Boltzmann law (cf. Sec. 5.5).

Let us now consider the situation from a quantum mechanical point of view. A microscopic state is one in which definite systems are assigned to definite energy states. If, for instance, systems 1 and 2 have energy $\mathcal{E}_0$, 3, 4, 5 have energy $\mathcal{E}_1$, 6 has energy $\mathcal{E}_3$, etc., the statefunction is

$$\psi_0(q_1)\psi_0(q_2)\psi_1(q_3)\psi_1(q_4)\psi_1(q_5)\psi_3(q_6) \cdots \psi_s(q_\nu). \tag{5}$$

In the designation of a macroscopic state we do not distinguish between the systems as individuals; we can permute all the q's and still have the same macroscopic state. If, therefore, all the ψ_i's were nondegenerate, we might conclude that each macroscopic state can be realized by $\nu!$ permuted functions of the type (5). In case ψ_i has a g_s-fold degeneracy, their number would be $\nu!\, g_0{}^{a_0} g_1{}^{a_1} \cdots g_s{}^{a_s}$. But we have forgotten the fact that a permutation of q's among ψ functions with equal indices does not produce a new function! Hence there are not $\nu!\, g_0{}^{a_0} \cdots g_s{}^{a_s}$ functions, but only

$$w_c = \frac{\nu!\, g_0{}^{a_0} g_1{}^{a_1} g_2{}^{a_2} \cdots}{a_0!\, a_1!\, a_2! \cdots}. \tag{6}$$

The degree of degeneracy of a macroscopic state agrees precisely with the weight (3) calculated previously. Hence, on this manner of reasoning, we should arrive at (4) and all the results of classical statistics.

In this consideration we have admitted all possible permutations of the q's in (5); in other words, we have not heeded the exclusion principle. To be in accord with it we should select from all linear combinations of such permutations the anti-symmetrical one if the individuals are elementary charged particles. But there is only one such function, and this, as we have seen, vanishes when any $a_\lambda > 1$. Hence we find that the weight of a macroscopic state, i.e., the number of possible microscopic ones composing it, is 1 or 0, according to whether all a_λ are equal to or smaller than 1, or any of them are greater than 1. The type of statistics based on Pauli's principle was first developed by Fermi and Dirac and bears their name. To indicate quantities pertaining to it we shall use the subscripts F.D. We may, then, state the result

$$w_{\text{F.D.}} = \begin{cases} 1 \text{ if all } a_\lambda \leq 1 \\ 0 \text{ if any } a_\lambda > 1 \end{cases}. \tag{7}$$

If, on the other hand, the only permissible Ψ functions are symmetrical, one can again form but one such combination from (5). But this combination, as we have seen, does not vanish when any a_λ is greater than 1. The statistics based on this selection was worked out by Einstein and Bose. It should be applicable to neutral atoms, and is also found to be valid for *photons*, a fact which cannot be deduced from the exclusion principle. We have found that, regardless of the values of the a_λ,

$$w_{\text{E.B.}} = 1. \tag{8}$$

The distribution laws which follow from the two non-classical choices (7) and (8) for the weights of macroscopic states are not the same as that of Maxwell-Boltzmann and will now be derived. To get all three distribution laws at once we use a method due to Fowler. We can write (6), (7), and (8) in a uniform way:

$$w = \text{constant} \cdot \prod_\lambda [\gamma(a_\lambda)] \tag{9}$$

if we put

$$\begin{aligned} \gamma_c(a_\lambda) &= \frac{g_\lambda^{a_\lambda}}{a_\lambda!} \\ \gamma_{\text{F.D.}}(a_\lambda) &= \begin{cases} 1 \text{ if } a_\lambda \leq 1 \\ 0 \text{ if } a_\lambda > 0 \end{cases} \\ \gamma_{\text{E.B.}}(a_\lambda) &= 1. \end{aligned} \tag{10}$$

Π in (9) is to be regarded as an infinite product over terms depending on λ. Instead of (4) we must then write more generally

$$\bar{a}_r = \frac{\sum_{(a)} a_r \prod_\lambda [\gamma(a_\lambda)]}{\sum_{(a)} \prod_\lambda [\gamma(a_\lambda)]} \equiv \frac{A}{C}. \tag{11}$$

For simplicity we call the numerator of this expression A and the denominator C. The meaning of the summation over a has been explained in Sec. 5.7. It is a manifold summation extended over all indices a_λ, but subject to (1). Let us first calculate C.

To free ourselves from the cumbersome conditions (1) we observe that C is the coefficient of $x^r z^E$ of the expression

$$M = \sum_{a_1 a_2 a_3 \ldots \lambda} \prod [\gamma(a_\lambda)] x^{\Sigma a_\lambda} z^{\Sigma a_\lambda \varepsilon_\lambda},$$

where the summation over the a's is now entirely unrestricted (except for the exclusion of negative values). But M can be simplified, as the reader will easily verify by expansion:

$$M = \sum_{a_1 a_2 \ldots \lambda} \prod \left\{ \gamma(a_\lambda) x^{a_\lambda} z^{a_\lambda \varepsilon_\lambda} \right\} \tag{12}$$

$$= \prod_\lambda \left\{ \sum_{n=0}^{\infty} \gamma(n) x^n z^{n \varepsilon_\lambda} \right\}.$$

Let us now introduce for

$$\sum_{n=0}^{\infty} \gamma(n) x^n z^{n \varepsilon_\lambda}$$

the abbreviation $f(xz^{\varepsilon_\lambda})$. Reference to (10) shows that f is summable in all three cases.

$$\left. \begin{array}{l} f_c = \displaystyle\sum_{n=0}^{\infty} \frac{g_\lambda^n}{n!} x^n z^{n \varepsilon_\lambda} = e^{g_\lambda xz^{\varepsilon_\lambda}} \\[2ex] f_{\text{F.D.}} = \hspace{4em} 1 + xz^{\varepsilon_\lambda} \\[2ex] f_{\text{E.B.}} = \displaystyle\sum_{n \neq 0}^{\infty} x^n z^{n \varepsilon_\lambda} = \frac{1}{1 - xz^{\varepsilon_\lambda}}. \end{array} \right\} \tag{13}$$

We have seen that C is the coefficient of $x^\nu z^E$ in M. Hence, by the theorem of residues (5.7–11),

$$C = \left(\frac{1}{2\pi i}\right)^2 \oint \oint \frac{dx\,dz}{x^{\nu+1} z^{E+1}} M. \tag{14}$$

On the other hand,

$$M = \prod_\lambda f(xz^{\varepsilon_\lambda}). \tag{15}$$

This is the form for C which we desire, for it can easily be evaluated by the method of steepest descent (cf. Sec. 5.7).

The next step is to calculate the numerator A of (11). For this purpose, let us calculate the expression $\dfrac{1}{\log z}\dfrac{\partial M}{\partial \varepsilon_r}$ from (12). We find

$$\frac{1}{\log z}\frac{\partial M}{\partial \varepsilon_r} = \frac{1}{\log z}\frac{\partial}{\partial \varepsilon_r}\sum_{a_1 a_2 \ldots} \prod_\lambda \left\{\gamma(a_\lambda) x^{a_\lambda} (e^{\log z})^{a_\lambda \varepsilon_\lambda}\right\}$$

$$= \sum_{a_1 a_2 \ldots} a_r \prod_\lambda \left\{\gamma(a_\lambda) x^{a_\lambda} z^{a_\lambda \varepsilon_\lambda}\right\}.$$

But A of (11) is just the coefficient $x^\nu z^E$ in this expression! Hence

$$A = \left(\frac{1}{2\pi i}\right)^2 \oint \oint \frac{dx\,dz}{x^{\nu+1} z^{E+1}} \frac{1}{\log z}\frac{\partial M}{\partial \varepsilon_r}.$$

The term $\dfrac{1}{\log z}\dfrac{\partial M}{\partial \varepsilon_r}$ can be thrown into a more convenient form. From (15)

$$\frac{1}{\log z}\frac{\partial M}{\partial \varepsilon_r} = \frac{1}{\log z}\left\{\frac{M}{f(xz^{\varepsilon_r})} f' \cdot xz^{\varepsilon_r} \cdot \log z\right\}$$

$$= Mx\frac{f'(xz^{\varepsilon_r})}{f(xz^{\varepsilon_r})} z^{\varepsilon_r}$$

$$= Mx\frac{\partial}{\partial x}\log f(xz^{\varepsilon_r}).$$

Hence

$$A = \left(\frac{1}{2\pi i}\right)^2 \oint \oint \frac{dx\,dz}{x^{\nu+1} z^{E+1}} Mx\frac{\partial}{\partial x}\log f(xz^{\varepsilon_r}). \tag{16}$$

We are now ready to calculate $\bar{a}_r = \dfrac{A}{C}$, using the method of steepest descent. Closer study would show that all the requirements for the applicability of this method are satisfied. The integrands of both (14) and (16) are small except for $x = \mu$ and $z = \vartheta$, two constants to be determined presently. We have already identified ϑ with $e^{-1/kT}$, T being the absolute temperature (cf. Sec. 5.7). The ratio of (16) and (14) will therefore have the value

$$\overline{a}_r = \mu \frac{\partial}{\partial \mu} \log f(\mu \vartheta^{\mathcal{E}_r}).$$

If we put (13), with $x = \mu$, $z = \vartheta$, and $\lambda = r$, into this expression, our final result becomes

$$\begin{aligned}(\overline{a}_r)_c &= \mu g_r \vartheta^{\mathcal{E}_r} = \mu g_r e^{-\mathcal{E}_r/kT} & (a) \\ (\overline{a}_r)_{\text{F.D.}} &= \mu \frac{\vartheta^{\mathcal{E}_r}}{1 + \mu \vartheta^{\mathcal{E}_r}} = \frac{e^{-\mathcal{E}_r/kT}}{\dfrac{1}{\mu} + e^{-\mathcal{E}_r/kT}} & (b) \\ (\overline{a}_r)_{\text{E.B.}} &= \mu \frac{\vartheta^{\mathcal{E}_r}}{1 - \mu \vartheta^{\mathcal{E}_r}} = \frac{e^{-\mathcal{E}_r/kT}}{\dfrac{1}{\mu} - e^{-\mathcal{E}_r/kT}}. & (c)\end{aligned} \quad (17)$$

The first of these is the classical (Maxwell-Boltzmann) distribution law; the second is the basis of the Fermi-Dirac statistics which should hold for electrons; the third underlies the statistics of light quanta and of certain atoms.

What is the significance of the number μ, the only parameter in our final results which is undetermined? Its meaning is clear when we consider the physics of our problem. $\overline{a}_r$ is the average number of systems in the rth energy eigenstate. By summing $\overline{a}_r$ over all r, or by integrating with respect to the energy in the case of continuous distributions, we obtain the total number of systems present. Thus μ is essentially a normalization parameter. The indicated summation can be easily carried out in (17a); in (b) and (c) it is best done by an expansion of the exponentials.[1] It is evident, however, that μ will invariably be a function of n, the total number of systems (per unit volume), and the temperature. In the Fermi-Dirac case, μ is a monotone increasing function of n. Thus, for small values of n, $\mu \ll 1$. But then (17b) reduces to (17a): electrons at low concentrations should obey classical statistics; only at high concentrations do the peculiarities of quantum statistics appear.

Sommerfeld showed that the concentration of "free" electrons [2]

[1] A. Sommerfeld, Z. f. Phys., 47, p. 1, 1928.
[2] Electrons in a metal are not strictly free; but theories based on this assumption are, within limits, singularly successful.

within metals is so great that $\mu \gg 1$, calling for an application of Fermi-Dirac statistics. As a result, he was able to improve considerably the theory of metals.

We have pointed out that (17c) applies to neutral atoms and light quanta. The latter case is particularly simple. In the case of light, the number of systems (photons) is not constant, hence conditions (1) are too stringent. The first, $\sum_\lambda a_\lambda = \nu$, need not be satisfied. But all that we have to do in the subsequent mathematical work in order to suppress this condition is to remove the variable x from M in (12). This, in turn, is equivalent to putting $x = \mu = 1$ in the final formulas (17). If we observe in addition that $\mathcal{E}_r = h\nu$, (17c) at once takes the form of Planck's radiation law (cf. Sec. 5.7).

We shall not concern ourselves with the application of these statistics; our principal aim has been to discuss their foundation, which could best be done by the actual process of deriving them. In concluding we merely state that all these results of the exclusion principle are in good accord with the facts of observation, and this indeed strengthens our belief in the correctness of so peculiar a postulate.

9.17. Dirac's Theory of the Electron. *The free electron.* Thus far, in formulating operator equations no account whatever has been taken of the requirements of the theory of relativity. Schrödinger's equation in particular, which appeared as the quantum mechanical modification of the classical energy equation, was based upon the non-relativistic form of the Hamiltonian function and could not, therefore, be expected to be relativistically correct. It will now be of interest to inquire into the possibilities of removing this inadequacy, and to construct a relativistic analogue of Schrödinger's equation:

For a detailed review of the demands of relativity the reader may be referred to Chapters VII and VIII. Suffice it to say at this time that, if relativity requirements are to be satisfied, all equations must be invariant under Lorentz transformations. Analytically, such invariance is most easily achieved if physical quantities are combined so as to form 4-vectors or, in general, tensors of higher rank. Physical laws must then be equations between the invariants formed from their components, or between similar components. The energy of a material point, for instance, appears in the theory of relativity as the fourth (ict) component of a 4-vector whose remaining components are the three momenta. The reader who is interested in these matters will observe that in the subsequent developments we are guided by these considerations; but we shall not make further explicit mention of them.

A thorough logician may indeed enter with some misgivings upon the task confronting us, for it is not altogether certain that quantum mechanics and relativity theory are compatible. The latter operates with the notion of definitely localized point events, the coordinates of which appear in Lorentz's transformation equations. But these equations involve also the velocity of the inertial system in which these events are observed. Does not this fact contradict the ever-present scattering of observations on canonically conjugate quantities required by quantum mechanics?[1] We shall not attempt to decide this question from the point of view of observability, for this can hardly be done without dogmatic extrapolation as to the ultimate form of quantum mechanical notions. But it is well to remember that relativity theory does more than regulate observable phenomena: it endows the equations of theoretical physics with symmetry and fundamental simplicity which alone are sufficient to dispose our minds favorably towards its basic axioms. And even if the facts predicted by the relativity theory could be derived from some other system of postulates, we feel that for the sake of its conceptual beauty most scientists would still be loath to surrender the doctrine in toto. It is principally for this reason that one is strongly tempted to recast the equations of quantum mechanics in a properly invariant form. The greatest success in this attempt has been achieved by Dirac, whose equation we shall now formulate.

There are numerous ways of arriving at a Lorentz-invariant equation which, when relativity corrections are neglected, will reduce to Schrödinger's. In order to find the most satisfactory one it is necessary to proceed by trial. The general method of attack is known. We must write down the classical relativity expression for the energy and then replace the dynamical variables appearing in it by suitable operators. There is no cogent reason why these operators should have the same form which they possess in non-relativistic quantum mechanics. For the present, let us restrict our consideration to the problem of a free electron.

Classically, its relativistic energy H can be expressed in two different but equivalent ways:[2]

$$H^2 = c^2 p^2 + m^2 c^4 \tag{1}$$

---

[1] For a more thoroughgoing discussion of this point see E. Schrödinger, *Sitz. Ber. d. Preuss. Akad.*, **3**, 63, 1931.

[2] According to (7.3-42), $T = mc^2[(1-\beta^2)^{-1/2} - 1]$. But $E = T + mc^2$, the last term being the rest energy of an electron. Hence $E = \dfrac{mc^2}{\sqrt{1-\beta^2}}$. Expressions (1) and (2) are seen to be equivalent to this if it is remembered that $p = \dfrac{mv}{\sqrt{1-\beta^2}}$.

and
$$H = \mathbf{v}\cdot\mathbf{p} + \sqrt{1-\beta^2}\,mc^2. \tag{2}$$

Here v is the velocity, p the momentum, and m the *rest* mass of the electron; c denotes the velocity of light, and $\beta = v/c$. It might seem natural to use the first of these expressions, take its square root, and substitute for p the usual operator $\dfrac{h}{2\pi i}\nabla$. But this procedure leads to difficulties, primarily because one cannot extract the square root of an operator in an unambiguous way. Dirac's equation results if we start with eq. (2), and make the following assignment of operators:

$$\left.\begin{aligned} \mathbf{p} &\to \frac{h}{2\pi i}\nabla \\ v_x &\to c\alpha_x, \quad v_y \to c\alpha_y, \quad v_z \to c\alpha_z \\ \sqrt{1-\beta^2} &\to \alpha_4. \end{aligned}\right\} \tag{3}$$

So far the nature of the operators α is undetermined. But let us require that the operator which, with these substitutions, results from (2) shall, if applied twice, be identical with the operator that results from (1). In symbols,

$$\left[\frac{h}{2\pi i}c\left(\alpha_x\frac{\partial}{\partial x} + \alpha_y\frac{\partial}{\partial y} + \alpha_z\frac{\partial}{\partial z}\right) + \alpha_4 mc^2\right]$$
$$\cdot\left[\frac{h}{2\pi i}c\left(\alpha_x\frac{\partial}{\partial x} + \alpha_y\frac{\partial}{\partial y} + \alpha_z\frac{\partial}{\partial z}\right) + \alpha_4 mc^2\right] = -\frac{h^2 c^2}{4\pi^2}\nabla^2 + m^2 c^4.$$

If now we assume that the α's do not contain coordinates and therefore commute with $\partial/\partial x$, $\partial/\partial y$, $\partial/\partial z$, we find at once from this equation:

$$\alpha_x^2 = \alpha_y^2 = \alpha_z^2 = \alpha_4^2 = 1;\; \alpha_x\alpha_y + \alpha_y\alpha_x = \alpha_y\alpha_z + \alpha_z\alpha_y$$
$$= \alpha_z\alpha_x + \alpha_x\alpha_z = \alpha_4\alpha_{x,y,z} + \alpha_{x,y,z}\alpha_4 = 0. \tag{4}$$

That is, each of the operators α, if iterated (applied twice), produces the result: multiplication by one; all α's "anticommute." The operator equation for the energy, $H\psi = E\psi$, takes the form

$$\left\{\frac{hc}{2\pi i}\left(\alpha_x\frac{\partial}{\partial x} + \alpha_y\frac{\partial}{\partial y} + \alpha_z\frac{\partial}{\partial z}\right) + \alpha_4 mc^2\right\}\psi = E\psi. \tag{5}$$

This is Dirac's equation for the free electron. But we have not as yet entirely specified the form of the α's; we only know their properties of commutation. As a matter of fact, it is not necessary to know more, since all results are independent of the special character, i.e., the mathematical form, of the operators α. It is conducive to clarity of thinking, however, to fix their form. This can be done in a variety of ways; customarily they are regarded as matrices. The simplest set of four matrices which can be made to satisfy eqs. (4) has four rows and four columns. It is not necessary for our purposes to write them down. If such a matrix, however, is to operate on ψ in accordance with eq. (5), then ψ itself must be a matrix. Indeed it must be a matrix of four rows, since the α's are applied to it from the left (cf. Sec. 9.8). Again we are at liberty to choose the number of its columns, and for the sake of simplicity we shall take this to be one. We are thus dealing in Dirac's theory, not with one eigenfunction, but with four, arranged as follows to form the matrix ψ:

$$\psi = \begin{Bmatrix} \psi_1 \\ \psi_2 \\ \psi_3 \\ \psi_4 \end{Bmatrix}. \tag{6}$$

Eq. (5) is equivalent to four equations.

In non-relativistic quantum mechanics, the mean value of an observable whose operator is P was given by $\int \psi^* P \psi d\tau$. How must this definition be modified in Dirac's theory? Let us take the simplest case, where the operator is the idem factor, and consider the normalizing relation $\int \psi^* \psi d\tau = 1$. If ψ^* were simply the matrix ψ of (6) with its components changed to $\psi_1^* \ldots \psi_4^*$, $\psi^* \psi$ would be meaningless, because the first factor of a matrix product must have as many columns as the second has rows. Clearly, in order to avoid this difficulty we must define:

$$\psi^* = (\psi_1^* \, \psi_2^* \, \psi_3^* \, \psi_4^*), \tag{7}$$

a matrix with one row and four columns.[1] With this modification we may retain the mean value relation: $\bar{p} = \int \psi^* P \psi d\tau$ in Dirac's theory. It is evident that the normalization condition, in expanded form, now reads:

$$\int (\psi_1^* \psi_1 + \psi_2^* \psi_2 + \psi_3^* \psi_3 + \psi_4^* \psi_4) d\tau = 1.$$

---

[1] We shall adhere to this convention. Thus when a star is added to a matrix symbol it signifies not only passage to complex-conjugate values of the elements, but also interchange of rows and columns.

9.17 DIRAC'S THEORY OF THE ELECTRON

Let us now return to (5) and attempt to find the permitted energies for a free electron. We observe that (5) is a set of *first* order differential equations, in contrast to Schrödinger's equation, which is of the second order. Without limiting the generality of the problem we may suppose the electron to have but one degree of freedom, and to be moving along x. The equation to be solved will then reduce to this form:

$$(E - \alpha_4 mc^2)\psi - \frac{hc}{2\pi i}\alpha_x \frac{d\psi}{dx} = 0.$$

It is to be observed that $d\psi/dx$ is the matrix having components $d\psi_1/dx, \ldots d\psi_4/dx$ and that, whenever ψ is multiplied by a constant, like E, we mean multiplication by E times the unit matrix. But this set of differential equations is easily seen to have the solutions

$$\psi = A e^{2\pi i/h \cdot px} \tag{8}$$

where p is a number which will soon be shown to be the momentum of the electron, and A is the matrix

$$\begin{Bmatrix} A_1 \\ A_2 \\ A_3 \\ A_4 \end{Bmatrix}$$

If we substitute this solution we get the matrix equation (no longer involving the variable x)

$$(E - mc^2\alpha_4 - cp\alpha_x)A = 0. \tag{9}$$

Let us operate on both sides of this equation with the matrix appearing on the left of A; taking care, of course, to preserve the order of the α's. We find

$$\{E^2 + m^2c^4\alpha_4{}^2 + c^2p^2\alpha_x{}^2$$
$$- 2E(mc^2\alpha_4 + cp\alpha_x) + pmc^3(\alpha_4\alpha_x + \alpha_x\alpha_4)\}A = 0.$$

But according to eq. (9), $(mc^2\alpha_4 + cp\alpha_x)A = EA$. If, in addition to this relation, we make use of (4), the result is simply

$$(-E^2 + m^2c^4 + c^2p^2)A = 0. \tag{10}$$

Clearly, all operators have disappeared from this equation, which is equivalent to four ordinary algebraic equations in which A is replaced, successively, by $A_1, \ldots A_4$. But the A_i's cannot all vanish; hence,

$$E^2 = m^2c^4 + c^2p^2.$$

The permissible values for E are therefore

$$E = \pm \sqrt{m^2c^4 + c^2p^2}. \tag{11}$$

Now it is easy to show that p is the average momentum of the electron along x. This is done by calculating

$$\frac{\int \psi^* \frac{h}{2\pi i} \frac{\partial}{\partial x} \psi dx}{\int \psi^* \psi dx}$$

on the basis of (8). The result is seen to be p. Eq. (11) then indicates that the electron may have energies $\geq mc^2$ as well as $\leq -mc^2$. The first set is to be expected, for mc^2 is merely the relativistic energy of a particle at rest. But the second set of energies appears most peculiar if one recalls that the energy of a free electron must be entirely kinetic. Dirac's theory thus leads to the conclusion that an electron can possess a negative kinetic energy smaller than $-mc^2$. To be sure, the existence of negative kinetic energies follows even from classical relativity considerations; but in classical theory it can be disregarded by supposing that no electrons exist in these states. This would then be a permanent situation, for, since an electron can never be found to have an energy intermediate between mc^2 and $-mc^2$, it could never reach these negative energy states. This argument breaks down in quantum mechanics, which permits discontinuous transitions. Even if at present all electrons had energies greater than mc^2, some would immediately pass into the negative energy states under emission of radiation, for the probability of transition-between the two sets of states, as computed by the principles of Sec. 9.11, does not always vanish. We shall return to this matter after we have studied some other interesting and perhaps equally paradoxical properties of the electron.

To exhibit the consistency of the theory, let us calculate the mean velocity of the electron whose state is described by (8). It is at once apparent that the mean of any one of the α matrices, $\frac{\int \psi^* \alpha \psi d\tau}{\int \psi^* \psi d\tau}$, is identical with $\frac{A^* \alpha A}{A^* A}$, where A^* is the matrix formed from A in the same manner as ψ^* is formed from ψ. Hence

$$\bar{v}_x = \frac{cA^*\alpha_x A}{A^*A}.$$

Now if A satisfies eq. (9), the matrix A^* must obey [1]

$$A^*(E - mc^2\alpha_4 - cp\alpha_x) = 0. \qquad (9)^*$$

Let us multiply (9) by $A^*\alpha_x$ on the left, and (9)* by $\alpha_x A$ on the right. Upon adding the resulting equations we find

$$2EA^*\alpha_x A - mc^2(A^*\alpha_4\alpha_x A + A^*\alpha_x\alpha_4 A) - 2cpA^*\alpha_x^2 A = 0,$$

and hence, by making use of (4),

$$\frac{A^*\alpha_x A}{A^*A} = \frac{cp}{E}.$$

Thus we have shown that

$$\bar{v}_x = \frac{c^2 p}{E}. \qquad (12)$$

If the reader recalls the familiar relativity formulas: $E = \dfrac{mc^2}{\sqrt{1-\beta^2}}$, $p = \dfrac{mv_x}{\sqrt{1-\beta^2}}$, he will observe that (12) is by no means surprising. But the result is interesting in this sense: let us suppose that the electron is in one of the negative energy states. Then E is negative, and $\bar{v}_x = -\dfrac{c^2}{|E|}p$; *the average velocity of the electron is opposite to its average momentum.* Physically, this is impossible to imagine. For this and other reasons one must conclude that the negative energy states can hardly represent states of an electron as we know it.

To find $\bar{v}_y = \dfrac{cA^*\alpha_y A}{A^*A}$, and $\bar{v}_z = \dfrac{cA^*\alpha_z A}{A^*A}$, we proceed in a similar fashion. We multiply (9) by $A^*\alpha_y$ from the left, and (9)* by $\alpha_y A$ from the right. On adding and using (4) we have: $A^*\alpha_y A = 0$. If in this procedure we substitute α_z for α_y the result is: $A^*\alpha_z A = 0$. Hence

$$\bar{v}_y = \bar{v}_z = 0, \qquad (13)$$

as we should expect.

---

[1] The reader will observe that, if A and B are matrices, and the star indicates interchange of rows and columns as well as change to complex conjugate elements, $(AB)^* = B^*A^*$. For Hermitean matrices, such as α_4 and α_x, $\alpha^* = \alpha$.

But on the other hand we know that $\overline{v_x^2}$, which is given by $c^2 \dfrac{A^*\alpha_x^2 A}{A^*A}$, must be c^2 in view of (4), and the same is true for $\overline{v_y^2}$ and $\overline{v_z^2}$. How shall we reconcile this fact with eqs. (12) and (13)? Evidently the electron has no average velocity along y and z, and yet the absolute value, or the root mean square, of its velocity along any of these directions is c. The only possible resolution of this apparent paradox is to assume that the electron performs, in a classical sense, a rapidly periodic movement with the speed of light, while it progresses uniformly along x in conformity with (12). Schrödinger [1] was the first to point out this peculiar trembling motion; its actual significance is not clearly understood.

The electron in an electromagnetic field. The triumph of the Dirac theory comes when it is generalized to the case of an electron moving in an electromagnetic field. This generalization is carried out in a perfectly natural manner, again on the basis of classical considerations. If we wish to include the action of a field of scalar potential ϕ and vector potential $\mathbf{A}$ in the classical energy expression (2), we must replace

$$\left. \begin{array}{c} \mathbf{p} \text{ by } \mathbf{p}' \equiv \mathbf{p} + \dfrac{e}{c}\mathbf{A}, \\[4pt] \text{and the total energy} \\[4pt] H \text{ by } H' \equiv H + e\phi. \end{array} \right\} \qquad (14)$$

Written explicitly, the operator equation $H\psi = E\psi$ now becomes

$$H\psi \equiv \left\{ c\left[\alpha_x\left(\frac{h}{2\pi i}\frac{\partial}{\partial x} + \frac{e}{c}A_x\right) + \alpha_y\left(\frac{h}{2\pi i}\frac{\partial}{\partial y} + \frac{e}{c}A_y\right) \right. \right.$$
$$\left. \left. + \alpha_z\left(\frac{h}{2\pi i}\frac{\partial}{\partial z} + \frac{e}{c}A_z\right) \right] + \alpha_4 mc^2 - e\phi \right\}\psi = E\psi. \qquad (15)$$

It is convenient to write this equation as though the matrices α_x, α_y, α_z constituted the components of a vector. If further we avail ourselves of the abbreviations (14), Dirac's equation takes the simpler form

$$(c\boldsymbol{\alpha}\cdot\mathbf{p}' + \alpha_4 mc^2 - e\phi)\psi = E\psi. \qquad (15')$$

This is again a set of four first order differential equations. We shall not solve (15). Suffice it to say that if the Coulomb potential e/r is substituted for ϕ and $\mathbf{A}$ is taken to be zero, the eigenvalues E of (15) should be the energy levels of the hydrogen atom. An actual calcu-

---

[1] Schrödinger, *Sitz. Ber. d. Preuss. Akad.*, **3**, 63, 1931.

9.17 DIRAC'S THEORY OF THE ELECTRON

lation [1] leads to complete agreement with experiment, even with regard to the fine structure of the energy levels. Sommerfeld's well-known fine structure formula results automatically from the calculation. Non-relativistic theory had been unable to produce such detailed agreement.

We are at present interested in knowing in what essential respects (15) differs from Schrödinger's equation. This is most easily seen from the following analysis. Let us consider (15) solved. Upon adding $e\phi\psi$ to both sides of the equation we obtain

$$H'\psi = E'\psi, \tag{16}$$

where $E' = E + e\phi$. (This relation is no longer an operator equation with E' as eigenvalue, for E' is not a constant!) Now multiply (16) on the left by E'. The result is

$$E'H'\psi = E'^2\psi.$$

But the left-hand side of this equation may be written: $H'E'\psi + (E'H' - H'E')\psi$, and the first term of this sum is $H'^2\psi$ because of (16). Hence our equation becomes:

$$[H'^2 + (E'H' - H'E')]\psi = E'^2\psi. \tag{17}$$

H' is the operator H of (15) without the last term, $-e\phi$. The reader will at once verify that [2]

$$H'^2\psi = \{c^2[p'^2 + \alpha_x\alpha_y(p_x'p_y' - p_y'p_x') + \alpha_x\alpha_z(p_x'p_z' - p_z'p_x')$$
$$+ \alpha_y\alpha_z(p_y'p_z' - p_z'p_y')] + m^2c^4\}\psi$$

and also that

$$(p_x'p_y' - p_y'p_x')\psi = \left(\frac{h}{2\pi i}\frac{\partial}{\partial x} + \frac{e}{c}A_x\right)\left(\frac{h}{2\pi i}\frac{\partial}{\partial y} + \frac{e}{c}A_y\right)\psi$$

$$- \left(\frac{h}{2\pi i}\frac{\partial}{\partial y} + \frac{e}{c}A_y\right)\left(\frac{h}{2\pi i}\frac{\partial}{\partial x} + \frac{e}{c}A_x\right)\psi$$

$$= \frac{he}{2\pi ic}\left[\frac{\partial}{\partial x}(A_y\psi) + A_x\frac{\partial \psi}{\partial y} - \frac{\partial}{\partial y}(A_x\psi) - A_y\frac{\partial \psi}{\partial x}\right]$$

$$= \frac{he}{2\pi ic}\left(\frac{\partial A_y}{\partial x} - \frac{\partial A_x}{\partial y}\right)\psi = \frac{he}{2\pi ic}\mathcal{H}_z\psi,$$

where $\mathcal{H}$ is the magnetic field strength. Two similar relations are

---

[1] A. Sommerfeld, "Wellenmechanischer Ergänzungsband."
[2] The p's in this expression are operators.

derived by cyclic permutation of the subscripts x, y, z. If, furthermore, we introduce a new set of matrices

$$\sigma_z = -i\alpha_x\alpha_y, \quad \sigma_y = -i\alpha_z\alpha_x, \quad \sigma_x = -i\alpha_y\alpha_z, \tag{18}$$

$H'^2\psi$ takes the simple form

$$H'^2\psi = \left[c^2\left(p'^2 + \frac{he}{2\pi c}\boldsymbol{\sigma}\cdot\mathcal{H}\right) + m^2c^4\right]\psi. \tag{19}$$

Finally, in order to evaluate (17), we must calculate $(E'H' - H'E')\psi$. We have

$$(E'H' - H'E')\psi = e(\phi H' - H'\phi)\psi = ce(\phi\boldsymbol{\alpha}\cdot\mathbf{p}' - \boldsymbol{\alpha}\cdot\mathbf{p}'\phi)\psi$$

$$= \frac{hce}{2\pi i}(\phi\boldsymbol{\alpha}\cdot\nabla\psi - \boldsymbol{\alpha}\cdot\nabla(\phi\psi))$$

$$= -\frac{hce}{2\pi i}\boldsymbol{\alpha}\cdot(\nabla\phi)\psi = \frac{hce}{2\pi i}\boldsymbol{\alpha}\cdot\mathfrak{E}\psi,$$

where $\mathfrak{E}$ is the electric field strength and it is assumed that $\dot{\mathbf{A}} = 0$. On substitution of these results in (17), the equation reads

$$(c^2p'^2 + m^2c^4 + \frac{hec}{2\pi}\boldsymbol{\sigma}\cdot\mathcal{H} + \frac{hec}{2\pi i}\boldsymbol{\alpha}\cdot\mathfrak{E})\psi = E'^2\psi. \tag{20}$$

We are now ready to make a comparison with Schrödinger's equation. For that purpose, let us put $E' = mc^2 + W + e\phi$, so that W is the non-relativistic energy of the electron. We also remember that $p' = \frac{h}{2\pi i}\nabla + \frac{e}{c}\mathbf{A}$. If, after making these substitutions, we divide eq. (20) by $2mc^2$ the result is

$$\left[\left(-\frac{h^2}{8\pi^2 m}\nabla^2 - e\phi\right) + \left(\frac{he}{2\pi icm}\mathbf{A}\cdot\nabla + \frac{e^2}{2mc^2}A^2\right)\right.$$
$$\left. + \left(\frac{he}{4\pi mc}\boldsymbol{\sigma}\cdot\mathcal{H} + \frac{he}{4\pi imc}\boldsymbol{\alpha}\cdot\mathfrak{E}\right)\right]\psi \tag{21}$$
$$= \left[W + \frac{(W + e\phi)^2}{2mc^2}\right]\psi.$$

By neglecting all terms involving $1/c$ this equation is reduced to Schrödinger's. ($c_1\psi_1 = c_2\psi_2 = c_3\psi_3 = c_4\psi_4$, and each component satisfies Schrödinger's equation.) But what is the meaning of the added terms? The two terms in the second parenthesis on the left describe the usual effects of a vector potential on an electron; they appear in all relativistic quantum theories and are not particularly interesting. The same is true for the second term on the right, which corrects for the relativistic change in the mass of the electron. The remaining two terms on the left distinguish Dirac's theory. If the electron had a magnetic moment μ and an electric moment $\mathcal{E}$, its additional energy in an electromagnetic field would be $\mu \cdot \mathcal{H} + \mathcal{E} \cdot \mathfrak{E}$. We see that (21) ascribes to the electron an energy as if it had a magnetic moment $\dfrac{he}{4\pi mc} \sigma$ and an electric moment $\dfrac{he}{4\pi imc} \alpha$. Moreover, the magnetic moment has exactly the required magnitude of $\tfrac{1}{2}$ Bohr magneton, previously explained by the electron spin. Dirac's theory, therefore, produces the spin properties without a special postulate, and this is its major achievement.

But eq. (21) also warns us not to take the electron spin too literally. The equation merely indicates in a formal way that the electron, if placed in a field, has an added energy part of which may be interpreted by saying that the electron spins. The electric moment is not simply accounted for by this supposition. To be sure, the electric moment appears to be imaginary, but this does not nullify the effect of the term upon the energy; indeed, the results would be wrong if the term containing $\alpha \cdot \mathfrak{E}$ were omitted. Furthermore, there is no term in (21) which could be interpreted as energy due to mechanical rotation. On the whole, then, the situation is more complex than would be in accord with the simple classical statement: the electron spins.

The matrices σ_x, σ_y, σ_z, defined by (18), are known as the "spin matrices." Their squares are 1, as follows from (4). Then it is necessary that their eigenvalues be $+1$ and -1. This fact affords justification for assuming that the magnetic moment of magnitude $he/4\pi mc$, which appears in (21), must be directed either parallel or anti-parallel to the field $\mathcal{H}$.

Among the things that mar the beauty of Dirac's theory is its apparently false prediction of negative energy states for the free electron. Other difficulties are encountered when one attempts to include radiation fields in the formalism. Nevertheless, its positive successes are so impressive that one may well hesitate to abandon the theory for the sake of avoiding these pitfalls.

The positron. Dirac has made a striking proposal to eliminate the negative energy difficulty. It must be admitted that an electron of energy $\leq -mc^2$ is unobservable because of its paradoxical physical properties. Dirac supposes that there exist an infinite number of such electrons, filling practically all available energy states. Pauli's principle, which forbids that one state be occupied by more than one electron, then becomes operative and prevents other electrons from falling into these deep energy levels. We must, in accordance with this picture, think of the universe as densely packed with unobservable electrons, all having energies below $-mc^2$.

If this infinite distribution of negative electricity is not observable, any hole in it will manifest itself; the absence of an electron from one of the negative energy states would be detectable. Indeed, such a hole, if free to propagate itself, would react with external agencies like a positive charge, and would exhibit the inertia of the absent particle, i.e., have the mass of an electron. Hence this hole would be a positron.

On the basis of this view an ordinary observable electron and a positron can combine to form an unobservable negative energy electron. In terms of Fig. 9.4 this process is equivalent to the passage of an electron from one of the upper levels to an unoccupied level below. In this transition, if energy is to be conserved, radiation (two quanta) of total energy $\geqq 2mc_2$ must be emitted. On the other hand, it is possible for sufficiently energetic radiation to lift an electron from one of the lower levels to one of the upper set. This process corresponds to the simultaneous production of an observable electron and a positron. Such "photoelectric" production of pairs with concomitant absorption of radiation is at present strongly supported by experiment. The striking facts are that pair production occurs frequently, but only under the action of γ rays of frequency greater than $2mc^2/h$, as this picture requires, and that the absorption of γ rays near atomic nuclei is stronger than it should be in the absence of pair production; in fact, absorption as a result of pair production would, as is claimed by some investigators, account fairly well for the discrepancies.

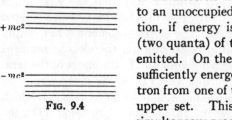

FIG. 9.4

The fact that the positron has so long escaped detection is also qualitatively explained by this hypothesis, for it may be shown that a hole in the distribution of negative kinetic energy electrons cannot long persist because of the eagerness of the ubiquitous ordinary electrons to fill the vacancy. Indeed, a positron moving slowly in the atmosphere would on the average be annihilated within about 3×10^{-7} sec.

It is a consequence of this view that positrons can be created or destroyed only in conjunction with electrons. Whether this is true is at present difficult to determine. There are other features of this theory of the positron which are not completely understood. Hence one cannot at this time pronounce reasonable judgment regarding the strange formalism, although it appears certain that Dirac's proposal is more than an idle speculation. The fact that it was made before the positron was discovered is of interest in this connection.

In surveying the situation one cannot fail to perceive that no existing theory is strictly applicable to phenomena which take place in a region of space smaller than the classical volume of an electron (strictly speaking, smaller than h^3/m^3c^3) and that, if we knew the laws regulating phenomena on this scale, the difficulties might, at least partly, resolve themselves. Perhaps the reader recalls that, prior to the advent of quantum theories, classical theory had surrendered before the problem of explaining the stability of an electron or of any aggregation of charge. In the joy of quantum theoretical achievements physicists then forgot their fiasco. It seems that fate now forces us to remember.

These last remarks make clear why quantum mechanics, even with its latest refinements, has hitherto been unsuccessful in explaining the structure of the nucleus. Our account of the foundations of modern physics would be decidedly incomplete, however, if it left unmentioned the spectacular experimental advances upon the mysteries of the nucleus that have been made within the past two years. Signal achievements are the discovery of the isotope of hydrogen, H^2, by Urey, Brickwedde, and Murphy, in 1932; Chadwick's discovery of the neutron (1932); and the discovery of the positron by Anderson, and by Blackett and Occhialini in 1933. Of these new entities only the positron is theoretically welcome, as is clear from our foregoing discussion. The question as to the ultimate constituents of matter has been raised anew. Until these recent discoveries the proton has been considered one of the primary building-stones of nuclei; it had been regarded as the electron's partner, although its different mass was a source of annoyance.

Now the positron claims this partnership, and the proton's status as an elementary entity has become doubtful. The view seems current among physicists that, of the proton and the neutron, one is elementary and the other complex. But whether the proton is a neutron combined with a positron, or whether the neutron is a proton combined with an electron, is a much-debated question. We have never been as far from achieving unity of explanation in the realm of ultimate constituents of matter as we are at present. To make the confusion complete there have arisen the advocates of the neutrino, a neutral entity of small or zero mass, invented mainly to account for the continuous β ray spectrum.[1]

The reader will observe that the trend of recent discoveries appears to be away from uniformity and simplicity of explanation. But we may rest assured that the list of discoveries is not yet complete and that, when their ring is closed and an adequate theory connects the facts, uniformity and simplicity will be restored. Antithesis must forever precede synthesis.

[1] The existence of a continuous range of velocities among the electrons emitted by radioactive substances has hitherto been impossible to explain. If quantum mechanics is correct, electrons within nuclei should be in definite energy states and should therefore be emitted with definite energies. Experiments show that they do not lose unquantized amounts of energy after expulsion from nuclei, thus rendering fairly definite the circumstance that they are actually expelled with continuous energies. This seems to indicate one of three things: (1) failure of energies to be conserved; (2) concomitant emission of unobserved "particles" of very small mass (neutrinos) which carry away the energy required to make the β ray spectrum discrete; (3) failure of the quantum theory to describe nuclear processes. One may well feel that the evidence for the last supposition is so great as to make the former two unnecessary.

CHAPTER X

THE PROBLEM OF CAUSALITY

This book began with very general considerations regarding the nature of physical theory. We shall bring it to a close by examining an equally general problem, and at the same time one which will cause us to survey in retrospection the various single theories of which we have given a brief account. In presenting an analysis of the problem of causality we are venturing on controversial grounds, where it is difficult to proceed entirely free from bias, and where it would be ill advised for an expositor to conceal his preference entirely. Nevertheless, our attempt will be to appraise the status of the question objectively.

The disagreement with regard to the validity of the principle of causality, existing today among scientists, has its roots in the diversity of definitions of the principle itself rather than in a problematic scientific situation. As far as the formulation of quantum theory is complete, its bearing upon philosophical questions can be fixed with precision, provided the questions are phrased intelligibly. But a question is intelligible from a scientific point of view only if it satisfies two conditions: (1) the meaning of its terms must be fixed; (2) it must be in accord with the conventions of the science to which the question is put.

The necessity of the first requirement is at once evident; if it were not satisfied, the question would have several correct but self-conflicting answers, such as those to which the undisciplined discussion of philosophical problems usually leads. The second requirement, however, reflects a particular weakness of philosophy. Within the domain of the latter, words have retained a variability of meaning which, to be sure, makes for beauty and flexibility of expression, but impairs precision of speech. This has already been commented on in our examination of physical theories in Chapter I. Indeed, the difference between philosophy and the specialized sciences began when the latter caused their concepts to crystallize and agreed to name them universally and with care. Even now the distinction between science and philosophy is best described in saying that science proceeds by making the acceptance of its terms obligatory for all its followers, whereas philosophy allows its advocates to coin largely their own phraseology. This individualism frequently causes confusion and indefiniteness of philosophical attitude. To illustrate: whether causality is a category or not is chiefly a matter of definition and nomenclature and may be answered correctly by yes and no, or in fact the question may be meaningless; proponents for each of these three answers are to be found among modern philosophers. However, the question: is mercury an element, has but one correct answer. Similarly, the term energy, which has a perfectly definite scientific meaning, is constantly used in phrases such as " mental energy " which signifies nothing unless ignorance of the laws of physics on the part of the speaker. Science has avoided such ambiguities. Hence we can effectively guard against them by using technical language wherever it is possible. In discussing causality it is necessary to formulate the problem in physical terms as far as they are available.

Considerations like these are common with all scientists and many philosophers today; they constitute no unsympathetic critique of philosophy, or of the methods employed in philosophical investigations. For the solution of numerous problems the methods of science, which, speaking figuratively, is nothing but the crystallized part of philosophy, are not yet at hand. It is also true that the philosophical terminology is more strongly subjected to popular misuse than scientific language, and therefore prevented from being standardized. None of these arguments is sufficient, however, to justify the use of vague philosophical phrases where definite scientific terms present themselves.

Let us then, with this critical attitude, examine some of the more outstanding modern interpretations of the problem at hand. Kant's formulations of the causality principle can hardly be regarded with favor from this point of view. He says in the first edition of his " Critique of Pure Reason ": " Alles, was geschieht (anhebt zu sein), setzt etwas voraus, worauf es nach einer Regel folgt" ; this statement is not scientifically clear. " Regel " is entirely undefined; it is possible to find a rule for everything that is susceptible of description. On the other hand, if the word is to be interpreted as a means of knowing subsequent events in advance, there arises the difficulty of who is to know and employ the rule, together with other inconsistencies which will be encountered shortly in connection with similar formulations. Another possibility of interpretation places the emphasis upon the first part of the sentence and neglects the last as an unessential explanatory phrase. But then, if " voraussetzen " is taken in its temporal sense, Kant's statement amounts to nothing more than the assertion that the universe has no beginning in time, which is plainly not identical with the causality postulate.

Kant's modified formulation, as it appeared in the second edition of the same book, reads: " Alle Veränderungen geschehen nach dem Gesetz der Verknüpfung von Ursache und Wirkung." Kant's concept of " law " is indeed well defined within his own philosophy, and different from that which we have discussed in Chapter I. If, however, we wish to deal with causality in its physical aspects (we are not here examining other possible interpretations), we are forced to understand the term law in its physical sense. But then it is found that physics knows of no law connecting cause and effect; indeed, there is no plausible way of defining these concepts. Moreover, as has already been noted in Chapter I, if the distinction between cause and effect is artificially impressed upon physical phenomena, physical laws do not even allow us to differentiate between the two. Newton's law of gravitation, for instance, sets up a relation between an observation on the rate of change of the radial velocity between two masses on the one hand, and the distance between them on the other. But it contains no criterion to determine the causal status of these observations. There is no law of connection between cause and effect known to science; these concepts are foreign to physical analysis. Nor is it of any avail to inject them externally, for the meaning usually conveyed by the words in question is expressed more adequately and precisely by technical terms like boundary condition, initial and final state. (Cf. Sec. 1.9.)

Most of the difficulties discussed so far are avoided in Laplace's statement [1] of what he and many later scientists consider to be the essence of causality. He postulates the existence of a universal formula according to which all happenings take place, and expresses this state of affairs as follows: " An intelligence knowing, at a given instant of time, all forces acting in nature, as well as the momentary positions of all things of which the universe consists, would be able to comprehend the motions of the largest bodies of the world and those of the smallest atoms in one single formula, provided it were sufficiently powerful to subject all data to analysis; to it, nothing would be uncertain, both future and past would be present before its eyes."

This is certainly an intelligible proposition; it is excellent in its clarity and precision. All its terms are well defined; the word force is to be understood in its accurate physical sense as the product of mass and acceleration, and " knowledge of a force " means knowledge of the

---

[1] Théorie analytique des probabilités (Paris, 3d ed., 1820).

differential equation which relates this product to a function of position, this function being also known. The spirit of Laplace's proposition pervades all of classical physics and has been eminently fruitful in the development of that science. Is the proposition true? In answering this question we shall find reason for abandoning this particular formulation of the causality principle.

Closer inspection shows Laplace's statement to be true, but unsuited to define a causal state of affairs. It is obviously not an easy matter to formulate the meaning of causality in a positive way. However, it appears less difficult to point to a typical situation which most people would agree to call non-causal. Imagine if possible a perfectly arbitrary universe with a god agitating it according to his ever-changing desires. Although we are still searching for a suitable definition of causality, it seems clear that this would constitute the model of a non-causal world. If we further suppose the happenings in this chaotic universe to be discernible and describable, then it must be possible also to represent them by means of equations. These equations will not necessarily contain analytic functions only, nor will they be differentiable. However, if things take place with reasonable smoothness and not too suddenly, if "natura non facit saltus," the functions will possess the property of differentiability. It is then clear that differentiation will, in general, simplify the equations, for it will cause additive constants to disappear. An intelligence powerful enough to know all these differential equations together with the values of coordinates and derivatives at a given time, and able to solve them, would have a complete survey of all events, future and past. In Laplace's world this survey must be possible on the basis of a knowledge of all forces, i.e., differential equations of the second order. This, in itself, constitutes no further restriction upon our arbitrary universe, since there is nothing to prevent us from differentiating the equations twice; but it expresses a preference which should manifest itself in a universe satisfying this particular causality postulate. In such a world, laws should take on an especially simple form if they are stated as second order differential equations. Thus it is seen that Laplace's postulate is not a stringent one; it is true for almost any imaginable universe. It imposes nothing upon a world which by itself runs smoothly, and is certainly a valid approximation to the course of events in the arbitrary, non-causal universe. In that sense the statement in question is true, but it does not seem to characterize causality. Whether or not nature is conveniently describable in terms of second order differential equations is an entirely different issue and must probably be affirmed—although there are cases of physical analysis where description by equations of higher order is customary.

The differential equations governing the processes in the arbitrary, lawless world which has been imagined will in general involve the time explicitly. Consequently the forces (which are always to be defined in terms of accelerations, not popularly as " pushes and pulls ") will change with time in an essential manner, and not only through the coordinates on which they depend. In order to predict the future, Laplace's demon not only would have to know the instantaneous values of all forces, he would require their complete form as functions of the space coordinates and of the time. An entirely different situation arises if we postulate that the forces be functions of space coordinates and possibly their time derivatives only. This would constitute a very definite limitation upon natural processes, a limitation indeed which the world agitated by a god would fail to exhibit. For now the forces have the same instantaneous values whenever the coordinates and their derivatives assume a given set of values; this endows nature with a regularity which, it appears to us, is very nearly what causality is meant to convey. But this important feature hardly follows from Laplace's formulation and may, at any rate, be expressed more directly as we shall see later.

The definition of the causality principle most widely accepted at present is closely related to that just considered. The argument runs: Introducing a superior intelligence is not permissible because it is man who makes his science and it must be he who is to judge whether his world is causal or not. A universal formula without an individual knowing it is a vague phrase. Let us therefore modify Laplace's proposition by substituting man in place of the demon. The causality principle is then valid if it is possible for the scientist, on the basis of known laws, to reconstruct the past and to project the future when the present state of the world, or of a portion of it, is completely known. We shall term this conception of causality the positivistic one, though it might equally well be called anthropomorphic or utilitarian. It is clearly in harmony with the present trend of eliminating things that have merely logical status but no concrete meaning in terms of physical operations.

Postponing for the present certain obvious shortcomings of this definition as stated, we cannot fail to observe that it enjoys the definiteness and directness which are typical of all positivistic pronouncements: it can at once be decided to be either applicable or inapplicable to the world as the scientist knows it. Moreover, both decisions have positive meaning. It is for this reason that the positivistic form of the causality principle has an immediate appeal to scientists and has found greatest favor with the working physicist. Let us, therefore, examine it more closely.

It is evident that our inquiry into the *general* validity of the principle encounters a natural obstacle in the fact that known physical laws describe only a small portion of the physicist's universe. Therefore, if our discussion is to have meaning, we must restrict our investigation to that portion of physical experience which is regulated by laws known at present. This limitation is, or should be, implied in the positivistic formulation of our principle. To choose an example: we know nothing about the internal structure of an electron; we can conjecture nothing about its past or future internal constitution. But this situation, which results from ignorance, is to be overlooked when we examine the validity of our principle.

If then we survey the fields of physics which we think we understand, do they justify us in asserting that, given the complete present state of a system, we are able to deduce its past and future behavior? The answer depends entirely on what is meant by "state." The reader will now do well to recall the discussion of Chapter IX where the principal meanings of this term have been reviewed. He will then observe that, if the classical definition of states in terms of positions and momenta of particles is adopted, the causality principle is valid in classical physics but breaks down in quantum mechanics, for, as was shown in Chapter IX, the formalism of quantum mechanics is such as to produce unavoidable related uncertainties in the positions and momenta of particles, permitting only statistical knowledge and predictions regarding these properties. Many physicists prefer to adhere to the classical definition of states and therefore declare causality void. Their contention is correct, of course, provided that we adopt both the positivistic formulation of causality and the classical definition of state.

On the other hand, the principle is valid if we agree to operate with quantum mechanical states (ψ functions, transformation functions) since, as we have seen in Chapter IX, the latter are completely predictable on the basis of our knowledge of present states.

These alternatives have been beautifully combined by Bohr in a new formulation, called by him the principle of complementarity. It is essentially equivalent to our last two conclusions, which Bohr phrases pregnantly in some such way as this: The ordinary description of nature in terms of space and time (classical definition of states) is necessarily acausal, whereas its description in terms of ψ functions and operators satisfies the causality principle. It is only by combining

THE PROBLEM OF CAUSALITY

these two complementary aspects that a true and complete picture of the physical world can be obtained. The reader will agree that this statement sums up the situation very clearly indeed; nevertheless, it should not be construed as preventing anyone who is willing to content himself with a purely quantum mechanical definition of all states, from asserting the validity of the causal principle.

The physicist's interest in the problem at hand may well stop at this point. The philosopher, however, who is eager to view the facts from an even higher plane, will continue his investigation and probe into the logical texture of the positivistic formulation itself. And there he will probably find some infelicities which we shall now discuss.

In the first place, it may be said that the meaning of the principle in the form last considered is not entirely unambiguous for the following reason. The term "possible for the scientist," or, if it is preferred, "possible for the human mind," permits various interpretations even if our attempt to predict is limited to fields of physics wherein the laws are supposedly well known. For who is to decide the capabilities of the scientist or the human mind? The causality principle is certainly not tantamount to asserting that there exists at the present time an individual clever enough to complete the analysis which leads to the correct prediction. Causality, it appears, deals with the fundamental workings of nature and should not be affected by accidental circumstances such as the extent of scientific knowledge or human ability. Of course it is far from certain that nature and our knowledge about nature can ever be separated—a circumstance which may render the reader immune to the previous argument. No one will deny, however, that, even though a present physical theory repudiates the possibility of knowing all data upon which a detailed analysis of the future depends, this possibility may be restored by later developments. In order to escape these and similar difficulties one is forced to modify the positivistic formulation of the principle by stating it in the form: Causality exists if no confirmed physical theory contradicts the acquisition of data by which a determination of future and past events can be made. But here we observe that the road is narrowing; our principle is gradually becoming more and more specific; as a result of this attenuation our chances of grasping the fundamental aspects of causality finally disappear. In fact, this last formulation has, to our knowledge, never been proposed; nor is it exempt from the criticism that follows.

The foregoing remarks do not touch the principal weakness of the positivistic causality formula. We shall try to show that the latter may be satisfied by a non-causal universe. The point is simply this: causality has nothing to do with the question whether future events may be known in advance; its prerequisite is not that the scientist turn prophet. Suppose that the god who agitates his universe according to his inscrutable desires and without restrictions by law or order should give the scientist exact forewarnings of his actions, so that the latter is able to prophesy with accuracy. Would this make his playland a causal universe? Of course there is no rigid answer since we are still searching for a suitable definition of causality. But if we interpret correctly the universal implication of this concept regardless of its various formulations we feel that one must answer negatively. It is commonly conceived, for instance, that the occurrence of miracles contradicts causality: the circumstance that many miracles have allegedly been predicted is hardly sufficient to reconcile their occurrence with the causality postulate. We conclude: the positivistic formulation with its main emphasis upon human ability to know in advance does not express the nucleus of what is understood by causality.

Let us now put an end to criticism and select a definition that will satisfy the outlined requirements more widely, and state the central part of the concept in question. It will be granted that the crucial feature which makes the arbitrary world non-causal is the irregularity arising from the whims of the god, whether this irregularity may be known or not. To be more specific it is the fact that in a non-causal world the force between two electric charges may vary, say, as the inverse second power of the distance today, but possibly as the inverse tenth power tomorrow. We feel that causality is violated when a given physical state A is not always followed by the same state B.[1] Crudely speaking, nature is consistent in the sense that she does not alter her ways; her reactions are independent of the time in an absolute manner. To avoid circumlocutions we shall refer to this property, which we wish to regard as the most adequate expression of the causality principle, as consistency of nature.[2]

In more accurate phraseology, and without reference to states, *consistency of nature*, i.e., causality, may be characterized by saying: *The differential equations in terms of which nature is described do not contain explicit functions of the time.* This statement is less definite

---

[1] There is indeed an indefiniteness in this statement because of the numerous possibilities of defining states, but when any one definition has been chosen the statement is unambiguous.

[2] For a more detailed discussion of these matters cf. H. Margenau, *The Monist*, XLII, 161 (1932); *Journ. Phil. Science*, I, 133 (1934).

than it seems because it does not contain directions as to the choice of variables appearing in the equations, a vagueness which is the counterpart of the indefinite meaning of " state." The existence of variables must of course be supposed since they form the condition under which description of nature is possible at all. But then, if nature is consistent in this sense, the integrated equations will not depend on time in an absolute manner; more specifically, if $x = f(t)$ is a solution, then $x = f(t - t_0)$ is another, so that the motion in question has an arbitrary beginning in time which becomes fixed only if accessory conditions are known. Furthermore, to use a previous example, the exponent in Coulomb's law of attraction will be invariable in time; if it is -2 today it will be -2 forever. The same should hold for all parameters appearing in the differential equations of physics if nature is consistent or causal.

It is to be remembered, however, that the occurrence of equations which do not satisfy this requirement is not at once a proof against causality. In fact, we often encounter such equations when a problem is not completely analyzed, for example, in the case of forced oscillations. But here, as well as in all similar instances, the explicit time dependence could be eliminated by including in the analysis those parts of the system which produce the varying force, i.e., by " closing " the system. " Impressed forces " always indicate that the physical system in question is an open one. In fact, a closed system is simply one to which causal analysis can be applied, that is, one which can be described by differential equations not containing the time explicitly.

Causality, if interpreted as consistency, thus banishes absolute time from the essential description of nature. It is true that time appears in most physical laws, but always in a manner which allows of arbitrary regulation by initial conditions (cf. Chapter I). The elimination of absolute time is necessary because the latter has no physical meaning. The theory of relativity with its rejection of absolute frames of reference is of necessity a causal one, though of course causality cannot be said to be identical with relativity. The consistency formula also enjoys the advantage of simplicity; nevertheless it involves everything implied in the usual conception of causality, such as the existence of unique laws and strict determinism of states (but not necessarily of measurements). Theorems of conservation (energy, momentum), generally felt to be in some way connected with the causality principle but unaffected by its customary formulations, follow at once as analytical consequences from the fact that the differential equations do not involve the time explicitly.

The principle of consistency, and hence of causality, has no meaning if it can be applied only to the universe as a whole, for then the number of describing variables would probably be infinite and the description in terms of differential equations loses its sense. Using the more intuitive definition according to which a state A is always followed by the same state B the failure is evident when we realize that state A may occur only once. Causality is then an empty phrase unless the universe is periodic. But the same would be true of all the laws of physics, in fact, of physics as a science, if its statements were inapplicable to small domains of nature. Hence the condition that causality shall have meaning is the same as that for the existence of science. As a matter of experience the universe *is separable* into smaller systems to which differential equations can be applied, and if these equations are of the type here postulated then nature is causal.

An observation of this kind carries little weight with those who feel too keenly that the processes in the universe are separable only to a rough approximation. But we note in the first place that it is by no means necessary for an arbitrary nature to be *roughly* or *approximately* separable into systems whose fates are independent; to say that it *is*, imposes a definite restriction even if the separation is not completely possible. But let us consider more exactly how it is performed.

Measurements on the force of attraction between two electric charges will not in general verify Coulomb's law. We observe that the force depends in some peculiar way upon the position of external charges, which suggests to us that the measured effects are not entirely due to the system in question, namely, the two test charges. This is expressed by stating that the system is not completely separated from its surroundings. Next we remove our system farther and farther from surrounding bodies and notice a gradual improvement in the consistency of the measurements. Now the situation would be very simple and satisfactory if progressive isolation produced better and better agreement between the observations, for then the condition of complete isolation, i.e., a closed system, could be defined by a simple limiting process much in the same manner as limits of functions are defined in mathematics. However, the situation is here of greater complexity, though still manageable. The agreement is improved only up to a certain point, and then further isolation fails to make itself felt. We have reached the limit of precision of our measurements. This limit of precision is a very definite thing which scientists have always considered very carefully. Quantum mechanics emphasizes it greatly and even renders its value in some instances calculable by means of the uncertainty relation (cf. Chapter IX). Nevertheless,

THE PROBLEM OF CAUSALITY

there is a well-known method of dealing with the ever-present divergence of observations: the theory of errors allows us to compute the most probable value of a measurement from any group of observations. It is the limit of this most probable value upon which we base the derivation of Coulomb's law, and not the limit to which actual observations tend. The latter does not exist, but the former can be defined. Hence it is possible to define a state of separation, not by actual physical operations but by blending experimentation with reasoning. We wish to emphasize in this connection that physical concepts need not—and cannot—be defined solely in terms of experimental operations or observations; it is both customary and proper in scientific investigations to characterize an abstract state of affairs by its logical consequences if they are more simply expressible, whether they are observable or not.

By its definition, a closed or independent physical system is a causal one, because we call it closed when the laws governing its behavior do not involve the time. But strictly speaking only closed systems are accessible to physical analysis. Thus it would seem that physics can never inform us of a failure of the causality principle. This brings us to a point of importance: is the causal principle a tautology? Here one is forced to make a large concession; an unbiased investigation must not fail to recognize its character as an analytic proposition. Kant, who thought of it as an *a priori* synthetic judgment, did not formulate it in a way in which its analytic character became apparent. It is certainly true that whenever a physical system does not appear to be closed, that is when the differential equations describing it contain the time explicitly (if it does not behave consistently), we conclude that the variables determining the state in question are not completely known. We then look immediately for hidden properties whose variation may have produced the inconsistencies, and whose inclusion in the analysis would eliminate them; moreover, if we do not find any we invent them. This procedure is possible because in the consistency formulation of causality the term "state" is undefined. The corresponding indefiniteness in the more abstract formulation lies in the absence of specifications as to the number of variables entering into the differential equations on the one hand, and of the order of the equations on the other. If the number of variables is increased indefinitely, or if an indefinite number of differentiations is permitted, time dependence can be ultimately eliminated no matter how complicated the processes of nature, provided only that the equations of motion can be differentiated a sufficient number of times. But this requires no more than a certain smoothness and continuity which nature certainly satisfies. On the whole, it seems, then, that the causality postulate reduces to a definition of what is meant by "state." It is an agreement to consider those quantities as composing the state of a system which enter into a time-free differential equation describing its behavior.

This line of reasoning leads at once to the inevitable question: why retain the proposition if it is merely tautological? Tautologies, as everybody knows, add nothing to the knowledge of nature; they expose a property already included in the term they are to explain; their careless use often produces vicious circles. All this is true in a sense; but to suppose that tautologies are always useless and to be avoided is fallacious. Every definition is a tautology except the first time it is stated. To many the word tautology conveys something more objectionable than the good old "analytic judgment." Really the two are synonymous, and one must free himself from any intuitive bias that may cling to the former term. The principle of conservation of energy is a tautology in the proper sense of the word, yet nobody doubts its fruitfulness, and it is even customary to speak of its validity as though it were an actual proposition about the world. The point in question is this: The general definition of the latter principle contains no restriction as to the number of different kinds of energy which may be transformed into one another and whose sum is constant. If this number became indefinitely great as a consequence of invention whenever a new type was needed the principle would hardly be applicable, it would merely define energies. As a matter of fact, however, nature permits us to get along with very few different types of energy, and its description in terms of energies is exceedingly useful and convenient. Therefore, while the energy principle in its logical formulation is a tautology which amounts to a definition of energies, the *analysis of nature in terms of this definition is advantageous*. Precisely the same is true with regard to causality. Its logical formulation is inevitably tautological and leads to a definition of physical "states," or a selection of variables in the differential equations. But this selection is useful and applicable, so that the causality principle, though tautological in its abstract form, does amount to a statement about nature. Moreover, it makes sense to say that nature is not causal, for this would be true if, in an attempt to describe nature in accordance with our definition of consistency, the resulting differential equations were found to contain very many variables or to be of very high order. One might, of course, arbitrarily restrict the latter and permit only equations of the second order, but we feel that this would do violence to the common conception of causality.

We have seen that the tautological character of the consistency postulate is no particular fault. Indeed, the postulate may be transformed into a synthetic statement if this be desired, but somewhat at the expense of its precision; it might be phrased: nature is so constituted that its description in terms of differential equations which do not involve the time explicitly is *convenient*. We merely state that Laplace's causality formulation is also tautological, while the one identifying causality of nature with human power to prophesy is not. But it stands to reason that the latter modification, missing, as it does, the central point of the causality concept, pays too large a price for its non-tautological form.

In formulating the principle of causality we have practically solved the question as to its validity. Classical physics was based upon it and therefore presents no argument against it. The effect of quantum mechanics upon its status may be sketched briefly as follows. Classical analysis had come face to face with experiences which its usual methods failed to describe; in fact, the treatment of certain problems threatened to become non-causal. At this very instant quantum mechanics achieved a revolutionary feat of great importance: it redefined the concept of physical states in a more abstract manner (in terms of mathematical functions satisfying certain requirements) and *thereby restored the causal character of physical analysis*.

The impression that quantum mechanics violates the principle now arises whenever the older classical conception of states (positions and velocities of the component parts of a system) is carelessly carried over into the new field of description in which it has little meaning. The uncertainty principle forbids ultimate extrapolation to the quantities defining a state in the classical sense, but it does not prevent an ultimate extrapolation to ψ functions. To be quite impartial, one should add that the trustworthiness of quantum mechanics even in questions of ultimate extrapolation to classical states is not entirely evident, for it is precisely in the very small domains of space (structure of the nucleus, structure of the electron, and its trembling motion) where its present axioms break down. In this sense, quantum mechanics does not constitute an argument against causality.

It seems difficult to envisage a situation which would compel physicists to surrender the consistency postulate. In general, strange new discoveries force us merely to alter the definition of states. There is an observation, however, which may, if verified, upset the causal principle completely. If the velocity of light, which is, at least at present, an essential parameter in the differential equations of physics, undergoes a variation in time, then our theory of electrodynamics is rendered definitely, and perhaps hopelessly, acausal, for absolute time would enter the equations through the parameter c.

One can see easily, from general considerations, that the description in time and space which is customary in classical physics, if coupled with the assumption of spatial continuity (cf. Chapter VI), must ultimately become contradictory to the principle of causality, for if the structure of nature's elements is continuous in space, and hence infinitely detailed, the equations representing the behavior of any of its parts will of necessity contain an infinite number of variables. Hence causal analysis, as it proceeds into finer and finer details of structure, will meet the same obstacle which prevented it from exploring an inseparable universe as a whole. Two means are available for avoiding this difficulty; both have been employed. One is to adopt a field theory which fixes minutely the values of a physical quantity at every point of space, but fixes it by means of simple functions so that the scientist is enabled to dominate in one grand sweep all the intricacies of spatial structure. There are reasons, however, why this procedure is unsatisfactory. Unless the field functions used in physical theory are periodic they imply singularities which, while they are insignificant as far as many properties of nature are concerned, certainly do not exist.

The other way of avoiding the difficulty without giving up classical description is to maintain continuity of spatial structure, but to endow finite parts of space with homogeneity so that those finite portions of space require no more elements of description than does a point. This was done in the early form of quantum theory by assuming the discrete existence of electrons, protons, energy quanta, and the like. The unsatisfactory consequences of this procedure are well known. But even logically the attempt is of no promise, because the postulation of indivisible entities is a command that reason must stop its inquiry at certain spatial confines, and reason refuses to be commanded.

But, after all, how can we know that the more abstract, and indeed more general, quantum mechanical procedure in vogue at present will not meet similar difficulties? How can we know that this world of ours is ultimately explorable? Is there a unique system of physical explanation? If there were, and the physicist were slowly learning it, his occupation would be that of a photographer who takes an enormous number of pictures in studying an object. If, however, there is no certainty about these questions, then his work is not photography; it is artistic creation. It seems that past experiences favor the latter alternative.

SELECTED READINGS FOR FOUNDATIONS OF PHYSICS

CHAPTER I

Braithwaite, R. B., *Scientific Explanation.* Cambridge University Press, 1953.
Bridgman, P. W., *The Nature of Physical Theory,* Dover, New York, 1936.
Cassirer, E., *The Problem of Knowledge,* tr. W. H. Woglom and C. W. Hendel, Yale University Press, 1950.
Duhem, Pierre, *The Aim and Structure of Physical Theory* (translated from the second French edition, 1914). Princeton University Press, 1954.
Lindsay, R. B., *Concepts and Methods of Theoretical Physics.* Van Nostrand, 1951.
Weyl, Hermann, *Philosophy of Mathematics and Natural Science.* Princeton University Press, 1949.

CHAPTER II

Jammer, Max, *Concepts of Space.* Harvard University Press, Cambridge, 1954.
Johnson, M. C., *Time, Knowledge and the Nebulae.* Dover, New York, 1947.
Mach, E., *Space and Geometry.* Open Court, Chicago, 1943.
Reichenbach, H., *Philosophy of Space and Time.* Dover, New York, 1956.
Schilpp, P., ed., *Albert Einstein: Philosopher-Scientist.* Evanston, Illinois, Library of Living Philosophers, Inc., 1949. Reprinted by Tudor, New York, 1951.

CHAPTER III

Bergmann, P., *Basic Theories of Physics. Mechanics and Electrodynamics.* Prentice Hall, 1949.
Corbin, H. B., and Stehle, P., *Classical Mechanics.* John Wiley and Sons, New York, 1950.
Goldstein, H., *Classical Mechanics.* Addison-Wesley Press, Inc., Cambridge, Mass., 1951.
Lindsay, R. B., *Physical Mechanics* (2nd edition). Van Nostrand, New York, 1950.
Truesdell, C., *Rational Fluid Mechanics,* 1687–1765 (Editor's introduction to Volume 12, Series II, of Euler's Works). Orell Füssli, Zurich, 1954.
Webster, A. G., *The Dynamics of Particles and of Rigid, Deformable and Fluid Bodies.* Teubner, Leipzig, 1904.
Whittaker, E. T., *Analytical Dynamics of Particles and Rigid Bodies.* Third Edition, Dover, New York, 1944.
Yourgrau, Wolfgang and Mandelstam, Stanley, *Variational Principles in Dynamics and Quantum Theory.* Pitman, London, 1955.

CHAPTER IV

Bures, C. E., "The Concept of Probability." *Philosophy of Science*, vol. 5, 1938.
Carnap, R., "Probability as a Guide in Life." *Journal of Philosophy*, vol. 44, 1947.
Carnap, R., *Logical Foundations of Probability*. University of Chicago Press, 1950.
Hopf, E., "On Causality, Statistics and Probability." *Journal of Mathematics and Physics*, vol. 13, 1934.
Jeffreys, H., *Theory of Probability*. Clarendon Press, Oxford, 1939.
Mises, R. von, *Probability, Statistics, and Truth*. Macmillan, New York, 1939 (originally 1928).
Nagel, E., "Principles of the Theory of Probability." Vol. I, No. 6 of *International Encyclopedia of Unified Science*, University of Chicago Press, 1939.
Reichenbach, H., *Theory of Probability*. University of California Press, 1949.

CHAPTER V

Chapman, S., and Cowling, T. G., *The Mathematical Theory of Nonuniform Gases*. University Press, Cambridge, 1939.
Cox, R. T., *Statistical Mechanics of Irreversible Change*. Johns Hopkins Press, Baltimore, 1955.
Donnan, F. G., and Haas, Arthur (Editors), *Commentary on the Scientific Writings of J. Willard Gibbs* (two volumes). Yale University Press, 1936.
Fowler, R. H., and Guggenheim, E. A., *Statistical Thermodynamics*. Cambridge University Press, 1949.
Lindsay, R. B., *Physical Statistics*. John Wiley and Sons, 1941.
Mayer, J. E., and Mayer, M. G., *Statistical Mechanics*. Oxford Press, 1949.
Rushbrooke, G. S., *Introduction to Statistical Mechanics*. Oxford Press, 1949.
Schrödinger, E., *Statistical Thermodynamics*. Cambridge University Press, 1948.
Ter Haar, D., *Elements of Statistical Mechanics*. Rinehart, New York, 1954.
Tolman, R. C., *The Principles of Statistical Mechanics*. Clarendon Press, Oxford, 1938.

CHAPTER VI

Frank, N. H., *Introduction to Electricity and Optics*. McGraw-Hill, New York, 1950.
Goldstein, H., *Classical Mechanics*. Addison-Wesley Press, Inc., Cambridge, Mass., 1951.
Harnwell, G., *Principles of Electricity and Electromagnetism*. McGraw-Hill, New York, 1949.
Jeans, J., *The Mathematical Theory of Electricity and Magnetism*. Cambridge University Press, 1948.
Lamb, H., *Hydrodynamics*. Dover, New York, 1906.
Loeb, L., *Fundamentals of Electricity and Magnetism*. John Wiley and Sons, New York, 1930.
Love, A. E. H., *Treatise on the Mathematical Theory of Elasticity*. Cambridge University Press, 1892.
Mason, M., and Weaver, W., *The Electromagnetic Field*. Dover. New York, 1929.
Milne-Thomson, L. M., *Theoretical Hydrodynamics* (3rd Edition). Macmillan, London, 1955.
Morse, P. M., and Feshbach, H., *Methods of Theoretical Physics*. McGraw-Hill, New York, 1953.
Page, L., and Adams, N., *Principles of Electricity*. Van Nostrand, New York, 1931.
Smythe, W. R., *Static and Dynamic Electricity*. McGraw-Hill, New York, 1950.
Stratton, J. A., *Electromagnetic Theory*. McGraw-Hill, New York, 1941.
Whittaker, E., *A History of the Theories of Aether and Electricity*. Phil. Lib., 1953.

CHAPTERS VII and VIII

Bergmann, P. G., *Introduction to the Theory of Relativity*. Prentice-Hall, New York, 1942.
Eddington, A. S., *The Mathematical Theory of Relativity*. Cambridge University Press, 1923.
Laue, M. V., *Das Relativitäts Prinzip*. Vieweg, Braunschweig, 1923.
Lorentz, H. A., *The Theory of Electrons and its Applications to the Phenomena of Light and Radiant Heat*. Dover, New York, 1909
Lorentz, H. A.. Einstein. A., and Minkowski H., *The Principle of Relativity*, Dover, New York, 1922.
Schrödinger, E., *Expanding Universes*. Cambridge University Press, 1956.
Weyl, H., *Space, Time, Matter*. Dover, New York, 1950.

CHAPTER IX

Bohm, D., *Quantum Theory*. Prentice-Hall, Inc., New York, 1951.
de Broglie, Louis, *The Revolution in Physics*. The Noonday Press, New York, 1953.
de Broglie, L., and Brillouin, L., *Selected Papers on Wave Mechanics*. Blackie and Sons, London, 1928.
Dirac, P. A., *Principles of Quantum Mechanics*.. Clarendon Press, Oxford, 3rd Edition, 1947.
Dushman, S., *Elements of Quantum Mechanics*. John Wiley and Sons, New York, 1938.
Houston, W. V., *Principles of Quantum Mechanics*. McGraw-Hill, New York, 1951.
Jordan, P., *Anschauliche Quantenmechanik*. J. Springer, Berlin, 1936.
Kemble, E. C., *The Fundamental Principles of Quantum Mechanics*. McGraw-Hill, New York, 1937.
Landé, Alfred, *Foundations of Quantum Theory*. Yale University Press, 1955.
Landé, A., *Quantum Mechanics*. Pitman Publishing Corp., New York, 1951.
Mott, N. F., and Sneddon, I. N., *Wave Mechanics and its Applications*. Oxford Press, 1948.
Neumann, J. v., *Mathematische Grundlagen der Quantenmechanik*. J. Springer, Berlin, 1932.
Pauling, L., and Wilson, E. B., *Introduction to Quantum Mechanics*. McGraw-Hill, New York, 1935.
Schiff, L. I., *Quantum Mechanics*. McGraw-Hill, New York, 1949.
Schrödinger, E., *Wave Mechanics*. Blackie and Sons, London, 1928.
Slater, J. C., *Quantum Theory of Matter*. McGraw-Hill, New York, 1951.

CHAPTER X

Bergmann, H., *Der Kampf um das Kausalgesetz in der jüngsten Physik*. Vieveg und Sohn, Braunschweig, 1929.
Bohr, N., "Kausalität und Komplementarität." *Erkenntnis*, vol. 14: 293, 1937.
Born, M., *Natural Philosophy of Cause and Chance*. Oxford Press, 1949.
Cassirer, E., *Determinism and Indeterminism in Modern Physics*, tr. T. Benfey, Yale University Press, 1956.
Frank, P., *Das Kausalgesetz und seine Grenzen*. Springer-Verlag, Vienna, 1932.
Jeffreys, H., *Theory of Probability*. Clarendon Press, Oxford, 1939.
Keynes, J. M., *Treatise on Probability*. Macmillan, London, 1921.
Lenzen, V. F., "Physical Causality." *University of California Pub. Philosophy*, vol. 15, Berkeley, 1942.

CHAPTER X continued

Lillie, R. S., "Biological Causation." *Philosophy of Science*, vol. 7: 314, 1940.
Margenau, H., *The Nature of Physical Reality*. McGraw-Hill, New York, 1950.
Rosenfeld, L., "L'Evolution de l'idée de causalité." *Mem. Soc. Roy. Sci. Liège*, 4th ser., vol. 6, 1942.
Winn, R. B., "The Nature of Causation." *Philosophy of Science*, vol. 7: 192, 1940.

INDEX

A

Absorption of light, 474
Abstraction,
    method of elementary, 29
Acceleration, 82
Ackermann, 8
Acoustic waves in a fluid, 38
Action and reaction, 86, 97
Action, principle of least, 128
    principle of varying, 144
Action variables, 154
Aggregate, probability, 161
    analytical nature of, 163
    derived, 162
    primary, 162
Alpha rays, emission of, as probability problem, 172
Anderson, 513
Angle variables, 154
Antisymmetry of state functions, 490
Appell, 112
Arithmetical mean, 186
Assembly, of several kinds of systems, 264
    of simple harmonic oscillators, 268
    of systems in Darwin and Fowler's statistical mechanics, 253 et seq.
Atomic structure, 493 et seq.
Averages, 168, 230
    deviations of properties from, 233
Axes, coordinate, 82

B

Bernoulli, D., 47
Bernoulli, J., 136
Bernoulli's distribution, 171
    dispersion of, 173
Bernoulli's problem, 169
Blackett, 513
Bohr, 88, 391, 454
    theory of atom of, 391, 454
Boltzmann, 227.
    constant of, 229, 242
Bolyai, 65
Born, 154, 454
Boundary conditions, 43, 48
    general, 52
    in wave motion, 43
    specific, 49

Boyle's law, 11
Bracket, Poisson, 148
Bradley, 324
Breadth of spectral lines, 476
Brickwedde, 513
Bridgman, 411
Brownian movement, 193
Brunn, T., 375
Bucherer's experiment, 323

C

Campbell, 20, 58
Canonical ensemble, 225
Canonical equations of motion, 145
Canonical transformation, 149
Canonically conjugate operators, 405
Canonically conjugate quantities, 148
Carnot cycle, 213
Causality, 515 et seq.
    as consistency of nature, 522
    Kant's formulation of, 516
    Laplace's formulation of, 517
    positivistic conception of, 519
Cause and effect, 18
Central field motion, in classical mechanics, 139
    in quantum mechanics, 436
Central field problem, 436
Chadwick, 513
Chance contrasted with determinism, 189
Charlier, 154
Chronon, 77
Classes, 9
Classical mechanics, method of, as distinguished from that of quantum mechanics, 394 et seq.
Clausius, 215, 319
Complementarity principle of Bohr, 520
Complete integral of Hamilton-Jacobi equation, 152
Completeness, of eigenfunctions, 446
    of function sets, 408
Compton effect, 390
Condensation, 316
Condon, 430, 467
Conduction of heat as irreversible process, 197
Conductor, Faraday's law for a perfect, 204

INDEX

Conjugate momentum, 144
Conservative system, 124
Consistency of nature, 523
Constant of radioactive decay, 191
Constants, method of variation of, 472
Constraint, principle of least, 112
Contact transformation, 456
Continuity, equation of, 30, 33
  in electromagnetism, 309
Continuous eigenvalues, 409
Continuum, concept of, 280
Contraction, Fitzgerald, 338
Coordinates, generalized, 136
  systems of, 81, 82
Coriolis acceleration, 331
Cosmical constant, 386
Coulomb's law, 308
Courant, 432, 442
Curtiss, 174
Curvature, constant, surface of, 67
  Gaussian, 384
  of space, 386
  radii of, 66
  total, of a surface, 66
Cylindrical coordinates, 82

D

D'Alembert, principle of, 103
  Mach's formulation of, 109
Dantzig, 5
Darwin, 252 et seq., 495
Data of physics, 1
Davisson-Germer experiment, 393, 399
De Broglie, 352, 393, 399
Decay constant, radioactive, 191
Deformable media, 292
Deformation tensor, 294
Degeneracy of quantum states, 256, 464
Degenerate case in perturbation theory, 464
Delta function, 408, 409
Density, as scalar field quantity, 288
  fluctuations, 177
  of a fluid, 32
  probability, 164
Density-in-phase, conservation of, 223
Description of experience, 2
Determination vs. chance, 189
Deviation, 168
  mean, 185
  of properties from averages, 233
  root mean square, 169
  standard, 169, 233
Diagonal matrix, 451
Dielectric constant, 307

Difference equations, 57
Diffuseness, elementary, 398 et seq.
Dirac, 29, 149, 472, 478, 488, 500 et seq.
  on electron in electromagnetic field, 508
Dirac's δ-function, 408
Dirac's equation for free electron, 503, 506
Dirac's theory, of electron, 501
  of positron, 512
Dispersion, statistical, 168 et seq.
  of error curve, 185
Displacement, 82
Displacements, principle of virtual, 102
Dissipation function, 126
Dissipative system, 126
Dissociation, 211
Distribution, arithmetical, 162, 168
  Bernoulli's, 171
  canonical, in-phase, 227
  continuous, 175
  density, of molecules, 178
  geometrical, 162, 168
  laws in quantum statistics, 500
  mean value of, 168
  microcanonical, 238
  moments of, 169
  of errors, 181
  of high energies in Sturm-Liouville problem, 444
  of minor planets, 246
  of velocities, Maxwell's law, 229
  probability, 162
Double pendulum as illustration of D'Alembert's principle, 110
Dynamical equation of an electron, 320
Dynamical laws, 191, 192
  application of, to thermodynamics, 206
  contrasted with statistical, 201
  examples of, 194
Dynamical theory of electric currents, 201
Dynamics, 85

E

Eckart, 458
Eddington, 76, 87, 385
Eigenfunctions, antisymmetry of, 490
  completeness of, 446
  general properties of, 440
  orthogonality of, 424
Eigenstates, 407
Eigenvalues, 406
  continuous, 409
  of angular momentum, 410
  of linear momentum, 409
  of stretched vibrating string, 407
  of x-coordinate, 408
  stationary character of, 442

INDEX

Einstein, 193, 330 et seq., 351 et seq., 358 et seq., 389
  —Bose statistics, 497
Electric currents, dynamical theory of, 201
Electric oscillations, 205
Electrokinetic momentum, 204
Electromagnetic energy, 305
Electromagnetic equations, in free space, 303
  interpretations of, 315
  solution of, 312
Electromagnetic ether, 302
Electromagnetic field, 302
Electromagnetic force equation, 307
Electromagnetic units, 309
Electromagnetic waves, 304 et seq.
Electromotive force, 308
Electron, dynamical equation of, 320
  electric moment of, 511
  magnetic moment of, 511
  numerical values of mass and charge, 323
Electron spin, 478
Electron theory, 319
  of Dirac, 501 et seq.
Electrostatic units, 309
Elementary abstraction, 29
Elementary diffuseness, principle of, 398 et seq.
Elementary weights, 255
Elements in a probability aggregate, 161
Emission of alpha rays as a probability problem, 172
Energy, average value of kinetic, 231
  average value of potential, 233
  concept of, 120
  kinetic, 121
  most popular value of, 236
  potential, 123, 284
  zero point, 433
Energy-momentum-stress tensor, 381
Ensemble, 220
  canonical, 225
  microcanonical, 238
Entropy, 210, 215, 217, 275
  thermodynamic definition of, 216
Equation of state, 213
Equations, canonical, 147
  difference, 55, 57
  integral, 55
Equilibrium, meaning of, in statistical mechanics, 224
Equipartition theorem, 232, 273
Error function, 177, 181

Errors, accidental, 182
  origins of, in measurements, 181
  probable, 185
  systematic, 182
  theory of, 181 et seq.
Euclidean geometry,
  axioms, 64
  definitions, 63, 64
  postulates, 64
Euclidean space, 63
Exclusion principle, 488 et seq.
Expected mean, in probability, 168
  expansion in terms of eigenvalues, 414
  of sequence of measurements in quantum mechanics, 411
Experiment, 4,
  symbolic description of, 11

F

Faraday, 302
  law of induction of, 202
Fechner, 161
Fermat's principle, 135
Fermi-Dirac statistics, 497
Field, concept of, 283
  electromagnetic, 302
  gravitational, 364
  of force, 283
    fictitious, 361
    permanent, 361
  scalar, 287
  vector, 287
Field equations, 303
Fitzgerald, 329, 338
Fizeau, 324
Fluctuation, density, 177
  law of, 179
  relative, 179
Force, 94
  electromotive, 308
  field of, 283
  mechanical, acting on a circuit, 204
  on an electron, 321 et seq.
Forces, common types of, 99
Four-vector, 348
Fourier integral theorem, 183, 313
Fourier series, 48
Fowler, 495, 252 et seq.
Frames of reference, 80
Free expansion of a gas, 199
Free particle in, quantum mechanics, 427
Frequency of harmonic wave, 44
Freundlich, 375

INDEX

Fry, 175
Function, 15
Functional, 56

G

Gas, ideal, 209, 271
Gauss, 112, 116
  function of, 177, 181
  integral of, 185
  law of, 284
  principle of least constraint of, 112
Gaussian curvature, 384
Generalized coordinates, 136
Generalized forces, 142
Generalized momenta, 144
Geodesic, 69, 364
Geometrical space, 62
Geometry, abstract, 362
  Euclidean, 63
  Lobatchevskian, 64
  natural, 362
  Riemannian, 65
Gerlach, 400
Germer and Davisson experiment, 393, 399
Gibbs's function, 217
Gibbs's method of statistical mechanics, 218 et seq.
  critique of, 250
Goudsmit, 479
Gouy, 193
Group, definition of, 9
Growth, law of organic, 190

H

Hamiltonian function, 145
Hamilton-Jacobi equation, 152
Hamilton's principle, 128, 131, 195, 365
Harmonic frequencies, 46
Harmonic functions, 36
Heat conduction as an irreversible process, 197
Heaviside units, 308, 319
Heisenberg, 24, 88, 420 ff., 453 ff.
Helmholtz, 188
Hereditary mechanics, 56
Heredity, coefficient of, 56
Hermitean matrices, 453
Hermitean operators, 422
Hermitean orthogonal functions, 432
Hertz, H., 85, 116
  mechanics of, 118 et seq.
  principle of the straightest path of, 120
Herzfeld, 229
Hilbert, 8, 432, 442
Hobson, E. W., 68

Holonomic systems, 140
Homaloidal space, 365
Hooke, Law of, 19, 56, 301
Hydrodynamic equations of motion, 36
Hypotheses, 23
  indifferent, 26
  natural, 25
  nature of physical, 25

I

Ideal gas, 271
Impulse, 98
Independence, linear, 425
Indeterminacy, 51; see also Uncertainty principle
Index of probability, 239
Indistinguishability of elementary particles, 491
Induction, Faraday's law of, 202
  mutual, 202
  self, 202
Inertia, 80, 92
Inertial system, 331
  primary, 81
Integral equations, 55
Integration, physical meaning of, 35
Intensity, of electric field, 303 et seq.
  of magnetic field, 303 et seq.
Invariance, 101, 287
  general, 356
  of field equations, 323, 326
Irreversibility, examples of, 197
Irrotational vector, 303
Isochore, reaction, 212

J

Jacobi, 152
James, 73
Jeffreys, 58, 302
Jevons, 8
Jordan, 454

K

Kelvin, 330
Kemble, 440
Kinematical concepts, 79
Kinetic energy, 121
  average value of, 231
  negative, 506
Klüber, 375
Kohlrausch, 174

L

Lagrange, 112, 136
  equations of, 138, 201 et seq.
  method of, 136

Lagrangian function, 201, 206, 365
average, 207
Laguerre polynomials, 439
Lamb, 111
Laplace's equation, 36, 282, 285
Laplace's formula, 174, 177 et seq.
Laplace's probability definition, 160
Laplacian, 285
Larmor frequency, 469
Law, Boyle's, 11
of induction, 202
relation to theory, 22
Laws, dynamical, 191 et seq.
examples, of, 194
of motion, 94, 98
of organic growth, 190
of thermodynamics, 214 et seq.
physical, 14
statistical, 191
examples of, 194
Least action, principle of, 128, 132
Least squares, 116
Legendre polynomials, 435
Leibniz, 60, 136
Lenzen, 58
Lewis, 76
L'Hopital, 136
Lie, 71
Light, absorption of, 474
Light ray, path of, in general relativity, 373
Line of force, 283
Linear differential equations, physical significance of, 47
Linear independence, 425
Liouville's theorem, 221
Lobatchevski, 64
Lorentz, 311, 319, 327
Lorentz transformations, 337
Love, 300, 302

M

Mach, 90, 93, 97, 107, 333
Macroscopic state, 254
Mass, definition of, 91 et seq.
longitudinal, of electron, 322
rest, of electron, 350
transverse, of electron, 322
Matrices, 448 et seq.
function of, 451
Hermitean, 453
spin, 483
Matrix, perturbation, 463
transformation, 457
unit, 450
Matrix mechanics, 453

Maupertuis, 133
Maxwell, 96, 310 et seq., 315, 318
Maxwell-Boltzmann law, 227, 264
Maxwell's law of distribution of velocities, 229
Mean, arithmetical, 186
expected, 168
Mean value of operators, 411, 413
Mean values in probability, 168
Measure of precision, 181
Measurement, 4
and indeterminacy, 51
Mechanical force acting on a circuit, 204
Mechanics, as illustration of physical theory, 21
foundations of, 79
statistical,
average values, 230
canonical ensemble, 225
deviations of properties from averages, 233
ensemble, 220
equilibrium, meaning of, 224
Liouville's theorem, 221
Maxwell-Boltzmann law, 227, 264
Maxwell's law of distribution of velocities, 229
method of Darwin and Fowler, 252
method of Gibbs, 218
terminology, 219
thermodynamic relations, 238
Media, deformable, 292
Michelson-Morley experiment, 324 et seq., 354
Microcanonical ensemble, 238
Microscopic state, 253
Miller, 324, 354, 377
Millikan, 323
Minimal concept, 115
Minkowski, 342 et seq.
four-dimensional representation, 343
Minor planets, distribution of, 246
Mises, v., 161
Molecule, rotating, 433
Momentum, 89, 98
eigenvalues of, 409
electrokinetic, 204
Motion, Newton's laws of, 85 et seq.
third law of, 97
Murphy, 513

N

Necessity, metaphysical implications of, 193
v. Neumann, 24, 408

Neutrino, 514
Neutron, 513
Newtonian postulates, 85
Newton's formula, 171, 173, 178, 179
Non-Archimedean geometry, 71
Non-conservative system, 125, 142
Non-degenerate case in perturbation theory, 461
Non-Euclidean geometries, 64
Non-Euclidean world of Poincaré, 68
Non-holonomic system, 140
Normal law of errors, 181
Normal value of a property in an assembly, 255
Normalization, 168, 413
Nucleus, 513
Number, 397
  cardinal, 5
  concept of, 6, 397
  ordinal, 5

O

Observables, 402
Occam's razor, 310
Occhialini, 513
Oersted, 302
Operations, 6
  symbolic representation of, 8
Operators, 404
  hermitean, 422
  theorem on permutable, 417
Orthogonality of functions, 407
Orthogonalization of functions, 425
Oscillations, electric, 205
Oscillator, in quantum mechanics, 430
Oscillators, assembly of, 268
Ostwald, 216

P

Page, 117, 285, 321, 355
Pair production, 512
Partition function, 261
Pauli, 480 et seq., 488 et seq., 495
Pauli principle, 488
Path, principle of straightest, 120
Pendulum, simple, 17
Periodic system, quantum mechanics description of, 494
Permeability, 307
Permutable operators, 417
Perpetual motion, 216
Perturbation matrix, 463
Perturbation theory, 460
Phase integral, 154
  of Gibbs, 230, 273
Phase space, 219

Photoelectric effect, 389
Physical law, 14
Physical space, 61
Physical symbol, 12
Physical system in quantum mechanics, 401
  state of, in quantum mechanics, 402
Physical theory, 20
Physical world, 1
Planck, 315, 388 et seq.
  radiation law of, 269
Planetary motion in general relativity, 364
Poincaré, 26, 58, 68, 71, 74, 91, 191
Poisson bracket, 148, 455
Poisson's equation, 285
Poisson's formula, 179
Polar coordinates, 81
Polarization rules, 456, 477
Position, specification of, 79
Positron, 511, 513
Potential, 284
  advanced, 315
  retarded, 313
  vector, 312
  velocity, 36
Potential energy, 123
Poynting flux, 306
Precision, measure of, 181, 185
Prediction, 16
Pokrowski, 77
Principle, D'Alembert's, 102
  exclusion, 488
  Fermat's, 135
  Gauss's, 112
  Hamilton's, 128
  of least action, 128
  of least constraint, 112
  of straightest path, 120
  of superposition, 47
  of varying action, 144
  of virtual displacements, 102, 107
  Pauli's, 488
Probability, *a posteriori* definition, 161, 165
- *a priori* definition, 160
  aggregates, 161, 173
    analytical nature of, 163
  density, 164
  index of, 239
  Laplace's definition of, 160
  mathematical, 168
  meaning of, 159
  predictions, 245
  thermodynamic, 218

Probable error, 185
Propagation of light, 134
Proper reference system, 321
Proper system, 321, 347
Pure case, 400

Q

Quantized systems in statistical mechanics, 258
Quantum conditions, 391
Quantum dynamics, 469
Quantum mechanics, 387
  axiomatic foundation of, 401 et seq.
  Bohr's theory of, 391, 454
  historical introduction to, 387
  stationary states in, 469
  time dependent states in, 469
  uncertainty in, 420
  use of matrices in, 450, 453
  use of operators in, 404
Quantum number, 439
Quantum of energy, 388
Quantum statistics, 495 et seq.

R

Radioactive constant, 191
Radioactive decay, 173, 191, 192
  law of, 191
Ramsauer effect, 393
Reaction, 97
Reciprocal reference system, 320
Reichenbach, 58
Relation between states and experience in quantum mechanics, 404
Relativity, general, 356
  abstract geometry, 362
  mechanics in, 377
  natural geometry, 362
  path of a light ray in, 373
  planetary motion in, 364
  shift of spectral lines, 375
  special, 330
    dynamics of, 347
    invariance in, 355
    kinematics of, 337
    Minkowski's interpretation of, 342
    transformation equations, 330
Rest mass, 350
Retardation of moving clocks, 339
Retarded potential, 313
Reversibility, as distinction between dynamical and statistical theories, 195
  definition of, 196
  examples of, 197

Riemann, 65, 67, 319
Riemann-Christoffel tensor, 363 et seq.
Ritz, 311
Rotating molecule in quantum mechanics, 433
Rowland, 329
Ruark, 57, 387
Rupp, 393

S

Scalar fields, 287
Scheffers, 67
Schmidt, 424
Schopenhauer, 29
Schrödinger, 502 et seq., 508 et seq.
Schrödinger's equation, 410
  for central motion, 436
  for free particle, 427
  for linear oscillator, 430
  for rotator, 433
  for time-dependent state function, 470
  general properties of, 440
  special forms of, 426 et seq.
Secular equation, 466
Selection rules, 477
Separable universe, 524
Shift of spectral lines, 375
Simple harmonic motion, 129
Simplicity, criterion of, 18
Simultaneity, 75, 334
Simultaneous eigenstates, 417
Singularity in a field, 286
Smoluchowski, 193
Solenoidal vector, 303
Sommerfeld 430, 478, 500
Space, Euclidean, 63
  geometrical, 62
  homaloidal, 365
  in physics, 59
  integral of force, 121
  properties of, 60
  psychological, 59
  public, 61
Spectroscopy, data of, 454
Spencer, 96
Spherical coordinates, 82
Spin degeneracy, 487
Spin function, 482
Spin in atomic problems, 486
Spin matrices, 483
Spin operators, 481
Spin theory, Dirac's, 511
  Pauli's, 480
Spinning electron, 478, 511
Stark effect, 463

State, in quantum mechanics, 402 et seq.
  macroscopic, 253
  microscopic, 253
  stationary, 469
State functions, 402
  antisymmetry of, 490
Statistical laws, 191
  contrasted with dynamical, 195, 201
  examples of, 194
Statistical mechanics, basic postulate, 224
  canonical ensemble, 224
  Liouville's theorem, 221
  method of Darwin-Fowler, 252
  method of Gibbs, 218
  terminology, 219
Statistics, Einstein-Bose, 497 et seq.
  Fermi-Dirac, 497 et seq.
  quantum, 495
Steepest descents, method of, 261
Stern-Gerlach experiment, 400
Stokes's theorem, 303
Stoner's rule, 494
Strain tensor, 293
Stress tensor, 293, 381
Structure, atomic, 493
Sturm-Liouville problem, 440, 478
Subshells of atoms, 494
Superposition, principle of, 47
Symbolism, 6, 12
  in logic, 7
  in physics, 10
Symmetry, of state functions, 490
  principle of, 26
System, closed, 525
  conservative, 125
  dissipative, 126
  holonomic, 140
  in Gibbs's statistical mechanics, 219
  inertial, 290
  non-conservative, 142
  proper, 321, 347
  reciprocal, 320
Swann, 311

T

Tensor, strain, 293
  stress, 293
Tensor equations, 300
  significance of, 300
Tensors, 292
  contravariant, 298
  covariant, 296
Theories, dynamical, 188
  mechanistic, 188
  statistical, 188

Theory, criteria for success of, 27
  mathematical development of, 29
  meaning of a physical, 20
  relation of law to, 22
Thermodynamic relations, 238
Thermodynamics, application of dynamical theories to, 206
  first law of, 214
  fundamental facts of, 212
  second law of, 215
Thirring, 58, 299
Thomson, J. J., 208
Thomson, G. P., 354, 393
Time, 72
  abstract, 74
  averages, 136
  continuity of, 76
  directionality of, 75
  physical, 73
  private, 72
  public, 73
Transformation, of coordinates, 289
  of tensors, 296
Transformation equations, Lorentz-Einstein's, 337
  of special relativity, 330 et seq.
Transformation matrix, 457
Transformation theory of mechanics, 149 et seq.
Transition probabilities, 455
True value, 181
  deviation from, 182

U

Uhlenbeck, 479
Uncertainty, 399, 420 et seq.
Uncertainty principle, 399, 418, 420 et seq.
Uncontrollable coordinates, 207
Uniformity, of experience, 1
  of a statistical distribution, 244
Uniqueness of determination, 189
Urey, 387, 513

V

Value, average, 230
  of kinetic energy, 231
  most popular, 236
  most probable, 186
  true, 181
    deviations from, 182
Variation of constants, method of, 472
Vector, definition of, 289
  four, 348
  irrotational 303
  solenoidal, 303

Vector fields, 287
Vector potential, 312
Vector product, 292
Vector representation, 83
Velocity, 82
Velocity potential, 36
Velocity transformations in relativity, 341
Virtual displacements, principle of, 102, 107
Volterra, 56

W

Wave, harmonic, 44
  meaning of, 41
Wave equation, 38
Wave length of harmonic wave, 44
Wave motion, 36 et seq.
Wave mechanics, de Broglie's, 352, 399
Wave velocity in a fluid, 42
Weber, 319
Weights, elementary, 255
Weyl, 24
Whittaker, 142
Wigner, 456
World line, 343
World point, 343

Y

Young, 64, 72

Z

Zeeman effect, 393, 467
Zero point energy, 433
Zustandssumme, 261